Sustainable Disease Management of Agricultural Crops

Sustainable Disease Management of Agricultural Crops

S.K. Biswas
S.R. Singh

2024

Daya Publishing House®
A Division of
Astral International Pvt. Ltd.
New Delhi – 110 002

First Published, 2011

Reprinted, 2024

ISBN: 978-81-7035-715-5 (HB)

Published by : **Daya Publishing House**®
A Division of
Astral International Pvt. Ltd.
– ISO 9001:2015 Certified Company –
4736/23, Ansari Road, Darya Ganj,
New Delhi-110 002
Ph. 011-43549197, 23278134
E-mail: info@astralint.com
Website: www.astralint.com

Printed at : **Replika Press Pvt. Ltd.**

Preface

The Indian agricultural scenario is undergoing in rapid change. The yield stagnation in crops is a major challenge for agricultural research in the 21st century. An analysis of agricultural scenario indicates that there are several concerns like biotic and abiotic factors which are responsible for lowering crop yield. The food grain production in India is impressive since independence. The food grain production increased from 50 million tones in 1950-51 to 234 million tones in 2008-09 which is unparallel in the globe. But at the same time, the population has also increased three fold and is expected to reach around 1.4 billion by 2025. The increase population alongwith improved living conditions will put greater demand for food and other agricultural commodities. We have to improve agricultural production and ensure food security in the coming era. The need based cultivation, problems solve in crop cultivation and production oriented research can take a leading role on changing scenario of Indian Agriculture which ultimately can help to reach our goal "Food, Nutrition & Environmental security".

The efforts have been made to compile invited book chapters on various issues in the present form. It is also an attempt to update the available information on various aspects related to disease résistance, integrated disease management, soil solarisation, transgenic crop, organic disease management and biopesticides in disease management. The chapters are devoted to the management of major diseases of vegetables, cereals, medicinal plants, pulses and fiber crops. The wide range of issues discussed in all chapters will help to generate new ideas towards disease management strategies. We hope that this will serve as a useful for all students, researchers and extension workers who are interested in research, teaching and extension activities.

We express our sincere thanks and appreciation to all contributors for their invaluable input and prompt response, which enabled to bring this book in this form.

We are highly grateful to Dr. G.C.Tewari, Vice chancellor, C.S. Azad University of Agriculture& Technology, Kanpur for giving us inspiration, encouragement and permission to edit the book.

We extend our gratefulness to Dr. Jitendra Singh, Professor & Head, Department of Plant Pathology & Dr. M. Rai, Professor for providing valuable suggestions, co-operation and special attention to complete the same. We also extend thanks to our students Mohd. Rajik, Kahkashan Arzoo & N. K. Pandey

We express our gratitude to Daya Publishing House, New Delhi for the interest and trouble taken in connection with publication of the book.

Lastly, but not in the least, we express thanks to our family members Mrs. Parwati Biswas, Chandana Biswas,, Sansthita Biswas, Shashi Mala, & Hira Devi for their constant support and encouragement in bringing out this book.

S.K. Biswas

S.R. Singh

Contents

List of Contributors

Arzoo, Kahkashan
Department of Plant Pathology, C.S. Azad University of Agriculture and Technology, Kanpur – 208 002

Bag, T.K.
Head, Central Potato Research Station, Silong

Banik, S.
Division of Plant Pathology, Indian Agricultural Research Institute, Pusa, New Delhi – 110 012

Benagi, V.I.
Department of Plant Pathology, UAS, Dharwad, Karnataka

Bhagawati, R.
ICAR Research Complex for NEH Region, Arunachal Pradesh Centre, Basar – 791 101

Biswas, C.
Central Research Institute for Jute and Allied Fibres, Barrackpore, 24-Parganas (N), Kolkata –70 0120

Biswas, S.K.
Department of Plant Pathology, C.S. Azad University of Agriculture and Technology, Kanpur – 208 002, U.P.

Chand, Gireesh
Department of Vegetable Science, N.D. University of Agriculture and Technology, Kumarganj, Faizabad – 224 229

Chandra, K.K.
Krishi Vigyan Kendra, Chandauli, N.D. University of Agriculture and Technology, Kumarganj, Faizabad – 224 229

Chaudhary, R.G.
Division of Crop Protection, Indian Institute of Pulses Research, Kanpur – 208 024

De, R.K.
Central Research Institute for Jute and Allied Fibres, Barrackpore, 24-Parganas (N), Kolkata –700 120

Dwivedi, Kusum
Department of Entomology, C.S. Azad University of Agriculture and Technology, Kanpur – 208 002, U.P.

Gangwar, R.K.
Krishi Vigyan Kendra, Chomu (Tankarda), Jaipur, Rajasthan

Gharde, Yogita
IASRI, New Delhi – 110 012

Gupta, P.K.
Division of Plant Pathology, Indian Agricultural Research Institute, New Delhi – 110 012

Kamalwanshi, R.S.
Department of Plant Pathology, C.S. Azad University of Agriculture and Technology, Kanpur – 208 002, U.P.

Kanwat, M.
ICAR Research Complex for NEH Region, Arunachal Pradesh Centre, Basar – 791 101

Kulkarni, Shripad
Department of Plant Pathology, UAS, Dharwad, Karnataka

Kumar, Dharmendra
Krishi Vigyan Kendra, Chandauli, N.D. University of Agriculture and Technology, Kumarganj, Faizabad – 224 229

Kumar, Sunil
Department of Plant Pathology, C.S. Azad University of Agriculture and Technology, Kanpur – 208 002, U.P.

Loganathan, M.
Division of Crop Protection, Indian Institute of Vegetable Research, Post Box No. 1, P.O. Jakhini, Shahanshapur, Varanasi – 221 305, U.P.

Mala, Shashi
Aligarh Muslim University, Aligarh

Mandal, Kunal
Directorate of Medicinal and Aromatic Plants Research, Boriavi, Anand – 387 310, Gujarat

Manzer, Abu
Assistant Professor, Pulses Research Substation, HABAK, SKUAST-K, Habak Crossing, Srinagar–190 006 (J&K)

Narain, Udit
Department of Plant Pathology, C.S. Azad University of Agriculture and Technology, Kanpur – 208 002, U.P.

Naresh, Prem
Department of Plant Pathology, C.S. Azad University of Agriculture and Technology, Kanpur – 208 002, U.P.

Pandey, N.K.
Department of Plant Pathology, C.S. Azad University of Agriculture and Technology, Kanpur – 208 002, U.P.

Pandey, R.K.
Division of ACB and PDM, Amity University, Noida – 201 303, U.P.

Pandey, Sudhakar
Division of Crop Protection, Indian Institute of Vegetable Research, Post Box No. 1, P.O. Jakhini, Shahanshapur, Varanasi – 221 305, U.P.

Prajapati, C.R.
Krishi Vigyan Kendra (Baghpat), S.V. Patel University of Agriculture and Technology, Modipuram, Meerut, U.P.

Prajapati, R.K.
Krishi Vigyan Kendra, Tikamgarh, M.P.

Rajik, Mohd.
Department of Plant Pathology, C.S. Azad University of Agriculture and Technology, Kanpur – 208 002, U.P.

Ratan, Ved
Department of Plant Pathology, C.S. Azad University of Agriculture and Technology, Kanpur – 208 002, U.P.

Reddy, C.N.L.
Assistant Professor, Department of Plant Pathology, Sericultural College, Chintamani – 563 125, Chikkaballapura, Karnataka

Sarkar, S.K.
Central Research Institute for Jute and Allied Fibres, Barrackpore, 24-Parganas (N), Kolkata – 700 120

Sarma, Pranabjyoti
ICAR Research Complex for NEH Region, Arunachal Pradesh Centre, Basar – 791 101

Shah, T.A.
Department of Plant Pathology, S.K. University of Agricultural Sciences and Technology-K, Srinagar (J&K)

Singh, A. Kirankumar
ICAR Research Complex for NEH Region, Arunachal Pradesh Centre, Basar – 791 101

Singh, Prabhas Chandra
Division of Crop Protection, Indian Institute of Vegetable Research, Post Box No. 1, P.O. Jakhini, Shahanshapur, Varanasi – 221 305, U.P.

Singh, S.P.
ICAR Research Complex for NEH Region, Arunachal Pradesh Centre, Basar – 791 101

Singh, S.R.
Krishi Vigyan Kendra, Belatal, Mahoba – 210 423, U.P.

Sinha, P.
Division of Plant Pathology, Indian Agricultural Research Institute, Pusa, New Delhi – 110 012

Srivastava, S.S.L.
Department of Plant Pathology, C.S. Azad University of Agriculture and Technology, Kanpur – 208 002, U.P.

Varma, P. Kishore
Assistant Professor, Department of Plant Pathology, Agricultural College, Aswaraopet – 507 301, Khammam, Andhra Pradesh

Sustainable Disease Management of Agricultural Crops (2011) *Pages* **1–26**
Editors: **S.K. Biswas and S.R. Singh**
Published by: **DAYA PUBLISHING HOUSE, NEW DELHI**

Chapter 1

Plant Pesticides: A Novel Ecofriendly Approach for Management of Plant Diseases

☆ ***S.K. Biswas, Kahkashan Arzoo, N.K. Pandey and Mohd. Rajik***
Department of Plant Pathology,
C.S. Azad University of Agriculture and Technology, Kanpur – 208 002

Man started practicing agriculture around 7500 BC. Neolithic agricultural revolution started around 7000 BC in West Asian countries Israel, Jordan, Iraq, Iran etc. where wild ancestor of wheat and barely was grown. Around 3000 BC irrigated agriculture started in Egypt from where it spread to Indus valley civilization (Dhaliwal and Arora, 2001). Pests and diseases of crops were developed side by side with development of agriculture.

During ancient times man evolved many cultural and physical practices to protect crops from ravage of insects and pathogens. This was followed by the use of botanicals.The earliest recommendation for the use of plant products was given by Democritus in 470 B.C. He recommended the control of plant blights by sprinkling plants with the olive grounds left after extraction of olive oil. Later Surapala in 9th century suggested the use of vidanga (*Embelia ribis*), mustard and sesame for curing tree diseases in his book "Vrikshayurveda".

As agriculture advanced man invented many new techniques for the control of pathogen. These included the use of inorganic and organic chemicals. The chemical control was so efficient in control of diseases that man started using it indiscriminately. This indiscriminate use continued until 1962 when Rachel Carson for the first time attracted attention towards the bad effect of chemicals in her book "Silent Spring".

Excessive use of chemicals has caused soil, air, surface and ground water pollution besides affecting the crop produce. The chemicals have entered our food chain also. Other detrimental effects of excessive use of chemicals are pesticide resistance in pathogen, contamination of food by toxic residues, resurgence of pest and detrimental effect on non target organisms and also cost of production. Judging the seriousness of this problem, attention has been paid into search for an effective alternative with practically no residual or toxic effects on ecosystem. Botanicals can serve as an alternative approach in this situation.

There is an increasing stimulus for the development of natural antimicrobials all over the world. "European union" took on the stimulus for the development of natural antimicrobial which they refer as "green chemicals". The USA "Environmental Protection Agency" has directed the registration of a new category of pesticides called "Plant Pesticides" in the year 1994. Infact the U.S.E.P.A. generally requires much less data to register a plant based chemical as conventional pesticides; registration costs are also low.

What are Plant Pesticides?

"Plant Pesticides" may be defined as substances of plant origin that have been used for pest control due to its pesticidal properties. Green plants appear to the reservoir of biotoxicants and constitute an inexhaustible source of innocuous pesticides (Swamminathan, 1978).

The presence of antifungal compounds in higher plants has long been recognized as an important factor for disease control (Mahadevan1982). Such compounds being biodegradable and selective in their toxicity are considered valuable for controlling plant diseases (Singh and Dwivedi 1987). The pesticidal compounds of plant origin are more effective and have little or no side effects in human beings in comparison to synthetic compounds (Kumar, *et al*; 1995).The natural compounds provide less phytotoxic, more systemic and easily biodegradable fungitoxic compounds (Saxena, *et al*; 2005).

Natural plant products have played a significant role in discovery of a new germicide, nematicide and to certain extent of viricide, either by their direct application to diseases or through their exploitation in the resilienting plant with optimum biological and physical properties.

In Prospects of good safety and public health, plant products satisfy following aims:-

1. They are harmless and environmental friendly.
2. Biodegradable
3. They can increase host metabolism.
4. They are slow in action of limited persistence.
5. Have a shelf life of 2-3 years.
6. Target specific.
7. Low mammalian toxicity.
8. Increase crop quality.
9. Systemic in nature.
10. Optimum bioactivity.
11. Increase benefit: cost ratio.
12. Novel mode of action.
13. Low toxicity.
14. Phytotoxicity of ecotolerance.
15. Less likely development of resistance.
16. Cost efficacy.

However, there are some drawbacks of botanicals control:

1. Loss of immediate killing action.
2. Higher dosage is require.
3. Some plant oils show phytotoxicity of lighter doses, for eg. *Azadirachta indica, Cymbopogon* sp.
4. Problems in preparation of formulation of spraying.
5. Some oils are non-edible, so their equipped toxicity problem also needs investigation
6. Shelf life is comparatibely less.

List of Promising Plants that have Pesticidal Properties

Plants are rich sources of bioactive organic chemicals. Of about 2,50,000-5,00,000 different plant species are known today that which have pesticidal properties but only <10 per cent have been examined chemically indicating that there is enormous scope for further work. Some of the most important families Meliaceae, Zingiberaceae, Apocynaceae, Compositae, Verbenaceae, Lamiaceae, Rutaceae, Solanaceae, Leguminaceae, Euphorbiaceae, contained plants with antifungal activity (Prasad and Ram, 2007 and Ray *et al.*, 2004) (Table 1.1).

Types of Antimicrobial Compound Present in Plants

Plants produce a number of secondary metabolites some of which have antimicrobial properties and help protect the plant against attack. The following important antimicrobial secondary metabolites are reported in plant (Agrios, 2005).

Flavonoides

These are phenolic compounds present in plant tissues especially the vacuoles. It is also present in various bark and heart wood. Flavonoides are inhibitory/toxic to pathogens and some of them act as phytoallexin which induces defense in plant against fungi eg: mediacarpin.

Glycosides

On wounding of plants, cynogenic glycosides and glycosinolates and their enzyme mingle and interact producing cyanide, isocyanates, nitriles and thiocynates. All these compounds are toxic to all organisms and also to fungi.

Saponins

Constitutive substances made from glycosylated steroidal triterpenoid compounds have antifungal membranolytic activity. Saponins form complexes with cell membrane, form pores and lead to loss of membrane integrity of pathogens. eg: tomatins in tomato, avenacin in oats.

Tannins

A group of compounds containing phenol, hydroxy acids and glucosides. They are phenolic oligomers. Condensed tannins are proanthocyanadins group of tannin formed by polymerization of flavonoids.

Phenols

Toxic phenolic compounds accumulate at faster rate after infection. *e.g.* Chlorogenic acid, caffeic acid and ferulic acid.

Table 1.1

Common Name	*Botanical Name*	*Family*	*Plant Part Used*	*Antifungal Compound*	*References*
Neem	*Azadirachta indica*	Meliaceae	Leaf, oil cake, bark, fruit pulp	Limonoid triterpenoid, Azadarachtin, tannin, tetranorterpenoid, protolimonoid.	(Lekhpal, *et al.,* 2008, Jat *et al.,* 2008, Singh 2008, Singh and Dewivedi, 1990.
Garlic	*Allium sativa*	Amaryllidaceae	Cloves	Diallyl Sulphide, Diallyl thiosulphide, Allicin	Burrow *et al;* 1998, Yamada and Azuma, 1971.
Onion	*Allium cepa*	Amaryllidaceae	Bulbs	Cycloallin and carbohydrate Propenyl Sulphonic acd, thiopropenol S oxide	Srivastava and Yadav, 2008, Sindhar *et al.,* 1999, Singh and Prithviraj 1997.
Ghrit Kumari	*Aloe vera*	Liliaceae	Leaves	Isoflavone glycosides, Hydroxanthra-quinone, alobin, aloesin, aloemodin, aloenoside-A, B and C, anglycone	Prasad and Ram, 2007
Custered apple	*Annona squamosa*	Annonaceae	Leaves	Diterpenoids anasquamosins, Annonacin, Annonelliptine, annonidines	Jat *et al.,* 2008, Srivastava and Yadav, 2008
Satavar	*Asparagus racemosus*	Liliaceae	Root	Saponins, Hydroxyecdyosone, Anthraquinone, Asparagic acid (S-containing), Steroidal glycosides asparaninl and asparanin B	Prasad and Ram, 2007, Mandal *et al.,* 2000
Kalmegh	*Andrographis paniculata*	Acanthaceae	Leaves, Stem, roots	Flavonoids, Phenols, Andrographosterol, andrographolide	Prasad and Ram
Bael	*Aegle mamrelos*	Rutaceae	Leaves and Fruit	Coumarine marmelosin, marmesin, marmin, xanthotoxol, methyl ether, pectin arabinose	Srivastava and Yadav, 2008
Babool	*Acacia nilotica*	Mimosaceae	Leaves, stem, bark	Triterpenoid, saponins, acaciasides A and B Sonnin	Srivastava and Yadav, 2008 Singh and Dwivedi, 1999
Adusa/ vasaka	*Adhatoda vasica*	Acanthaceae	Leaves	Alkaloids-vasinin, beouine, vaskin, vasicinone, vasicins anisotine, vasicid	Singh, 2008, Srivastava and Yadav, 2008
Vidhara	*Agorgyreia seciosa*				Srivastava and Yadav, 2008
Apamarga	*Achyranthus aspersa*	Amranthaceae	Leaves and seeds	Saponins, pentatriaconteins, triacontane, alkaloid-Achyranthine, betain	Srivastava and Yadav, 2008
Anethium	*Anethum graveolens*	Apiaceae			Srivastava and Yadav, 2008
Brahmi	*Bacopa monnieri*	Scrophulariaceae	Leaves, whole plant	Triterpenoid, saponins, bacosides, brahmine, hespertine, monnierin, flavonoids, glycosides	Mamatha and Rai, 2004

Contd...

Table 1.1–Contd...

Common Name	*Botanical Name*	*Family*	*Plant Part Used*	*Antifungal Compound*	*References*
Palas/Dhak/ Tesu	*Butea monosperma*	Leguminaceae	Stem/bark	Mediacarpin	Prasad and Ram, 2007
Baugainvillea	*Bongenvillea spectabilis*	Nyctaginaceae	Leaves		Allice and Rao 1984
Turmeric	*Curcuma longa*	Zingiberaceae	Leaves and rhizomes	Polypropanoid-diphenyl heptanoids cucurmin, cucurminoid, eugenol, phytosterol	Srivastava and Yadav, 2008
Amahaldi	*Curcuma amada*	Zingiberaceae	Leaves and rhizomes	Cucurmin; cucurminoid, eugenol, Phytosterol	Srivastava and Yadav, 2008
Dalchini	*Cinnamomum zeylanicum*		Bark	Eugenol, benzaldehyde,	Prasad and Ram 2007, Ray *et al.*, 2004
Chenopodium	*Chenopodium ambrosioides*	Chenopodiaceae	Leaves	Saponins, isoflavone, flavone	
Lemon grass	*Cymbopogon flexuosus*	Poaceae	Leaves	Citrol (75-85 per cent) linalool, geraniol, Citronellon terpenoids	Dhaliwal *et.al.* 2002
Raja grass	*Cymbopogan martini*	Poaceae		Geraniol	Prasad and Ram, 2007
Sadabahar	*Cartharanthus pusillium*	Apocynaceae		Pentacyclic alkaloid serpentine	Prasad and Ram 2007, Ray *et al.*, 2004
Citronella	*Cymbopogan winterianus*	Poaceae	Leaves	Terpenoid like linalool, eugenol, menthol, cineole and geraniol)	Dubey *et al.*, 1983; Agarwal *et al.*, 2000
Cumin	*Cuminum cyminmum*	Apiaceae	Seeds	Cuminaldehyde, cuminol	Dhaliwal *et.al.* 2002
Periwinkle	*Catharanthus roseus*	Apocynaceae	Leaves, flowers	Alkaloids like- ajamalicine, cantharanthine, akummacine, mannoside anthoxanthine, choline	Sreelatha and Bhagyanarayan 2008
Cannabis	*Cannabis sativa*	Cannabinaceae	Leaves, flowers	Tetrahydrocannabinol, Cannabidiol	Srivastava and Yadav, 2008
Papaya	*Carica papaya*	Carcicaceae	Leaves, flowers		Srivastava and Yadav, 2008
Amaltas	*Cassia fistula/ C.occidentalis*	Caesalpineaceae	Leaves, pods and seeds	Tannins, anthraquinone, emodin, steroids, sennoside, proanthocyanidin	Sharma and Basandari 2004, Srivastava and Yadav, 2008
Cleome	*Cleome viscosa*	Capridaceae	Leaves	Isothiocynate	Srivastava and Yadav, 2008
Coleus	*Colius aromaticus*	Lamiceae	Foliage	Diterpenoid, sitosterol, Rosaminic acid	Sreelatha and Bhagyanarayan 2008

Contd...

Table 1.1–Contd...

Common Name	*Botanical Name*	*Family*	*Plant Part Used*	*Antifungal Compound*	*References*
Crinium	*Crinum asiaticum*				Srivastava and Yadav, 2008
Callistemon	*Callistemon lanceolatus*	Myrteaceae	Leaves		Chaudhary *et al.*, 1986, Pandy *et al.*, 1986
Cissus	*Cissus quadrangularis*	Vitaceae	Stem		Srivastava and Yadav, 2008
Datura	*Datura metel, D.stromonium*	Solanaceae	Leaves	Steroid withametelin	Ray *et al.*, 2004, Lekhpal *et al.*, 2008
Yam	*Dioscoria bulbifera*	Dioscoreaceae	Root tuber rhizomes	Dioscorin, furastanol, glycosides (protodiosci and deltoside), saponins	Ray *et al.*, 2004
Desmostachya	*Desmostachya bipinnata*	Graminae	Leaves, flower		Srivastava and Yadav, 2008
Eucalyptus	*Eucalyptus citridoa*	Myrtaceae	Leaves and bark	Citronellol	Sindhar and Indra Hooda 2008, Dhaliwal *et al.*, 2001
Aonla	*Emblica officinalis*	Euphorbiaceae	Leaves and fruits	Phyllemblin Tannins (5 per cent), Salicylic acid	Singh, 2008
Chotidudhi, baridudhi	*Euphorbia microphyla, E.pulcherima*	Euphobiaceae	Leaves	Terpenoids	Sreelatha and Bhagyanarayan, 2008
	Ixora coccinia	Rubiaceae	Leaf		Srivastava and Yadav, 2008
	Inula viscosa				Ray *et al.*, 2004
Jatropa	*Jatropa gossypifolia, J. laminarioides*	Euphorbiacceae	Leaves	Methyl 2-propanamide 2 methoxycarbonylbenzoate	Sreelatha and Bhagyanarayan, 2008
Bhasmpatti	*Kalanchoe heterophylla*	Labiataeae	Leaves		Kumar *et al.*, 1979
	Lantana camara	Verbinaceae	Leaves	Triterpenoid-Camarinic acid, flavonoid glycosides-(Lantanoside, lantanosoid)	Lekhpal *et al.*, 2008, Mamtha and Rai, 2004
	Lawsonia inermis	Lythraceae	Leaves	2-hydroxy-1, 4-napthoquinone (lowsone)	Srivastava and Yadav, 2008, Bhatnagar *et al.*, 2004
Pundina	*Mentha arvensis*	Labiateae	Leaves		Dikshit *et al.*, 1980, Kaur and Singh, 1991
Murraya/ mitha neem	*Murraya konengii*	Rutaceae	Leaves	Koenimbine, Cadinene, Cary ophylens, Murryanol	Srivastava and Yadav, 2008

Contd...

Table 1.1–Contd...

Common Name	*Botanical Name*	*Family*	*Plant Part Used*	*Antifungal Compound*	*References*
Melia	*Melia azadirachta*	Meliaceae	Leaves	Limonoid glycoside, Tetranortriterpenoid-meliandiol, melianone, meliantriol, nimbolidin A	Ray *et al.*, 2004
Shajana	*Moringa olifera*	Moringaceae	Leaves, pods root and flower	Benzyl isothiocynnate, phenyl acetonitrile	Srivastava and Yadav, 2008
Tea tree	*Melaleuca alternifolia*			Thymol, eugenol	Ray *et al.*, 2004
Majorana	*Majorana hortensi*	Labiateae			Yadav and Saini, 1991; Keltawi *et al.*, 1980
Mint	*Mentha spp.*	Labiateae	Leaves	Methanol-docineoleole, geroniol	Prasad and Ram, 2007
Tulsi	*Ocimurr sanctum*	Lambiaceae	Leaves	Eugenol, methyl eugenol, Caryaphyllene, Terpenoides (linalool, eugenol, methanol, geraniol) triterpenoid oleanalic acids.	Sreelatha and Rhagyanaryan, 2008, Singh, 2008
Ocimum	*C canum giatisimum*	Lambiaceae	Leaves	Ethyl cinnamate	Lakhpale, *et al.*, 2008
Oxalis	*Oxalis quatosella*	Oxalidaceae	Leaves		Srivastava and Yadav, 2008
Marigold	*Tagetues erecta, T. pa ula*	Compositae	Whole plant		Prasasd and Ram, 2007
Thuja	*Thuja c iantalis*		Leaves and bark	β-sitosterol	Ray *et al.*, 2004
Tamrind	*Tamrinds indicus*	Caesalpineaceae	Seed		Mishra and Dixit, 1979
Arjun	*Terminalia arjuna*	Combretaceae	Leaves, bark	2α, 19αdihydroxy-3-oxo-olean -12 en 28 acid-28–o–β-D-glucopyranoside, Casuranin(tannin)	Ray *et al.*, 2004
Fenugreek	*Trigonella foenum granatum*	Papionaceae	Seeds	Flavnoids	Ray *et al.*, 2004
Tiliacora	*Tiliacora racemosa*		Roots	Tiliacorin	Ray *et al.*, 2004
Thevitia	*Thevetia peruvina*	Apocynaceae	Leaves seed, roots	Flavanone, flavonol and iriod glycoside	Srelatha and Bhagyanarayan, 2008, Srivastava and Yadav, 2008
	Thabernaemontana diveraricata	Apocynaceae			Ray *et al.*, 2004
Khas	*Viterveria zizanoides*	Poaceae	Roots	Kusimol, vetisnenol, β eudesmol	Gangada, *et al.*, 1990
Verbina	*Verbena offcinalis*		Leaves		Gangada, *et al.*, 1991

Contd...

Table 1.1–Contd...

Common Name	*Botanical Name*	*Family*	*Plant Part Used*	*Antifungal Compound*	*References*
Ashwagandha	*Withamia somnifera*	Solanaceae	Leaves, green berries, seeds and roots.	Alkaloid and steroidal compounds-withanolides, tropin, pseudotropins	Rao and Balasubramanyam, 1992
Sadabahar	*Vinca rosea*		Leaves		Singh, *et al.*, 1993
Portulaca/ Kulfa	*Portulaca oleracea*	Portulacaceae			Ohk *et al.*, 2000, Ray *et al.*, 2004
Patchauli	*Pogastemon patchaulii*	Labiateae	Leaves		Osawa *et al.*, 1990, Ray *et al.*, 2004
Bachochii	*Psoralea corylifolia*		Leaves, seeds	Phenolic meroterpine bakuchiol	Newton *et al.*, 2002, Ray *et al.*, 2004
Carrot grass	*Parthenium hysterophorus*	Astoaceae	Leaves, stems root, inflorescence	Parthenin, sesquiterpine lactone	Lekhpal, *et al.*, 2008
Polyalthia	*Polyalthia longifolia, P.suberosa*	Anonaceae	Leaves	Lanuginosine, oxostephanine, pendulamine A, Pendulamine B	Ray *et al.*, 2004, Srivastava and Yadav, 2008
Chitrak	*Plumbago zeylanica*		Root	Plumbagin	Durga *et al.*, 1990, Ray *et al.*, 2004, Gopal *et al.*, 1986,
Isabgol	*Plantago ovata*	Plantoginaceae	Seeds, husk	Mucilage, polysaccharides (20-30 per cent), Arabinoxylans (85 per cent), arabinose, xylan	Bhatnagar *et al.*, 2004
Sarpagandha	*Rauwolfia serpentina*	Apocynaceae	Root, stem, leaves, bark	Alkaloid, serpentine, reserpine, ajmaline	Prasad and Ram, 2006
Castor	*Ricinus communis*	Euphorbiaceae	Seeds and leaves	Recinin	Sreelatha and Bhagyanarayan, 2008
Giant Knot	*Reynoutria sachalinensis*				Ray *et al.*, 2004
Kava	*Piper methysticum*	Piperaceae	Roots		Singh, 2008
Sunflower	*Helianthus annus*	Asteracaea/ Compositae	Receptacle	Sesquiterpine lactone, 6-acetyl-2, 2-dimethyl-1, 2-benzopyran.	Shiv Puri *et al.*, 1997, Ray *et al.*, 2004
Sida	*Sida cordifolia*	Malvaceae	Leaves		Alam *et al.*, 1991, Srivastava and Yadav, 2008

Contd...

Table 1.1–Contd...

Common Name	*Botanical Name*	*Family*	*Plant Part Used*	*Antifungal Compound*	*References*
Ber	*Zizyphus jujube*	Rhamnaceae	Leaves and flower	Oleanolic acid and ursolic acid	Srivastava and Yadav, 2008
Hyptis	*Hyptis suveolens*	Lamiaceae	Roots	α-peltoboykinolic acid	Mishra *et. al.*, 1981
Nerium	*Nerium oleander*	Apocynaceae	Leaves	Nerifolin,	Ahmad *et. al.*, 1993
Eelgrass	*Zosetra marina*	Poaceae	Leaves	Zosetric acid p-sulpho-oxy-cinnamic acid	Singh, 2005
Clove	*Syzygium aromaticum*	Myrtaceae	Flower buds	Eugenol	Beg and Ahmad, 2002
Ginger	*Zingibar officinalis*	Zingberaceae	Rhizome	Sesquiterpines β-sesquiphellandrene	Sindhan *et al.*, 1999, Singh and Prithvi Raj, 1997

Terpenoids

It is a 10 carbon compound, a major component of essential plant oils and inhibitory effect against predators and pathogens.

Alkaloids

There are several types of alkaloids present in plant which have strong antifungal and antibacterial properties against number of diseases.

All these antifungal compounds are produced either constitutively or produced in response to attack by pathogen. Compound produced constitutively are called phytoanticipins while those produced in response to attack by pathogen are called phytoallexins.

The constitutive antifungal compounds belong to following major classes of secondary metabolites:

1. *Terpenoids*: *e.g.* Iridoides, sesquiterpines, saponins etc.
2. *Nitrogen and Sulphur containing compounds*: *e.g.* alkaloids, amides amines etc.
3. *Aliphatic*: long chain alkanes and fatty acids.
4. *Aromatics*: *e.g.* phenol, flavinoids, stilbinoids, biobonyls, xanthones, bimoguimones etc.

Again constitutive antifungal compound can be divided into following groups:

1. *Pre-inhibition*: In this type of inhibition, antifungal compounds are already present in plant from the very beginning. These include constitutive antifungal compounds or the phytoanticipines.
2. *Inhibition*: These inhibitory compounds are present in plant from the beginning but initially they are present at very low concentration, soon after infection it concentration increase enormously.
3. *Post inhibition*: These are antifungal metabolites produced by plant in response to infection. Post inhibitors are normally present in inactive form but are converted to active form after infection by means of start of simple biochemical reaction. For *e.g.* enzyme hydrolyses. Cynogenic glycosidase release toxins after wounding of leaves and change the inactive inhibition to active inhibition.

Fungal Diseases Management

Plant disease causes enormous amount of crop losses. However, maximum amount of loss causes due to fungal pathogen. The management of the disease can be done through cultural, chemical, biological, use of resistance varieties etc. But recently use of botanical fungicide are getting popularity in fungal disease management. Some examples are showing in Table 1.2.

Fungicidal Properties of Oil

Oil sprays protect against fungi probably by helping to repel the water that is needed for germination of spore and fungal growth. The following essential oils have been recommended for control of several diseases.

Petroleum Based oil

The orchardist, even before 1970s used to protect their trees from ravages of insect–pest by spraying with heavy dormant oils during spring. Now, a day's different type of horticultural products are available. As for example–Sun Spray is a refined horticultural oil product which is effective against

Table 1.2

Name of Diseases	*Causal Agent*	*Botanical Control*	*References*
Aphanomyces root rot of pea	*Aphanomyces euteiches*	Green manuring with oat, rape and sweet corn, soil application of paper mill residues.	Williums Woodward (1997)
Downy mildew of pearl millet	*Sclerospora graminicola*	Seed treatment with crude extract of *Vinca rosea, Ocimum sanctum, Allium sativum, Parthenium hysterophorus, Datura stromonium, Datura metel, Azadirachta indica, Thuja sinensis*	Ray *et al.*, 2004
Powdery mildew of pea	*Erysiphe pisi*	Nemadole a neem product, Extracts of *Allium cepa, Allium sativum,* rhizome of ginger and neem	Sindhar *et al.* (1999) Singh and Prithviraj (1997)
Powedry mildew of grapes	*Uncinula necator*	Oils of *Allium sativum, Cyprus scariousus, Cuminum cyminum, Cympopogon, Eucalyptus camaldulensis*	Dhaliwal *et al.* (2002)
Powdery mildew of apple	*Podosphaera leucotricha*	Oils of sunflower, olive maize soyabean, rape-seed.	Northover and Schneider, 1993
Ergot of sorghum	*Claviceps*	Garlic extract (12 per cent conc)	Singh and Navi (2000)
Common bunt/stinking smut of wheat	*Tilletia caries, T. foetida*	Yellow mustard meal	Varshrey (2001)
Strip diseases of barley	*Drechslera graminea*	Diluted neem and mariegold extracts	Varshrey (2001) Paul and sharma (2002)
Brown leaf spot of rice	*Drechslera oryzae*	Aquous extract of neem	Amadioha (2002) Beg and Ahamad (2002)
Blast of rice	*Pyricularia oryzae/ P. grasea*	Eelrass (Zosetra marina) NSKE or water extract of neem	Ahmad *et al.* (1993)
Mango anthracnose	*Colletotrichum gloeosporioides*	5 per cent leaf extract of *Ocimum sp.*	Kumar *et al.* (1997)
Anthracnose of bean	*C.lindemuthianum*	Oil of eelgrass (Zosteric acid)	Stanley *et al.* (2000)
Charcoal rot of soyabean/cluster bean	*Macrophomina phaseolina*	Pearlmillet, residue composet, cauliflower leaf compost	Lodha and Aggarwal (2002)
Rhizoctonia stemcanker and black scurf of potato	*Rhizoctonia solani*	Soil application of neem cake @ 25 g/hac.	Singh *et al.* (1972)
Sheath blight of rice	*Rhizoctonia solani*	Application of *kava* (*Piper methysticum*) root @ 1 ton/hac; Oil cake of margosa and coconut @1 ton/hac Rice husk	Meena and Muthuswamy (1999)
Fusarium wilt of tomato	*Fusarium oxysporum fsp lycopersici*	Flower extracts of *Desmostachya bipinnata* L.	Srivastava and Yadav (2008)
Fusarium wilt of cumin	*Fusarium oxysporum* fsp. *cumini*	Dhatura and Isabgol extracts	Bhatnagar *et al.*, 2004
Damping off of chilli	*Pythium aphanidermatum* (Edson Fitz)	*Crinium asiaticum, Ardisia crenulata, Eucalyptus totigoni, Parthenium hysterophorus, Allamanda catharlica* leaf extracts	Ramanathan *et al.*, 2004

Contd...

Table 1.2–Contd...

Name of Diseases	*Causal Agent*	*Botanical Control*	*References*
Banana fruit rot	*Colletotrichum gloiosporioides*	Leaf extracts of neem, marigold, custard apple, heena and lemongrass	Jat *et al.*, 2008
Karnal bunt of wheat	*Neovossia indica* (Mitra) Mundkar	*Azadiracta indica, Cassia fistula, Eucalyptus tersticornis*	Sharma and Basandrai, 2004
Black scurf of potato	*Rhizoctonia solani*	Extacts af garlic, datura, eucalyptus, ginger, onion and *Ocimum sanctum*	Shinde and Patel, 2004
Leaf blight of *Terminalia catappa*	*Fusaium solani*	Leaf extracts of *Lantana camara, Azadiracta indica, Acalypha indica* and *Baccopa monnieri*	Mamtha and Rai, 2004
Brown leaf spot of pearl miller	*Curvularia pennisiti*	Extracts of *Emblica officinalis, Azadiracta indica, Adhatoda vasica* and *Osimum sanctum*	Singh, 2008
Alternaria blight of *Brassica juncea*	*Alternaria brassici*	Foliar spray with bulb extracts of *Allium sativum*	Ray *et al.*, 2004

powdery mildew and sometimes against black spot of rose. However, the following care should be taken before applying such of oils-

They should not be used on:

1. Drought stressed plants
2. Plants weakened with diseases
3. When temperature exceed 85°F

Vegetable Oil

Emulsified vegetable oil sprays of sunflower, olive, conola, peanut, soyabean, corn, grapeseed or safflower can control powdery midew on apple trees, roses and other plants. Cotton seed oil has protectived value against powdery mildew. Vegetable oil are biodegradable and do not cause any long term problem.

Essential Oils

Many aromatic plant species eg. Eucalyptus, clove, mint, cumin fenugreek, basil, barera, sweet orange etc yield essential oils with potential fungicidal properties against broad spectrum of fungi. Some examples are given in Table 1.3.

Use of Neem in Fungal Disease Management

The multiferous use of neem is known from ancient time, but its scientific renaissance regarding insecticidal, fungicidal, nematicidal, antiviral properties came into existence quite recently. The several use of neem have been listed in the Table 1.4.

Management of Bacterial Diseases

The common practice to control bacterial disease is the use of antibiotics and chemicals. However, excess use of antibiotics may lead to development of acquired resistance in bacteria and use of broad spectrum antibiotics may lead to killing of non target pathogen. All these limitation of control has forced to search an alternative control measure. Therefore, bacterial disease management through botanical are getting popularity in the field of agriculture. It has been found that parts of some plant like root, bulb, rhizome, leaf, flower, seed, trunk etc. contain antibacterial substances and is using for disease management.

Table 1.3: Some Example of Essential Oils and their Use

Common Name	Botanical Name	Family	Part Used	Effective Against
Cumin	*Cuminum cyminum*	Umbelliferae	Seed	Sugarcane rot
Clove	*Syzygium aromaticum*	Myrtaceae	Flower bud (cuminaldehyde)	Sugarcane rot
Basil	*Ocimum sanctum*	Labiaceae	Leaves	Soil borne pathogen
Mint(fungal stop-commercial	*Mentha sp.*	Labiateae	Leave	Soil borne pathogen
Sweet orange	*Citrus sinensis*	Rutaceae	Leaves (linonene)	Bacterial disease
Bael	*Aegel marmelos*	Rutaceae		
Ajwain		Umbellifereae	Seed	
Barera	*Hyptis suavedens*	Lamiaceae	Leaves	

Table 1.4: Effect of Various Neem Products on Fungal Pathogens

Sl.No.	Neem Product	Effective Against Pathogen	References
1.	Neem oil cake	*Sclerospora sacchari* in maize.	Ray *et al.*, 2004
2.	Fruit pulp and Neem product	*Rhizoctonia solani* (supress formation of sclerotia)	Ray *et al.*, 2004; Biswas and Raychoudhary, 2003
3.	Extract oil	Inhibition of germ tube growth of *Erysiphe polygoni* (Singh and Singh, 1981)	Ray *et al.*, 2004
4-	Extract	Mycelial growth and spore germination of *Curvularia lunta, Alternaria alterneta, Drechslera turcica*	Bhowmick and Vardhan, 1982; Ray *et al.*, 2004
5-	Neem oil	*Fusarium monoliforme, M. phaseolina, D. rostrtatum,* powdery mildew, Black spots.	Ray *et al.*, 2004
6.	Leaf extract	*Pythium aphanidermatum*	Singh and Pandey, 1967
7.	Extract of leaf, trunk bark, fruit pulp and oil	*R. solani, S. rolfsi, F. o.* f sp.*ciceri*	Singh *et al.*, 1972
8.	Leaf extract	*Curvularia* sp. *Drechslera turcica*	Bhowmick and Vardhan, 1982
9.	Leaf extract	*F.eqiseti, F.semitictum* and *Curvularia lunata*	Krisna, *et al.*, 1986
10.	Leaf extract	*Mycosphaerella Berkeleyi* and *Puccinia arachidis*	Ray *et al.*, 2004
11.	Aqueous neem extract	*Aspergillus niger* and *Aspergillus flavus*	Bhowmick and Vardhan, 1982
12.	Ethenol extract of neem	*Pyricularia oryzae, R. solani, C.lunata, A. niger* and *F. moniliforme*	Mishra, and Tewari 1999

Antibacterial Principle

Although many plant products has been screened against bacterial plant pathogens, specific compounds responsible for antibacterial activity has not yet been eluted. However, in some plant products with antibacterial principles have been identified.

Kotasthane and Gupta (1986) reported the antimicrobial activity of essential oil. They also found that the skeleton of essential oil is terpenoids. The six selected terpenoids *viz.* citral, citronellol, citronellal, eugenol, and nerol were evaluated against all bacteria. Terpenoids are non toxic to plants and are highly volatile.

Inhibitory plant extracts from *Juglans nigra, Berberis vulgaris* and *Rhus typhira* as aquous solution was compatable with the antibiotic streptomycin (17ppm).

Effect of Neem in Bacterial Disease Management

Antipathogenic chemicals present in neem are effective in controlling many bacterial diseases. Kotasthane and Gupta (1986) reported that spraying of 10 per cent leaf extracts of *Prosopis guliffera, Acacia arabica*, NSKE and 2 per cent *Bacillus subtilis* showed BLB intensity grade of 2.88, 2.88, 3.0, and 2.0 against 5.70 under field condition.

Eswaramorthly *et al.* (1995) evaluated 25 per cent neem oil (emulsifiable concentrate, E.C.), 25 per cent kernel wettable powder and neem extract. Neemark also found effective against green pepper leaf spot incited by *X.c.* pv.*vasicatoria*.

Table 1.5: Bacterial Diseases Managed by Botanicals

Bacterial Disease	*Causal Organism*	*Botanical Control*	*Reference*
Citrus canker	*X.c.pv. citri*	Margosa, neem seed cake spray prepared by soaking 25 kg cake in 100 lit water and spraying @ 80 kg/acre	
Black arm of cotton	*X.c.pv. malvacearum*	Neem based formulation *viz.* plantolyte and agricare	Hulloli *et al.* (1998)
BLB of rice	*X. c.* pv *oryzae*	Leaf extracts of *Adhatoda vasica*, neem, turmeric and bulb extract of *Allium cepa.*	Madhiazhagan *et al.* (2002)
Bacterial spots of chillie and tomato	*X. c.* pv. *vasicatoria*	Week spray of aqous suspension of neem @ 0.5 (v/v)	Abbassi *et al.*, 2003
Bacterial canker of tomato.	*Clavibacter michiganensis*	Fragin an antibiotic isolated and purified from soluble fraction of strawberry leaf extract.	
Common scab of potato	*Streptomyces scabies*	Soil amendment with soyameat.	Abbassi *et al.*, 2003

Viral Disease Management

Viral diseases are transmitted by many means of mechanical transmission through sap, vegetative propagating materials, seeds, pollens and through vectors like insect, mites, nematodes, fungi etc. When virus is transmitted mechanically through sap, instense sanitation methods are carried out for the control but when it is transmitted by vectors, the best method is to control the vector.

Several antiviral proteins obtained from various plant parts have been found effective in various virus host system (Sendhilvel *et al.*, 2004) as shown in the Table 1.6.

When insects act as vector of plants viruses it damages the plant in two ways:

1. *Directly*–by feeding sap from plants.
2. *Indirectliy*–by transmitting virus from one plant to another.

Table 1.6

S.No.	Antiviral Protein	Source	Virus-Host System	Host
1.	Abrin	*Abrus precatorius*	TMV	*N.glutinosa* seeds
2.	Agrotins	*Agrostemma githago*	TMV	*N.glutinosa* seeds
3.	Bryodin	*Bryonia dioica* leaves and roots	TMV	*N.glutinosa* seeds
4.	Ca-SRI	*Cleodendrum aculeatum* leaves	TMV	*N. tabaccum/Cymopsis tetragonoloba, N.glutinosa* seeds
5.	Dianthins	*Dianthus cryophyllus*	TMV	*N.glutinosa* seeds
6.	EHL(*Eranthis hyemalis* Lectin)	*Eranthis hyemalis* bulb	AMV	*Vigna vigna*
7.	Gelonin	*Gelonium multiflorum*	TMV	*N.glutinosa* seeds
8.	MAP	*Mirabilis jalapa* roots	TMV	*N.glutinosa* seeds
9.	Modeccins	*Adenia digitata* roots	TMV	*N.glutinosa* seeds
10.	Momordin	*Momordica charantia*	TMV	*N.glutinosa* seeds
11.	PAP	*Phytolacca americana* seeds leaves	TMV	*N.glutinosa* seeds
12.	:995	*Phaseolous vulgaris*	TMV	*P. vulgaris*
13.	AIMV	*Chenopodium amranticolar*	PVX	*Gmphrema globosa*
14.	RICIN	*Ricinus communis*	ACMV CaMV Brassica compertris	*N. benthaminana*
15.	Soponins	*Saponaria officinalis*	TMV	*N.glutinosa* seeds
16.	YLP (*yucca recurvifolia* protein)	*Yucca recurvifolia* leaves	TMV	*C. amranticolour*

A number of plant extracts have shown to have excellent pest controlling properties against many insect hosts of plants pathogen which is indirect means of virus disease management.

Mechanism

The extact mechanism of action of plant product against virus diseases is still not known but it has been assumed that virus acquisition and transmission is interfered by influencing vector host relationship and feeding behavior with antifeedent chemicals. The possible mechanism of action of plant product are as follows:

Antifeedent Activity

As an effect of botanical pesticides, insect do not feed properly. Some example narrating antifeedent activities are:

1. Cabbage and cauliflower crops sprayed with NSKE aqueous extracts on leaf showed antifeedent activity against *Bravicornea brassicae* (cabbage aphid)
2. Pure azadirachtin, a limonoid terpenoid acts as insecticide or antifeedent against other insects but is weakly reactive against aphids.
3. Botanical product RA Repellin is highly repellent against pea aphid at concentration of 1 per cent, 4 per cent or 10 per cent which transmit Zucchini Yellow Mosaic Virus.

Effects on Growth and Physiology of Insects

Plant products affect the physiology of insects by its toxic affect and also by acting as growth regulator. It has been found that when azadirachtin solution applied at juvenile stage of insect, their growth was arrested. Insect may produce malformed minute adults or may be killed before reaching the adult stage.This may be due to delayed release of morphogenetic hormone. There is also evidence that azadirachtin interfere with neuro endocrine system which control ectysome of juvenile hormones, but how it is accomplished it is still unclear (Singh, *et al.*, 1993)

There are also several reports on direct toxic affect of neem against a number of insect vector.

Oviposition Deterrent

Application of plant extracts over foliage causes reduction of egg laying, viability of oviposition, prolonged period of larval mortality resulting in delay further spread of viruses from one plant to another

Induced Resistance

Antiviral proteins identified in many plants species can stimulate the active defense mechanism against viruses. Activation of such mechanism results in plants becoming highly resistant to subsequent infection. As for example flower extracts of *Azadirachta indica, Euphorbia milli* induce resistance against PVX infection in hypersensitive host *Chenopodium amranticolor*

Table 1.7: Some Plant-Products and their Efficacy

Sl.No.	*Plant Product*	*Virus*	*Vectors*
1.	Neem oil(5 per cent spray)	RTV	*N.virecens*
2.	Neem cake	"	"
3.	Custard apple oil	"	"
4.	Custard + neem oil	"	"
5.	Achras sapota	Ragimosaic	"
6.	Basella	Nubra	"
7.	Mirabilis jalapa	"	"

Nematode Management

Phytoparasitic nematodes are very difficult to control because most of them spend their lives in soil or within plant roots and the cuticle and other surface structure of nematodes are impermeable to many organic molecules. Therefore, nematologists are not optimistic about the chemical ways of nematode control. So attempts are made in the direction of plant products and phytochemical based strategies.

Research with crude plant extract has lead to the development of "Sincocin", the trade name of recently developed product containing a mixture of extracts from prickly pear (*Opuntia engelmanni lindheimeri*, Cacteaceae), southern red oak (*Quercus falcota*, Fegaceae), the sumac (*Rus aromatica*, Anacardiaceae) and the mangrove (*Rhizophora mangle*, Rhizophoraceae).In the field situation "Sincocin" has been found effective in controlling *Tylenchulus semipenetrans* (citrus nematode) on orange, *Rotylenchulus reniformis*(reniform nematode) on sunflower and cyst nematode *Heterodera schactti* on sugarbeet (Chitwood, 2002).

Table 1.8: Plant Produced Chemicals and Nematode Control

Common Name	*Botanical Name*	*Family*	*Nematode Antagonistic Phytochemical*	*Controlled Nematode*
Mariegold	*Tagetes* spp.	Compositae	Thiophene,α-terthienyl Tetrachloro-thiophene (Feldmesser *et al.*, 1985; Uhlenbroek and Bijloo, 1958)	*M.incognita, Pratylenchus penetrans, Globodera rostochinensis, Anguina tritici, Ditlenchus dipsaci*
Sudangrass	*Sorghum sudanensis*	Poaceae	Glucoside dhurrin (Widmer and Abawi, 2000)	*Meloidogyne halpa*
Cassava roots	*Mannihot esculenta*	Euphorbiaceae	Linamarin (*Cynogenic glucosides*) (Pant *et al.*, 1996; Sena and Pant, 1982)	*Meloidogyne* sp.
Susan	*Rudbeckia hista*		Thiorubrines, Polythenyles (San Cheze *et al.*, 1998)	*M. incognita, P. penetrans*
Helenium spp.	*Helenium* spp.	Apiaceae	Tridec-1-en-3,5,7,9,11 pentayne (Gommer, 1973, 1981)	*P. penetrans*
Angelica	*Angelica pubescens*	Fabaceae	Heptadeca-1,9-diene-4,6-diyene-3,8-diol (Munokata, 1983)	*Pratylenechus coffea, C. elegans*
Calabar bean	*Physostigma venenosum*	Brassicaceae	Physostigmine (Bijloo, 1965)	*D. dipsaci*
Black Mustard	*Brassica nigra*	Brassicaceae	Ally iso thiocyanates, thioglucoisonates (Euenby, 1945; Morgan, 1925)	*G. rostrochinensis*
Rapeseed/ canola	*B. napus*	Malvaceae	Isothiocyanates 2 phenyl ethyl glucosinolates (Halbrendp, 1996; Potter *et al.*, 1999)	*C. elegans, Xiphinema americanus, H. schachtti, G. rostrochinensis, Pratylenchus neglectus*
Cotton	*Gossipium hirsutum*	Lamiaceae	Hemigossypol, 6-methoxygossipol (sesquiterpenoids) (Cveech and Mcchure, 1977)	*M. incognita*
Oregana	*Origanum* spp.	Umbellifereae	Carvacrol and Thymol (Terpenoid) (Oka *et al.*, 2000)	*M. javanica*
Fennel	*Foeniculum vulgare*	Iridaceae	Transanethole (Terpenoid) (Oka *et al.*, 2000)	*M. javanica*
Iris	*Iris japonica*	Poaceae	Nyristic, Palmitic, Oleic, Myristic acids (Munakata, 1983)	*Many nematodes*
Timothy	*Phlenum prantense*	Umbelliferae	Butaric acid (Sayve, *et al.*, 1965)	*M. incognita, P. penetrans*
Caraway	*Caravi*	Geraniaceae	Carvone (terpenoid) (Oka *et al.*, 2000)	*M. javanica*
Pelargonium	*Pelargonium graveolans*	Papeveraceae	Linalool, geraniol, citronellol (Terpenoids) (La, *et al.*, 1992; Rao *et al.*, 1996)	*M. incognita*
Bacconia	*Bacconia cordata*	Liliaceae	Chelerythrin, Sangerinarine, Bacconine (alkaloids) (Onda, *et al.*, 1970; 1965)	*Rhabditis* spp., *Panagrolaimus* spp.

Contd...

Table 1.8–Contd...

Common Name	Botanical Name	Family	Nematode Antagonistic Phytochemical	Controlled Nematode
Gloriosa	*Gloriosa superba*	Fabiaceae	Cochicine (Nidiry, *et al.*, 1993)	*M. incognita*
Crotolaria	*Crotolaria spectabilis*	Poaceae	Pyrrolizidine alkaloid, Monocrotalin (Fassuliopis and Skucas, 1969)	*M. incognita*
Sorghum	*Sorghum bicolar*	Apocynaceae	Glycoside Dhurrin (Widmer and Abawi, 2000)	*M. hapla*
Madagaskar periwinkle	*Cartharanthus roseus*	Fabiaceae	Pentacyclic alkaloid serpentine (Rao *et al.*, 1996; Chandra Badana *et al.*, 1994)	*M. incognita*
Sophora	*Sophora flavescence*	Fabiaceae	N-methyl cystine, anagyrine, Matrine, sophocarpin and sophoramide (Matsuda, *et al.*, 1989, 1991)	*Bursaphelunchus xylophilus*
Velvet bean	*Mucana* sp.	Liliaceae	1-triacontanol and triacontanyl teteracosanoate (Nogueira, *et al.*, 1996)	*M. incognita*
Allium	*Allium grayri*	Papeveraceae	1-octanol (Tada, *et al.*, 1988)	*M.incognita*
Katili	*Argemone mexicana*	Iridaceae	Triglyceride (Sn–glycerol-1-eicosa-9,12-dienoate-2-palmitoleate-3-linoleate) (Salch, *et al.*, 1987)	*M. incognita*
Groundnut	*Arachis hypogea*	Lamiaceae	Di-n-butyl succinate (Kimura, *et al.*, 1981)	*P.coffiae*
Clove	*Eugenia caryophyllata*		Eugenol (Sangwar, *et al.*, 1990)	*A tritice, M. japonicum, T. semipenetrans, H. cajani*
Pepermint	*Mentha piperatum*		Menthol, Cineole, Geraniol (Sangwar, *et al.*, 1990)	*A. tritici, M. javanica, T. semipenetrans, Hctorodra cajani*

Mechanism

Numbers of nematode antagonistic compounds are obtained from many plants. They inhibit or control the nematode population by following ways:

Inhibition of Hatching of Eggs

The isothiocyanate compounds of black mustard seed was inhibited hatching of eggs of *Globodera rostochenensis*.

Inhibition of Motility

A combination of linalool and methyl chavicol isolated from essential oil of *Pelargonium graveolens* (Geraniaceae) inhibited motility of *H. cajani* and *M. incognita.*

Inhibition of Enzyme Action

All members of Brassicaceae produce thioglucose conjugate called glucosinolates which are hydrolysed either in soil or by mammal to form isothiocyanates.These isothiocynates react with sulphahydryl group of protein or enzymes and metabolism is disturbed.

Repellant

Heartwood extract from *Pinus massoniana* (Pinaceae) containing sesquiterpene humulene repelled *B. xylophilus. M. incognita* is repelled by high cucurbitacin produced by cucumber.

Induction of Mortality of Juveniles/Direct Toxicity

Linalool, geraniol and citronellol isolated from essential oil of *Pelargonium gravealens* (Geraniaceae) induced mortality in *M.incognita* juveniles. A triterpenoid caumeric acid from *Lantana camara* can directly kill *M.incognita* juvenile at 1 per cent concentration.

Inhibition of Penetration of Host

Three Quassinoids from seeds of *Hannoa undulata* (Simarubaceae)-chaparrinone, Klaineanone and glaucaubolone inhibited penetration in tomato roots by *M.incognita* at juvenile stage.

Inhibition of Host Finding Juvenile Ability

α-terthienyl from marigold inhibit host finding and induce mortality in entamopathogenic nematode *Steinernemaglaseri.*

Some commercial product of botanical pesticides:

1. Funga stop Commercial formulation of mint oil.
2. Sun Spray Refined horticultural oil.
3. Triology Neem oil formulation.
4. Rose defence Neem oil formulation.
5. Triact Neem oil formulation.
6. Milsana Knotweed (*Reynoutria sachalinensis*)

Conclusion

The pesticidal compounds of plant origin, are much more effective and have little or no side effects on human, still a lot of work needs to be done before large scale utilization of botanical pesticides. Once a potentially useful plant species is identified, intensive breeding and selection work will have to be undertaken so that economic production, from such plant of antagonistic compounds is possible. The safety and selectivity of such botanicals needs to be tested.

People are not interested at large in mode of action than in development of environmentally safe, inexpensive agonomicaly useful compounds. The international agronomical industry is well aware of both public perception of chemical hazards and customers deires for alternative methods to protect their liver hood. This has stimulated a wide range of technologist to complement synthetic chemicals with other cost effective natural products for all types of farmers in future.

There is likely that current global thrust on botanical pesticide may yield new leads and prototype modules to guide future research programme in plant protection.

References

Abbsi, P. A., Cuppels, D. A. And Lazarovits, G. (2003). Effect of foliar application of neem oils and fish emulsions on bacterial spot and yield of tomato and pepper. *Can. J. Pl. Path.* 25 (1): 41

Agarwal, K. K., Ahmad, A., Santha Kumar, T.R., Jain, N., Gupta, V. K., Kumar, S. and Khanuja, S. P. S. (2000). Antimicrobial activity spectra of Pelargonium grveolens, L. and Cymbopogon winterianus. Jowitt oil constituents and acyl derivatives. *J. Med. Arom. Plant Sci.* 22 (1B): 544-548.

Agrios. G. N. (2005). Plant Pathology, 4th edn. Academic Press, California, USA, pp. 635.

Ahmad, B. A., Sulayman,K. D., Abd, A.A. and Rashan, L.J. (1993). Antibacterial activity of the leaves of Nerium oleander. *Fitoterapia*, 64 (3): 273-274.

Akhtar M. (2000). Nematicidal potentials of neem trees *Azadirachta indica* (A.Juss). *Integrated Pest Management*. Rev.5: 57-66.

Alam, M., Joy, S. and Ali, U. S. (1991). Antimicrobial activity of *Sida cordifolia* Linn. *Sida Rhomboidea* Roxb and *Triumfetta rotundifolia* Lam. *Indian Drugs*. 28 (12): 570-572.

Alice, D. (1984). *Studies on antifungal activities of some plant extracts. M. Sc. Thesis, TNAU, Coimbatore.*

Allen, E. H., Feldmeser, J. (1971). Nematicidal effect of alfa tomatine on *Panagrellus redivivus*. *Phytopathology*. 60: 1013.

Allen, E. H., Feldmesser, J. (1971). Nematicidal activity of alfa chaconine: effect of hydrogen ion concentration. *J. Nematol*. 3: 58-61.

Amadioha, A.C. (2000). Controlling rice blast in vitrro and in vivo with extracts of Azadirachta indica. *Crop Protection*. 19(5): 287

Amadioha, A.C. (2002). Fungitoxic effects of extracts of Azadirachta indica against Cochliobolus miyabeanus causing brown spot disease of rice. *Arch. Phytopath. Plant Prot*. 35(1): 37.

Ames, B. N., Profet, M., Gold, L. S. (1990). Nature's chemicals and synthetic chemicals: comparative toxicology. *Prac. Natl. acad. Sci. U. S. A*. 87: 7782-86.

Anke, H., Stadler, M., Mayer, A., Sterner,O. (1995). Secondry metabolites with nematicidal and antimicrobial activity from nematiphagous fungi and Ascomycetes. *Can. J. Bot*. 73: 5932-39.

Ansari, M. M. (1995). *Indian Phytopath*. 48: 268-270.

Baldridge,G. D., O'Neill, N. R., Samac,D.A. (1998). Alfa-alfa (*Medicago sativa*, L.) resistance to root lesion nematode, *Pratilenchus penetrans;* defence response gene mRNA and isoflavonoid phytoallexin levels in roots. *Plant. Mol. Biol*. 38: 999-1010.

Bauske, E. M., Rodriguez-Kabana, R., Estaun, V., Kloepper, J. W., Robertson, D.G. (1994). Management of *Meloidogyne incognita* on cotton by use of botanical aromatic compounds. *Neatropica*. 24: 143-50.

Beg A.Z. and Ahmed, I. (2002). In vitro fungitoxicity of the essential oil of *Syzygium aromaticum*. World Jour. Microbiol. Biotechnol. 18(4): 317

Bhatnagar, K., Sharma, B. and Cheema, H. S. (2004). Efficacy of plant extract against *Fusarium oxisporum* f. sp. *cumini* wilt in cumin. *Journal of Mycology and Plant Pathology*. 34(2): 360-361.

Bhore, S. T., Thakare, C. S. and Kadam, V. (1995). Efficacy of some plant extracts and commercial plant products against test fungi. *J. Maharashtra Agric. Univ*. 20: 251-254.

Bhowmik, B. N. and Vardhan (1982). Antimicrobial activity of leaf extracts of some medicinal plants on *Drechslera turicica* (Pers) Subram. and Jain *Biol. Bullt*. India. 4 (1): 58-60

Bijloo, J. D. (1965). The Pisum test: a simple method for the screening of substances on their therapeutic nematicidal activity. *Nematologica*. 11: 643-644.

Bisht, S. and S. L. Srivastava. (1996). Effects of angiospermic extracts on the in vitro growth of different seed borne pathogens. *Annual Proc of Indian Nat sci con. Sec* 8: 115.

Biswas, A. and Roychoudhry, B. (2003). Relative efficacy of some botanicals against sheath blight disease of rice. *Journal of Mycopathological Research*. 41(2): 163-165.

Burrows, P.R., Barker, A.D.P. Newell, C.A., Hamilton, W.D.O. (1998). Plant derived enzyme inhibitors and lectins for resistance against plant parasitic nematode in transgenic crops. *Pestc. Sci.* 52: 176-183.

Chandravadna, M. V., Nidiry, E. S. J., Khan, R. M., Rao, M. S. (1994). Nematicidal activity of serpentine against *Meloidogyne incognita*. *Fundam. Appl. Nematol.* 17: 185-92.

Chitwood, D.J. (2002). Phytochemical based strategies for nematode control. *Annual Riview of phytopatology.* 40: 221-49.

Chowdhary, A. K. and Saha, N. K. (1985). Inhibition of urd bean leaf crinkle virus by different plant extracs. *Indian Phytopathol.* 38 (3): 566-568.

Dhaliwal, G. S. and R. Arora. (2001). Intensive agriculture and pest problems.*Integrated Pest Management.* Kalyani Publishers.

Dhaliwal, H.S., Thind, T.S., Mohan, C. and Chhabra, B.R. (2002). Activity of some essential oil against *Uncinula necator* causing powdery mildew of grapevines. *Indian Phytopath.* 55(4): 329.

Dixit, A., Singh, A. K. and Dixit, S. N. (1980). Fungitoxic activity of some essential oils against *Helminthosporium oryzae*. *Indian Perfum.* 24 (4): 222-223.

Dubey, G., Rao, T.S.S. and Nigam, S. S. (1983). Antifungal efficacy of some Indian essential oils. *Indian Perfum.* 27 (1): 5-8.

Dubey, S. C. (1998). *J. Mycol.Pl. Pathol.* 28: 266-269.

Durga, R., Sridhar, P. and Polasa, H. (1990). Effects of plunbagin on antibiotic resistance in bacteria. *Indian J. Med. Res.* 91: 18-20

El- Keltawi NEM, Megalla, S.E. and Ross, S.A. (1980). Antimicrobial activity of some Egyptian aromatic plants. *Herba Pol.* 26 (4): 245-250.

Ellenby, C. (1945). The influence of crucifers and mustard oil on the emergence of potato root eelworm, *Heterodera rostochinsis* Wollenweber. *Ann. Appl. Bio.* 32: 367-70.

Eswarmurthy, S., Muthusamy, M., Muthusamy, S. and Pariappan, V. (1989). Inhibitory effect of neem, Acacia, Prosopis and Ipomea extract on *Sarocladium oryzae* and *Fusarium oxysporium* f. sp. *cepae*. Neem News Lett. 6 (1): 4-5

Fassuliotis, G., Skucas, G. P. (1969). The effect of pyrrolizidine alkaloid ester and plants containing pyrrrolizidine on *Meloidogyne incognita acrita*. *J. Nematol.* 1: 287-288 (Abst).

Feldmessa, J., Kochanky, J., Jaffe, J., Chitwood, D. (1985). Future chemical for control of nematodes. In: *Agricultural Chemical of the future* (eds.) J. L. Hilton, Pp 327-44.

Gangrade, S. K., Shrivastava, R. D., Sharma, O. P., Jain, N. K. and Trivedi, K. C. (1991). *In vitro* antifungal effect of essential oils. *Indian Perfum.* 35 (1): 46-48.

Gangrade, S. K., Shrivasthava, R. D., Sharma, O. P., Moghe, M. N. and Trivedi, K. C. (1990). Evaluation of some essential oils for some essential oils for antibacterial properties. *Indian Perfum.* 34 (3): 204-208.

Gerard, E.J., V. Chandrshekhar and V. Kurucheve. (1994). *Indian phytopathology. 47: 183-185.*

Gommere, F. J. (1973). Nematicidal principles in Compositae. Diss. Agric. Univ., Wageningen, Neth. 73 pp.

Gommere, F. J. (1981). Biochemical interaction between Nematode and plants and their relevance to control. Helminthol. Abstr. 50B: 9-24.

Gopal, R. H. and Purushothamam, K. K. (1986). Effect of few plants isolates and extracts of bacteria. *Bull. Med. Ethnobot. Res.* 7 (1-2): 78-83.

Halbrendt, J. M. (1996). Alleopathy in managemnent of plant parasitic nematodes. *J. Nematol.*, 28: 8–14.

Hooda, K. S. and Srivastava, M.P. (1998). Biochemical response of scented rice as influenced by fungitoxicant and neem products in relation to rice blast. *Indian J. Pl. Pathol.* 16: 64-66.

Hulloli, S.S., R.P.Singh and J.P Verma (1998). Management of bacterial blight of cotton by Xanthomonas axonopodis pv.malvacearum with the use of neem based formulations. *Indian Phytopath.*51: 21.

Jain, T., and Sharma, K. (2007). Antifungal potential of *Polyalthia longifolia* Benth and Hook leaves. *Proc. Nat. Acad. Sci. India.*77 (1): *105-109.*

Jat, B. L., Gour, H. N. and Sharma, P. (2008). Effficacy of phytoextracts and bioagents against *Colletotrichum gloeosporioides* causing banana fruit rot. *J. Mycol. Pl. Pathol.*38 (3): 635-638.

Jayalakshmi, C. and Seetharam, K. (1998). Biological controls of fruit rot and die back of chillie with plant products and antagonistic microorganism. *Pl. Dis. Res.* 13: 46-48.

Jobling, J. (2000). Essential oils: A new idiea for post harvest disease control. *Good fruit and vegetables magazine.* 11(3): 50.

Kimur, Y. Mori, M., Hyeon, S. B., Suzuki, A., Mitsui, Y. (1981). Rapid and simple methods for assay of nematicidal activity and its application to measuring the activity of dicarboxylic acid. *Agric. Biol. Chem.* 45: 249-51.

Kotasthan, S. R. and Gupta, O. M. (1986). Management of soil borne diseases of crop plant-A seminar, Department of Plant Pathology, CPPS. TNU

Kumar, A., Srivastava and Bihari Lal. (1997). Studies on biofungicidal properties of leaf extracts of some plants. *Indian Phytoopathology.*50 (3): 408-411.

Lakpale, N., Khare, N. and Trimurthy, V. S. (2008). Evaluations of botanicals of antimicrobial properties. *Journal of Mycology and Plant Patology.* 38(3): 614-617.

Latif, M. A., Saleh, A. K. M., Khan, M. A. I., Rahman, H. and Hossain, M. A. (2006). Efficacy of some plant extracts in controlling seed borne fungal infections of mustard. *Bangladesh J. Micro Biol.* 23 (2): 168-170.

Leela, N. K., Khan, R. M., Reddy, P. P., Nidiry, E. S. J. (1992). Nematicidal activity of essential oil of *Pelargonium graveolens* against the root knot nematode. *Meloidogyne incognita. Nematol. Medit.* 20: 57-58.

Lodha, S. and Aggarwal, R.K. (2002). Inactivation of *Macrophomina phaseolina* propagule during composting and effect of compost on dry root rot severity and on seed yield of cluster bean. *Eur.J.Plant Pathol.* 108 (3): 253

Madhiazhagan, K., Ramadoss, N. and Anuradha, R. (2002). Effect of batanicals on bacterial blight of rice. *J. Mycol. Pl. Pathol.* 32(1): 68

Mahadevan, A. (1982). Biochemical aspects of plant disease resistance Part 1. Preffered inhibitory substance prohibition. *Today and tomorrow*. Printer and Publishers, New Delhi.

Mamtha, T. and Ravishankar Rai, V. (2004). Evaluation of fungicides and plant extracts against *Fusarium solani* leaf blight in *Terminalia cattappa. J. Mycol. Pl. Pathol.*34 (2): 306-307.

Mandal, S.C., Nandy, A., Pal, M., and Saha, B.P. (2000). Evaluation of antibacterial activity of *Asparagus racemous* willd. root, *Phytother. Res.* 14 (2): 118-119.

Matsuda, K., Kimura, M., Komai, K., Hamada, M. (1989). Nematicidal activities of (-)–N- methyl cysisine and (-)-anagyrine from Saphora flavescens against pine wood nematode. *Agric. Biol. Chem.* 53: 2287-2288.

Matsuda, K., Yamada, K., Kimura, M., Hamada, M. (1991). Nematicidal activity of matrine and its derivatives against pine wood nematodes. *J. Agric. Food. Chem.* 29: 189-191.

Meena, b. and Muthusami, M. (1999). Effect of organic amendments against rice sheath blight. *Indian Phytopath.* 52: 92.

mhtml: file: //L: Natural disease control-A common sense approach to plant first aid. mht.

mhtml: file: //L: /www_infonet_biovision org_Plant extract garlic. mht.

Misra, S. B. and Dixit, S.N. (1979). Antifungal activity.of leaf extracts of some higher plants. *Acta Bot. Indica.* 7 (2): 147-150.

Misra, T. N., Singh, R. S., Ojha, T. N. and Upadhyay, J. (1981). Chemical constituents of *Hyptis suaveolens.* Part-1. Spectra and biological studies a triterpine acid. *J. Nat. Prod.* 44 (6): 735-738.

Morgan, D. O. (1925). Investigations on eelwarm in potatos in south. *Lincolnshiri J. Helminthol.* 3: 185-92.

Munakata, K. (1983). Nematicidal natural products. In Natural Products for Innovative Pest Management, Ed. D. L. Whitehead, W.S. Bowsers, pp 293-310. Oxford: Pergamon.

Nene, Y. L., Thalpiyal, P. N. (1971). History. *Fungicide in Plant Disease Control.*

Newton, S. M., Lau, C., Gurcha, S. S., Berra, G. S., Wright, C. M. (2002). The evaluation of forty three plant species for *in vitro* antimycobacterial activities: Isolation of active constituents from *Psoralea corylifolia* and *Sanguinaria Canadensis. J. Ethopharmacol.* 79 (1): 57-67.

Nidiry, E.S. J., Khan, R. M., Reddy, P. P. 1993. *In vitro* nematicidal activity of *Gloriosa superba* seed extract against *Meloidogyne incognita. Nematol. Medit.* 21: 127-128.

Nogueira, M. A., de Oliveira, J.S., Ferraz, S. (1996). Nematicidal hydrocarbons from *Macuna aterrima. Phytochemistry*. 42: 997-998

Nogueira, M. A., de Oliveira, J.S., Ferraz, S., Dos Santos, M. A. (1996). Nematicidal constituents of *Mucuna aterrima* and its activity on *Meloidogyne incognita* race 3. *Nematol. Medit.* 24: 249=252.

Northover, J.and Schnieder, K. E. (1993). Activty of plant oils on diseases caused by *Podosphaera leucotricha, Venturia inaequalis* and *Albugo occidentalis. Plant Dis.* 77: 152.

Oh, K. B., I.M. Chang, K. J. Hwang and Mar, W. (2000). Detection of antifungal activity of portulaca oleraca by a single-cell bioassay system. *Phytother. Res.* 14 (5): 329–332.

Oka, Y., Nacar, S., Putievsky, E., Ravid, U., Yaniv, Z., Spiegel, Y. (2000). Nematicidal activity of essential oils and their components against root knot nematode. *Phytopathology*. 97: 710-15.

Onda, M., Abe, K., Yonezawa, K., Esumi, N., Suzuki, T. (1970). Studies on the constituents of *Bocconia cordata* 2. Bacconine. *Chem. Pharm. Bull.* 18: 435-439.

Onda, M., Takiguchi, K., Hirakuna, M., Fukushima, H., Akagawa, M., Naoi, F. (1965). Studies on constituents of *Bocconia cordata. Pt. I. On Nematicidal alkaloids Nippon Nagcikaguka Kaishi.* 39: 168-170.

Osawa, K., Matsumoto, T., Muruyama, T., Takiguchi, T., Okuda, K. and Takazoe, I. (1990). Studies of the antibacterial activity of plant extacts and their constituents against preodontopathic bacteria. *Bull. Tokyo. Dent. Coll.* 31 (1): 17-21.

Pandey, B. P. and Mohan, J. (1986). Inhibition of turnip mosaic virus by plant extracts. *Indian Phytopathol.* 39 (3): 489-491.

Paul, P.K. and Sharma, P.D. (2002). *Azadirachta indica* leaf extact induces resistance in barley against leaf stripe disease. *Physiol. Mol. Plant Pathol.* 61(1): 3.

Ponte, JJda, Cavada, B. S., Silveira-Filho, J. (1996). Teste com lectina no controle de *Meloidogyne incognita* em tomateiro. *Fitopatol. Bras.* 21: 489-491.

Potter, M. J., Vanstone, V. A., Davies, K. A., Kirkegaard, J. A., Rathjen, A. J. (1999). Reduced susceptibility of *Brassica napus* to *Pratylenchus neglectus* in plants with elevated root levels of two phenyl ethyl glucosinolates. *J. Nematol.* 31: 291-298.

Prasad, J. and Ram, D. (2007). Handbook of medicinal and aromatic plants. Department of Horticulture, N.D.U.A. and T, Faizabad pp. 64

Ramanathan, A., Marimuthu, T. and Raguchander, T. (2004). Effects of plant extracts on growth of *Pythium aphanidermatum. J. Mycol. Pl. Pathol.*34 (2): 315-316.

Rao, K. and Balasubramanian,K. A. (1992). Management of pearl millet downy mildew by seed treatment with plant extract. Abstr. *Indian phytopath.* 44 (Suppl).

Rao, M. S., Reddy, P. P., Mittal, A., Chandravadna, M. V., Nagesh, M. (1996). Effect of some secondry plant metabolites as seed treatment agents against *Meloidogyne incognita* on tomato. *Nematol. Medit.* 24: 49-51.

Ray, A. B., Sharma, B. K. and Singh, U. P. (2004). Medicinal Properties of Plants; Antifungal, Antibacterial and Antiviral Activities. IBDC Publishing Division.

Saleh, M. A., Abdel Rahman, F. H., Ibrahim, N. A., Taha, N. M. (1987). Isolation and structure determination of new nematicidal triglyceride from *Argimone Mexicana. J. Chem. Ecol.* 13: 1361-70.

Sanchez de Viala, S., Brodie, B. B., Rodriguez, E., Gibson, D. M. (1998).The potential of thiarubrine C as a nematicidal agent against plant parasitic nematodes. *J. Nematol.* 30: 192-200.

Sangwan, N. K., Verma, B. S., Verma, K. K., Dhindsa, K. S. (1990). Nematicidal activities of some essential plant oils. *Pestic. Sci.* 28: 331-335

Satinder Kaur, and Sinha, G. K. (1991). *In vitro* antifungal activity of some essential oils. *J. Res. Ayur. Siddha.* 12 (3-4): 200-205.

Sawant, G.G. (1999). Studies on biofungicidal properties of some plant extracts on growth of *Alternaria solani* and *Colletotrichum lagenarium* in vitro. *Indian J. Environment and Toxicol.* 9: 46-48.

Saxena, A. R., Sahni, R. K., Yadav, H. L., Upadhyay, S. K. and Saxena, M. (2005). Antifungal properties of some higher plants against *Fusarium oxisporum* f. sp. *pisi.*

Sena, E. S. And Ponte, JJda. (1982). A manipueria no controle da meloidoginosa da cenoura. *Rev. Soc. Bras. Nemato.* 6: 95-96.

Sharma, B. K. and Basandrai, A.K. (2004). Efficacy of fungicides and plant leaf extracts for the control of Karnal bunt of wheat.*J. Mycol. Pl. Pathol.* 34(1) 102-104.

Shinde, G. R. and Patel, R. L. (2004). Evaluation of plant extracts against *Rhizoctonia solani* incitant of black scurf disease of potato. *J. Mycol. Pl. pathol.* 34 (2): 284-285.

Shivpuri, A., O. P. Sharma and S. L. Jhamaria (1997). *J. Mycol. Pl. Pathl.* 27: 29-31.

Sindhan, G. S. and Hooda, Indra. (2004). Control of white rust of mustard with bioconrol and plant extracts. *J. Mycol. Pl. Pathol.* 34(2): 286-287.

Sindhan, G. S., Hooda, I. and Parashar, R. D. (1999). Evaluation of plant extracts for the control of powdery mildew of pea. *Indian Phytopath.* 53: 257.

Singh, H. N. P., M. M. Prasad and K.K. Sinha. (1993). *Lett in appli. Microbiol.* 17: 289-291.

Singh, R. K. and R. S. Dwivedi. (1987). Effects of oils on *Sclerotium rolfsii* causing root rot to barley. *Indian Phytopathology*.40: 531-533.

Singh, R. K. and R. S. Dwivedi. (1990). Fungicidal properties of neem and baboolgum against *Sclerotium rolfsii. Acta Bot. Indica.* 18: 260-262.

Singh, R. S. (2005). *Plant Diseases.* Oxford & IBH Publishing Co., New Delhi, pp. 564.

Singh, R.S., Chaube, H. S. and Singh, N. (1972). Studies on control of potato black scurf disease. *Indian Phytopathol.* 25: 343.

Singh, S. D. and Navi, S. S. (2000). Garlic as a biocontrol agent for sorghum ergot. *J. Mycol. Pl. Pathol.* 30: 350.

Singh, U. P. and Prithviraj, B. (1997). Neemazol-a product of neem (*Azadiracta indica*) induces resistance in pea against *Erisiphe pisi. Physiol. Mol. Plant Pathol.* 51: 181.

Sreelatha, R. and Bhagyanarayan, G. (2008). Biocontrol efficacy of botanicals on anamorphic fungi. *J. Mycol. Pl. Pathol.* 38(3): 647-649.

Srivastava, D. K. and Yadav, H.L. (2008). Antifungal activity of some medicinal plants against *Fusarium oxisporum* f. sp. *lycopersici. Indian Phytopathology.* 61(1): 99-102.

Stanley, M. S. Callow, M. E., Perry. (2002). Inhibition of fungal spore spore adhesion by zosteric acid as the basis for the novel non toxic crop protection technology. *Phytopathology.* 92(4): 378.

Swaminathan, M. S. (1978). Inaugural address, Fruit, Bot. Conference, Meerut, India, pp. 1-31.

Syre, R. M., Patrick, Z. A., Thorpe, H. J. (1965). Identification of a selective nematicidal component in extracts of plant residues decomposing in soil. *Nematologica.* 11: 263-68.

Tada, M., Hiroe, Y., Kiyohara, S., Suzuki, S. (1988). Nematicidal and antimicrobial constituents from *Allium grayi* Regel and *Allium fistulosum* L. var *caespitosum. Agric. Biol. Chem.* 52: 2383-2385.

Uhlenbroek, J. H. and Bijloo, J.D. (1959). Investigations on nematicides II. Structure of second nematiidal principle isolated from Tagetes root *Rec. Trav. Pays. Bas.* 78: 382-90.

Varshey, V. (2001). Effect of plant extracts on *Drechslera graminea*, the causal agent for stripe disease of barley.*Indian Phytopathol.* 54: 88.

Veech, J. A., Mc Clure, M. A. (1977). Terpenoid aldehydes in cotton roots susceptible and resistant to the root knot nematode, *Meloidogyne incognita. J. Nematol.* 9: 225-229.

Widmer, T. L. and Abawi, G. S. (2000). Mechanism of suppression of *Meloidogyne hapla* and its damage by a green manure of sudan grass. *Plant Diseases*. 84: 562-568.

Williams–Woodward, J.L. (1997). Green manures of oat, rape and sweet corn for reducing common root rot in pea (*Pisum sativum*) caused by *Aphanomyces euteiches. Plant Soil.* 188 (1: 43)

Yadava, R. N. and Saini, V. K. (1991). Antimicrobial efficacy of essential oils of *Majorana hortensis* and *Anisomeles indica.* (Linn) Kunteze. *Indian Perfum.* 35 (1): 58-60.

Yamada, Y. and Azuma, K. (1977). Evaluation of the *in vitro* antifungal activity of allicin. *Antimicrob. Agents. Chemother.* 11 (4): 743-749.

Sustainable Disease Management of Agricultural Crops (2011) *Pages* **27–49**
Editors: **S.K. Biswas and S.R. Singh**
Published by: **DAYA PUBLISHING HOUSE, NEW DELHI**

Chapter 2

Recent Advance Approaches in Biological Management of Soil Borne Diseases of Pulse Crops

☆ *R.K. Prajapati*[1], *S.R. Singh*[2] *and R.K. Gangwar*[3]
[1]*Krishi Vigyan Kendra, Tikamgarh, M.P.,*
[2]*Krishi Vigyan Kendra, Belatal, Mahoba – 210 423, U.P.*
[3]*Krishi Vigyan Kendra, Chomu (Tankarda), Jaipur, Rajasthan*

Pulses are an integral component of Indian farming. Its importance is signified in terms of ecology, soil health and human and animal nutrition. Out of a dozen of pulse crops grown in this country, only few such as chickpea [*Cicer arietinum* L.], pigeonpea [*Cajanus cajan* (L.) Millsp.], mungbean [*Vigna radiata* (L.)Wilczek], Urdbean [*V. mungo* (L.) Hepper], lentil [*Lens culinaris* Medik.], fieldpea [*Pisum sativum* L.] and dry beans or rajmash [*Phaseolus vulgaris*] are popularly grown in different seasons. The green revolution, which boosted production of cereal, and soil seed crops many fold, could not accelerate pulses production much. The present level of total pulse production of 13.35 million tonnes in 1999-2000 from 21.29 million ha of land provides its productivity level of 630 kg/ha against the potential yield of 2000 kg/ha. It is so because of its cultivation on marginal lands, without giving almost no crop input, lack of irrigation and high proneness to biotic stresses especially the insect pests and the diseases. Conservative estimate show around 8-10 per cent annual yield loss in pulses due to the attack by diseases. Resistant varieties have been developed only against few major diseases that too have not yet reached the farmers all around. Therefore, their eco-friendly, cost effective management is the need of the day and the use of biocontrol agents is certainly an important and inspiring approach.

Important Disease Problems

There is a heavy toll of pathogens that attack the pulse crops. Among them, fungal pathogens, particularly the soil borne causing root and stem diseases are economically most important. Many of them are also seed borne and play active role in their dissemination distribution and annual recurrence. Vishwa Dhar *et al.* (2000) has reviewed the diseases of pulse crops. Besides, reviews on diseases of chickpea (Haware, 1998; Nene and Reddy, 1987; Gurha *et al.*, 2003; Gurha, 2004), pigeonpea (Reddy and Vishwa Dhar 1998, Reddy *et al.*, 1990; Vishwa Dhar and Chaudhary, 1998, 2004; Vishwa Dhar *et al.*, 2005), mungbean and urdbean (Singh and Naimuddin, 2004) lentil (Agrawal and Prasad, 1997; Saxena *et al.*, 1998; Chaudhary, *et al.*, 2004), peas (Kraft *et al.*, 1998; Vishwa Dhar and Chaudhary, 1998; Chaudhary and Naimuddin, 2000; Chaudhary *et al.*, 2004) and rajmash or bean (Allen *et al.*,, 1998; Chaudhary *et al.*, 2004) are available. Fortunately, only limited number of fungal pathogens, which infect root, and stem of pulse crops in India (Table 2.1) are important and have the potential to be managed through biocontrol approaches.

Table 2.1: Fungal Pathogens Infecting Root and Stem of Pulse Crops in India

Name of the Disease	Causal Organism	Crop	Nature of Pathogen	Problematic Areas
Wilt	*Fusarium oxysporum* f.sp. *ciceri*	Chickpea	Seed and soil borne	NEPZ, NWPZ, CZ,SZ
Wilt	*Fusarium oxysporum* f.sp. *lentis*	Lentil	Seed and soil borne	NEPZ, CZ
Wilt	*F. udum*	Pigeonpea	Seed and soil borne	NEPZ,CZ,SZ
Black root rot	*F. solani*	Chickpea, lentil, pea, rajmash	Seed and soil borne	NEPZ, CZ, NWPZ, NHZ
Root rot	*Fusarium* spp.	All crops	Soil borne	ALL ZONES
Wet root rot	*Rhizoctonia solani*	All crops	Soil borne	ALL ZONES
Dry root rot	*R. bataticola*	All crops	Soil and seed borne	NEPZ, CZ, SZ
Stem canker	*Macrophomina phaseolina*	Pigeonpea	Soil and seed borne	NEPZ, CZ, SZ
Collar rot	*Sclerotium rolfsii*	All crops	Soil borne	ALL ZONES
White rot/stem rot	*Sclerotinia sclerotiorum*	Chickpea, pea, rajmash, lentil	Soil borne	NHZ, NEPZ, NWPZ
Blight	*Ascochyta pisi, A. pinodes, A. pinodella*	Pea	Seed and Soil borne	NHZ, NWPZ, NEHZ
Blight	*A. rabies*	Chickpea	Seed and soil borne	NWPZ
Blight	*Phytophthora drechsleri* f.sp. *cajani*	Pigeonpea	Soil borne	NEPZ, NWPZ, CZ
Root rot/ damping off	*Pythium* spp.	All crops	Soil borne	NEPZ, NHZ, NEHZ

Economics and Biocontrol

The increasing value of pulse crops, severity and economic losses caused by the fungal pathogens in these crops, cost of application of biocontrol agents and economic gains by their use are the important and deciding factors for farmers acceptance of recommended bioagents. The incentives to optimum would be high if the crop value is high, the biocontrol agent is stably effective and the disease is economically important. Studies showed that seed treatment prior to sowing are highly cost effective but many a times they fail to deliver the desired results primarily because of the adverse edaphic factors for the bioagent and secondly its quality in terms of effective strain or quantity of viable propagules in the formulation. Soil application though bit costly, it provides more effective disease control and better sustainability over the crop seasons and years. In addition to short term gains through disease control in the current crop, microbial biocontrol agents show long term carryover in subsequent crops and manage their diseases in the years to come (Adams and Ayers, 1982). It is also noticed that these bioagents, particularly the *Trichoderma* spp., significantly increase the plant emergence, growth of seedlings and nodulation in pulses (Baker *et al.*, 1984; Chang *et al.*, 1986; Kloepper *et al.*, 1988; Siddiqui *et al.*, 1998; Chaudhary *et al.*, 2003). However, plant pathologists have largely ignored this role of microbes in plant growth.

The environmental and social gains of bioagents use are largely over looked while considering the economics. The adverse effects of pesticides in terms of soil, water and air pollution, detrimental effects on other beneficial organisms and creatures, health hazards to human being and domestic animals and the development of pesticide resistant pest population are the established and known facts. The environmental gains of these costs can be reduced by replacing pesticides with bioagents. Besides, small production units for the bioagents can be established in the nearby areas of the end user (farmer) while for pesticides it is not. Largely all these economic/social aspects are often ignored by all.

Present Status

Biological control of plant pathogens differs from the biocontrol of insect pests in many ways. Firstly, the management occurs at microbial level typically in a plant ecosystem such as root, stem, leaf or pod/fruit surfaces. Secondly, bioagents of pathogens act in three different ways. While the parasites of pathogens function almost the same way as do the natural enemies (parasitoids) or orthopods, the competitors function by occupying and using resources in a non pathogenic manner and disallow the pathogenic microorganism in colonizing the tissues and many others act through the strong mechanism of antibiosis. Although many mycoparasites and antagonists have been reported as potential bioagents, only few of them have been successfully exploited for the effective management of plant diseases in green house and field. The development of microbial bioagents for commercial use has been quite slow in the past. The bioagents seem to be a viable alternatives to chemicals provided they could be established in the same ecosystem where pathogens exist and this is possible through their augmentation either naturally or through manipulation and with this process they out compete the pathogens in these niches.

Among the antagonistic fungi *Trichoderma* species are the most studied bioagents in pulse eco-system. They are successfully used to control wilt, root rots, collar rot and stem rot diseases incited by different *Fusarium* spp., *Rhizoctonia solani, R. bataticola, Sclerotium rolfsii, Sclerotinia sclerotiorum, Ascochyta* spp. and *Pythium* spp., in different pulses and other field crops. Other fungi reported to suppress the fungal diseases in pulses are *Aspergillus candidus, A. niger, Pacilomyces lilacinus, Glomus mosseae, Verticillium chlamydosporum, Absidia cylindrospora, Coniothyrium minitans, Talaromyces flavus* and *Trichothecium roseum.*

Two bacterial bioagents *viz., Pseudomonas fluorescens* and *Bacillus subtilis* are also commercially exploited for the management of pulse crops diseases. Pathogen wise studies conducted in the management of pulse crop diseases have been listed in Table 2.2.

Mode of Action and Delivery Systems

The basic philosophy of biocontrol of diseases is that the bioagent should be in active state in a right place and at a right time *i.e.* it should be active in the infection court at the time of active period of the pathogen. A single application technology cannot be effective in all the situations. The strategies used for the biocontrol of soil borne pathogens include protection of the infection sites, impeding the

Table 2.2: Bio-control Agents Effective Against Soil Borne Pathogens Infecting Pulse Crops

Name of Bioagent	Effective Against	Crop	References
T. harzianum	***Sclerotium rolfsii***	Pea Bean	Elad *et al.* (1980) Siveira (1994)
T. koningii		Chickpea	Mukhopadhyay (1998)
T. harzianum, T. viride		Bean	Robert *et al.* (1996)
Gliocladium virens		Bean	Mukherjee and Tripathi (2000)
T. harzianum, T. virens, T. Viride		Chickpea	Prasad R.D., Rangeshwaran (2001)
T. harzianum, T. koningii, T. viride, G. virens, A. candidus, P. lilacinus, Bacilus sp.		Lentil	Iqbal *et al.* (1995)
T. viride	***Rhizoctonia bataticola (M. phaseolina)***	Mungbean	Raghuchander *et al.* (1993) Kehri *et al.* (1991)
T. viride, T. harzianum and *Bacillus subtilis*		Bean	Majumdar *et al.* (1996)
T. koningii, T. harzianum, T. harzianum, A. niger		Cowpea	Adekunle *et al.* (2001) Pandey *et al.* (2000), Braga *et al.* (2003)
T. viride, Chaetomium, Globosum		Pigeonpea	Pradeep Kumar *et al.* (2000)
T. viride, T. harzianum		Cowpea	Ushamalini *et al.* (1997)
T. viride, T. harzianum, T.longi brachiatum		Cowpea	Mathur and Srivastava (2005)
T. viride, P. fluorescens, Bacillus subtilis		Pigeonpea	Kaswate *et al.* (2003)
T. viride, T. harzianum, A. flavus, A. niger, A. ochraceus, Pencillium citrinum, Bacillus subtilis		Chickpea	Sindhan *et al.* (2002)
T. harzianum, T. viride		Bean	Singh *et al.*, 1993
T. harzianum	***Rhizoctonia solani***	Bean	Elad *et al.* (1980), Strashnov *et al.* (1985), Noranha *et al.* (1996)
T. harzianum, Gliocladium		Lentil, chickpea	Mukhopadhyay (1988)
		Chickpea	Prasad and Rangeshwaran (1999, 2006, 2002)
		Pea	Elad *et al.* (1982)
T. viride		Bean	El-Farnawany *et al.* (1996)
T. harzianum, G. virens, P. Lilacinus		Bean	Enteshmanl *et al.* (1992)
T. harzianum, G. virens		Bean	Mathew and Gupta (1998)
T. koningii		Bean	Lacicowa (1990)
T. hamatum		Pea	Harman *et al.* (1980)
T. viride, C. globosum		Pigeonpea	Pradeep Kumar *et al.* (2000)
T. viride, T. Harzianum, T. Virens		Bean	Harzarika and Das (1998)
		Lentil	Sadowskis and Kryzia (1991)
		Pea	Velikanov *et al.* (1994)

Contd...

Table 2.2–Contd...

Name of Bioagent	Fffective Against	Crop	References
T. harzianum	***Fusarium udum***	Pigeonpea	Prasad and Rangeshwaran (2002)
Aspergillus nidulans		"	Upadhyay (1987)
T. harzianum, T. Viride and *G. Virens*		"	Chaudhary and Kumud Kumar (1999); Pandey and Upadhyay (1999, 2000)
Bacillus subtilis		"	Vasudeva (1949) Vasudeva and Roy (1950)
Bacillus bruis		"	Sangita Bapat *et al.* (2000)
T. viride, T. harzianum and *T. koningii*		"	Bhatnagar, (1996)
T. viride, A. niger and *Penicillium* sp.		"	Gaur and Sharma (1991)
T. viride		"	Bidari and Gundappa (1997), Mahalinga *et al.* (2004)
T. harzianum, T. koningii, T. viride and *T. hamatum*		"	Somasekhara *et al.* (1996)
Glomus mosseae, T. harzianum, Verticillium chlamydosporum		"	Siddiqui and Mahmood (1996)
B.subtilis, Bradyrhizobium Japonioum		"	Siddiqui and Mahmood (1995)
T. koningii, T. viride, T. lamatum, T. Pseudokoningii		"	Jayalashmi *et al.* (2003)
T. harzianum, A. niger, G. Virens		"	Chaudhary and Prajapati (2004)
P, fluorescens, Bacillus subtilis		"	Shazia *et al.* (2005)
T. hamatum, T. longiconis, P. koningii		"	Biswasand Das (1999)
T. viride, P. fluorescens, Bacillus subtilis		"	Madukeshwara and Seshadri, (2001)
T. viride, P. fluorescens		"	Gholve Mudkar (2002)
T. harzianum, T. viride, T. virens, Bacillus subtilis		"	Singh *et al.* (2002)
Trichoderma sp	***Fusarium oxysporum* f.sp. *ciceri***	Chickpea	Mukhopadhyay (1988)
T. harazianum		"	De *et al.*, (1996)
T. harzianum, T. viride and *G. Virens*		"	Khot *et al.* (1996), Singh *et al.*, (2003), Harichand *et al.*, (2005)
P. fluorescens		"	Vidhyasekaran and Muthamilan (1995)
P. lilacinus, A. niger		"	Akhtar *et al.* (2002)
T. viride, P. fluorescens		"	Shamaro-Jahagirdar (2002)
B. subtilis, P. fluorescens		"	Landa *et al.* (2004)
T. harzianum, Rhizobium ciceri		"	Poddar *et al.*, (2004)
T. harzianum, P. fluorescens		"	Khan *et al.*, (2004)
T. harzianum, T. viride, P. fluorescens, B. subtilis		"	Dhedhi *et al.* (2001), Agrawal *et al.*, (2002)
T. harzianum, T. hamatum, P. fluorescens, Rhizobium, T. Viride		"	Sonawane and Pawar (2001)

Contd...

Table 2.2–Contd...

Name of Bioagent	Effective Against	Crop	References
T. viride	***F. oxysporum* f.sp. *lentis***	Lentil	De and Chaudhary (1999), Haware *et al.*, (1999) Abha Agrawal *et al.*, (1999)
T. lamatum		"	El. Hassan and Gowen (2003)
T. harzianum, G. virens, Pseudomonas fluorescens		"	De *et al.* (2003)
T. harzianum	***Sclerotinia***	Pea	Singh (1991)
T. koningii	***sclerotiorum***	Bean	Lacicowa (1990)
T. harzianum, Absidia, Cylendrospora		Chickpea	Sharma *et al.* (1999)
Coniobhyrium minitans, Talaromyces flavus, T. virens and *Trichothieim roseum*		Bean and Pea	Huang *et al.*, (2000)
T. harzianum, T. viride and G. Virens	***Pythium* spp.**	Pea	Velikanov *et al.* (1994)
		Chickpea	Shahriazry *et al.* (1996)
T. harzianum and G. virens		Pea	Steinmetz and Schoubeck (1994)
T. koningii and T. harzianum, Rhizobacteria		Pea	Hadar *et al.*, (1984), Xi *et al.*, (1996)
	Other *Fusarium* spp.		
T. harzianum, T. viride and *G. Virens*		Bean	Velikanov *et al.* (1994)
G. virens		Bean	Mukherjee and Tripathi (2000)
G. roseum	***Mycosphaerella Pinodes***	Pea	Xue (2000)

progress of pathogen in soil and inactivation of surviving structures of the pathogen. Bioagents that work through high competitive ability or antibiosis are employed for protecting the infection court and restrict the growth of pathogen in the soil where as the mycoparasites are often used for the destruction of surviving structures (Lumsden *et al.*, 1995) and the bioagents such as *Trichoderma* spp., which act through all three modes (Nigam *et al.*, 1997; Sumeet and Mukherji, 2000) are more versatile and better effective. Dennis and Webster (1971b) isolated various volatile and non-volatile antifungal antibiotics from different species of *Trichoderma* and attributed them for fungicidal activity against the pathogens. Widely studied two bacterial bioagents *B. subtilis* and *P. fluorescens* can suppress *Fusarium* species (Goudar and Kulkarni 2000; Khan and Akram, 2000; Khan and Khan, 2002; Naik, 2003), *Sclerotium rolfsii*(Bhatia *et al.*,2005 and *R. solani* (Khan and Sinha, 2005). Application of *B. subtilis* (Podile and Laxami, 1998) and *P. fluorescens* (Gholve and Kurkundar, 2003) significantly reduced pigeonpea wilt. Incidence of chickpea wilt could be effectively controlled through the use of *P. fluorescens* (Rangeswaran *et al.*, 2001; Jahagirdar *et al.*, 2002). The mechanism involved in the biocontrol of plant pathogens through bacterial antagonists is mainly the production of antimicrobial compounds. *B. subtilis* has been reported to produce bulbiformin (Brannen, 1995), iturir A (Asaka and Shoda, 1996), surfactin (Edwards and Seddan 1992), mycocerein (Wakayama *et al.*, 1984), Agrocin (Kim *et al.*, 1997) mycobacillin (Sengupta *et al.*, 1971), mycosubtilins (Peypoux and Michel, 1976) and fungimycin

(Vanittakam and Lowfller, 1986). Similarly, *P. fluorescens* is reported to produce Pyoluteorin (Whisler *et al.*, 2000), tryptophan side chain oxidase (Laville *et al.*, 1991), phenazine (Schoonbeek *et al.*, 2002), Oomycin A (Gulterson *et al.*,1988), IAA (Mordukhova *et al.*, 2000), siderophores (Perez *et al.*, 2001), alginate, HCN, Pseudomonic acid, ovafluorin, fluopsin C and F, sorbistin A and B, salicylic acid (Johri *et al.*, 1997), lipodepsipeptides, syringotoxins, syringomycins (Woo *et al.*, 2002), N-butylbenzene sulphonamide (Kim *et al.*, 2000) and 2,4-Diacetyl phloroglucinol (Mazzola *et al.*, 2002). Above antimicrobial compounds act on pathogenic fungi through fungistasis, inhibiting spore germination, causing lysis of mycelia or by inducing other fungicidal effects. Strains of *P. fluorescens* synthesize certain compounds (siderophores) that have very high affinity to ferric ions (Boruah and Kumar, 2002) which in turn is reduced to ferrous ions and thus reduce its availability to pathogens. Similar activity has been reported for *B. subtilis* also (Glick, 1995). Bioagents introduced once or occasionally if established in the ecosystem will have long lasting effect. However, if the ecosystem does not favor and the introduced bioagent do not persist, it will be less effective and frequent augmentations are necessary for the pathogen management. The techniques employed for the delivery of bioagents for the soil borne pathogens are seed treatment, soil amendment and drenching. Among these, seed treatment is more economical and easy method. Seed inoculation helps in managing externally seed borne and soil borne inoculums. Spore suspension as well as dry powder of *Trichoderma* spp., have been used for seed coating in pulses in India (Mukhopadhyay and Mukherjee, 1991; De and Chaudhary, 1999; Chaudhary and Kumar, 1999; Pandey and Upadhyay, 1999; Chaudhary and Prajapati, 2004). Numerous efforts have been made to manage different soil borne diseases such as *R. solani*, *S. rolfsii*, *S. sclerotiorum*, *Pythium* spp., *Phytophthora* spp. and *F. udum* by incorporating bioagent colonized natural substrates into the soil. However, *Trichoderma harzianum* as wheat bran powder granules has been used in pigeonpea and chickpea (Prasad and Rangeshwaran, 1999). Soil drenching has not so far used in pulses though drenching of soil with aqueous suspension of *T. harzianum* has been used against *Colletotrichum lindemuthianum* in cowpea, *S. rolfsii* in groundnut and *F. oxysporum* in chrysanthemum. It is observed that bioagent distribution in the soil is not even with this method. However, the author observed good control of pigeonpea wilt with the application of *T. harzianum* liquid culture through irrigation water in the standing crop.

Mass Production

With the increasing interest in developing the biopesticides as alternatives to fungicides, mass production of bioagents with quality has become the focus of research and development in recent years. The production strategies for fungal and bacterial bioagents differ considerably.

Fungal Biogents

The major work on the multiplication of fungal bioagents across the world centered around *Trichoderma* species. The two approaches employed for inoculum production of fungal bioagents are given below.

Solid State Fermentation

Solid state fermentation is the most common method of mass production of *Trichoderma* spp. For this purpose various cheap agricultural wastes and byproducts are utilized. This fermentation results in a product that is used directly rather than formulated further. The substrates used for solid state multiplication are grain bran, wheat straw, wheat bran, wheat bran sawdust, sorghum grain, wheat bran, saw dust modified medium, sand sorghum medium, tapioca rind, tapioca refuse, FYM, press mud, sand corn meal, pea or its combinations, FYM and coconut coir pith, coffee bery husk, wheat bran and biogas manure, cow dung, compost, oil cakes, wood bark, groundnut shell medium, rice bran, rice straw, pigeonpea leaves and urdbean straw etc. But none proved ideal for commercial formulation.

Liquid Fermentation

Liquid fermentation is largely being used by the bioagent producing industry for past 40 years. In industrialized countries deep tank fermentation using less expensive media such as molasses and brewers yeast for large scale production of *Trichoderma* is being employed. It is very important that the bioagents are produced in a cost effective and time efficient manner and should be suitable for each processing step such as fermentation, harvesting, drying, formulation, storage and delivery. Jin *et al.* (1991) amended modified Richards medium (RM 8) with polyethylene glycol (PME) and obtained desciction tolerant propagules of *T. harzianum*. In their later studies in 1996 glycerol amended RM 8 medium gave maximum biomass and viable propagules in a fermenter. Similar results were obtained by using potato dextrose broth, molasses yeast medium and V8 juice in laboratory fermenter. Maximum biomass and propagules were obtained in 96 hours with aeration, agitation, temperature, pH and antifoam controls than in shake flask culture. Various media used for liquid fermentation are molasses, brewers yeast (Papavizas *et al.*, 1984), RM8, Czapeks Dox broth, V8 broth, modified RM 8, potato dextrose broth, Richards broth, V8 juice and molasses yeast and molasses-soy medium (Prasad and Rangeshwaran, 2000a).

Bacterial Bioagents

Mass production of bacterial bioagents such as *B. subtilis* and *P. fluorescens* etc., involves the growing of the organism in suitabie medium using rotary shaker or fermenter. Bacteria require adequate nutrient supplement for proper growth. Kings B medium is most ideal. Other media like Tryptic Soya broth or nutrient broth can also be used. These bacteria are cultured in large flasks at 28°C for 48 hrs using rotary shaker. For large scale industrial multiplication, batch or continuous type fermenter is employed. The antagonist is grown for a definite period and the cell mass is separated or concentrated. Raw materials such as sugar beet, potato, sweet potato, tapioca, apple, rasins, grapes, sweet corn, molasses, sorghum, wheat, barley, malted cereals, honey mapple etc., have been used to grow bacteria by different workers. Since the aim is to get maximum viable cells quickly, the optimum conditions need to be standardized before the mass multiplication is taken up.

Development of Formulation

Biopesticide formulations are required to have many desirable characters such as easy in preparation and application, stability during transport and storage, good shelf life, sustainability of efficacy, acceptable cost and adequate market potential. Generally, the formulation from solid-state fermentation does not require sophisticated procedures. These products are dried, ground and added with the carriers for use. However, such preparations are bulky take longer time and are therefore prone to contamination. The liquid fermentation is devoid of such problems. The large quantity of propagules can be produced within few days, and the biomass can easily be formulated into dust, granule, pellet, wettable powder or emulsion. Various carriers used in *Trichoderma* formulations are celatom, vermiculite, talc, diatomaceous earth, molasses, gypsum, lignite, kaolin, peat, wheat bran, clay, perlite, sodium alginate, calcium chloride, bentonite, rice flour, gluten and pyrex (Papavizas *et al.*, 1984; Papavizas and Lewis, 1989; Zones *et al.*, 1984; Moody and Gindrant, 1977; Kelly, 1976; Khan *et al.*, 2001). For bacterial bioagents, diatomacious earth talc, calcium alginate, lignite, peat, FYM, cow dung cake and wood charcoal have been used by different workers. A brief review on the formulation of fungal and bacterial biopesticides has been given by Khan (2005).

Improving Efficiency of Bioagent

An efficient bioagent is able to control the pathogen in the right place and at a right time. The efficiency of the bioagent depends upon the biological equilibrium of pathogen and bioagent in the soil. If it is manipulated in favour of bioagent, it becomes detrimental to pathogen survival and its activity. Biological control often fails due to lack of thorough understanding between bioagent-pathogen-host-environment interactions and their impact on disease development. In many cases the failures occur because of our poor understanding about the ecological requirement of the bioagent strain for proper colonization and disease management mechanism. These aspects determine the method of application, quantity of bioagent and critical stage of application. Altering the edaphic conditions through cultural methods, cropping systems, soil amelioration, maintaining optimum moisture, providing organic source for bioagent and its application combined with fungicides etc., further improve the efficacy of bioagents.

Limitations and Future Thrusts

The effective amount of bioagents needed to be incorporated to counteract the pathogens should be 10^6 cfu/g soil which is a large quantity and cannot be achieved unless the antagonist establish itself in the ecosystem and multiplies. However, in order to economize the bioagents application, broadcasting in the field can be replaced by furrow application. Coating of seeds and priming is efficient technique especially against seed rots and pre and post emergence plant mortality caused by seed and soil borne pathogens. However, its efficacy for managing root diseases at reproductive stage of pulse crops has always been variable. The level of bioagent population in the rhizosphere for months together depends upon the availability of food base, soil type, pH, and temperature and moisture conditions of the soil. Also, the quality of bioagent formulation and efficiency of the antagonist strain is utmost important. It is measured in terms of inoculum potential (propagules/unit weight and the aggressiveness), shelf life, and ease of application and purity of the formulation. Generally, the formulations available in the market are of poor quality, which need to be regulated by the quality control agencies.

Lack of knowledge of handling, storage and marketing of microbial pesticides is another big limiting factor. There is a need to educate the dealers and the growers on storage conditions, shelf life and mode of action of biocontrol formulations. A grower who is tuned to a rapid results obtained with chemical pesticides, needs to be explained of slower but sustainable response by the biocontrol agents besides the environmental safety and other positive effects of microbes on plant and soil.

The antagonistic activity of microbes was discovered more than 70 years ago. However, their potential in managing the plant diseases could gain momentum worldwide during last 25 years and according to reports major soil borne diseases of pulse crops caused by *Fusarium spp., R. solani, R. bataticola, S. rolfsii, S. sclerotiorum, Ascochyta* spp.,*Pythium* spp., *Botrytis cineria* and many others can be effectively managed with bioagents. Since, the efficacy of bioagents, especially the *Trichoderma* spp., against soil borne diseases varies from isolate to isolate and location to location, the thrust for future research should be the selection of efficient strains which have the potential against wide range of pathogens under wide range of environments. Genetic engineering can provide a path to isolate lytic genes, their cloning and synthesizing into the antagonists, to produce a more efficient biocontrol agent.

Majority of the area of pulses in India lies under rainfed, moisture stress, high temperature and above normal soil pH (7.5–8.5). Therefore, there is a strong need to search the antagonists effective under such harsh ecological environments. Besides, these antagonists must be compatible with other bio-inoculants and agrochemicals applied to the pulse crops. In addition to develop superior strains of bioagents, easy, efficient and cheaper methods for their multiplication, increasing their shelf life and viability at least up to one year need elaborate researches particularly on nutrition, formulations and carriers, packaging, storage, delivery systems and survival/multiplication in the infection court.

References

Adams, P.B. and Ayers, W.A. (1982). Biological control of *Sclerotium* in the lettuce field *Sporidesmium sclerotiorum. Phytopathology* 72: 485-488.

Adekunle, A.T., Cardwell, K.F., Florini, D.A., Ikotun, T. (2001). Seed treatment with *Trichoderma* species for control of damping off of cowpea by *Macrophomina Phaseolina Biocontrol Science and Technology* 11: 4 449-457.

Agarwal, Abha, Tripathi, H.S. and Agarwal. A. (1999). Biological and chemical control of Botry gray mould of chickpea. *Journal of Mycology and Plant Pathology* 29(1): 52-56.

Agrawal, S.C. and Prasad, K.V.V. (1997). *Diseases of lentil.* Mohan Primlam for Oxford and B.M. Publishing Co. Pvt. Ltd., New Delhi. 155 pp.

Agrawal, S.C., Surendra-Sharma, Prasad, K.V.V. (2002). Efficacy of biological components in wilt management of chickpea. *Indian Journal of Pulses Research* 15: 2 177-178.

Akhtar-Haseeb, Shukla, P.K. and A. Haseeb (2002). Management of wilt disease of chickpea by the application of chemical, biopesticides and bioagent under field conditions. *Current Nematology* 13: (1-2): 61-63.

Allen, D.J., Bruchara, R.A. and Smithson, J.B. (1998). Diseases of common bean. *The Pathology of Food and Pasture Legumes* (Eds. D., J. Allen and J. M. Lenne). CA International, Wallingford, U.K. and ICRISAT, Patancheru, India, pp 179-266.

Asaka, K. and Shoda, P.M. (1996). The antibiotic action of *Bacillus subtilis* in relation to certain parasitic fungi with special reference to *Alternaria solani. Canadian Journal of Botany* 39: 220-8.

Baker, R., Elad, Y. and Chet, I. (1984). The controlled experiment in the scientific methods with special emphasis on biological control. *Phytopathology* 74: 1019-1021.

Baldwin, J.E., Derome, A.E., Field, L., Gallagher, P.T., Taha, A.A., Thaller, V., Brewer, D and Taylor, A. (1981). Biosynthesis of a cyclopentyl dienyl isonitrile acid in culture of fungus *Trichoderma hamatum. Journal of chem. Soc. Chem. Commn.* 1: 227-1229.

Bayaa, B. and Erskine, W. (1998). Diseases of lentil. Pages 423-471 *In: The Pathology of Food and Pasture Legumes*(Eds., d.J. Allen and J.M. Lenne), CAB International, Wallingfod, U.K. and ICRISAT, Patancheru, India.

Beagle-Ristaino, J.E. and Papvizas, G. C. (1985). Survival and proliferation of *Trichoderma* spp. and *Gliocladium* (*Trichoderma*) *virens* in soil and in plant rhizospheres. *Phytopathology* 75: 729-32.

Bhatnagar, H. (1996). Influence of environmental conditions on antagonistic activity of *Trichoderma* spp. Against *Fusarium udum. Indian Journal of Mycology and Plant Pathology* 26(1): 58-63.

Bidari, V.B. and Gundappa, R.C. (1997). *Trichoderma viride* in the integrated management of pigeonpea wilt under dryland cultivation. *Advances in Agricultural Research in India* 7: 65-69.

Biswas, K.K., Das, N.D. (1999). Biological control of pigeonpea wilt caused by *Fusarium udum* with *Trichoderma* spp. *Annals of Plant Protection Sciences* 7: 1 46-50.

Bjorkam-T, Blanchard-L-M., Harman, G.E. (1998). Growth enhancement of Shrunkenz (Sh2) sweet corn by *Trichoderma harzianum* 1295-22 effect of environmental stress. *Journal of the American. Society for Horticultural Science* 1: 35-40.

Boruah, H.P.D. and Kumar, B.S.D. (2002). Biological activity of secondary metabolites produced by a strain of *Pseudomonas fluorescens*. *Folia-Microbiologica* 47: 359-363.

Braga-M.A., Pessoa, M.N.G., Teofilo E.M. (2003). Biological and chemical treatment of cowpea. *Vigna uniguniculata* L. Walp. Seeds for the control of *Macrophomina Phaseolina* Tass Gold, *Revista-Ciencia, Agronomica*, 34 (2): 193-199.

Brannen, R. (1995). Production of antibiotics by *Bacillus subtilis* and their effect on fungal colonists of various crops. *Transactions of British Mycological Society* 65: 203.

Chang, Y.C., Baker, R., Kleifeld, O. and Chet, I. (1986). Increased growth of plants in the presence of the biological control agent *Trichoderma harzianum*. *Plant Disease* 70: 145-148.

Chaudhary, R.G. (2004). Biological control of soil borne disease of pulse crops. In "*Pulses in New perspective* (eds. Ali *et al.*), Indian society of Pulses Research and Development, IIPR, Kanpur, India, pp. 345-361.

Chaudhary, R.G. and Kumar, K. (1999). Potentials of Biocontrol agents against wilt in Pre rabi pigeonpea crop. *Indian Journal of Plant Pathology* 17 (1 and 2): 67-69.

Chaudhary, R.G. and Naimuddin, (2000). Pea diseases in Indian Perspective and their Economic Management. In "Advances in Plant Disease Management" (eds. Udit Narayan *et al.*). *Advance Publishing Concept*, New Delhi, pp. 47-60.

Chaudhary, R.G. Prajapati, R.K., Saxena, H and Dar, M.H. (2003). Synergistic effect of *Trichoderma harzianum* on nodulation and wilt management in pigeonpea. In: 6th *International, PGPR Workshop, session III. Integrated Biological Systems. Abstracts and Short Papers*, 5-10 October2003. Calicut, India, pp. 266-270.

Chaudhary, R.G., Naimuddin, De, R.K. and Rai, A.B. (2004). Masoor, matar evam rajma ke rog aur keet tatha unka prabandhan. In "Dalhan" (eds. Ali *et al.*), Bhartiya Dalhan Anusandhan Sansthan, Kanpur, India, pp. 237-267.

Chaudhary, R.G., Prajapati, R.K. (2004). Comparative efficacy of fungal, biological agents against *Fusarium udum*. *Annals of Plant Protection Sciences* 12(1): 75-79.

Chavan, A.M. (1996). Effect of fungal. Pigment on hydrolytic enzyme activity. *Flora and Fauna* Jhansi 2: 29-30.

Chet, I, Harman, G.E. and Baker, R. (1981). *Trichoderma hamatum*, its hyphal interaction with *Rhizoctonia solani* and *Pythium* spp. *Microbiology and Ecology* 7: 29-38.

Cipriano, T., Cirvilleri, G and Cartia, G. (1989). In-vitro activity of antagonistic microorganisms against *Fusarium oxysporum* f.sp. *lycopersici*, the causal agent of tomato crown root-rot. *Informatore Fitopatologico* 39(5): 46-48.

Cotes-A. M., Thonart-P, Lepoivre-P (1994). Relationship between the protective activities of several strains of *Trichoderma* against damping off agents and their ability to produce hydrolylic enzymes activities in soil or in synthetic media. 46th International Symposium on Crop Protection 59: 3a 931-941.

Daecon, J.W. (1997). Fungal interactions: mechanisms, relevance and practical exploitation. In: *Modern Mycology*. Third Edition. Blackwell Science Ltd., Oxford. pp. 205-223.

De, R.K. and Chaudhary, R.G. (1999). Biological and chemical seed treatment against lentil wilt. *Lens-Newsletter* 26 (1-2): 28-31.

De, R.K. and Mukhopadhyay, A.N. (1994). Biological control of tomato damping off by *Gliocladium (Trichoderma) virens*. *Journal of Biological Control* 8(1): 34-40.

De, R.K., Chaudhary, R.G. and Naimuddin. (1996). Comparative efficacy of biocontrol agents and fungicides for controlling chickpea wilt caused by *Fusarium oxysporum* f.sp. *ciceri*. *Indian Journal of Agricultural Sciences* 66(6): 370-373.

De, R.K., Dwivedi, R.P., Udit Narain. (2003). Biological control of lentil wilt caused by *Fusarium oxysporum* f.sp. *lentis*. *Annals of Plant Protection Sciences* 11: 1 46-52.

Dennis, C. and Webster. J. (1971). Antagonistic properties of species groups of *Trichoderma*.I. Production of non-volatile antibiotics. *Transactions of British Mycological Society* 57(1): 25-39.

Dennis, C. and Webster.J. (1971b). Antagonistic properties of species groups of *Trichoderma*. II Production of volatile antibiotics. *Transactions of British Mycological Society* 57(1): 41-48.

Dhedhi, B.M., Om Gupta, Jharia, H;.K., Gupta, O. (2001). Management of fusarial wilt of chickpea through antagonistic Microorganisms. *JNKVV Research Journal Puli* 35: 1-2 38-42.

Dutta-P, Das-B.C. (1999). Control of *Rhizoctonia solani* in soybean (Glycine max) by Farmyard manure culture of *Trichoderma harzianum*. *Indian Journal of Agricultural Sciences* 69(8): 596-598.

Edwards, K. and Seddon, P. (1992). Biological interactions. In: Microbial Ecology: A Conceptual Approach. K. Edwards and P. Siddow (eds.) John Wiley and Sons, New York, pp. 171.

Elad, Y., Chet, I and Katan, J. (1980). *Trichoderma harzianum* a biocontrol effective against *Sclerotium rolfsii* and *Rhizoctonia solani*. *Phytopathology* 70: 119-121.

Elad, Y., Chet, I., Doyle, P. and Henis, Y. (1982). Parasitism of *Trichoderma* spp. On *Rhizoctonia solani* and *Sclerotium rolfsii*-scanning microscopy and fluorescent microscopy. Phytopathology 72: 85-88.

Elad, Y., Kalfon, A. and Chet, I. (1982). Control of *Rhizoctonia solani* in cotton by seed-coating with *Trichoderma harzianum* spp Spores. *Plant and Soil* 66: 279-281.

Elad, Y., Katan, J. Aharonson, N. Cohen, E. Rubin, B and Mathews, G.A. (2000). Biological control of foliar pathogens by means of *Trichoderma harzianum* and potential modes of action. *XIVth International Plant Protection Congress*, Jerusalem 19: 709-714.

El-Hassan-S.A., Gowen, S.R. (2003). Mechanism in the biological control of lentil. Vascular wilt (*Fusarium oxysporum* f.sp. *lentis*) by *Trichoderma hamatum*. *Crop Sciences and Technology* Volumes 1 and 2 proceedings of an International Congress 651-654.

Enteshamul-Haque-S, Hashmi R.Y., Ghaff (1992). Biological Control of root rot disease of lentil. Lens 19:2 43-45.

Erskine W, Bayaa B, Dholi M. (1990). Effect of temperature and sove media and biotic factors on the growth of *Fusarium oxysporum* f.sp. *lentis* and its mode of seed transmission. *Arab Journal of Plant Protection* 8: 134-137.

Farnawany-M, El, Shamas (1996). Biological control of Rhizoctonia solani affecting bean seedlings damping off. *Alexandria–Journal of agricultural Research* 41 (1) 253-260.

Fravel, D.R., Marois, J., J., Dunn, M.T. and Papavizas, G.C. (1985). Compatibility of *Talaromyces flavus* with potato seed piece fungicides. *Soil Biology and Biochemistry* 17: 163-166.

Gaur, R.B., Sharma, Singh, Vinod (2005). Manipulations in the mycoparasite application techniques against Rhizoctonia root rot of cotton. *Indian Phytopathology* 58 (4): 402-409.

Gaur, V.K. and Sharma, L.C. (1991). Microorganisms antagonistic to *Fusarium udum* Butler. *Proceedings of Indian National Science Academy, Part B, Biological Sciences* 57 (1): 85-88.

Gholve, V.M. and Kurundkar, B.P. (2003). Efficacy of *Pseudomonas fluorescens* isolates against wilt of pigeonepa. *Journal of Maharashtra Agricultural University*. 27: 327-328.

Gholve-V.M., Kurundkar, B.P. (2002). Biological control of pigeonpea wilt with *Pseudomonas. Fluorescens* and *Trichoderma Viride*. *Indian Journal of Pulses Research* 15(2): 174-176.

Glick, B.R. (1995). The enhancement of plant growth by free-living bacteria. *Cannadian Journal of Microbiology* 41: 109-117.

Goel-A.K., Sindhu, S.S., Dadarwal, K.R. (2000). Pigment diverse mutants of *Pseudomonas* spp. Inhibition of fungal growth and stimulation of growth of *Cicer arietinum*. *Biologia Plantarum* 43: 4 563-569.

Gotfredsen, W.O. and Vangedal, S. (1965). Trichodermin, a new sequiterpene antibiotic. *Acta, chem. Scand*. 19: 1088.

Goudar, S.B. and Kulkarni, S. (2000). Bioassay of antagonists against *Fusarium udum* the causal agent of pigeonpea wilt. *Karnataka Journal of Agricultural Sciences* 13: 64-67.

Gulterson, N., Ziegle, J.S., Warren, G.J. and Layton, T.J. (1988). Genetic determinants for catabolite induction of antibiotic biosynthesis in *Pseudomonas fluorescens* Hv37a. *Journal of Bacteriology* 170: 380-385.

Gupta, O.M, Sharma, N.D., Gupta, O. (1995). Effect of fungal. Metabolites on seed germination and root length of black gram (*Phaseolus mungo*). *Legume Research* 18(1): 64-66.

Gurha, S.N. (2004). Chana mein rog prabandhan. In "Dalhan" (eds. Masood Ali *et al.*). Bhartiya Dalhan Anusandhan Sansthan, Kanpur, India, pp. 201-212.

Gurha, S.N. and Vishwa Dhar (2005). *Chickpea disease and their management*. Chickpea Production and Productivity Constraints (eds. Amerika Singh *et al.*). NCIPM, IARI, New Delhi, India pp. 99-110.

Gurha,S.N., Gurdip Singh and Sharma, Y.R. (2003). *Diseases of chickpea and their management*. In *Chickpea Research in India* (eds. Masood Ali *et al.*). Indian Institute of Pulses Research, Kanpur, India, 195-227.

Hadar, Y., Harman, G.E. and Taylor, A.G. (1984). Evaluation of *Trichoderma koningii* and *T. harzianum* from new York soils for biological control of seed rot caused by *Pythium* spp. *Phytopathology* 74: 106-110.

Haggag, W.M. and Mohamed, H.A.A. (2002). Enhancement of antifungal metabolites production from gamma-ray induced mutants of some *Trichoderma* species for control onion white rot disease. *Plant Pathology Bulletin* 11: 145-156.

Hari-Chand, Surender Singh (2005). Control of chickpea wilt *Fusarium oxysporum* f.sp. *ciceri*. Using bioagents and plant extracts. *Indian Journal of Agricultural Sciences* 75(2): 115-116.

Harman, G.E., Chet, I. and Baker, R. (1980). Seed treatment with *Trichoderma harzianum* as a means of preventing seed and seedling disease induced by *Pythium* spp. Or *Rhizoctonia solani. Phytopathology* 70: 1167-1172.

Harman, G.E., Hayes, C.K., Lorito, M., Broadway, R.M., Di Pietro, A., Peterbauer C and Tronsmo, A. (1993). Chitinolytic enzymes of *Trichoderma harzianum*: purification of chitobiosidase and endochitinase. *Phytopathology* 83: 313-318.

Harman, G.F., Chet, I. and Baker, R. (1981). Factors affecting *Trichoderma namatum* applied to seeds as a biocontrol agent. *Phytopathology* 71: 569-572.

Harnam, G.E., Jin, X., Stasz, T.E. Peruzzotti, G, Leopold., A.C. and Taylor, A.G. (1991). Production of conidial of *Trichoderma harzianum* for biological control. *Biological Control* 1: 23-28.

Haware, M.P. (1998). Diseases of chickpea. *The Pathology of Food and Pasture Legumes* (Eds. D.J. Allen and J.M. Lenne), CAB International, Wallingfod, U.K., ICRISAT, Patancheru, India, pp. 473-516.

Haware, M.P., Mukherjee, P.K., Lenne, J.M., Jayanthi, S., Tripathi, H.S and Rathi, Y.P.S. (1999). Integrated biological and chemical control of botrytis gray mould of chickpea. *International Phytopathology* 52(2): 174-176.

Hazarika, D.K. and Das, K.K. (1998). Biological management of root rot of French bean *Phaseolus vulgaris* caused by *Rhizoctonia solani. Plant Disease Research* 13(2): 101-105.

Howell, C.R., Stipanovic, R.D. and Lumsden, R.D. (1993). Antibiotic production by strains of *Gliocladium virens* and its relation to the biocontrol of cotton seedling disease. *Biocontrol Science and Technology* 3, 435-41.

Huang, H., Chang, Y.C., Erickson, R.S. and Huang, H.C. (2000). Soil treatment with fungal agent control of apothecia of *Sclerotinia sclerotiorum* in bean and pea crops. *Plant Pathology Bulletin* 9(2): 53-58.

Iqbal, S.M., Bakhshi, A., Hussain, S and Malik, B.A. (1995). Microbial antagonism against *Sclerotium rolfsii* the cause of collar rot of lentil. *Lens Newsletter* 1-2: 48-49.

Itoh, Y., Kodama, K., Furuya, K., Takahash, S., Heneishi, J., Taki-guchi, Y and Arai, M. (1980). A new sesquiterpene antibiotic heptelidic and producing organisms. Fermentation, Isolation and Characterization. *Journal of Antibiotics*. 33: 468-473.

Jahagirdar, S., Yenjarappa, S.T., Ravikumar, M.R. and Jamadar, M.M. (2002). Field evaluation of biocontrol agents against chickpea wilt. *Legume Research* 25(4): 299-300.

Jangid, P.K., Mathur, P.N., Kusum. (2004). Role of chitinase and B,1,3 glucauce elicitation in the *Trichoderma harzianum* induced systemic resistance in capsicum. *Indian Phytopath* 57 (2): 217-221.

Jayalaxshmi, S.K., Sreeramali, K., Benigi V.L. (2003). Efficacy of *Trichoderma* spp. against pigeonpea wilt caused by *Fusarium udum* Butler. *Journal of Biological Control* 17(1): 75-78.

Jayaraj-J, Ramobadrau R. (1996). Evaluation of certain organic substrates and adjuvants for the mass multiplication of *Trichoderma harzianum* Rifai. *Journal of Biological Control* 10: 1-2 129-131.

Jin, X., Harman, G.E. and Taylor, A.G. (1991). *Biological Control* 1: 237-243.

Jin, X., Taylor, A.G. and Harman, G.E. (1996). Development of media and automated liquid fermentation method to produce desiccation-tolerant propagules of *Trichoderma harzianum*. *Biological Control* 7: 267-274.

Johri, B.N., Rao, C.V.S. and Goel, R. (1997). Fluorescent *Pseudomonad* In plant disease management. In: Biotechnological Approaches in Soil Microorganisms for Sustainable Crop Production, (ed. K.R. Dadarwal), Scientific Publisher, Jodhpur, India, pp. 193-223.

Kapat-A, Zimand G., Elad, Y. (1998). Effect of two isolates of *Trichoderma harzianum* on the activity of hydrolytic enzymes. Produced by *Botrytis cinerea. Physiological and molecular–Plant Pathology* 52 (2): 127-137.

Kaswate, N.S; Shinde, S.S., Rathod, R.R. (2003). Effect of biological agents against different isolates of *Rhizoctonia bataticola*. Butler India17(2): 167-168.

Kaur, N.P. and Mukhopadhyay, A.N. (1992). Integrated control of chickpea wilt complex *Trichoderma* and chemical methods in India. *Tropical Pest Management* 38: 272-275.

Kehri, H.K. and Chandra, S. (1991). Antagonism of *Trichoderma viride Macrophomina phaseolina* and its application in the control of dry root rot of mungbean. *Indian Phytopath.* 44(1): 60-63.

Kelley, W.D. (1976). Evaluation of *Trichoderma harzianum* impreganated clay granules as a biocontrol of damping off of pine seedlings caused by *Phytophthora cinnamomi. Phytopathology* 66: 1023-7.

Khan M.R. and Khan, S.M. (2002). Effect of root-dip treatment with certain phosphate solubilising microorganisms on Fusarial wilt of tomato. *Bioresource Technology* 85: 213-215.

Khan, A.A. and Sinha, A.P. (2005). Comparative antagonistic potential of some biocontrol agents against sheath blight of rice. *Indian Phytopath.* 58(1): 41-45.

Khan, M.R. and Akram, M. (2000). Effect of certain antagonistic fungi and rhizobacteria on wilt disease complex caused by *Meloidogyne incognita* and *Fusarium oxysporum* f.sp. *lycopersici* on tomato. *Nematologia Mediterranea.* 28: 139-144.

Khan, M.R. and Khan, S.M. (2001). Bio management of *Fusarium* wilt of tomato by the soil application of certain phosphate solubilizing microorganisms. *International Journal of Pest Management* 47(3): 227-231.

Khan, M.R., Khan, N. and Khan, S.M. (2001). Evaluation of Agriculture materials as substrate for mass culture of fungal biocontrol agents of Fusarial wilt and root knot nematode diseases. *Annals of Applied Biology* (TAC-21 suppl.) U.K. 22: 50-51.

Khan, M.R., Khan, S.M., Moniddin, F.A. (2004). Biological control of *Fusarium* wilt of chickpea through seed treatment with the commercial formulation of *Trichoderma harzianum*. Pseudomonas Fluorescens. *Phytopatholigia Mediterranea* 43 (1): 20-25.

Khan, M.S., Almas-Zaidi, Zaidi, A. (2002). Plant growth promoting Rhizobacteria from rhizosphere of wheat and chickpea. *Annals of Plant Protection Sciences* 10(2): 265-271.

Khare, M.N. (1981). Diseases of lentil. Pages 163-171. *In: Lentils* (Eds., G. Webs and G. Haware) CAB Farnham. Royal, U.K.

Khot, G.G. Tauro. P. and Dadarwal, K.R. (1996). Rhizobacteria from chickpea rhizosphere effect in wilt control and promote nodulation. *Indian Journal of Microbiology* 4: 217-222.

Kim, B., Bulit, J. and Bugaret, Y. (1997). The production of antifungal volatiles by *Bacillus subtilis. Journal of Applied Bacteriology* 74: 119-126.

Kim, K.K., Kang, J.G., Moon, S.S. and Kang, K.Y. (2000). Isolation and identification of antifungal N-butylbenzene-sulphonamide produced by *Pseudomonas* spp. *Journal of Antibiotics* 53:2, 13-6.

Kloepper, J.W., Hume, D.J., Seher, F.M., Singleton, C., Tipping, B., Laliberte, M., Frauley, Kutchaw, T., Simonson, C., Lifshitz., R., Zaleska, I. and Lee, L. (1988). Plant growth promotion *Rhizobacteria* on canola (rapeseed). *Plant Disease* 72: 42-46.

Koch E, (1999). Evaluation of commercial products for microbial. Control of soil borne plant disease. *Crop Protection* 18(2): 119-125.

Kraft, J.M., Larsen, R.C. and Inglis, D.A. (1998). Diseases of peas. Pages 179-266. In: *The Pathology of Food and Pasture Legumes* (Eds. D.J. Allen and J.M. Lenne), CAB International Wallingfod, U.K. and ICRISAT, Patancheru, India.

Krishna, Murthy, Niranjana, S.R., Shetty, M.S., Murthy, K. (2003). Effects of chemical fungicides and biological agents on seed quality improvement in pulses. *Seed Research* 31(1): 121-124.

Kumar Sreerama, P., Palakshappa, M.G. and Leena Singh. (2005). Effect of Polyethele glycol on the biomass characteristics of *Trichoderma harzianum* and *T. viride* in liquid culture. Indian Phytopathology 58(4): 466-469.

Kumar-BSD, Dabe-HC (1992). Seed bacterization with a fluorescent *Pseudomonas* for enhanced. Plant growth yield and disease. *Soil Biology and Biochemistry* 24(6): 539-542.

Lacicowa, B. (1999). Protective effect of *Trichoderma koningii* Oud and *Gliocladium catenulatum* Gilman et Abbot used for seed treatment of bean. *Ochrona Roslin* 34 (11-12): 35-36.

Lakshmi, Tewari, Chandra Bhanu, Tewari K., Bhanu C. (2003). Screening of various substrates for sporulation and mass multiplication of biocontrol agent *Trichoderma harzianum* through solid State Fermentation. *Indian Phytopath.* 56(4): 476-478.

Landa, B.B., Navas Cortes-J.A., Jimenez Diaz R.M. (2004). Influence of temperature on Plant Rhizobacteria interactions related to biocontrol. Potential for suppression of *Fusarium* wilt of chickpea. *Plant Pathology* 53(3): 341-352.

Landa, B.B., Navas-Cortes-J.A., Jimenez-Diaz, R.M. (2004). Integrated management of *Fusarium* wilt of chickpea with sowing date host resistance and biological control. *Phytopathology* 94(9): 346-960.

Laville, J.C., Voisard, C., Keel, C., Maurhofer, M., Defago, G. and Haas, D. (1991). Global control in *Pseudomonas fluorescens* mediating antibiotic synthesis and suppression of black rot of tobacco. Proceedings of National Academic Science. USA 89:1562-66.

Lumsden, R.D., Locke, J.C. and Walter, J.F. (1991). Approval of *Gliocladium virens* by the U.S. Environ. Protec. Agency for biological control of *Pythium and Rhizoctonia damping-off. Petrial: 138.*

Lumsden, R.D., Lewis, J.A. and Fravel, R.D. (1995). Bioradiational pest control agents formulation and delivery (Eds. R. Franklin, W. Hall, A. John, C.S. Barry) Symposium Series 595. Amer. Chemical Society, Washington, USA. 306 pp.

Madukeshwara, S.S. and Seshadri, V.S. (2001). Variation and management of *fusarium* wilt of pigoenpea (*Cajanus cajan* L. Mill spp.) *Tropical Agricultural Research* 13: 380-394.

Mahabeer Singh, Majumdar, V.K., Singh, M. (1995). Antagonistic activity of *Trichoderma* spp. *Macrophomina phaseolina* Goid *in vitro*. *Environment and Ecology* 13(2): 481-482.

Mahalinga, D.M., Jayalakhsmi, S.K., Gangadhara,G.C. (2004). Management of pigeonpea wilt through integration of bioagents and resistant source. *Plant Disease Research*, Ludhiana 19(2): 181-182.

Majumdar, V.L., Jat, J.R. and Gour, H.N. (1996). Effect of Biocontrol agents on the growth *Macrophomina phaseolina* the incitant of blight of moth bean. *Indian J. Myco. Pl.Path.* 26(2): 202-203.

Malik, G., Dawar, S., Sattar, A., Dawar, A. (2005). Efficacy of *Trichoderma harzianum* after multiplication on different substrates in the control of root rot fungi. *International Journal of Biology and Biotechnology* 2(1): 237-242.

Mandhare, V.K., Suryawanshi, A.V. (2005). Standardization of storage conditions to increase the shelf life of *Trichoderma* formulation. *Agriculture Science Digest* 25(1): 71-73.

Marfori, E.C., Kajiyama, S., Fukusaki, E. and Kobayashi, A. (2002). Trichosetin, a novel tetramic acid antibiotic produced in dual culture of *Trichoderma harzianum and Catharanthus roseus callus.* Zeitschrift-fur-Naturforschung. *Section-C,-Biosciences* 57: 465-470.

Mathew, K.A., Gupta, S.K. (1998). Biological control of root rot of French bean caused by *Rhizoctonia. Solani. Indian J. Myco. Pl.Path.* 28 (2): 202-205.

Mathur, A.C. and Srivastava, M.P. (2005). *In vitro* inhibition studies of *Macrophomia phaseolina* by local *Trichoderma spp.* The cause of charcoal rot of cowpea. *Annals of Biology* 21(1): 79-81.

Mazur S, Nawrocki J, Kucmierz, J. (2003). Fungi isolated from chickpea Plants (*Cicer arietinum*) and their effects on growth of chickpea Pathogens. *Sodininkyste in Darzininkyste* 22(3): 281-289.

Mazzola, M., Van Veen, J.H., Laanbroek, H.J. and de Vos, W.M. (2002). Mechanisms of natural soil suppressiveness to soil borne diseases. Proceeding of the 9th International symposium on Microbial Ecology, Amsterdam, Netherlands August 2001. *Antonie-van-Leeuwenhoek* 81 (1-4): 557-564.

Moody, A.R. and Gindrat D. (1977). Biological control of Cucumber black rot by *Gliocladium roseum. Phytopathology* 67: 1159-1162.

Mordukhova, E.A., Sokolov, S.L., Kechetkov, V.V. Koshelva, I.A., Zelenkova, N.F. and Boronin, A.M. (2000). Involvement of naphthalene dioxygenase in indole-3-acetic acid bijosynthesis by *Pseudomonas putida. FEMS-Microbiology-Letters* 190: 279-285.

Muhammad-S, Amusa N.A. (2003). *In vitro* inhibition of growth of some seedling blight inducing pathogens by compost, inhabiting microbes., *African Journal of Biotechnology* 2(6): 161-164.

Mukerjee, P.K., Haware, M.P., Raghu, K. (1997). Induction and evaluation of benomyl tolerant of *Trichoderma viride* for biological control of Botrytis grey mold of chickpea. *Indian Phytopathology* 50 (4): 485-489.

Mukherjee, S. and Tripathi, H.S. (2000). Biological and chemical control of wilt complex of French bean. *Indian J. Myco. Pl.Path.* 30(3): 380-385.

Mukhopadhyay, A.N. (1988). Biocontrol efficacy of *Trichoderma* spp. In controlling damping off in tomato and egg plants and wilt complex of lentil and chickpea. *Proceedings of the 5th International Congress of Plant Pathology*, Kyota, Japan.

Mukhopadhyay, A.N. and Mukherjee, S. (1991). Biological seed treatment for control of soil-borne plant pathogens. *FAO Plant Protection Bulletin* 40 (1 and 2): 21-30.

Naik, M.K. (2003). Challenges and Opportunities for Research in Soil borne Plant Pathogens with Special Reference to *Fusarium* Species. *Indian J. Myco. Pl.Path.* 33: 1-14.

Nene, Y.L. and Reddy, M.V. (1987). Chickpea diseases and their control. *In: The Chickpea* (Eds., M.C. Saxena and K.B. Singh), CAB International, Wallingfod, U.K. pages 233-270.

Nigam, N., Kumar, R.N., Mukerji, K.G. and Upadhyay, R.K. (1997). Fungi-a-tool for Biocontrol. In *IPM system in Agriculture, Vol.II. Biocontrol in Emerging Bio Technology*. R.K. Upadhyay, K.G. Mukerjii and R.L. Rajak (eds). Aditya Books Pvt. Ltd., New Delhi, pp. 503-526.

Noronua, M.A., Antunes Sobrinno-S, Silveira, N.S.S., Michereff, S.J., Marianno, R.L.R., de (1996). Selection of *Trichoderma* spp., isolates for *Rhizoctonia solani* control of beans. *Summa Phytopathological* 22 (2): 156-162.

Pandey, K.K. and Upadhyay, J.P. (1999). Comparative study of chemical, biological and integrated approach for management of *Fusarium* wilt of pigoenpea. *J. Myco. Pl.Path.* 29(2): 214-216.

Pandey, K.K. and Upadhyay, J.P. (2000). Micorbial population from rhizosphere and non-rhizosphere soils of pigeonpea-Screening for resident antagonist and mode of mycoparasitism. *Indian J. Myco. Pl.Path.* 30(1): 7-10.

Pandey, K.K., Pandey, P.K. and Upadhyay, J.P. (2000). Selection of potential isolate of biocontrol agents based on biomass production, growth rate and antagonistic capability. *Vegetable Science* 27(2): 194-196.

Papavizas, G.C. and Lewis, J.A. (1989). Effect of *Gliocladium* and *Trichoderma* on damping off and blight and snap bean caused by *Sclerotium rolfssi* in the green house. *Plant Pathology* 38: 278-286.

Papavizas, G.C. and Lewis, J.A. (1989). Effect of *Gliocladium* and *Trichoderma* on damping-off of snapbean caused by *Sclerotium rolfsii* in the green house. *Plant Pathology*. 38: 277-286.

Papavizas, G.C., Dunn, M.T., Lewis, J.A. and Beagle-Ristaino, J. (1984). Liquid fermentation technology for experimental production of biocontrol fungi. *Phytopathology*. 74: 1171-1175.

Patel, T.C., Nafade, S.D., Ghetiya, L.V. and Thakor, K.J. (2001). Management of important diseases of pigeonpea. *Pesticide Research Journal* 13(1): 92-95.

Perez, C., Fuente, L-de-La., Arias, A., Altier, N. and De-la-Fuente, L. (2001). Use of native fluorescent *Pseudomonas* for controlling seedling diseases of *Lotus cornicalatus* L. *Agrociencia-Montevidea* 5:4: 41-47.

Peypoux, F. and Michel, G. (1976). Structure de la mycosubtilin, antibiotique isole de *Bacillus subtilis. European Journal of Biochemistry* 63: 391-398.

Poddar, R.K., Singh, D.V., Dubey, S.C. (2004). Management of chickpea wilt through combination of fungicides and bioagents. *Indian Phytopath.* 57 (1): 39-43.

Podile, A.R. and Laxmi, V.D.V. (1998). Seed bacterization with *Bacillus subtilis* AF 1 increases phenylalanine ammonia-lyase and reduces the incidence of fusarial wilt in pigeonpea. *Journal of Phytopathology* 146: 255-259.

Pradeep Kumar, Anuja and Kumud Kumar. (2000). Biocontrol of seed borne fungal pathogens of pigeonpea *Cajanus cajan* (L.) Millsp. *Annals Pl. Protec. Sci.* 8: 30-32.

Prasad R.D., Rangeshwaran, R. (1999). Granular formulation of *Trichoderma* and *Gliocladium* spp.-biocontrol of *Rhizoctonia solani* of chickpea. *Indian J. Myco. Pl.Path.* 29 (2): 222-226.

Prasad R.D., Rangeshwaran, R. (2001). Biological control of root and collar rot of chickpea caused by sclerotium rolfsii. *Ann. Pl. Protec. Sci.* 8 (2): 287-303.

Prasad, R.D. and Rangeshwaran, R. (2000a). An improved medium for mass production of the biocontrol fungus *Trichoderma harzianum*. *Indian J. Myco. Pl.Path.* 30: 233-235.

Prasad, R.D. and Rangeshwaran, R. (2000b). Effect of soil application of a granular formulation of *Trichoderma harzianum* on seed rot and damping off of chickpea incited by *Rhizoctonia solani*-saprophytic growth of the pathogen and bioagents proliferation. *Indian J. Myco. Pl.Path.* 30: 216-220.

Prasad, R.D., Rangeshwaran, R. (2000). Effect of soil application of a granular formulations of *Trichoderma harzianum* on *Rhizoctonia solani* incited seed rot and damping off chickpea. *Indian J. Myco. Pl.Path.* 30 (2): 216-220.

Prasad, R.D., Rangeshwaran, R., Hegde, S.V., Anuroop, C.P. (2002). Effect of soil and seed application of *Trichoderma harzianum* on pigoenpea wilt caused by *Fusarium udum* under field conditions. *Crop Protection* 21: 4 283-287.

Prasad, R.D., Rangeshwaran, R. (2000). Self life and bio-efficacy of *Trichoderma harzianum* formulated in various carrier materials. *Plant Disease Research* 15(1): 38-42.

Prasad, Rangeshwaran, R., Anuroop, C.P., Phanikumar, P.R. (2002). Bioefficacy and self life of conidial and chlamydospore formulation of *Trichoderma harzianum*. *J. Biol. Control* 16(2): 145-148.

Prasad, R.D., Rangeshwaran, R. (2000). Effect of soil application of a granular formulation of *Trichoderma harzianum* on *Rhizoctonia solani* incited seed rot and damping off chickpea. *Indian J. Myco. Pl.Path.* 30(2): 216-220.

Prasad, R.D., Rangeshwaran, R., Anuroop, C.P., Phanikumar, P.R. (2002). Bioefficacy and shelf life of conidial and chlamydospore formulation of *Trichoderma harzianum* Rifai. *Journal of Biological control* 16(2): 1451-48.

Praveen, Shabana and Kumar, R. Vijay (2004). Antagonism by *Trichoderma viride* against leaf blight pathogen of wheat, *J.Myco. Pl. Patho.* 34(2): 220-222.

Priya-rani Agarwal and Mehrotra, R.S. (1998). Establishment of nursery technology through *mosseae*, *Rhizobium* sp. and *Trichoderma harzianum* on better biomass yield of IPro. Cinearia *Linn. Proceedings of the National Academy of Sciences–India. Section Biological Science* 68 (3-4): 301-305.

Pyke, T.R. and Dietz, A. (1966). U-21, 963, a new antibiotic I, Discovery and biological activity. Applied Microbiology 14: 506-509.

Raguchander, T., Sumiappan, R. and Arjunam, G. (1993). Biocontrol of Macrophomina root rot mungbean. *Indian Phytopathology* 46: 379-382.

Rakesh Kamar, Jalali, B.L., Harichand (2004). Interaction between V.A. Mycorrhizal fungi and soil borne plant pathogens of chickpea legume research 27(1): 19-26.

Rangeshwaran, R., Prasad, R.D. and Anuroop, C.P. (2001). Field evaluation of two bacterial antagonists, *Pseudomonas putida* (PDBCAB-19) and *Pseudomonas fluorescens* (PDBCAB-2) against wilt and root rot of chickpea. *Journal of Biological Control* 15(2): 165-170.

Rao, C.V.S., Sachan, I.P., Johri, B.N. (1999). Influence of fluorescent Pseudomonas on growth and nodulation of lentil in *Fusarium* infected soil. *Indian Journal of Microbiology* 39(1): 23-29.

Reddy, M.V. and Vishwa Dhar. (1998). Current approaches and new dimensions in the management of pigoenpea diseases. Abstract *In: National Symposium on Management of Biotic and Abiotic Stresses in Pulse Crops* 26-28 June 1998, ISPRD and IIPR, Kanpur, India, p 2.

Reddy, M.V., Sharma, S.B. and Nene, Y.L. (1990). Pigeonpea diseases management. *In: The Pathology of Food and Pasture Legumes* (Eds. D.J. Allen and J.M. Lenne), International, Wallingford, U.K. and ICRISAT, Patancheru, India, pp 517

Robert, R., Flori, P. and Pisi, A. (1996). Biological control of soil borne *Sclerotium rolfsii* infection by treatment of bean seed with species of *Trichoderma. Petria* 6(2): 105-116.

Rudresh, D-L., Shivaprakash, M.K., Prasad, R.D. (2005). Effect of combined application of Rhizobium Phosphate solublizing bacterium and *Trichoderma* spp. On growth nutrient. Uptake and yield of chickpea. Applied soil ecology 28(2): 139-146.

Sadowski, S., Kryzia, K.A. (1991). Trial using *Trichoderma harzianum* and *Trichoderma lignorum* to control root diseases of lentil. *Phytopathologia Polonica* 14: 83

Sangita, B., Shah, A.K. and Bapat, S.A.D. (2000). Biological control of *Fusarium* wilt of pigeonpea by *Bacillus brevis. Journal of Microbiology* 46(2): 125-132.

Saxena, D.R., Saxena, Moly and Khare, M.N. (1998). Diseases of lentil and their management. *In: Diseases of Field Crops and their Management* (Ed., T.S. Thind), National Agricultural Technology Information Centre, Ludhiana, India, pp 271-283.

Schoonbeek, H., Raaijmakers, J.M., Waard, M.A.de and de Waard, M.A. (2002). Fungal ABC transporters and microbial interactions in natural environments. *Molecular Plant Microb Interaction* 15 (11): 1165-1172.

Sengupta, S., Banerjee, A.B. and Bose, S.K. (1971). Glutamyl and D-orL-petide linkages in mycobacillin, a cyclic peptide antibiotic. *Biochemical Journal* 121: 839-846.

Shahriary, D., Okhovvat, M., Rouhaw, H. (1996). Biological control of *Pythium ultimum* through the causal agent of chickpea. Seed rot and damping off disease by antagonistic fungi. *Iranian Journal of Agricultural Sciences* 27: 3 1-7 pp 8.

Shamarao-Janagirdar, Yenjerappa S.T., Ravi Kumar M.R., Jamadar, M.M. (2002). Field evaluation of biocontrol agents against chickpea wilt. *Legume Research* 25(4): 299-300.

Shanmugam, V., Sriram, S., babu, S., Nandakumar, R., Raguchander, T. and Balsubramanian, P. (2001). Purification and characterization of an extracellular alphaglucosidase protein from *Trichoderma viride* which degrades a phytotoxin associated with sheath blight disease in rice. *Journal of Applied Microbiology* 90: 320-329.

Sharma, S.K., Verma, B.R. and Sharma, B.K. (1999). Biocontrol of *Sclerotinia sclerotiorum* caused stem rot of chickpea. *Indian Phytopathology* 52(1): 44-46.

Shazia-Siddiqui, Siddiqui-Z.A., Iqbal, Ahmad. (2005). Evaluation of Flourescent *Pseudomonas* and *Bacillus* isolates. For the biocontrol of a wilt disease. Complex of pigeonpea. *World Journal of Microbiology and Biotechnology* 21(5): 727-732.

Shivanna-M.B., Meera, M.S., Kageyama, K, Hyakhmachi M. (1996). Growth promotion ability of Zoysia grass, rhizosphere fungi 14. Consecutive Plantings of Wheat and Soyabean Mycoscience 37(2): 163.

Siddique, Z.A., Singh, L.P. (2004). Effects of soil inoculants on the growth transpiration and wilt diseases of chickpea. *Zeitschrift-Fur-Pflanzenkrankheiten Uud-Pflanzen Schutz* 111(2): 151-157.

Siddiqui, Z.A. and Mahmood, I. (1995). Biological control of *Heterodera cajani* and *Fusarium udum* by *Bacillus subtilis, Bradyrhizobnium japonicum* and *Glomus fasciculatum* pigoenpea. *Fundamental and Applied Nematology* 18(6): 559-566.

Siddiqui, Z.A. and Mahmood, I. (1996). Biological control of *Heterodera cajani* and *Fusarium udum* on pigeonpea by *Glomus mosseae, Trochoderma harzianum. Verticillium Chlamydosporium. Israel Journal of Plant Sciences* 44(1): 49-56.

Siddiqui, Z.A., Mahmood, I., Hyat, S. and Mahmood, S. (1998). Biocontrol of *Heterodera cajani* and *Fusarium udum* on pigeonpea using *Glomus mosseae, Paecilomyces lilacinus, Pseudomonas flouorescens. Thai Journal of Agricultural Science* 31(3): 310-321.

Silveria, N.S.S., Michere, F.A., Menezes, S.J., Campos, M. and Takaki, G.M. (1994). Potential *Trichoderma* spp. Isolates on the control of *Sclerotium rolfsii* on beans. *Summer Phytopathologica* 20(1): 22-25.

Sindhan, G.S., Indra Hooda, Karwasra, S.S., Hooda-I (2002). Biological control of dry root rot of chickpea caused by *Rhizoctonia bataticola. Plant disease research* 17(1): 68-71.

Singh, B.K., Mukesh Srivastava, Udit Narain, Srivastava, M. (2003). Evaluation of bioagents against *Fusarium oxysporum* f.sp. *ciceri* causing chickpea wilt. *Farm Science Journal* 12(1): 48-49.

Singh, D.P. (1991). Genetics and Breeding of Pulse Crops. Kalyani Publishers. Ludhiana 371 p.

Singh, G, Singh US, Mukopadhyay, A.N. (2001). Mass production of spore. Powder of *Trichoderma virens* and its viability on coated lentil seed. *Ann. Pl. Protec. Sci.* 9(1): 16-18.

Singh, G. and Sharma, Y.R. (1998). Diseases of chickpea and their management. Pages 155-177. *Diseases of Field Crops and their Management* (Ed., T.S. Thind), National Agricultural technology Information Centre, Ludhiana, India.

Singh, R.A. and Naimuddin. (2004). Mung Evam Urd mein rog prabandhan. In "Dalhan" (eds. Ali *et al.*) IIPR, Kanpur, pp. 225-235.

Singh, R.S., Daljeet Singh, Singh, H.V., Singh, D. (1997). Effect of fungal antagonists on the growth of chickpea. Plants and wilt caused by *Fusarium oxysporum* f.sp. *ciceri. Plant Disease Research* 12(2): 103-107.

Singh, R.S., Narinder Singh, Kaur, M.S. and Singh, N. (1993). Rhizosphere mycoflora mungbean and their interaction with *Macrophomina phaseolina. Plant Disease Research* 8(1): 25-28.

Singh, S.K., Singh, R.H. and Dutta, S. (2002). Integrated management of pigeonpea wilt by biotic agents and biopesticides. Annals of Plant Protection Sciences 10: 323-326.

Sivan, A. and Chet, T. (1989). Biological control effects of a new isolate of *Trichoderma harzianum on Pythium aphanidermatum. Phytopathology* 74: 498.

Skrobakova, E. (1995). Effect of growth Regulators on yield formation in garden pea. *Agrochemia Bratislava* 35(3-4): 61-62.

Somasekhara, Y.M., Kumar, A., Siddaramaiah, T.B. and Ali, A.L. (1996). Biocontrol of pigeonpea (*Cajanus cajan* L. Millps.) wilt *Fusarium udum.* Butler. *Mysore Journal of Agricultural Sciences* 30(2): 159-163.

Somasekhara, Y.M., Siddarmaiah, A.L., Anil Kumar, T.B. (1998). Evaluation of *Trichoderma* isolates and their antifungal extracts as Potential, biological control agents against pigeonpea wilt pathogen. *Fusarium udum* (Butler) current research-university of Agriculture Science, Bangalore 27: 158-160.

Sonawane, S.S., Pawar, N.B. (2001). Studies on biological management of chickpea wilt. *Journal of Maharashtra Agricultural Universities* 26(2): 215-216.

Steinmetz, J. and Schonbeck, F. (1994). Conifer bark as growth medium and carrier for *Trichoderma harzianum* and *Gliocladium roseum* to control *Pythium ultimum* on pea. *Zetschrift-Fun-Pflan, Zeukrankheiten–und- Pflanzenchutz* 101: 200-211.

Strashnov,Y., Elad, Y., Sivan, Y., Rudich and Chet, I. (1985). Integrated control of *Rhizoctonia solani* Kuhn by methylbromide and *Trichoderma harzianum* Rifai Agriculture *Plant Pathology* 34: 146-151.

Sumeet and Mukerjii, K.G. (2000). Exploitation of Protoplast Fusion Technology in Improving Biocontrol. Potential.: In Biocontrol Potential and its exploitation in sustainable agriculture Vol. 1: Crop disease, weeds and nematodes, pp. 33-48.

Tiwari, A.K., Krishna-Kumar, Razdan, V.K., Rather, T.R. (2004). Mass production of *Trichoderma viride* on indigenous Substrates. *Ann. Pl. Protec. Sci.* 12(1): 71-74.

Ulhoa, C.J. and Peberdy, J.F. (1993). Effect of carbon sources on chitobiose production by *Trichoderma harzianum*. *Mycological Research* 97: 45-48.

Upadhyay, R.S. (1987). Tolerance of higher temperature by *Aspergillus nidulans* and its possible implication on biological control of wilt disease of pigeonpea. *Plant and Soil* 97: 77.

Ushamalini,C, Rajappan K, Kousalya Gangadharan, Gangadhara, KI. (1997). Management of charcoal rot of cowpea. Using biocontrol agents and plant products. *Indian Phytopathology* 50 (4): 504-507.

Vanillakam, N and Lowffler, W. (1986). Fengycim a novel antifungal lipopeptide antibiotics produced by *Bacillus subtilis* R. 29-3. *Journal of Antibiotics* 39: 888-901.

Vanillakam, N. and Lowffler, W. (1986). Fengyeim a noval antifungal lipopeptide antibiotics produced by *Bacillus subtilis* R. 28.3. *Journal of Antibioch.* 39: 888-901.

Vasudeva, R.S. (1949). Soil borne plant diseases and their control. *Current Science* 18: 114-115.

Vasudeva, R.S. and Roy, T.C. (1950). The effect of associated soil microflora on *Fusarium udum* Butler, the fungus causing wilt of pigeonpea (*Cajanus cajan* L. Millsp.). *Annals of Applied Biology* 37: 169-178.

Velikanov, L.L., Cukhonosenko, E., Yu., Nikolaevu, S.I. and Zavelishko, I.A. (1994). Comparison of hyper parasitic and antibiotic activity of the genus *Trichoderma* Pers.: Friend *Gliocladium virens* Miller. Giddens at foster isolates towards the pathogens causing root rot of pea. *Mikologiya Fitopathologia* 28: 52-56.

Vidyasekara P, Muthamilan M. (1995). Development of formulation of *Pseudomonas fluorescens* for control of chickpea wilt. *Plant Disease* 78(8): 782-786.

Vishwa Dhar and Chaudhary, R.G. (1998). Diseases of pigeonpea and fieldpea and their management *In: Diseases of Field Crops and their Management* (Ed., T.S. Third). National Agricultural Technology Information Centre, Ludhiana, India, pp 217-238.

Vishwa Dhar and Chaudhary, R.G. (2004). Arhar mein rog prabandhan. In "Dalhan" (eds.) Ali *et al.*. Bhartiya Dalhan Anusandhan Sansthan, Kanpur, India, pp. 213-223.

Vishwa Dhar, Reddy, M.V. and Chaudhary, R.G. (2005). Major diseases of pigeonpea and their management. In "Advances in pigeonpea research (eds. Ali, M and Shiv Kumar). Indian Institute of Pulses Research, Kanpur, India, pp. 229-261.

Wakayama, N.S., Ishikawa, F. and Oishik, (1984). Mycocerein a novel antifungal peptide antibiotic produced by *Bacillus cereus*. *Antimicrobial Agents and Chemotherapy*. 26: 939-940.

Wakelin, S.A., Ryder, M.H. (2004). Plant growth promoting inoculants in Australian Agriculture. *Crop Management* 1-5, March.

Wells, H.D. (1988). *Trichoderma* as a biocontrol agent. Pages 71-82 *In: Biocontrol of Plant disease*, Vol. 1 (Eds. K.G. Mukherji and K.L. Garg), CRC Press Inc., Boca Raton, Florida, USA.

Whistler, C.A. Stockwell, V.O. and Loper, J.E. (2000). Lon protease influences antibiotic production and UV tolerance of *Pseudomonas fluorescens* Pf-5. *Applied and Environmental Microbiology* 66: 2718-2725.

Woo-S.R., Van Veen, J.H., Laanbroek, H.J. and de Vos, W.M. (2002). Synergism between fungal enzymes and bacterial antibiotics may enhance biocontrol. Proceeding of the 9th International An. Leeuwenhoek. 2002, 81 (1-4): 353-356.

Xi, K., Stephens, J.N.G. and Verma, P.R. (1996). Application of formulated rhizobacteria against root rot of field pea. *Plant Pathology* 45(6): 1150-1158.

Xue, A.G.A.D. (2000). Effect of seed borne *Mycosphaerella pinodes* and seed treatment on emergence, foot rot severity and yield of fieldpea. *Can. J. Pl. Path.* 22(3): 248-253.

Yamano, T., Schroth, M.N. and McCain, A.H. (1985). Reduction of *Fusarium* wilt of carnation with suppressive soils and antagonistic bacteria. *Plant Diseases.* 69: 1071-1075.

Sustainable Disease Management of Agricultural Crops (2011) *Pages* **50–61**
Editors: **S.K. Biswas and S.R. Singh**
Published by: **DAYA PUBLISHING HOUSE, NEW DELHI**

Chapter 3

Disease Management in Cucurbitaceous Vegetables

☆ *T.K. Bag*[1], *M. Loganathan*[2] *and Sudhakar Pandey*[2]
[1]*Head, Central Potato Research Station, Shillong*
[2]*Division of Crop Protection, Indian Institute of Vegetable Research, Post Box No. 1, P.O. Jakhini, Shahanshapur, Varanasi – 221 305, U.P.*

Introduction

Cucurbits belong to the family Cucurbitaceae and the family consists of about 118 genera and 825 species. Most of the cucurbits are nutritionally rich and consumed in various forms of salad, pickles and sweets. Some of them (*e.g.* bitter gourd) are well known for their unique medicinal properties. In India, several major and minor cucurbits are cultivated, which shares about 5.6 per cent of the total vegetable production. The cucurbits are cultivated in wide range of environment from tropical to temperate regions and they are affected by wide range of biotic and aboitic stresses including fungi, bacteria, viruses, phytoplasma and nematodes. According to American Phytopathological Society report, cucurbits are affected by 51 fungal diseases, 7 bacterial diseases, 17 virus diseases, 1 phytoplasma disease, 10 nematode diseases and 8 physiological disorders (Martyn *et al.*, 1993). Although, all the cucurbit diseases are not equally important and cause considerable damage, some are really destructive world wide and cause significant colossal loss and they need special attention for management with integrated holistic approaches. In this article, some important cucurbit diseases are discussed with their symptoms, diagnostics and management approaches.

Anthracnose

Anthracnose, caused by *Colletotrichum orbiculare* var. *lagenarium* (formally known as *Colletotrichum lagenarium*)), is a destructive disease of cucurbits occurring in warm and rainy seasons. The disease can cause significant damage to cucumber, muskmelon and watermelon unless resistant varieties are grown. Bottle gourd, cucumber, muskmelon and watermelon are the most susceptible host while bitter gourd is least affected, and pumpkin and squash are rarely affected.

Diagnostic Disease Symptoms

Disease symptoms are found on all above ground plant parts started from cotyledonary leaf to fruits. Symptoms differ according to crop species. In most of the cucurbits, symptoms on leaves are observed as water soaked small yellow spots that enlarge and turn to brown but black on the watermelon. The necrotic portion dries and shatters. On cucumber and muskmelon the spots turn brown and can enlarge considerably. Symptoms on the stems are, elongated water soaked sunken lesions and the lesions become light yellow to brown due to abundant sporulation. Similarly, when fruits are at juvenile stage, small, sunken, light brown and cracked spots are produced. The most striking diagnostic symptoms are visible circular, black and sunken cankers on the fruits. On watermelon the spots may be of 6 to 13 mm in diameter. In moist weather, the black centres of the lesions are covered with a slimy mass of salmon colored spores. Cankers lined with this characteristic color can never be mistaken for any other disease. Similar fruit lesions are also produced on muskmelon and cucumber.

Management Strategies

Use of commercially produced disease free seeds and seed treatment with carbendazim @ 2.5 g/kg must be practiced. Follow crop production with bower system to avoid soil contact. Field sanitation by burning of crop debris and proper drainage in the field should be followed. Seed production should be preferably carried out in summer season since summer crop is often free from pathogen. Carbendazim @ 0.1 per cent or chlorothalonil @ 0.2 per cent spraying can be done soon after infection. Use of resistant varieties/hybrid could be a successful practice for an instance watermelon cv. 'Yarilo' reported as a fairly resistant to *Colletotrichum orbiculare* in Russia (Dyutin *et al.*, 1982). Similarly, in water melon, the cultivars like Fair, Charleston gray, Congo and PI 189225 are resistant to anthracnose but Arka Manik is multiple resistant to downy mildew, anthracnose and powdery mildew, and can be grown in the epidemic area of these diseases (Rai *et al.*, 2008).Cucumber hybrid 'Zhongnong 14' has been identified as resistant to downy mildew (*Pseudoperonospora cubensis*), powdery mildew (*Sphaerotheca fuliginea*), bacterial angular leaf spot (*Pseudomonas syringae pv. lachrymans*) and cucumber mosaic virus in China (Gu *et al.*, 2002).

Downy Mildew

Downy mildew disease is very common in all cucurbits; however, it is severe problem in sponge gourd, ridge gourd, muskmelon and cucumber. The other gourds like bitter gourd, bottle gourd, and snake gourd are moderately susceptible where as pumpkin and summer squash are the less susceptible to downy mildew. Besides, these crops this pathogen also infects watermelon, round melon and pointed gourd. The disease is very common in northern India and it is serious during later part of rainy season. The disease is caused by *Pseudoperonospora cubensis*. The downy mildew fungus exists as pathotypes varying in ability to infect the various cucurbit types. Some can infect all types while others are able to infect cucumber and cantaloupe but not watermelon, squash or pumpkin. Races and strains have been described within pathotypes based on variation in virulence and fungicide sensitivity.

Diagnostic Disease Symptoms

Symptom appears as irregular, numerous, small and yellow areas surrounded by green tissues scattered all over the leaf lamina. It also appears as just like definite mosaic pattern particularly in cucumber. The yellow areas are angular and bounded by veins. Symptoms on bottle gourd are light brown colour while grayish brown on bitter gourd without prominent yellowing. Early lesions of downy mildew in melon usually appear as water-soaked on the underside of leaves. Affected tissue in pumpkin can be more orange than yellow. Initially, on the underside of leaf typically water-soaked spots appears later, an extensive defoliation occurs when conditions are favorable. Leaf petioles often remain green and upright after the leaf blade has died and drooped. In high humid weather, faint white downy growth of fungus is observed on lower side of leaves.

Management Strategies

Dense cropping and shading should be avoided in the planting sites to provide good air movement and light. Overhead irrigation must be avoided in early morning when leaves are wet from dew or late in the day when leaves will not have an opportunity to dry before dew forms. Application of judicious amount of nitrogen fertilizers, field sanitation and bower system of cropping normally reduces the disease incidence. Seed production should be preferably carried out in summer season to obtain disease free seed. Protective spray of mancozeb @ 0.25 per cent at seven days interval gives good disease control. In severe case, one spray of metalaxyl + mancozeb @ 0.2 per cent may be given. There are certain multiple disease resistant varieties of cucurbits which can be exploited in disease prone areas for an example in USA, cucumber inbred line 'USDA 6632E' showed high degree of resistant to several diseases *viz. Pseudomonas syringae* pv. *lachrymans, Colletotrichum orbiculare, Pseudoperonospora cubensis, Sphaerotheca fuliginea, Cladosporium cucumerinum* and cucumber mosaic virus (Staub and Crubaugh, 2001). In China, cucumber cv. 'Henan Cucumber No.2' was reported to be highly resistant to downy mildew (*Pseudoperonospora cubensis*), *Fusarium* wilt (*F. oxysporum f. sp. cucumerinum*), powdery mildew (*Sphaerotheca fuliginea*) and anthracnose (*Colletotrichum orbiculare*) diseases (Li *et al.*, 1999). Muskmelon (cv. B-159 and B-184) and bitter gourd (cv. VBRT-39 and NIC-12285) have been identified as resistant to downy mildew in India (Rai *et al.*, 2006). Similarly, melon cv. 'Punjab Rasila' was resistant to downy mildew where as, cv. Arka Manik was shown multiple disease (downy mildew, anthracnose and powdery mildew) resistance (Rai *et al.*, 2008).

Powdery Mildew

All the cucurbits are susceptible to powdery mildew disease but it is severe in crops like bottle gourd, cucumber, ridge gourd and sponge gourd. Pumpkin and bitter gourd are less infected. This is more severe in winter season and green house crops. Two pathogens *Sphaerotheca fuliginea* and *Erysiphe cichoracearum* are associated with this disease.

Diagnostic Disease Symptoms

The powdery mildew on the foliages and green stems is characterized by the appearance of tiny, white to dirty gray spot (sometimes with a reddish brown tinge). They become powdery as they enlarge. The superficial powdery mass may ultimately cover the entire green surface of the leaf. Because of thick powdery growth on the aerial parts of plants, there is heavy reduction in photosynthetic area. Plant may wither and die. Growth of plant and fruits seized. Transpiration rate is increased in infected leaves. On early stages plant tissues appears normal, but finally the spot turn brown and dry. In India, downy mildew generally comes in winter months. The effect of the severe infection may cause premature defoliation and death of the vines. Fruits also get covered with the white powdery mass but this is not common, caused undersized and deformed fruits.

Management Strategies

Field sanitation and foliar spray of penconazole 0.05 per cent or tridemorph 0.1 per cent or carbendazim 0.1 per cent gives good control of the disease. The melon variety 'Tabolinka' bread in USSR by crossing Kolkhoznitsa 749 and Rio Gold has reported as resistant to powdery mildew (Dyutin, 1981). Muskmelon cv. Milong is highly resistant to powdery mildew (*Erysiphe cichoracearum*) in China (Li *et al.*, 2000). Melon cv. 'Arka Rajhans' and Arka Manik are resistant to powdery mildew diseases (Rai *et al.*, 2008).

Gummy Stem Blight

Gummy stem blight (GSB) caused by *Phoma cucurbitacearum* (perfect stage: *Didymella bryoniae*) is a common disease of all major cucurbits. GSB symptom is most common on the stem where as black rot is associated with symptoms on fruits. This disease is now becoming severe in muskmelon, bottle gourd, ridge gourd and sponge gourd as well.

Diagnostic Disease Symptoms

Different types of symptoms occur on foliar parts of the cucurbit plants which make diagnosis difficult. On leaves of the muskmelon variety 'Early gold', symptoms appear as water soaked lesion on the leaf margin, interveinal necrotic scorch, and randomly distributed irregularly shaped circular lesions. Sometimes, lesions may be surrounded by a yellow halo, and the lesions often crack when they dry up. On pumpkin, a non-descript marginal necrosis is followed by larger and wedge-shaped necrotic areas. Leaf symptoms on cucumber and squash are in frequent but are similar in appearance to those on pumpkin. The asexual fruiting bodies (pycnidia) appear on affected leaves as small black specks, but if the tissue is rapidly killed, as on muskmelon, these diagnostic signs will not be evident on the foliage.

On the stem first symptoms appear as water soaked lesions which later turn tan. On older stems, pycnidia develop on the affected tissue, particularly in musk melon and cucumber. Stem lesions often emit gummy, reddish-brown or black beads to exude which can be confused with *Fusarium* wilts and injury caused by insect feeding. In the latter cases, however, pycnidia are not present.

Management Strategies

Cucumber cv. Homegreen 2 and an exotic collection, PI 200818 were found resistant against GSB in USA (Wyszogrodzka *et al.*, 1986). Watermelon cv. 'AU-Jubilant' and 'AU-Producer' developed from the backcross and inbreeding breeding programme were resistant to multiple diseases *viz., D. bryoniae, C. lagenarium* race 2 and *F. oxysporum f. sp. niveum* in the USA (Norton and Cosper, 1985). Muskmelon cv. 'Aurora' developed by crossing between Southland and PI-140471 was found highly resistant against *Pseudoperonospora cubensis, Sphaerotheca fuliginea* and *D. bryoniae* in the USA (Norton *et al.*, 1985). In Japan, Sakata *et al.* (2000) identified several accessions of *C. melo* var. *agrestis viz.*, MEX-2, KOR-33, J-3 and JMu-15 as highly resistant to the gummy stem blight pathogen.

Use of disease free seeds, seed treatment with carbendazim @ 2.5 g/kg and crop rotation with non cucurbitaceous crops for 2 years will help to control the disease. Apart from these, some additional management practices like controlling cucumber beetles and aphids to reduce predisposing the disease, and foliar spray of captafol or maneb or mancozeb or carbendazim or thiophanate methyl controls secondary infection, can be followed.

Fusarium Wilt

The vascular wilt caused by *F. oxysporum* is one of the most common soil borne diseases worldwide. Several species of *Fusarium* are responsible for the vascular wilt of cucurbits. *F. oxysporum* is known to cause wilt in cucumber, whereas *F. oxysporum* f. sp. *melonis* reported as wilt pathogen of cantaloupe, melon, honeydew melon and musk melon. *F. oxysporum* f. sp. *niveum* induces wilt in citron, squash and watermelon. *F. solani* f. sp. *cucurbitae* affects pumpkins, squash and other cucurbits. Some *Fusarium* spp. are host specific, for an example, *F. oxysporum* f. sp. *melonis* infects *C. melo* (cantaloupe, crenshaw, honeydew melon and muskmelons), but not gourd, pumpkin, squash and watermelon.

Diagnostic Disease Symptoms

Symptom is characterized by yellowing in lower leaves, which gradually progresses on upper leaves. Immediately after infection, the plant starts drooping followed by wilting. Later on wilting becomes permanent. Characteristic symptom was observed after splitting of roots and lower portion of stem as vascular browning. Severity of disease is more if soil is infected with root-knot nematode.

Management Strategies

The disease is seed borne as well as soil borne in nature. Seed treatment with carbendazim (0.25 per cent) before sowing is advisable. Disease can be minimizing by soil application of *Trichoderma* (Bioderma) at the rate of 5-8 kg/ha. Rotation to a non-host crop for 3-4 years is effective for most of the diseases caused by *Fusarium* spp.

Host plant resistance is the most effective for control of Fusarial wilt of watermelon (Martyn and McLaughlin, 1983) and muskmelon (Zink *et al.*, 1983). In India, muskmelon cultivars Durgapur Madhu and Punjab Sunehri were resistant to *F. oxysporum* and *F. solani* (Radhakrishnan and Sen, 1985). Besides, variety Tavrichanka from Ukraine, UC Top from USSA are also resistant to *Fusarium* wilt. In melon, breeding line MR-1 found resistant to races 0, 1 and 2 of *F. oxysporum* f. sp. *melonis*. Watermelon cultivars, Summit, Conqueror, Charleston gray, Dixilee, and Crimson sweet were resistant to *Fusarium* wilt and are been grown in wilt prone areas (Rai *et al.*, 2008).

Uses of biological control agents have been found very effective in inducing disease resistance (Sharma, 2004). Potential biocontrol agents *viz.*, *P. chlororaphis* strain 63-28, *T. harzianum* (Root shield drench), *Streptomyces griseoviridis* (Mycostop), *Gliocladium catenulatum* (Prestop WP, Prestop Mix), and *G. virens* (Soil gard) and a bio-molecule, crab/shrimp shell chitin, have been evaluated against the disease.

Phytophthora Blight

Phytophthora blight of cucurbits caused by the *P. capsici* is one of the most serious threats to production of cucurbits (cucumbers, melons, pumpkins, and squash) in worldwide (Babadoost, 2000 and Hausbeck, 2004). Recently, the incidence of *Phytophthora* blight on cucurbits has significantly increased in Illinois (Tian and Babadoost, 2004) and other cucurbit-growing areas in the world (Hausbeck, 2004), causing up to 100 per cent yield loss. Cucurbit industries, particularly processing industries, are seriously threatened by heavy crop loss resulting from *Phytophthora* blight. The *Phytophthora* blight in cucurbit has little impact in India and cucurbit blight caused by *P. parasitica* and *P. capsici* has been reported from some isolated part of the country in very rare case. But, cucumber late blight and root rot caused by *P. melonis* and squash blight caused by *Phytophthora capsici* have not reported from India.

Diagnostic Disease Symptoms

P. capsici can strike cucurbit plants at any stage of growth. The infection usually appears first in low areas of the fields where the soil remains wet for longer periods of time. The pathogen infects seedlings, vines, leaves, and fruit (Babadoost, 2004). The pathogen causes pre and post emergence damping-off in cucurbits in wet and warm (20 to 30°C) soil conditions (Zitter *et al.*, 1996). In seedlings, a watery rot develops in the hypocotyls at or near the soil line, resulting in plant death. Mature plants show symptoms of crown rot and sudden wilting. Normally wilting symptom is visible from the base to top of plant vines. On stem, initially the stem near to the soil turns light to dark brown becomes soft and water-soaked later the infected stems collapse and die.

Vines can be affected at any time during the growing season. Water-soaked lesions develop on vines. The lesions may be of dark olive in the beginning and become dark brown in a few days later. On leaves, development of leaf spots ranging from 5 mm to more than 5 cm in diameter chlorotic at first and then become necrotic with chlorotic to olive-green borders in a few days can be noticed. Fruit rot can occur from the time of fruit set until harvest. Fruit rot generally starts on the side of the fruit that when in contact with the soil. Symptoms on the upper surface of the fruit may also develop following rain or overhead irrigation, which provides splashing water for pathogen dispersal. *Phytophthora* foliar blight and fruit rot may result in total loss of the crop (Zitter *et al.*, 1996).

Management Strategies

A combination of all available measures should be practiced to reduce the damage caused by *P. capsici* on cucurbits (Zitter *et al.*, 1996; Hausbeck, 2004; Islam and Babadoost, 2004). Any single method can not provides adequate control of *Phytophthora* blight in any crop. The most effective practice in controlling *P. capsici* is preventing the movement of the pathogen from infected field to new field. The following practices can help to manage *Phytophthora* blight in cucurbit vegetables. Use of well-drained fields, selecting the field which has no history of *Phytophthora* blight infection, avoiding excessive irrigation, scouting the field for the *Phytophthora* symptoms, especially after major rainfall, and particularly in low areas to early management, and harvesting the seeds from *Phytophthora* blight free plots or fields. Seed treatment with metalaxyl can protects seedlings of cucurbits until 5 weeks after sowing seed (Babadoost and Islam, 2003). There are reports that continuous use of Ridomil Gold causes the development of resistance against *P. capsici in* North Carolina and in New Jersey hence judicious application of the fungicide should be followed.

Cucurbit Scab

The disease is caused by the fungus, *Cladosporium cucumerinum* and it affects muskmelon and cucumber. Symptoms appear as dry corky spots on cucumber and green spots on muskmelon fruits. Under moist conditions a dark olive-green velvety growth appears on the spot. This growth distinguishes scab from angular leaf spot on cucumber fruit. Spots also develop on young terminal stem growth and on petioles. When spots girdle young stems and petioles, the growth beyond the spot dies up. Affected areas on very young leaves are irregular in shape and the infected areas soon become dried. The disease causing fungus survives on seed and in residue from diseased plants.

Management Strategies

The resistant cultivars of cucurbits may be cultivated to avoid the disease. In Poland, pickling cucumber (*Cucumis sativus*) hybrid 'Polan' has been released in 1972 as high yielding as well as resistance to scab (*Cladosporium cucumerinum*) and CMV (Cucumber mosaic virus). Another set of three more hybrids, 'Aladyn', 'Cezar' and 'Parys', which have multiple disease resistance to scab, CMV, powdery mildew (*Sphaerotheca fuliginea*), anthracnose (*Colletotrichum orbiculare*), angular leaf spot (*Pseudomonas lachrymans*) and especially downy mildew were released in 1992. Therefore, new F1 hybrids 'Atlas', 'Cyryl', 'Bazyl', 'Hermes' and 'Izyd' were developed and released in 1996-99. They are very prolific and have high level of downy mildew resistance (Doruchowski *et al.*, 2000). In China, cucumber hybrid 'Jinyou No. 31' has been reported to be highly resistant to *Fusarium* wilt, downy mildew, powdery mildew and scab (Li *et al.*, 2004). Application of blitox 50 or phytolon or blue copper @ 3 g/l has also been recommended for control of this disease.

Alternaria Blight

Causal organism of the disease is *Alternaria cucumerinum* and it infects melon, squash, cucumber, pumpkin and watermelon especially under wet weather and moderate temperatures of 60-90 °F. Among the cucurbits melon is quite susceptible and most commonly infected.

Diagnostic Disease Symptoms

Spots develop first on older leaves near the centre of the hill. The spots are circular and appear somewhat water soaked. They enlarge rapidly up to 1/2 inch in diameter and turn light brown on melons, cucumbers and squash and dark brown to black on watermelon. Spots on the upper leaf surface may develop concentric rings giving them a target pattern. Rapid defoliation occurs when weather favors the disease. Occasionally the fruits of melon, squash, cucumber and watermelon become infected where sunken spots are there and often these spots are covered with an olive to black mold. Severe attack occurs when plants are being weakened by poor soil fertility, poor growing conditions, heavy fruit set and/or other diseases.

Management Strategies

Crop rotation and destruction of crop refuse are important. Use of good quality seed and treatment of the seed with a fungicide such as thiram or captan are also advisable. Sprays of fungicide like mancozeb (dithane M-45 or manzate 200) also help to control the disease.

Fruit Rot

Fruit rot is more common in rainy season than summer crop. Generally lower portion of fruits which touches ground are mostly affected. The disease is caused by species of several soil borne fungi namely *Pythium, Phytophthora, Rhizoctonia, Phomopsis* and *Colletotrichum*. Fruits of pointed gourd are severely infected by *Phytophthora cinnamomi* during rainy season. It also causes rotting of vines and leaves of pointed gourd and the infected crop is blighted within three to four days. In *Pythium* fruit rot, initially water soaked and soft rotting of the fruits starts from lower portion, which are in soil contact and in the early morning white, fluffy, mycelial growth is seen obviously on the affected portion. *Rhizoctonia* fruit rot is most severe in muskmelon showing rotting and cracking of infected portion of fruit. *Phomopsis* fruit rot caused by *Phomopsis cucurbitae* was first recorded on ash gourd in India. Symptoms are mostly observed on matured fruits showing comparatively dry rotting with characteristic pycnidia over it.

Management Strategies

To avoid fruit contact with ground soil, crop can be grown on bower or trained on stakes. Proper drainage should be provided in the field to avoid stagnation of water so that there will be less chance of *Pythium* and Phytophthora rots. Green manuring along with soil application of *Trichoderma* formulation (Bioderma) @ 5 kg/ha is very effective in checking most of the fruit rotting. Rotten fruits should be collected and burnt to reduce the built up of primary inoculums in field.

Leaf Spots

Cucurbits are affected by several types of leaf spots. The associated pathogens with leaf spot diseases are *Cercospora citrullina, Didymella bryoniae, Alternaria cucumerinum* and *Corynespora melonis*.

Often these leaf spots diseases are more pronounced at maturity stage. Spots are circular and variable in size and are observed mostly on the leaf lamina. They are light brown to dark brown with white centre in *Cercospora* initiated spots. Papery with rhythmic large spots followed by shot hole is observed in *Didymella*. Black small dot like black fruiting structures are also observed on the old spots. White fungal growth is visible on outer margins of the spots in the morning. The severity of the disease is increasing every year in vegetables *viz.*, pointed gourd, pumpkin and sponge gourd.

Management Strategies

Field sanitation, selection of healthy seeds and crop rotation are the normal practices followed to reduce the disease incidence. Alternate spray of mancozeb @ 0.25 per cent and hexaconazole @ 0.05 per cent is recommended to control the pathogen. Seed production should preferably be carried out in summer season to obtain disease free seed.

Bacterial Wilt

Bacterial wilt of cucurbits caused by *Erwinia tracheiphila* is a common and destructive disease of cucurbits. The disease is spread primarily by the striped cucumber beetle, *Acalymma vittata*, and the spotted cucumber beetle, *Diabrotica undecimpunctata howardi*. However, Mathew *et al.* (2002) reported that bacterial wilt of cucurbits such as bitter gourd (*Momordica charantia*) and ridge gourd (*Luffa acutangula*) is caused by *Ralstonia solanacearum* race 1 and biovar 3, and they claimed that this is the first report of cucurbit wilt by *R. solanacearum* from India. It has also been reported in United States of America, Canada, Europe, South Africa and Japan. Cucumber is the most susceptible host followed by muskmelon, squash and pumpkin while watermelon is least susceptible.

Diagnostic Disease Symptoms

The first sign is the wilting of individual leaves, followed by drooping and flaccid in sunny weather. Later, the wilting occurs eventually on the entire plants. Affected stem becomes soft and gets macerated but later on plant dry up permanently. If a transverse cut is made through the stem and gluey substance will be found is a very characteristic symptom of the disease, which forms a thread if the two parts of the stems are separated very gently. Cucumber, muskmelon, squash and pumpkin are most susceptible among cucurbit hosts.

The bacteria survive in the body of adult cucumber beetles such as red striped and spotted beetles. Primary infection is initiated when beetles feed and cause deep wounds on the young leaves and/or cotyledons of cucurbits plants. The bacteria are deposited in such wounds with the faces of the insects. This is the only means of natural spread known. This pathogen is one of the unusual organisms in which the bacterium is completely dependent upon the insect for its survival. After infection, the bacteria swim through the droplets of sap present in the wound. It enters in the xylem vessels and move rapidly to all parts of the plant. The bacterium does not penetrate the host through stomata. The bacteria present in the vessels of infected plants die within 1 or 2 months as and when the plants died. The survival period of these bacteria is very short in infected crop debris.

Management Strategies

The most effective method of disease control is immediate removal of cucumber beetles. The beetles can transmit squash mosaic virus as well as bacterial wilt and can cause severe damage by feeding on the leaves. Ploughing of soil is to be carried in summer to expose all the stages of beetles. For management of cucumber beetles in the soil at initial stage of infestation, soil application of neem cake is recommended. Control of cucumber beetles at high infestation level may be carried out by applying granular insecticides like carbofuran in surrounding soil of the plants or in the pits followed by dusting and foliar spray of carbaryl.

Control should start when the first beetle is seen or when the first cucurbit seedlings emerge. Some farmers plant "trap crops" of cucurbits thickly over a small area a few days before they plant the main crop. Trap crops are meant to attract beetles which can be sprayed with insecticide. All crop refuse must be removed and destroyed to reduce other cucurbit diseases and to reduce sites for beetle hibernation. Tolerant cucurbit varieties with restricted use of exotic cucumber lines should be grown. Most of the exotic hybrids of cucumber are susceptible to the disease. However, China has developed new cucumber F1 hybrid 'Zhongnong 201' which was highly resistant to powdery mildew, angular leaf spot, *Fusarium* wilt and was moderately resistant to downy mildew of cucumber (Sheng *et al.*, 2001).

Angular Leaf Spot

The causal agent of angular leaf spot is *Pseudomonas syringae* pv. *lachrymans* (Smith and Bryan) and it is reported on cucumber, gherkin, muskmelon, pumpkin, squash, watermelon and related species worldwide. In India, it is a serious disease in the cool climatic areas (Jindal, 1994).

Diagnostic Disease Symptoms

The bacterium affects both leaves and fruit. On the leaves, lesions begin as small water-soaked circular black spots surrounded by a yellow halo. The centres of the spots may become white. As they enlarge, lesions become angular in shape and may involve on entire lobe of the leaf. On fruits, lesions begin as very small circular water-soaked areas which enlarge with age to cover larger portions of the melon fruit surface. Lesions do not penetrate deeply into the fruit. On large lesions the cuticle ruptures and peels free from the melon fruit surface. On squash and watermelon, lesions are brown and black, with a yellow halo. The bacterial ooze forms a crust on them and dead tissues may peel-off leaving shot holes.

Management Strategies

Preventive measures like use of clean equipment and tools, avoid working in fields when they are wet, locating cull piles away from production fields and use of disease free seeds, will certainly prevent infection. Avoiding overhead irrigation, field sanitation and crop rotation may also reduce the disease incidence. Spraying of a fungicide containing cupric hydroxide plus zinc (0.5–1.0 per cent) gave good control of *P. syringae* pv. *lachrymans* on cucumber in Romania. Copper fungicides are useful but care should be taken not to spray when temperature is high or on young plants.

Bacterial Leaf Spot

The bacterial leaf spot of cucurbits caused by *Xanthomonas campestris* pv. *cucurbitae* infects *Cucurbita maxima*, *C. pepo*, *C. moschata* and *Citrullus lunatus*. The symptoms of this disease is minute water-soaked lesions on the lower surface of leaf with corresponding yellowing on upper surface later they become brown and surrounded with halo, the spots remain restricted between midrib and veins, and some times spots may coalesce to form large dry patches.

Management Strategies

Nursery site should be changed frequently to avoid seedling infection. Detach the lower infected leaves in afternoon after drying of bacterial ooze and burn them. Crop rotation with cereals is advised to minimize the infection. Use antagonistic bacteria *viz. Pseudomonas* spp and *Bacillus* spp. in the soil to minimize the inoculums. Spraying streptocycline @ 150-200 ppm or kasugamycin @ 0.2 per cent at 10-15 days interval is beneficial. Also, spraying of copper oxychloride @ 0.3 per cent at 10 days interval or mixture of streptocycline 100 ppm and copper oxychloride @ 0.3 per cent with sticker @ 0.1 per cent gives good control of the disease.

Cucurbit Mosaic

Cucurbit mosaic is caused by cucumber mosaic virus and squash mosaic virus. The disease appears as a mosaic pattern of light and dark green areas on the leaves and fruits. In addition to the mosaic pattern, the edges of the leaves turn down, and light yellow colour knobs on the fruits. The infected leaves become small and puckered, and plants become severely stunted. Fruits develop warts and become misshaped.

The cucumber mosaic virus is transmitted by mechanical as well as through aphids. The virus infects cucumber, melon, squash, pumpkin, pepper, spinach, tomato, and other vegetables, flowers, and weeds.

Squash mosaic is caused by the squash mosaic virus. It is transmitted from plant to plant by cucumber beetles. The virus infects squash, cucumber, melon, and occasionally watermelon. The virus may sometimes be seed borne.

Watermelon Mosaic Virus

Watermelon mosaic virus (WMV) is the second most important cucurbit virus in USA and other countries. Besides watermelon, this virus can infect and produce symptoms on all other commercially grown cucurbits. This is an aphid-transmitted virus and causes milder symptoms on the foliage of most infected squash plants. Fruit distortion and color breaking are still a problem on varieties like yellow straight neck of squash. Uses of variety such as 'Multipik' since the fruits are marketable in spite of foliar symptoms.

Pumpkin Mosaic Virus

The infected leaf initially shows vein-clearing symptoms. Serration is noticed in some foliage in which tips of veins protrude beyond the edges of the leaf; some times giving toothed appearance. Diseased leaves become smaller than normal. Fruits are small with less carotene content. The virus can be transmitted mechanically through sap and naturally by aphids.

Yellow Vein Mosaic of Pumpkin

Infected plant shows yellowing of veins and vein clearing on young leaves with light green or yellow colour patches. Normally the infected leaves become smaller than the normal leaves. Similarly, fruits become small and deformed. The virus is transmitted mechanically by sap and persistent transmission through white fly (*Bemisia tabaci*).

Zucchini Yellow Mosaic Virus (ZYMV)

This is a recently described virus disease of cucurbits, first identified in Europe in 1981. It has since been reported from most southern and southwestern states and was found in New York State in 1983. This virus has also been reported from India. Foliar symptoms consist of a prominent yellow mosaic, necrosis, distortion, and stunting. Fruits remain small, greatly malformed, and green mottled. The virus has characteristics very similar to WMV-1 and WMV-2 and like WMV-2; its host range is not limited to cucurbits. Currently, none of the genetic factors confer resistance to WMV-1 or WMV-2 virus strain.

Management Strategies

Virus diseases are managed by using good quality seeds and controlling aphids and cucumber beetles throughout the season. Diazinon can be used for aphid and the beetle control. Apart from that carbaryl (sevin) and methoxychlor also provide better control of the beetle. Insect control should be started as soon as plants emerge from the soil. Planting of cucurbits should be avoided near woods, brushy areas, or other weed prone areas. To avoid aphid infestation grow several rows of corn around the cucurbit patch, or at least plant corn on the windward side. To avoid viral spreads remove and destroy diseased plants as soon as mosaic appears and also after handling diseased plants, wash the hands with detergent and water.

References

Babadoost, M. (2000). Outbreak of *Phytophthora* foliar blight and fruit rot in processing pumpkin fields in Illinois. *Plant Disease* 84:1345.

Babadoost, M. and Islam, S. Z. (2003). Fungicide seed treatment effects on seedling damping-off of pumpkin caused by *Phytophthora capsici*. *Plant Disease* 87:63-68.

Doruchowski, R. W., Lakowska, R. E. and Katzir, N. (ed.), Paris-HS. (2000). F1 hybrid pickling cucumbers developed for increased yield, earliness and resistance to downy mildew (*Pseudoperonospora cubensis*). *Cucurbitaceae*. Proceedings of the 7th EUCARPIA meeting on cucurbit breeding and genetics, Ma'ale Ha Hamisha, Israel, 19-23 March, 2000. *Acta Horticulturae*. (2000, No. 510, 45-46.

Dyutin, K. E. (1981). Sources of resistance to powdery mildew in cucurbits. *Mikol.-i-fitopatol* 15: 299-301.

Dyutin, K. E., Bogoyavlenskaya, S. M., Mascheva, N. I. and Turchenkov, G. I. (1982). New watermelon varieties. *Kartofel'-i-Ovoshchi* 10: 36.

Gu, X. F., Fang, X. J. and Zhang, T. M. (2002). A new cucumber hybrid–'Zhongnong 14'. *Acta Horticulturae Sinica* 29: 398.

Hausbeck, M. (2004). *Phytophthora* lessens learned: Irrigation water and snap beans. *The Vegetable Growers News* 38: 2829.

Islam, S. Z. and Babadoost, M. (2004). Evaluation of selected fungicides for control of *Phytophthora* blight of processing pumpkin, *Fung. and Nemat. Tests* 59:V129.

Jindal, K. K. (1994). Occurrence of angular leaf spot bacterium *Pseudomonas syringae* pv. *lachrymans* on cucumber plant. *Plant Disease Research* 9: 66-67.

Li, J. W., Sun, S. R., Ma, C. S., Xu, X. L., Shi, X. J. (1999). 'Henan Cucumber No. 2'–a new cucumber hybrid variety. *Acta Horticulturae Sinica* 26: 3.

Li, J. W., Zhang, W. Z. and Wang, J. (2004). A new cucumber F1 hybrid–'Jinyou No. 31'. *China Vegetables* 5: 16-18.

Li. X. X., Duan, A. M., Lu, J. G. (2000). A new muskmelon F1 hybrid–Milong. *China Vegetables*. No. 6, 22-24.

Martyn, R. D. and Maclaughlin, R. J. (1983). Effect of inoculum concentration on the apparent resistance of watermelons to *Fusarium oxysporum f. sp. niveum*. *Plant Disease* 67: 263-266.

Martyn, R. D., Miller M. E. and B. D. Bruton. (1993). Diseases of Cucurbits. In: *Common names of Plant diseases*. American Phytopathological Society. Plant Pathology on line. APSnet.com

Mathew, S. K., Girija, D., Abraham, K.and Binimol, K. S. (2002). Occurrence of bacterial wilt of bitter gourd and ridge gourd in Kerala. *Bacterial Wilt Newsletter* 17: 21

Norton, J. D., Cosper, R. D. (1985). Breeding watermelons for disease resistance. *Phytopathology* 75: 1178.

Norton, J. D., Cosper, R. D., Smith, D.A., Rymal, K.S. (1985). 'Aurora' muskmelon. *Hort Science* 20: 955-956.

Radhakrishnan, P. and Sen, B. (1985). Efficacy of different methods of inoculation of *Fusarium oxysporum* and *Fusarium solani* for inducing wilt in muskmelon. *Indian Phytopathology* 38: 70- 73.

Rai, M., Pandey, S. and Kumar, S. (2008). Cucurbit research in India: a retrospect. In: Cucurbitaceae, Proceedings of the IXth EUCARPIA meeting on genetics and breeding of Cucurbitaceae (Pitrat M, ed), INRA, Avignon (France), May 21-24th, 2008. 1

Rai, M., Singh, M., Pandey, S., Pandey, K. K., Singh, J., Kumar, S. and Singh, B. (2006). IIVR: A Decade of Accomplishment. Indian Institute of Vegetable research, P.O. Jakhini (Shahanshapur), Varanasi 221 305, Up, India

Sakata,Y., Wako, T., Sugiyama, M., Morishita, M., Katzir, N. (ed.), Paris, H.S. (2000). Screening melons for resistance to gummy stem blight. *Cucurbitaceae* 2000. Proceedings of the 7th EUCARPIA meeting on cucurbit breeding and genetics, Ma'ale Ha Hamisha, Israel, 19-23 March, 2000. *Acta Horticulturae.* (2000), No. 510, 171-177.

Sharma, P. (2004). Induced resistance V/S biological control against plant pathogens. In *Advances in Plant Protection Sciences* pp230-243 (Ed.) D. Prasad and Amerika Singh.

Shen, D., Yin, Y., Qi, C. Z., Li, X. X. and Song, J. P. (2001). A new cucumber F1 hybrid — 'Zhongnong 201'. *Acta Horticulturae Sinica* 28: 481.

Staub, J. E. and Crubaugh, L. K. (2001). Cucumber inbred line USDA 6632E. *Cucurbit Genetics Cooperative.* No.24, 6-7.

Tian, D., and Babadoost, M. (2004). Host range of *Phytophthora capsici* from pumpkin and Pathogenicity of isolates. *Plant Disease* 88:485-489.

Wyszogrodzka, A. J, Williams, P. H., Peterson, C. E. (1986). Search for resistance to gummy stem blight (*Didymella bryoniae*) in cucumber (*Cucumis sativus* L.). *Euphytica* 35: 603-613.

Zink, F.W., Gubler, W.D. and Grogan, R.G. (1983). Reaction of muskmelon germplasm to inoculation with *Fusarium oxysporum f. sp. melonis* race 2. *Plant Disease* 67: 1251–1255.

Zitter, T. A., Hopkins, D. L., and Thomas, C. E. (1996). Compendium of Cucurbit Diseases. American Phytopathological Society, St. Paul, MN.

Sustainable Disease Management of Agricultural Crops (2011) *Pages* **62–79**
Editors: **S.K. Biswas and S.R. Singh**
Published by: **DAYA PUBLISHING HOUSE, NEW DELHI**

Chapter 4

Diseases of Jute and Mesta: Present Status and Management Options

☆ ***C. Biswas, S.K. Sarkar and R.K. De***
Central Research Institute for Jute and Allied Fibres, Barrackpore, 24-Parganas (N), Kolkata – 700 120

Jute (*Corchorus olitorius*, jute; *Corchorus capsularis*,white jute) and mesta (*Hibiscus cannabinus, H. sabdariffa*) are two important bast fibre crops. Jute is mainly grown in South East Asian countries such as India, Bangladesh, Nepal, China, Indonesia, Thailand, Myanmar, and some South American countries. In India, jute is cultivated in the Eastern region covering an area of little over 0.8 million hectare. The area under mesta is more or less 2.5 lakh hectares and the major growing areas of *H. cannabinus* are in West Bengal, Assam and Tripura and that of *H.sabdariffa* are in Andhra Pradesh, Orissa and Maharashtra. Both the crops are attacked by number diseases. The major diseases of jute are stem rot (*Macrophomina phaseolina*), anthracnose (*Colletotrichum corchorum, C.gloesporiodes*) black band (*Botryodiplodia theobromae*) and soft rot (*Sclerotium rolfsii*).Mesta crop is mainly affected by foot and stem rot (*Phytophthora parasitica* var.*sabdariffae*) and mesta yellow vein mosaic diseases.

Stem Rot of Jute

Stem rot caused by *Macrophomina phaseolina* is the most economically important disease of jute (both *white* and *tossa* jute) in all jute growing areas. Though commonly known as the stem rot, the symptom is not restricted to the stem only. The pathogen attacks any part of the plant at any stage of growth thus causing seedling blight, damping off, collar rot, stem rot and root rot. Average yield loss due to this disease has been estimated at around 10-15 per cent but in case of severe infection may lead to 35-40 per cent loss in fibre yield.

Symptom

Even just after germination the cotyledons are infected, turns brown to black and drop off. Under favourable condition infection spread from such seedlings to the stem and root resulting in damping

Figure 4.1: Stem Rot of Jute

off of seedlings. Damping off is seen only in young seedlings but seedling blight may occur in young as well as older seedlings up to 6-8^2 height. Well grown plants are also infected at the collar region *i.e.* the stem just above the ground level. Necrotic wounds develop at the site of infection and such plants either die or survive developing adventitious roots in lying position. Stem rot may start at any stage of plant growth. Seedlings of 6-8^2 height may show light to dark brown lesions with innumerable pycnidia on leaf surface as well as at the junction of the petiole and the stem causing stem rot followed by death of the seedling. In mature plants also innumerable minute dots, brown in colour appear on lamina and the margin of the leaves. Infection proceeds to the stem through the petiole or directly from lamina if the infected leaves hang down and touch the stem. The stem develops brown to black necrotic spots which vary in size and shape. Heavily infected stem of susceptible variety often breaks at the lesion site while comparatively less affected ones remain standing in the field showing shredded fibre tissue at the lesion. In case of root infection young rootlets or branch roots or even the main roots are marked by light brown areas which indicate the site of infection. The plants show wilting symptoms from the top. The disease is seed as well as soil born.

Pathogen

Macrophomina phaseolina (Tassi) Goid is the causal organism of stem rot in jute, the sclerotial stage of which is *Rhizoctonia bataticola*. The perfect stage is *orbilia obscura*. The pycnidial and sclerotial stage are primarily responsible for the disease. The fungus has more than 500 host range. *Major cultivated hosts include: Arachis hypogaea* (peanut), *Beta vulgarius, Brassica oleracea* (Cabbage), *Capsicum annuum* (pepper), *Cicer arietinum* (chick pea), *Citrus spp. Corchorus sp., Cucumis spp., Fargaria sp., Glycine Max* (soybean), *Gossypium sp., Helianthus annuus* (sunflower), *Ipomoea batatas* (sweet potato), *Medicago sativa* (alfalfa), *Phaseolus spp., Pinus spp., Prunus spp., Sesamum indicum* (sesame), *Solanum tuberosum* (potato), *Sorghum bicolor* (sorghum), *Vigna unguiculata* (bean), and *Zea mays* (corn). The fungus isolated from various hosts from different parts of the world showed wide range of variation in morphological shape and size that ranged form less than 100 to above 400μ, and based on this size, Haige (1935) classified into four groups *viz.* A, B, C and D. Isolates from jute collected from different jute growing areas in India belonged to Group C of Haige. But within this group also there is morphological and pathological variation and they have been further classified in to four sub-groups depending on their sclerotial morphology and pathogenicity (Mandal *et al.*, 1998). Eight physiologic races of the fungus were identified by workers at JRI, Dhaka, while four strains were recognized in terms of virulence in India (Ahmed *et al.*, 1969 and Ghosh and Sen 1973).

Pathogenesis

Pectinolytic and cellulolytic enzymes play a significant role in pathogenesis of seedling blight. Both the enzymes produced by *M.phaseolina* constitutively and inducibly. Distinct lesions were observed in 14 day old jute seedlings kept in enzyme solution at 21C for 48 hours (Chattopadhyay and Raj, 1978). Of the pectic enzymes contained in three strains of *Macrophomina phaseolina*, MP-3 was found to be more virulent in causing disease and retting jute indicating a relationship of pectin enzyme with stem rot disease and rotting of jute (Myser Ali *et al.*, 2005).

Disease Cycle and Epidemiology

M. phaseolina survives as microsclerotia in the soil and on infected plant debris. The microsclerotia serve as the primary source of inoculum and have been found to persist within the soil up to three years (Ahmad, 1968)(4). The microsclerotia are black, spherical to oblong structures that are produced in the host tissue and released in to the soil as the infected plant decays. These multi-celled structures allow the persistence of the fungus under adverse conditions such as low soil nutrient levels and temperature above 30°C. Microsclerotial survival is greatly reduced in wet soils surviving no more than 7 to 8 weeks and mycelium no more than 7 days. Seeds may also carry the fungus in the seed coat. Infected seed do not germinate or produce seedlings that die soon after emergence. Germination of the microsclerotia occurs throughout the growing season when temperatures are between 28 and 35°C. Microsclerotia germinate on the root surface, germ tubes form appresoria that penetrate the host epidermal cell walls by mechanical pressure and enzymatic digestion or through natural openings. The hyphae grow first intercellularly in the cortex and then intracellularly through the xylem colonizing the vascular tissue. Once in the vascular tissue *M. phaseolina* spreads through the taproot and lower stem of the plant producing microsclerotia that plug the vessels. The rate of infection increases with higher soil temperatures and low soil moisture will further enhance disease severity. The pathogen has a poor saprophytic compatibility in mycelia form and involvement of dormant sclerotia is the survival of the fungus, therefore, growing of Macrophomina prone crop in succession increases the inoculums in the soil (Bhattacharya and Samaddar, 1976). Environmental and soil factors like soil moisture, relative humidity and temperature influence the development of stem rot. Susceptibility to stem rot of jute increased with age irrespective of varieties and maximum disease was observed at the time of harvest. Overcast cloudy condition, heavy rainfall resulting in near field capacity of soil moisture, high atmospheric humidity, air temperature below 32°C and soil temperature below 30°C favour the infection (Rao, 1979).

Management

Varietal Resistance

True resistant variety against the disease is not available among either of the cultivated species, at least at the cultivators level. But the degree of resistance/susceptibility varies from variety to variety and also within the vagaries from place to place, which might be due to the pathotypes present in a particular area as well as the soil and climatic conditions of the place. The most ruling varieties JRO 524 and JRO 632 of *C. olitorius* and JRC 212 and JRC 321 of *C. capsularis* showed differential reactions at different places and also at the same place with pathogen isolates form different places (Mandal *et al.*, 2001). When the whole set of germplasm available at CRIJAF were assessed against the pathogen, quite a good number were totally free from the disease at Barrackpore and Budbud field condition (Saha *et al.*, 1994 and Mahapatra *et al.*, 1994). But when the short listed *C. capsularis* were further exposed to more vulnerable situation at Sorbhog, most of them were severely affected by the disease. Only 9 accessions out of 196 showed tolerant reaction (Mandal *et al.*, 2000). Some accessions of wild species of *Corchorus* like *C. aestuans, C. fascicularis, C. pseudo-olitorius, C. tridens* and *C. trilocularis* have shown very high degree of tolerance against the disease under Barrackpore condition (Palve *et al.*, 2003 and 2004). In *olitorius* jute germplasm OIN 221, OIN 157, OIN 651 and OIN 154 have been found to be resistant against stem rot on the basis of their PDI (5.0 or less).

Cultural Control

Selection of land: Heavy soil with low permeability are so compact when wet that the crop lacks proper aeration around the root zone. Such crops remain stunted and suffer from stem rot and therefore drainage supports a good, disease free crop (Ghosh and Basak 1965). K_2O/CaO ratio in leaf and stem of *C. capsularis* is believed to be an indicator of susceptibility to infection by *M. phaseolina* or *R. batatoticola* as assessed in a study on comparatively more resistant varieties JRC 918, JRC 212 and D 154 along with the highly susceptible JRC 412 (Mandal *et al.*, 1976). Since the pathogen is seed, soil as well as air borne, it remains in soil as either living on the dead soil-organic matter and jute stubbles as inactive sclerotia or the weed population harboring its active phase as mycelium or pycnospores (Ghosh 1983, Mandal, 2001). Even when the treated seeds are sown, they cannot protect the seedlings for very long time if the inoculum density in the soil is very high and seedling blight, damping off and collar rot symptoms are observed. The fungus from the soil attacks root system also. When only the branch roots are attacked, the plants may remain alive with reduced vigour showing rusty brown branch roots, but if the tap root is infected the plant generally dies (Som 1976). Jute crop in acid soils with pH 6.0 and below is more readily attacked than a crop at neutral soil. With continuous cultivation of jute, soil is depleted of calcium, potassium and other base substances whereas, the presence of iron, manganese and aluminium increases due to development of acidity. But under acidic conditions with low availability of potash, the mere addition of lime only does not help to reduce disease, and a simultaneous application of potash is imperative ((Mandal *et al.*, 1976). Selection of land, therefore, plays a very important role. Loam or clay loam, medium to high land with organic matter and having pH 6.5–7.0 or 7.5 is suitable. Depending upon pH level, application of lime or dolomite @ 2–4 t ha-1 at least 3–4 weeks before sowing once in 3–4 years is helpful in acidic soil. Application of FYM @ 7–8 t ha-1 is to be done in organic matter deficient soil (Ghosh and Basak, 1965).

Selection of Variety

Though resistant variety as such is not yet available, the cultivated varieties showed differential reaction to the pathogens of different places (Mandal, 1998) and accordingly, the varieties should be selected considering their status of susceptibility in a particular area.

Spacing

Close spacing is conducive to stem rot infection (Ghosh, 1963). Optimum spacing (line sowing in *C. capsularis* row to row 30 cm and plant to plant 5–7 cm and *C. olitorius* row to row 20–25 cm and plant to plant 5–7 cm, and broadcast sowing for both *C. capsularis* and *C. olitorius* at 10–15 cm plant to plant) results in less disease incidence and good yield. The pathogen has a very wide host range and a number of weeds serve as the source of inoculum. Therefore, weeding and thinning at 20–25 DAS and 35–40 DAS and clean cultivation is essential for a good healthy crop. Since water logging increases stem rot and root rot infection good drainage is to be provided to avoid such a situation.

Fertilizer Application

Jute needs a substantial quantity of nitrogen for required growth and yield, but application of N beyond 80 kg ha-1 and in new alluvial soils even above 60 kg ha-1 promotes incidence of stem rot, root rot and collar rot (Ghosh, 1983). Application of potash @ 50–100 kg K_2O ha-1 as basal dose can check the disease to a considerable extent (Cheng and Tu, 1970). In the era of intensive cropping system a number of other crops are cultivated both before and after jute. It is important to recognize that many familiar crops are susceptible to *Macrophomina phaseolina/Rhizoctonia bataticola* (Ghosh, 1983). Therefore, non host crop should be selected for rotation. Rice/wheat/mustard crop to follow jute but not potato or other solanaceous vegetables for more than consecutive three years (Mandal and Bandopadhyaya, 2000).

Chemical Control

The pathogen is seed borne, soil borne as well as air borne. Once introduced into the soil through the infected seed, the fungus can remain in soil for many years (Mandal 2001). Seed treatment is an economic and most effective step to reduce the losses due to this disease. Elimination of the pathogen from seed is possible by the treatment with organo-mercuric compounds (Mukherjee and Basak, 1972) but these compounds have been established to be hazardous from several considerations. It is for this fact some very effective non-mercurials were searched out for seed treatment. Bavistin 50 WP (carbendazim) @ 2.0 g kg^{-1} or Dithane M 45 (mancozeb) @ 5.0 g kg^{-1} and Captan are very effective compounds for this purpose (Som, 1977, Mandal, 1997). Spraying of Blitox 50 WP or Fytolan 50 WP (copper oxychloride) @ 5.0–7.0 g l^{-1} or Bavistin 50 WP (carbendazim) @ 2.0 g l^{-1} of water is helpful to keep the disease under control. In case of severe infection, alternate sprays of blitox and bavistin at 10 days intervals (total three sprays) are recommended (Mandal, 1997).

Biological Control

Health hazards and environmental pollutions have restricted the wide spread use of pesticides in plant disease control. In fact a good number of effective fungicides and insecticides have been banned. Comparatively, lesser toxic chemicals are still in uses and at the same time, continuous search for biological control measures are in progress. A good number of fungal antagonists and plant growth promoting rhizobacteria were tested in the past. The same are being tried also at present for disease control in jute and allied fibre crops. *Trichoderma viride, Aspergillus niger* AN 27 and some species of fluorescent *Pseudomonas* have been established as very effective biocontrol agents for stem and root rot in jute (Bandopadhyay and Bandopadhyay, 2003, Bandopadhyay *et al.*, 2004). Seed treatment with *T. viride* and soil drenching with the same antagonist has controlled the root disease in jute under different agro-ecological conditions (AINP, 2002-2004). Different combinations of *Trichoderma, Azotobacter, Aspergillus* and fluorescent *Pseudomonas* have effectively controlled root disease in jute and increased plant biomass and fibre yield (Bandopadhyay, 2004).

Anthracnose

This is also a very important disease mainly of *C. capsularis*. Two different species of *Colletotrichum* are responsible for anthracnose in the two different species of *Corchorus*. In *C. capsularis* it is *C. corchorum* and in *C. olitorius* it is *C. gloeosporioides*. However, in *C. olitorius* infection takes place at a very later stage than *capsularis* jute.

Economic Importance

It is more serious in *C. Capsularis* jute. Olitorius jute is rarely affected. In Assam, severity of anthracnose in olitorisus jute was also noticed. At Barrackpore the disease was also noticed in seed crop of Olitorius jute. In capsularis jute numerous spots appears on the stem and plant die in many cases. In case where the plant survives, the fibre is specky or knotty leading to cross bottom fibre. It causes considerable damage to the fibre quality as well as loss of yield.

Figure 4.2: Anthracnose

Geographical Distribution

The disease is of regular occurrence in *capsularis* belt of India *viz.* Assam, North Bengal, Bihar, and Uttar Pradesh. The disease is prevalent in Bangladesh also (Ghosh, 1957; Mandal, 1990). In all probability. anthracnose caused by *Colletotrichum corchorum* entered India during thirties along with jute germplasm from Southeast Asia, particularly Taiwan unknowingly. There was no mention of this disease in India prior to 1945 when it was first observed on "Jap-Red', a capsularis introduction from Formosa (Taiwan) at Dacca Farm, now in Bangladesh (Ghosh 1999). Then from Dacca it spread to other parts of Bangladesh. Later it entered India through Assam. Continuous rain, high humidity and temperature around 35ºC are congenial for this disease. Epidemic of anthracnose was noticed in exotic variety, Japanese Red at Chinchura, West Bengal during 1950-51 (Ghosh, 1957). By preventing release of hybrids involving South Asian susceptible or selections there of, India was enjoying comparative freedom from anthracnose while in Bangladesh, anthracnose established itself firmly. In recent years, the disease is appearing in very severe form in *capsularis* jute belt at Bahraich areas in Uttar Pradesh and also in some places of North Bengal, Bihar and Assam (Anonymous, 2002-04)

Symptom

In *C. capsularis* at seedling stage, the disease appears on leaf and stem as brownish spot and streaks following drying up to the entire stem. On mature plants, at first light yellowish patches can be seen on stem which turns to brownish depressed spots. The spots are irregular in shape and size. Several spots may coalesce causing deep necrosis showing crakes on the stem and exposing the fibre tissues. The coalescing spots may often girdle the stem and in such cases the plants break at that point and die. Affected plants when survived show necrotic wounds all over the stem. Fibres extracted from such affected plants are specky and knotty and fall under very low grade (cross bottom). Pods of diseased plants are also affected showing depressed spots and seeds collected from such fruits are also infected. The infected seeds are lighter in colour, shrunken and germination is poor. In *C. olitorius* incidence occurs at the later stage of crop growth. Generally it starts at mid July and continues up to September. Brown to black sodden spots appears on the stem.

Disease Cycle and Epidemiology

The mycelium enters through the epidermis and attacks the parenchymatous tissues between the wedges of bast fibre bundles. The entire parenchymatous tissue of cortex gradually disintegrates.

Under favourable condition the mycelium attack the thin walled phloem tissue in between the phloem bundles. The phloem parenchyma disintegrates and the bundles are exposed. The mycelia often reaches the cambial layer but seldom attack the wood. During humid days acervuli appear on the surface of the affected tissue and are visible with nacked eye; under a magnifying lens they appears bristled hemispherical or slightly lenticular eruption. The pod of the seeds is also affected and seed inside the pods are affected. The pathogen is also seed borne. The disease starts in the hot and very humid months of jolly when the crop is about 8-10 weeks old. The damage is most severe near about the harvest time. Continuous rain, high relative humidity and temperature of around 35°C are congenial for the faster development of this disease. *White* jute is more susceptible and *tossa* jute is rarely affected.

Pathogen

Colletotrichum corchori for *capsularis* jute and *Colletotrichum gloeosporioides* for *olitorius* jute *Acervuli*: black, superficial (errumpant) scattered, stroma: subglobose to hemispherical, sometimes lenticular, Setae: 10-30 in number, mostly 20-30 in each acervulum; deep dark brown scattered all over the stroma, septa 2-5; 52-138μ mostly 80 μ, 3.9-5.2 at the base. Conidiophores: simple straight hyaline bearing conidia singly, 25-25-30μ. conidia: acrogenous, falcate, hyaline to subhyaline, guttulate, 17-24 x 2.6-3.25μ mostly 20 x 3.5μ, germinate forming appresoria, in mass dirty flesh coloured to warm buff. The fungus grows very slowly on PDA. Growth starts with pale grayish-white aerial mycelium. The final mycellial growth is appressed, slightly dodden. Innumerable acervuli developed on the surface. Colour of the mycelia matt is masked by formation of abundant acervuli which when mature are blakish–brown to black. In culture the acervuli are arranged in concentric zones. In about a week's time at a temperature of 32°C a pale dirty whitish cream coloured mass covers the acervuli. This mass is the conidial accumulation in a semi viscid matrix. In older cultures the colour of the conidial mass changes to warm buff to dirty flesh tint.

Management

Cultural Control

Removal of affected plants from the field and clean cultivation.

Chemical Control

Seed treatment with carbendazim @ 2 g kg^{-1} of seed or captan @ 5 g kg^{-1} of seed to eliminate primary source of infection. Spraying carbendazim @ 2 g l^{-1} or captan @ 5 g l^{-1} or dithane M 45 @ 5 g l^{-1}. Seeds having 15 per cent or more infection should not be used even after treatment.

Hooghly Wilt

Geographic Distribution and Economic Importance

This disease was mostly prevalent in the areas where jute is followed by potato or other solanaceaous crop. The disease was observed in the districts of Hooghly, parts of Howrah, North 24 Parganas, Burdwan and Nadia of West Bengal, India. This was reported from Bangladesh also. *Olitorius* jute suffers from it. Still there is no report of *capsularis* jute being attacked by this disease. During late seventies to end eighties, more than 40 per cent infection was recorded in Hooghly district of India. (Mandal, 2002, Ghosh and Mukherjee, 1964, Chattopadhyaya and Sarma, 1970). But presently, this disease is not a concern.

Economic Importance

During late forties and early fifties, a typical wilt disease in jute (*Corchorus olitorius*) was observed in Tarakeswar areas of Hooghly district of West Bengal. The *malady* was so wide spread and the damage to crop was so severe that a new Pest and Disease Control Centre had to be established at Tarakeswar in the heart of Hooghly district (Ann. Rep. JARI, 1949). The disease was later widely spread in other areas of the district and also adjoining areas of surrounding districts where jute used to be followed by potato in winter. The benchmark survey estimated 30–34 per cent loss of jute crop each year between 1950-1954 (Ann. Report. JARI, 1949-56). During late eighties and early nineties, 5-37 per cent disease was recorded in Kamarkundu area of Hooghly district and 2-20 per cent in some areas of Nadia and North 24-Parganas districts (Mandal,1986, Mandal and Mishra, 2001). The disease was recorded in *olitorius* jute in North Bengal (Coochbehar) also though *capsularis* jute were disease free (Mandal and Khatua, 1986).

The Pathogen

Ghosh (1961) coined the name Hooghly wilt and isolated seven fungi and a bacterium from the diseased or dead plant. It appears that the term 'Hooghly wilt' has been accepted by the working pathologists (Chattopadhyay *et al.*, 1970). Later on by repeated experimentation it was established that *Ralstonia solanacearum* (= *Pseudomonas solanacearum*) is primary pathogen and *R. bataticola* and *M. incognita* are associated pathogens. Presence of these root rot pathogen (*R. bataticola / M phaseolina*) and root knot nematode (*M. incognita*) increases the disease since these create wound in the root, and thus facilitate the entry of the primary bacterial pathogen, *R. solanacearum* (Mandal *et al.*, 1984, Mandal and Mishra, 2001).

Management

Cultural Control

Potato or other solanaceaous crops in the rotation are to be avoided as per as practicable. Jute: Paddy: Paddy or Jute: Paddy: Wheat are the most effective rotation. In case where solanaceaous crop is the main crop of the area in the *rabi* season, it should be replaced by paddy or wheat at least after two years. Removing wilt affected plants from the field, burning of the dead plants of solanaceaous plants and burning rejected and rotten potato tubers (Mandal, 1986, Ghosh and Mukherjee 1970, Sarkar *et al.*, 1998) are important cultural practices to keep the disease under control. By adopting cultural practices particularly the appropriate crop rotation in Hooghly district the disease came down to 1-2 per cent compared to above 40 per cent in the late eighties (Mandal, 1986, Mandal and Ghosh, 2002).

Chemical Control

Seed treatment with Carbendazim @ 2 g kg^{-1} of seed and spraying the same fungicide @ 2 gl^{-1} of water is effective for management of disese. This fungicide helps to reduce root rot incidence which favours the entry of the bacteria.

Jute Mosaic

The disease was first observed in 1917 undivided Bengal province of India by Finlow, 1917. Since then it is variously known as jute yellow mosaic, jute leaf mosaic and jute golden mosaic (Ahmed, 1978, Ahmed *et al.*, 1980). Recently the incidence of the disease has increased from 20-40 per cent (Ghosh *et al.*, 2007a).

Symptomatology

The disease is characterized by appearance of small yellow flecks on leaf lamina in the initial stage, which gradually increased intermingled with green patches and produced a yellow mosaic appearance. Leaves in some cases produce small enation along the mid vein. In extreme cases the infected plant gets stunted and leads to reduced plant height to the extent of 20 per cent (Das *et al.*, 2001). The incidence was about 50 per cent on some leading cultivar such as JRC 7447 and JRC 212 (Ghosh *et al.*, 2007b).

Geographic Distribution

The disease was reported in *capsularis* jute from different jute growing belts of India and Bangladesh. Recently, from Vietnam some sequence of Jute golden mosaic virus has been deposited in GenBank, but no details is available. In India, the disease was reported in *capsularis* jute from West Bengal (Roy *et al.*, 2006, Ghosh *et al.*, 2007b) and Assam (Pun and Pathak, 2005).

Transmission

The causal virus was reported to be transmitted by white fly (*Bemisia tabaci*) (Ahmed, 1978, Roy *et al.*, 2006, Pun and Pathak, 2005). Some worker also reported that the virus was transmitted by seeds (Das *et al.*, 2001). Adult whitefly can acquire the virus within 30 minutes of its access to a disease plant and acquisition become maximum after 8 hours of access. A viruliferous whitefly can inoculate the virus to the test plant within 30 minutes of its inoculation. A viruliferous vector can retain the virus upto 10 days. A typical symptom of the disease appeared on the plants of cv. JRC 7447 and JRC 212 after 10 days of whitefly transmission with 60 per cent transmission efficiency when acquisition and inoculation access period were 24h and 12 h respectively.

Particle Morphology

As jute produce lots of mucilaginous substances, observation on virus particle is very difficult. So long no report on the purification of this virus is available and possibly that become a great hindrance to identification of particle from purified virus material. Presence of mucilage also became a great factor in viewing the particle morphology through dip method.

Detection, Identification and Phylogenetic Relationship

Based on sympotomatology and transmission, the virus was found to be a member of Begomovirus under family Geminiviridae. PCR based detection with Begomovirus group specific primer amplified a 1.2 kb DNA A fragment of the virus (Ghosh *et al.*, 2007a) Cloning and sequencing of this amplicon revealed that it consisted of 1263 nucleotides (Accession No EU 047706) and shared the highest nucleotide sequence identity (91.2 per cent) with Corchorus golden mosaic Vietnam virus (DQ 641688). The nucleotide sequence at the origin of replication was found to be CATTATTAC instead of conventional TAATATTAC. Such unique feature was also noticed in case of Corchorus golden mosaic Vietnum virus. Phylogenetic analysis with other begomovirus revealed that the begomovirus from jute grouped with other begomovirus reported to be associated with Corchorus species and clustered with new world begomoviruses. Two other primers have been designed to amply the complete DNA A and DNA B component of the viral genome which gave expected 2.7 kb amplicon in each case.

Management

No detailed work has been carried out. In general, rouging of diseased plant and spraying of imidacloprid, thiamethoxan and acetamedrid could prevent the spread of the disease. Studies in Bangladesh revealed that a combination of collection and use of seeds from healthy plants, one insecticidal spray around 30 DAE, field sanitation with rouging several times during growth period and application of an extra booster dose of nitrogen at around 45DAE reduced the spread (Hoque *et al.*, 2003).

Figure 4.3: Black Band and Soft Rot

Black Band and Soft Rot

Black band is caused by *Botryodiplodia theobromae*. The pathogen is seed borne as well as air borne and has a very wide host range. It was a minor disease in the past, but the incidence is gradually increasing now a days.

Soft rot is a disease of minor importance caused by *Sclerotium rolfsii*. This is primarily a soil borne fungus. It grows in the litter of fallen jute leaves. When the weather is hot and the soil is wet, it grows and initiates infection in the collar region. Both *tossa* and *white* jute are equally susceptible.

Integrated Disease Management for Jute

Resistant Varieties

Resistant varieties are not available at the cultivators level for managing the *Macrophomina*. However, six *white* jute accessions *viz.*, CIM 036, CIM 064, CIN 109, CIN 360, CIN 362 and CIN 386 were identified as resistant. Four *tossa* jute accessions OIN 125, OIN 154, OIN 651 and OIN 853 were categorized as moderately resistant to *Macrophomina phaseolina*. These accessions will be utilized for the breeding programmes to develop disease resistant varieties of jute.

Clean Cultivation

Deep ploughing during summer and clean cultivation by removing the leftover of all the previous crops are essential to prevent primary source of infection by *Sclerotium rolfsii,* the causal agent of soft rot disease.

Soil Amelioration

Acid soils are in general preferred for most pathogens. If the soil is acidic, then lime @ 2-4 tons/ha should be applied depending upon the soil pH. Lime application needs to be done at least a month before the sowing otherwise the seedlings will be damaged. A neutral soil having pH between 6.5 and 7.5 is preferable for healthy crop. Indiscriminate use of fertilizers the nitrogenous fertilizers, in particular makes the soil acidic.

Healthy Seed

Infected seeds serve as the primary source of inoculum. Therefore, healthy seeds should be used after treating them with recommended fungicides. Seed treatment with carbendazim @ 2.0 g/kg or mancozeb @ 5.0 g/kg or *Trichoderma viride* @ 10.0 g/kg of seed may be recommended for managing the *Macrophomina* disease complex, anthracnose, blackband and Hooghly wilt. Seeds with 15 per cent or more infection should be avoided even after treatment to contain the spread of anthracnose incidence.

Sowing Time

The crop can be escaped from some diseases by slight adjustment in the sowing time. Jute sown after second week of April suffers less from stem rot and root rot. Sowing time however, needs adjustment depends on the preceding and the succeeding crops in the cropping sequences.

Fertilizer Application

Application of nitrogen beyond 80 kg/ha and in new alluvial soils even above 60 kg/ha promotes the incidence of stem rot, root rot and collar rot. Thus judicious application of nitrogenous fertilizers is recommended.

Spacing

Closer spacing is conducive for the spread of the stem rot, but optimum spacing of 30´5-7cm for *white* jute and 20-25´5-7cm for *tossa* jute results in lesser incidence of the disease.

Weeding

Macrophomina phaseolina and *Sclerotium rolfsii* have a very wide host range including a number of weeds. Timely weeding should be done to avoid the spread of disease from the weeds.

Crop Rotation

Non-host crop should be selected for rotation to reduce the incidence of *Macrophomina* disease complex as well as Hooghly wilt of jute. Rice/wheat/mustard may be selected, but not potato or other solanaceous vegetables for more than two consecutive years. Jute-paddy rotation in rainfed and jute-paddy-wheat sequence under irrigated condition is recommended to contain the spread of the disease. By adopting cultural practices particularly appropriate crop rotation in Hooghly district, the disease has come down to 1-2 per cent as against above 40 per cent in the late eighties.

Fungicides

- ✰ Foliar spraying of carbendazim (0.2 per cent) or copper oxychloride (0.75 per cent) is suggested for the management of *Macrophomina* disease complex, anthracnose, blackband and soft rot.
- ✰ Integration of control measures consisting of maintaining soil pH 5.6–6.0, application of organic manure @ 5 t/ha, seed treatment with carbendazim @ 2g/kg of seed, foliar spraying of carbendazim @1.5 g/lit. of water when disease incidence reached 2 per cent and soil application of *T. viride* @12×10^6 spores/ml during sowing could reduce the incidence of stem rot to a greater incidence.
- ✰ In general, roguing of diseased plant and spraying of insecticides such as imidacloprid, thiamethoxam and acetameprid could prevent the spread of the viral diseases.

Major Diseases of Mesta

Foot and Stem Rot

Economic Importance

Foot and stem rot is the most important disease of mesta in India (Gosh, 1983) causing loss to the extent of 10–25 per cent. The disease attacks the plant at very early stage and continues upto maturity affecting quality and quantity of fibre. It is more serious in roselle (*Hibiscus saffdariffa*) than kenaf (*H. cannabinus*). In severe cases, more than 40 per cent crop loss in roselle was observed (Mandal *et al.*, 1995; De and Mandal, 2007a)

Figure 4.4: Foot and Stem Rot

Symptom

It is prevalent in all the mesta growing areas of India. In foot and stem rot the symptom appears on the stem generally a few inches above the ground level but the spots may be seen at higher or lower level also. The spots are deep brown to blackish in colour with variable size. Larger spots very often girdle the whole stem and as a result the plant breaks at the point of infection and is dead. No fibre is obtained from such plants.

Pathogen

The disease is caused by *Phytophthora parasitica* var. *sabdariffae*. It is favoured by high temperature and continuous drizzling. Water stagnation which is common during mesta growing season in west Bengal predispose the plants to infection. Continuous drizzling, high rainfall, cloudy condition from May to September may be responsible for epidemic in which the mean monthly maximum and minimum aerial temperature and relative humidity during the period were 33-0C and 93-70 percent. Average monthly rainfall 181-227 mm was distributed over 16 rainy days.

Management

Varietal Resistance

None of the cultivated varieties are resistant to the disease, but comparatively AMV 1, Roselle Type 1 and AP 481 were observed to be moderately resistant (Majumdar and Mandal, 2000a, b). Red bristled *H. saffdariffa* lines are more resistant than others (Majumdar and Mandal, 1994).

Chemical Control

Seed treatment with Dithane M 45 (mancozeb) @ 5.0 g kg^{-1} followed by soil drenching (0.2 per cent) twice and spraying of Blitox 50 WP or Fytolan 50 WP (copper oxychloride) @ 5.0–7.0 g l^{-1} or

Bavistin 50 WP (carbendazim) @ 2.0 g l^{-1} of water (Mukherjee and Basak, 1973, Raju *et al.,* 1989) is effective for management of disease. De and Mandal (2007a) reported that pre-sowing seed treatment with copper oxychloride was more effective than Carbendazim, mancozeb, metalxyl, ediphenphos, carboxim, hexaconazole and thiophenate methyl. Copper oxychloride reduces the disease indicence by 50.3-45.5 per cent at 30 days after sowing and at maturity, respectively and increased 20.6 per cent fibre yield of HS 4288.

Biological Control

Disease control was reported by the application of *Trichoderma viride* + *Azotobacter* and also by *Pseudomonas* + *Azotobacter* (Bandopadhyay, 2004).

Yellow Vein Mosaic of Mesta

Yellow vein mosaic of mesta was found in a devastating form in all mesta growing areas in India. It is reported from eastern (Chatterjee *et al.,* 2005) and northern part of India (Ghosh *et al.,* 2007b).

Symptom

The disease was first observed in few plant from Andhra Pradesh and then it was reported from west Bengal and Uttar Pradesh. The disease is characterized by yellowing of veins and veinlets

Figure 4.5: Yellow Vein Mosaic of Mesta

followed by complete chlorosis of the leaves of affected plants at and advance stage of infection. The disease is developed one or both halves of the leaves or with the appearance of erratic chlorotic flecks in the vein and veinlets (Chatterjee *et al.*, 2006). In severe cases, those flecks gradually increase in size and yellow netting. Along with the vein yellowing the leaf lamina started showing foliar discolouration. This discolouration is green yellow at first stage which in advance forms yellow mosaic, chlorotic yellowing and complete chlorosis of foliage in sequential chronology. Occasionally the stem of diseased plant becomes yellow, partially or totally. The flowers and fruits are malformed causing low seed yield. If the plants are infected by the virus at their early stage of growth they do not flower even. Infected plants in general show stunted growth with reduced leaf size. Survey revealed that 90 per cent incidence of the disease is yield loss due to this disease alone was found to be 12.78- 17.45 per cent with respect to fibre yield and 18.91-23.83 per cent with respect to seed yield (Roy *et al.*, 2007)

Transmission

Viruliferous whitefly could effectively transmit the virus with 78-85 per cent transmission efficiency (Chatterjee *et al.*, 2005). Three whitefly? per plant were found to be effective for disease transmission. Typical symptom appeared after a minimum incubation period of nine days under glass house condition. Minimum acquisition and inoculation access period for the vector was observed to be 12h and 4 h respectively.

Host Range

The virus has a very narrow host range. Besides the two species of mesta, it could be experimentally transmitted to *Vigna unguiculata* and *Vigna umbellate* where a mid crumpling symptom was observed (Chatterjee *et al.*, 2007).

Particle Morphology

Transmission electron microscopy with 2 per cent uranyl acetate stains form typical symptomatic leaves of *H. cannabinus* revealed that the disease is associated with geminivius. The size of the geminate particle is 20nm x 30 nm (Chatterjee *et al.*, 2005).

Detection, Identification and Phylogenetic Relationship

Positive hybridization signal in Southern blot using radio labeled probe of cotton leaf curl Rajasthan virus DNA A and α-DNA confirmed the involvement of a Begomovirus with the disease (Chatterjee *et al.*, 2005). From the northern India a strain of δ-DNA containing tobacco leaf curl New Delhi virus has been reported to be associated with this disease. Characterization of coat protein gene of associated Begomovirus and the satellite DNA Beta molecule from six different geographical isolates of eastern and northern India revealed isolates obtained from eastern and northern India formed two distinct groups indication the existence of two distinct Begomovirus complexes associated with the yellow vein mosaic disease of mesta in two different geographical location in India (Roy *et al.*, 2007). Nucleic acid based diagnostic have been developed for identification of these complexes (Chatterjee *et al.*, 2007). Characterization of DNA molecule of east India isolate revealed its close similarity with beta DNA molecule associated with cotton leaf curl disease (Chatterjee and Ghosh, 2007a). The full length DNA A homologue (-2.7 kb) of begomovirus have been characterized from two east Indian isolates which revealed that these two isolates consisted of 2728 and 2752 nucleotide, respectively (Chatterjee and Ghosh, 2007b, Roy *et al.*, 2007) and shared 84.3 per cent sequence identity with cotton leaf curl Bangalore virus followed by Malvestrum yellow vein virus from China. It has been noticed that this DNA A molecule shared 89 per cent sequence identity with any known begomovirus sequence and hence it is proposed as a new species of begomovirus with a tentative name mesta yellow vein mosaic virus. Individual ORF analysis revealed that this DNA A evolved as a recombinant molecule from at least three distinct begomovirus from north, south and China.

Management

Initial record on vector control revealed that imidachlorprid and Thiamethoxam were effective against reducing the population of whitefly.

Integrated Disease Management for Mesta

- ☆ None of the cultivated varieties are resistant to the foot and stem rot disease, but AMV 1, Roselle Type 1 and AP 481 were observed to be moderately resistant. Red bristled *H. sabdariffa* lines are more resistant than others.
- ☆ Seed treatment with copper oxychloride (5g/kg of seed) is more effective in managing *Phytophthora parasitica* var. *sabdariffae*, the causal organism of foot and stem rot disease.
- ☆ Spraying copper oxychloride @ 5-7 g/lit. of water, directed towards base of the plants helps in managing the diseases of mesta.
- ☆ The vector of yellow vein mosaic of mesta (*B. tabaci*) can be managed effectively by imidaclorprid 17.8 SC (3ml/10lit. of water) and thiamethoxam 25 WG (5g/20 lit. of water).

References

Ahmed *et al.* (1982). *FAO Field Document.* BGD/74/018, Dacca.

Ahmed, N (1978). *FAO, Plant Protection Bulletin* 26: 169-171

Ahmed, N. and Ahmed Q.A. (1969). *Mycopath. et. Mycologia Applicata.* 39: 129-138.

Ahmed, Q. A. (1968). *Jute and Jute Fabrics Pakistan.* 7 (8): 147-151.

Ahmed, Q.A., Biswas, A.C, Farukuzzaman, A.K.M., Kabir, M.Q and Ahmed, N. (1980). *Jute and Jute Fabrics, Pakistan.* 6: 9-13.

Annon. *Annual Report JARI.* 1949-1956

Annon. *Annual Report, AINPJAF,* 2001-2004.

Annon.*Annual Report, CRIJAF* 2007-08

Bandopadhyay, A.K. (2004). *Annual Report, CRIJAF*: 2004-05.

Bandopadhyay, A.K. and Bandopadhyay, A. (2004). *Indian Phytopath.* 57(3): 356-357.

Bandopadhyay, A. K. Prakash, G. and Som, D. (1982). *Pestology*7 (7): 21-23.

Bhattacharya, M and Samaddar, K.R (1976). *Plant and Soil* 44: 27-36.

Chatterjee, S., Mukherjee, N., Khatua, D.C. and Mandal, R.K. (1998) In: Proc. of National Seminar on Jute and Allied fibres: Agriculture and processing, held at CRIJAF, Barrackpore, (Eds. P. Palit, S. Pathak and D.P. Singh.) pp.221-235

Chatterjee, A. and Ghosh, S.K. (2007a). *Virus Genes* DOI 10. 1007/s 11262-007-0160-6.

Chatterjee, A. and Ghosh, S.K. (2007b). *Arch. Virol.* DOI 10. 1007/s 00705-07-1029-7

Chatterjee, A., Roy, A and Ghosh, S.K. 2006. In: Characterization, diagnosis and management of plant viruses (Eds. G.P.Rao, S.M. Paul Khurana and S.L. Lenardon) Vol. 1 Industrial crops Studium Press, Texas, USA. Pp 497-505.

Chatterjee, A., Roy, A., Padmalatha, K.V., Malathi, V.G and Ghosh, S.K, 2005. *Australasian Plant Pathol.*34: 609-10.

Chatterjee, A., Sinha, S.K., Roy, A., Sengupta, D.N. and Ghosh, S.K. 2007. *J. Phyto. Pathol.* 155: 683-689.

Chattopadhyay, S. B. and Raj, S.K. (1978). *Madras Agric. J.* 65(5): 320-324.

Chattopadhyay, S. B. and Sarma, B. D. (1970). *Indian. J. Mycol. Res.* 17: 55-64.

Cheng, Y.H. and Tu, C.C. (1970). Res. Rep. Tainan Fibre Exp. Sta. TARI, N.S. No. 14 (2).

Das, S. Khokon, A.R, Haque, M.L. and Ashrafuzzaman, M. (2001). *Pakistan J.Biol. Sci.* 4(12): 1500-1502.

De, R.K. (2004), *Annual Report, CRIJAF*, 2004-05.

Finlow, R.S. (1917). *Agric. J. India.* 12: 3-29.

Ghosh, R., Paul, S., Das, S., Palit, P., Acharyya, S., Mir, J.I., Roy, A. and Ghosh, S.K. 2007a. In: Proceedings of Xth Plant Virus Epidemiology Symposium. October 15-19, ICRISAT, India. pp 115.

Ghosh, R., Paul, S., Roy, A. Mir, J.I., Ghosh, S.K., Srivastava, R.K. and Yadav, U.S. 2007b. Online *Plant Health Progress* DOI 10. 1094/PHP-2007-0508-01-RS

Ghosh, S. K. and Sen, C. (1973). *Indian. Nat. Sci. Acad.* 39B (2): 221-227.

Ghosh, T. (1957). *Indian Phytopath.*, 10: 63-70.

Ghosh, T. (1961), *Ph.D. Thesis, Calcutta University.*

Ghosh, T. (1963). *Jute. Bull.* 26(6).

Ghosh, T (1999). In: Jute and Allied Fibres–Agriculture and Processing (Eds. P. Palit, S. Pathak and D.P.Singh), Central Research Institute for Jute and Allied Fibres, Barrackpore, India.

Ghosh, T. and Mukherjee, N. (1967). *Proc. 1st International. Symposium on Plant Path. Ind. Phytopath.Soc.* (Ed. S. P. Raychoudhury.) pp 363-369.

Ghosh,T and Basak, M. (1961). *Jute Bull.* June.

Ghosh,T and Basak, M. (1965). *Indian J. Agric. Sci.* 35(3): 90-100.

Ghosh, T. (1983). *Handbook on jute, FAO publication.*

Haige, J. C. (1930). *Peradeniya Ann. Rep. Bot. Gard.* II. 23: 213-239.

Mahapatra, A.K., Mandal R., Saha, A.K. Sinha M.K., Guha Ray, M.K., Kumar, D and Gupta, D. (1994) Information Bulletin, CRIJAF, Barrackpore, pp 1-31.

Majumdar, M and Mandal, N. (1988) *International J. Tropical Plant Diseases* 6: 73-76.

Majumdar, M and Mandal, N. (1998) *Indian Agric.* 42(2): 127-130.

Majumdar, M and Mandal, N. (2000). *Indian Agric.* 44(1-2): 75-78.

Majumdar, M and Mandal, N. (1994). *Indian Agric.* 38(3): 219-223.

Majumdar,M. and Goswami, K.K. (1998). *Indian J. Mycopath. Res.* 36(1): 1-6.

Mandal,R.K. and Ghosh T. (2002). *J. Mycopathol. Res.* 40(1): 67-69.

Mandal, R. K, Mahapatra, A.K. and Saha, A. (1995) Information Bulletin, CRIJAF, Barrackpore, pp 1-25.

Mandal, R,K, Mishra, C. D and Mishra, C. B. P. (1984). *Ann. Rep. JARI*, 1984.

Mandal,R.K, Saha, A. K. and Mahapatra, A.K. (1995) Information Bulletin, CRIJAF,Barrackpore, pp 1-29.

Mandal, A. K. (1976). *Bull. Indian. Soil. Sci.*, 10: 278-284

Mandal, R.K. and Bandopadhyay, A.K. (2000). Proc. of National Training Progamme on Jute production Technology held at CRIJAF, Barrackpore

Mandal, R.K. (1984). *Sabuj Sona*(Bengali), Sept. 1984,

Mandal, R.K. (1986). *Sabuj Sona* (Bengali) March, 1986.

Mandal, R.K. (1987) Proc. of Training Programme in Seed Pathology under Indo-Danish Project for staff of seed testing laboratories held from 17.11.87 to 31.12.87 at Division of Seed Science and Technology, IARI, New Delhi. (Ed. S.A. Siddique)

Mandal, R.K. (1990) In: Proc. National Workshop cum Training on jute, mesta, sunnhemp and ramie held at CRIJAF, Barrackpore.

Mandal, R. (1997) In: Proc. Training Programme for Senior Management/Research/Extension officers held at CRIJAF, Barrackpore from 4th to 6th Nov, 1997. (Ed. V. N. Saraswat.)

Mandal, R. (1997). Proceedings of IJO/CRIJAF Training Course on Field Techniques for Varietal Improvement of Jute and Mesta, 20-22 March,1997 at CRIJAF. (Ed. S. K. Hazra)

Mandal, R. K. (1978), Annual Report, JARI.1978,

Mandal, R. K. (2001). *J. Interacad* 5(3): 402-405.

Mandal, R. K., Sarkar, S and Saha, M. N (2000). *Environment and Ecology*, 18(4): 814-818

Mandal, R. Khatua, D. C. (1986) *Jute Dev. Journal*. April-June, 1986

Mandal, R. Prakash, S. (1987). Proc. 74th Indian Sci. Cong.

Mandal, R., Sinha, M.K., GuhaRay, M.K. and Chakrabarty, N.K. (1998). *Environment and Ecology* 16(2): 424-426.

Mandal, R.K. and Mishra C.D. (2001). *Environment and Ecology* 19(4): 969-972.

Mukherjee,N. and Basak,S.L. (1973), *Indian Phytopath* 26(3): 567-577

Myser Ali, M., Sayem, A.Z.M., Alam, S. and Ishaque, M (2005). *Mycopathologia* 38(3): 289-98.

Palve., S. M., Sinha, M.K. and Mandal, R.K. (2003). *Plant Genetic Resource Newsletter* No. 134: 10-12

Palve., S. M., Sinha, M.K. and Mandal, R.K. (2004). *Trop. Agric.* (Trinidad). 81 (1): 23-27.

Pun, K.V and Pathak, S. (2005). *Indian J. Virol.* 16: 53.

Rao, P.V. (1979). *Agri. Met.* 16: 107-112

Rao, P.V. (1980). *Ind. J. Agri. Sci.* 50(8): 608-611

Roy, A. De, R.K and Ghosh, S.K. (2008). In: Jute and allied fibre updates (Eds. P.G. Karmakar, S.K.Hazra, T. Ramasubramanian, R.K.Mandal, M.K.Sinha and H.S.Sen). Central Research Institute for Jute and Allied Fibres, Barrackpore, India, pp 217-241.

Roy, A., Ghosh, R., Paul, S., Das, S., Palit, P., Acharyya, S., Mir, J.I., and Ghosh, S.K. (2007). In: Proceedings of Xth plant virus epidemiology symposium. October 15-19, ICRISAT, India. Pp 85.

Roy, A., Mir, J.I., Ghosh, R., Paul, S. and Ghosh, S.K. (2006). Jal News–CRIJAF News letter 4(1): 11.

Saha, A. K., Mandal R., Mahapatra, A.K., Sinha M.K., GuhaRay, M.K., Kumar, D and Gupta, D. (1994) Information Bulletin, CRIJAF pp 1-54.

Sarma, B. K., Bandopadhyaya, A.K., Mandal, R.K. and Ghosh, K.L. (1981) *Farmer and Parliament*, September, 1981.

Sarma, M.S. (1962). *Field Crops.* Abstt. 22(4): 324-333.

Sustainable Disease Management of Agricultural Crops (2011) *Pages* **80–88**
Editors: **S.K. Biswas and S.R. Singh**
Published by: **DAYA PUBLISHING HOUSE, NEW DELHI**

Chapter 5

Integrated Diseases Management of Rice

☆ *S.P. Singh[1], R. Bhagawati[3] and S.R. Singh[3]*
[1]ICAR Research Complex for NEH Region, Arunachal Pradesh Centre, Basar – 791 101
[3]Krishi Vigyan Kendra, Belatal, Mahoba – 210 423, U.P.

Important Diseases of Rice

In Rice crop so many diseases are occured which hamper the production of crop, here we are discussing the some important diseases of rice and their management. The following diseases are mentioned below.

Rice Blast

Occurrence

- ☆ Earliest known plant disease
- ☆ Also known as rotten neck or rice fever
- ☆ Reported from 80 rice-growing countries. First recorded in China (1637). Later from Japan (1704), from Italy (1828), from USA (1876), from India (1918).
- ☆ Expected grain loss: 70 to 80 per cent

Three Type of Blast

1. Leaf blast
2. Nodal blast
3. Neck blast

Symptoms

The lesions are more or less boat or eye- shaped spots with grey or dark brown margin appear on the leaf and leaf sheath. The central portion of the young spot become pale- green or dull-grayish green

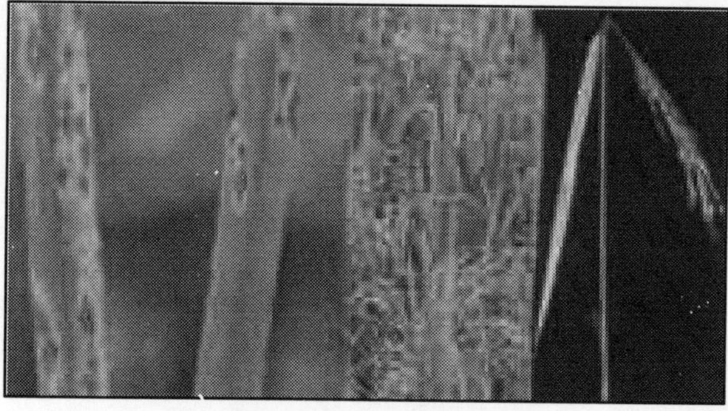

Figure 5.1: Leaf Blast

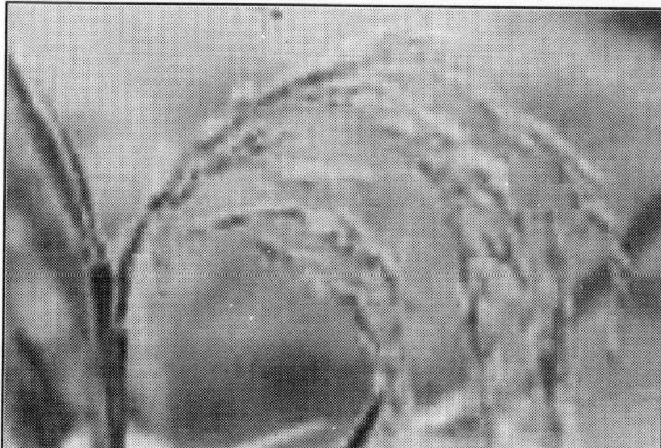

Figure 5.2: Internode Blast

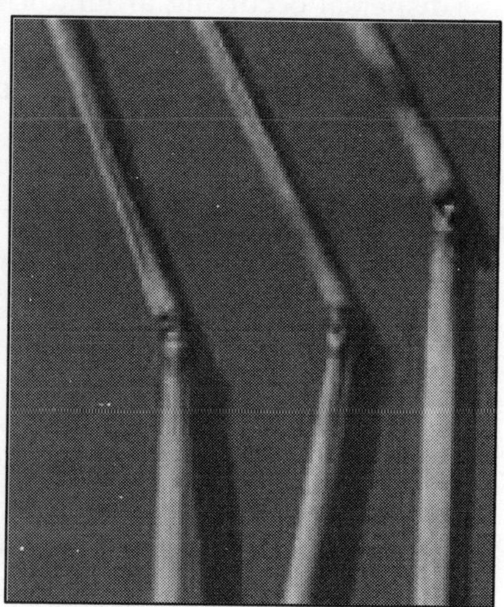

Figure 5.3: Neck Blast

and appears water-soaked but in old spots the central portions turn grey or straw colored. Fully developed spots reach a length of 1.0 to 1.5 cm and width of 0.3 to 0.5 with brown margins. The lesion on stem and grains are darker.

In case of neck infection, the stem below the earhead gets infection and turns into brown to black spots, which may cover the entire stem. The neck infection results in dropping of ears. Affecting grain are partially filled up. Due to necrosis of neck tissue the earheads break and fall off which results in maximum damage.

Disease Cycle

- ✰ In tropics, air borne conidia are present all round the year, because of the availability of collateral host grasses such as Setaria intermecia, Digitaria marginata, Panicum repens and Leersia hexandra.
- ✰ Some of these grasses may be acting as primary source of inoculum.
- ✰ The infection may be sometimes through infected seed and infected plant debris.
- ✰ The conidia produced on the leaves of nursery seedlings become wind-borne and cause secondary spread of the disease.

Epidemiology

- ☆ Day temperature (30°C), night temperature (20°C) and day light (14 hours) found to predispose the plants to infection.
- ☆ Relative humidity (92 per cent) and free water required for conidial germination and infection.
- ☆ Spores do not germinate in direct sun light
- ☆ Cloudy overcast weather, dew drops encourage blast spread
- ☆ Conidia exhibit nocturnal pattern of diurnal periodicity with peak concentration of spore dispersal occurring around 4AM favored by night temperature (25–27°C) and relative humidity (86 -98 per cent)
- ☆ Conidia could remain viable under snow to over winter period and 4–6 months after harvest.

Management

Cultural Practices

- ☆ Use seed from a disease free crop
- ☆ Destruction of wild collateral hosts
- ☆ Timely removal of weed hosts
- ☆ Destruction of infected plants
- ☆ Avoid excess N–fertilizer application
- ☆ Use of tolerant varieties (Penna, Pinakini, Tikkana, Sreeranga, Simphapuri, Palghuna, Swarnamukhi, Swathi, Prabhat, IR–64, Jaya, IR–36, MTU 9992, MTU 1005, MTU 7414)
- ☆ Burning of straw and stubbles after harvest

Chemical Management

- ☆ Seed treatment at 2.5 gm/kg seed with Captan or Carbendazim or Thiram or Triclyclazole.
- ☆ Spraying of Triclyclazole at 0.6 gm/liter of water or Edifenphos at 1 ml/lit of water or Carbendazim at 1.0 gm/lit. in 100 liter of water/hectare.
- ☆ 3 to 4 sprays each at nursery; Tillering stage and panicle emergence stage may be required for complete control.
- ☆ Light Infestation–Spray carbendazim or Edifenphos @ 0.1 per cent.

Pre-tillering to Mid-tillering Management

- ☆ Light at 2 to 5 per cent disease severities–Apply Ediphenphos, Carbendazim or 1 BP 48 @ 0.1 per cent. when infection is seen. Panicle initiation to booting delay top dressing of N fertilizers.
- ☆ At 2 to 5 per cent leaf area damage spray Edifenphos or carbendazim or Pyroquilon or 1 BP 48 @0.1 per cent.
- ☆ At 5 per cent leaf area damage or 1 to 2 per cent neck infection Spray Edifenphos, carbendazim@0.1 per cent or Triclyclazole @ 1 gm/lit of water.

Brown Spot: Caused by (*Helminthosporium oryzae*)

Occurrence

- ☆ Also called as sesame leaf spot or fungal blight
- ☆ Mostly seen in West Bengal, Orissa, and Andhra Pradesh and Tamil Nadu

Symptoms

The development of innumerable dark, brown, elliptical spots on the leaves stems and glume size 2-8 mm x 0.5mm on the. On the leaves, circular or oval, dark-brown to purplish-brown spots are found. At the time of flower emergence typical brown, water-soaked lesions appear at the principal node below the rachis or the 'neck'.

Disease Cycle

- ☆ Over winters mainly in infected plant parts
- ☆ Seedlings from diseased seed (primary source of inoculum)
- ☆ Seedlings show infection symptoms soon after germination
- ☆ Disease occurs on as many as 20 species of Oryzae
- ☆ Spores on seedlings cause secondary infection (wind–borne)

Epidemiology

- ☆ Optimum temperature is 25–30°C for condia germination
- ☆ Around 90 per cent humidity favours infection
- ☆ Darkness more favoured for fungus spread over sunlight
- ☆ Flowering phase is more susceptible
- ☆ Monsoon rains cloudy days favour the disease
- ☆ Heavy dose of nitrogen application aggravate the disease
- ☆ Low pH soils, deficient in essential and trace elements (especially low available K) favourble to disease
- ☆ Older leaves are more susceptible over younger ones

Management

Cultural Practices

- ☆ Use disease free seed (since it is seed borne)
- ☆ Treat the seed with Carbendazim (1g/kg seed) or Thiram or Mancozeb (2.5g/kg seed)
- ☆ Correct potash deficiency
- ☆ Avoid excess N–application

Chemical Control

- ☆ Spray Mancozeb (2.5g/lit) or Carbendazim (1g/lit) or Ediphenphos (1ml/lit)–2 to 3 times at 10–12 day intervals before appearance of initial symptoms
- ☆ Spray preferably during early hours or afternoon hours at flowering and post–flowering stages.

Stem Rot (*Sclerotium oryzae*)

Occurrence

- Earlier this disease was considered as minor.
- But now it has assumed importance with its sporadic occurrence in some places of Andhra Pradesh.

Symptoms

The disease appears after transplanting as small, black irregular lesions at the water line on leaf sheath. Fungus gradually enters the stem and develops large number of black, smooth, shining sclerotia in the stem. Such infected stem rot and falls off. The spikelets remain chaffy or partially filled.

Disease Cycle

- Primary source of inoculum is by means of infected plant debris.
- Secondary spread is by means of air borne conidia produced on the leaf sheath.
- Conidia are colorless, single celled and cylindrical in shape.

Epidemiology

- Night temperatures of 20°C, dew (mist), Cold weather, high humidity, more use of Nitrogen fertilizers are favorable for development of this disease.

Management

- Destroy the infected stubbles by burning and/or deep ploughing.
- Avoid standing water in the field for a longer period. Drain the standing water from time to time.
- Spraying of Mancozeb at 2.5 gm or Carbendazim at 1.0 gm/lit or Benomyl 0.5 gm/lit of water at flowering stage may be taken up.

Sheath Blight L. (*Rhizoctonia solani*)

Symptoms

The chief symptoms develop as 2-3 cm long greenish-grey lesion with dark brown margin on the leaf and leaf sheath. The lesion turns to straw colour and surrounded by bluish grey narrow bands. The lesion increase in size and girdle the stem. The leaf blade of the affected sheath dries up from the tip downwards. The grains are shriveled and poorly field. Hemispherical or spherical grayish-black sclerotia are formed on the lesion which falls on the ground. Sclerotia form even on the grain in case of severe infection.

Management

- Selection of healthy seeds, application of split (3-4) nitrogenous fertilizers, weed control in bunds and fields reduces the disease incidence.
- Spray application of Propiconazole @1ml/lit or hexaconazole 2ml/lit or validamycim at 2ml/lit twice at 15days interval when the first symptoms are noticed or at maximum tillering stage effectively controls the Sheath blight disease.

Bacterial Blight (*Xanthomonas oryzae* pv. *oryzae*)

Occurrence

- ✰ Earliest known plant disease in rice.
- ✰ Also known as rotten neck or rice fever.
- ✰ Reported from 80 rice-growing countries. First recorded in China (1637). Later from Japan (1704), From Italy (1828), from USA (1876), from India (1918). Expected grain loss: 70 to 80 per cent.

Symptoms

Lesions usually start as water soaked look strips along the margins of the upper parts of leaf blades. The lesion enlarges and turns yellow within a few days. Lesion may start at one edge or at both edges of the leaves. As the turning it white. On susceptible varieties lesion extend to the leaf sheath and may reach its lower ends.

'*Kresek*' symptoms may appear 2-3 weeks after transplanting. The infected leaves become grayish green and begin to roll along the mid rib.

Disease Cycle

- ✰ *Inoculum source*–seed husk, seed endosperm, soil, plant stubbles, debris, and collateral host grasses (*Leersia* spp. and *Cyprus* spp.)

Epidemiology

- ✰ Rainy (cloudy) weather, dull windy days, temperature between 22–26°C is conducive. Close planting enhances the disease incidence

Management

Cultural Practices

- ✰ Secure disease free seed
- ✰ Grow nurseries preferably in isolated upland conditions
- ✰ Balanced of fertilizers, avoid excess N–application
- ✰ Skip N–application at booting (if disease is moderate)
- ✰ Drain the field (except at flowering stage of the crop)
- ✰ Destruction of wild collateral hosts
- ✰ Avoid flow of water from affected fields
- ✰ Seed soaking for 12 hours and treating in hot water at 53°C for 30 minutes will make the seed free from bacterium
- ✰ Grow tolerant varieties (Swarna, Ajaya, Deepti, Badva mashuri, MTU-9992).

Chemical Control

- ✰ Soak the selected seed in 10 litres of water containing 1 g Streptocycline for 8-10 hours before sowing.
- ✰ Spray the crop with a mixture of copper oxychloride 500g and Streptocycline 7.5g in 500 litre/ha.

Bacterial Leaf Streak (*Xanthomonas campestris* pv *oryzicola*)

Symptoms

The primary infection occurs from the infected seeds. The early symptom is the appearance of interveinal long or short line, water soaked and grayish in colour on leaves. Minute yellowish or amber coloured beads may be found on the lesion. The lesion extend and coalesce to form large patches and become yellow due to death of cells. Eventually the entire leaf may become yellow or dirty white.

Management

- ☆ Use disease free seeds from the reliable sources.
- ☆ Soak the selected seed in 10 litres of water containing 1 g Streptocycline for 8-10 hours before sowing.
- ☆ Spray the crop with a mixture of copper oxychloride 500g and Streptocycline7.5g in 50 litre/ha.
- ☆ Seed soaking for 12 house and treating in hot water at 53°C for 30 minutes will make the seed free from bacterium.
- ☆ Grow resistant varieties like IR 42 and IR 20 etc.

False Smut

Occurrence

- ☆ Also known as green smut
- ☆ First reported from Tamil Nadu in 1878
- ☆ Later reported from Japan, USA, Phillippines, Indo–China, Burma, Srilanka
- ☆ Severe in Eastern States of India
- ☆ Not a serious disease in A.P
- ☆ Disease more prevalent in seasons of favourable growth and high yield
- ☆ Indication of good harvest.

Symptoms

Occurs on ear heads. Ovaries are transformed into large, velvety green masses round to oval, irregular sclerotial bodies. Few spikelets affected. Glumes are not affected. Sclerotial bodies–slightly flattened, smooth, yellow–covered by a membrane. Membrane bursts leads to further growth.

Disease Cycle

- ☆ Grasses and wild rice species are alternate hosts
- ☆ Spores–air–borne (main source of inoculum)
- ☆ Ascospores produced from sclerotia–source of primary infection
- ☆ Chalmydospores–secondary infection, air–borne, abundant at heading stage

Epidemiology

- ☆ Rainfall accompanied by cloudy days are favourable
- ☆ Appear between flowering and maturity of grain

Management

Cultural Practices

- ☆ Use of certified seed
- ☆ Collect and destroy diseased grains–> check secondary spread of disease–> helps in reducing inoculum for the next year

Chemical Control

- ☆ Spray Carbendazim (1g/lit) or Copper Oxychloride (2.5g/lit) at panicle initiation stage and 7–10 days later.

Rice Tungro Virus Disease

Occurrence

- ☆ Virus diseases are neglected initially with an impression of physiological disorders.
- ☆ Two types of viruses *i.e.* Tungro and Grassy stunt are reported. Tungro is severe in A.P.
- ☆ Reported in India in 1967. Appeared in epidemic form in N–E states of India (1969 and 1973) and Kerala (1964).
- ☆ Yield loss–38–71 per cent (less susceptible varieties)–84–100 per cent (highly susceptible varieties).

Symptoms

Disease can infect paddy at all growth stages and all aerial parts of plant (Leaf, neck and node). Among the three leaf and neck infections are more severe. Small specks originate on leaves–subsequently enlarge into spindle shaped Spots(0.5 to 1.5cm length, 0.3 to 0.5cm width) with ashy center. Several spots coalesce -> big irregular patches. Stunted growth, leaf colour is yellow to orange. Mottled appearance and slight twisting in young leaves. Rusty blotches in old leaves, discolored rusty blotches spread downwards from leaf tip. Delayed flowering (less susceptible varieties). Leads to death before flowering (highly susceptible varieties). Virus transmitted by green leaf hopper (Nepholettix virescens).

Disease Cycle

- ☆ Wild collateral grasses–*Eleusine indica, Echinochloa colonum* are the primary sources of inoculum.
- ☆ Green leaf hoppers are the secondary source of infection (female hopper is more efficient over male hopper)
- ☆ Virus is non–persistent in the vector
- ☆ Four strains of virus reported in India
- ☆ Virus particles are spherical (27.3 to 44.5mμ diameter)

Epidemiology

- ☆ Mineral nutrition and N–fertilization had marked influence on development of disease
- ☆ September to November and March to April the insect vector is more active and thus disease is more prevalent

Management

- The isolated plants having virus infection symptoms in the beginning and destroy them by burning so that the insect does not get inoculum to spread the disease.
- The green stubbles, voluntary plants should be uprooted and burnt after harvest.
- Adopt balance fertilizer application.
- Destroy weeds both in field and on bunds.
- Leaf yellowing can be minimized by spraying 2 per cent urea mixed with Mancozeb at 2.5 gm/lit.
- Instead of urea foliar fertilizer like multi-K (potassium nitrate) can be sprayed at 1 per cent which impart resistance also because of high potassium content.
- Grow tolerant varieties like MTU 9992, MTU 1002, MTU 1003, MTU 1005, Surekha, Vikramarya, Bharani, IR 36 etc.
- In epidemic areas follow rotation with pulses or oil seeds.

Chemical Control

- Green jassids acting as vectors are to be controlled effectively in time by spraying Monocrotophos at 1.6 to 2.2 ml/lit or Edifenphos @1.5 ml/lit or by applying carbofuron 3 G @ 10 kg/acre.
- In nursery when virus infection is low, apply Carbosulfan granules @ 1 kg a.i./ha to control vector population.
- During pre-tillering to mid-tillering when one affected hill/m is observed apply carbofuran granules @ 1 kg a.i./ha or spray monocrotophos @ 1.6 to 2.2ml/Lit to control insect vector.

References

Annonymous (2010). www.ikisan.com/links/ap_rice Detailed Study of Diseases.shtml

Saha, L. R. and Dhaliwal, G.S. (2006). *Handbook of Plant Protection*. Second Revised Edition, Kalyani Publishers.

Sustainable Disease Management of Agricultural Crops (2011) *Pages* **89–117**
Editors: **S.K. Biswas and S.R. Singh**
Published by: **DAYA PUBLISHING HOUSE, NEW DELHI**

Chapter 6

Phytophthora Fruit Rot of Tomato (*Lycopersicon esculentum* Mill.) and its Management

☆ *T.A. Shah[1], S.R. Singh[2] and C.R. Prajapati[3]*

[1]*Department of Plant Pathology,*
S.K. University of Agricultural Sciences and Technology-K, Srinagar (J&K)
[2]*Krishi Vigyan Kendra, Belatal, Mahoba – 210 423, U.P.*
[3]*Krishi Vigyan Kendra (Baghpat), S.V. Patel University of Agriculture and Technology, Modipuram, Meerut, U.P.*

The fruit rot of tomato (*Lycopersicon esculentum* Miller) is known to be caused by various fungal pathogens in different parts of the world and diverse strategies for management of the disease have, therefore, been suggested. In the present write-up, the information of the disease has been comprehended.

Status of Disease

The tomato fruit rot caused by *Phytophthora* spp. has been reported from all tomato growing areas of the world where high relative humidity coupled with warm temperature is prevalent (Obrero and Aragaki, 1965; Singh 1999). For the first time, the disease was recorded in the year 1915 at Goulds in Florida, USA (Sherbackoff, 1917).On the basis of various observations, Sherbackoff (1917) coined the name buckeye rot to this disease owing to the characteristic lesions with broad zonation on the fruit. Since then it has been reported from Argentina, Australia (Flett, 1986), Bulgaria (Elenkov, 1977), Canada (Conners, 1938; Richardson, 1941), China (Wei and Cheo, 1944), Cuba (Weimer, 1920), Croatia (Cvjetkovic and Jurjevic, 1993), England (Bewley, 1921), Germany (Skadow *et al.*, 1984), Greece, Hawaii, Holland (Voorst *et al.*, 1987), Indonesia, Ireland, Italy (Pane *et al.*, 2000), Japan (Matsuzaki *et al.*, 1980), Korea (Kim *et al.*, 1998), Malaysia, Mexico, Rhodesia, South Africa (Wager, 1935; Ferreira *et al.*, 1991), Sri Lanka (Williams *et al.*, 1937), Taiwan (Chang, 1983), Tasmania, Trinidad (Baker, 1939) Zambia

(Andersson, 1987) and the States of Indiana, Florida, California, Mississippi, Tennessee, Texas and Arkansas of USA (Cristopoulos, 1954; Fulton, 1954).

In India major fungal tomato rot due to *Phytophthora palmivora* (Butler) has been reported from Coimbatore area of Tamil Nadu (Ramakrishnan and Soumini, 1947). Later, from Bombay Patel *et al.* (1996) reported *P. parasitica* pathogenic on fruits of tomato. Jain *et al.* (1961) were first to report the natural occurrence of buckeye rot of tomato from India in Solan (Himachal Pradesh). Subsequently, it has been reported from Haryana, Jammu and Kashmir, Karnataka, Punjab, Tamil Nadu and Uttaranchal (Sokhi and Sohi, 1974; Sohi, 1981; Srivastava, 1981; Sharma, 1992; Sharma *et al.*, 1992; Gupta and Paul, 2001). Dar *et al.* (2004) while working on fungal fruit rots of tomato revealed association of three species of Phytophthora, identified as *P. infestans*, *P. capsici* and *P. nicotianae* var. *parasitica* in Kashmir province of Jammu and Kashmir State. However, the last one was more damaging. Moreover, the fungus has been reported from Uttar Pradesh and incites damping-off of tomato seedlings (Singh and Srivastava, 1966). From Maharashtra it was found as a disease of storage and transit (Rao, 1984). The disease has been observed to be serious at elevations ranging from 500 to 1700 m above sea level (Sohi and Sokhi, 1972).

Incidence and Losses

The economic losses due to fruit rot, particularly buckeye rot are direct and has assumed an alarming proportion as infected fruits fall down prematurely. Several workers have reported varying degree of losses in different parts of the world. Sherbakoff (1917) observed that this disease affects up to 15 per cent of the fruit in field and up to 10 per cent of the fruit in transit. Losses to the extent of 66 per cent have been reported by Rosenbaum (1920). Sherf and Macnab (1986) reported crop losses in the range of 25-50 per cent in USA. Later, losses up to 20 per cent have been estimated from Florida in USA (Engelhard and Pluetz, 1979). A crop loss of 16 per cent from Papua New Guinea was due to *P. parasitica* (Dodd, 1979). Great losses have also been reported from Nigeria.

In India, losses from this disease ranged from 30 to 60 per cent (Jain *et al.*, 1961; Rattan and Saini, 1979; Kohli *et al.*, 1996). Sharma (1974) and Kakar *et al.* (1980) reported losses from 35 to 45 per cent in Solan (HP). However, an incidence of 65 to 68 per cent and 65 per cent due to fruit rot disease in ripe tomatoes were recorded from Himachal Pradesh (Thapa and Sharma, 1976; Dodan *et al.*, 1995). Sohi and Sharma (1989) recorded up to 35 per cent net loss in fruit yield in field and 18 per cent in storage under Bangalore climatic conditions, while in another study from the same area losses from 18 to 35 per cent were recorded (Sokhi and Sohi, 1982). Sharma and Sohi (1982) observed the average loss of 50 per cent. Severe incidence of disease in some hybrid genotypes in Punjab caused 15 to 20 per cent crop loss (Sharma and Sharma, 1982). In Haryana buckeye rot incidence varied from 20 to 30 per cent (Srivastava, 1981; Thareja *et al.*, 1989). Maximum fruit rot infection in susceptible varieties was about 30 per cent and in absence of fungicidal treatment up to 35 per cent (Sherf and Macnab, 1986). Under Kashmir agro climatic conditions, a yield loss of 40 per cent in tomato fruit has been reported due to buckeye rot alone (Dar *et al.*, 2004).

Jones and McCarter (1974) reported 1.4 to 3.6 per cent fruit losses while studying the relative prevalence of different rots. Behar *et al.* (1993) from Punjab recorded average disease incidence of 38 per cent in early season crop and 3.4 per cent in main season crop. In Michigan (USA) anthracnose and Phytophthora together caused 92 per cent fruit rots in tomato (Byrne, 1997). Fageria (1994) found combined losses due to buckeye rot and alternaria blight disease in the mid hills of HP as high as 70 to 90 per cent. The disease of Phytophthora fruit rot ranged from 39.0 to 45.0 per cent in IPM plots (Deepak, *et al.*, 2003).

Host Range

The pathogen *Phytophthora nicotianae* var. *parasitica* has a very wide host range and associated with damping-off, root rot, crown rot, stem canker, tip blight, leaf blight and fruit rot in 72 genera of over 58 families (Jones *et al.*, 1997; Bose *et al.*, 2002). In addition to tomato the fungus has been recorded on aster, banana, beans, betelvine, bougainvillea, brinjal, capsicum, carnation, chillies, chrysanthemum, citrus, coconut, grapes, guava, hibiscus, jack fruit, maize, okra, onion, pears, pumpkin and sesame in India (Matherton and Matejka, 1990; Nema and Sharma, 1997; Nema and Sharma, 1999; Singh, 1999; Gupta and Paul, 2001). Moreover, the pathogen can infect a large number of fruit vegetables belonging to diverse families. Fruits of bitter gourd, cowpea, cucumber, okra, pointed gourd, ridge gourd, snake gourd besides tomato were found to be susceptible to the pathogen and developed rot symptoms within 6-10 days after inoculation (Roy, 1997).

Variability

Variation among different species of Phytophthora is inherent as observed by Erwin *et al.* (1983). Sharma and Sharma (1982) studied the growth of pathogen on 7 solid and 7 liquid media and also worked out the pH, carbon and temperature requirement (Cameron, 1962). Based on these studies, they concluded the presence of different physiological forms in the pathogen.

Obrero and Aragaki (1965) reported that *Phytophthora parasitica* isolates from tomato plants were more pathogenic to tomato fruits than isolates from pineapple, parsley, hibiscus, carnation, eggplant and a soil isolate. An isolate from papaya was non-pathogenic. Cultural and pathogenic variability among 6 different isolates of *P. parasitica* isolated from genetically different host plants have been reported (Siradhana *et al.*, 1968; Sharma and Sharma, 1987). However, Sehgal and Prasad (1966) had found no morphological and physiological differences among isolates of *P. nicotianae* var. *parasitca* that varied in pathogenicity. Sharma (1971) also studied the cultural and pathogenic variation among 20 isolates collected from buckeye rot affected tomato fruits and reported that significant differences were exist among the isolates; but could not establish any correlation between mycelial growth and virulence. Variation among 81 fungal cultures assigned to *P. nicotianae* var. *parasitica* was analysed under defined conditions and PCR was used to differentiate species (Hall, 1993; Ersek *et al.*, 1994).

Dodan and Shyam (1994) studied the pathogenic variation in 26 isolates of the pathogen collected from different agro-climatic zone of Himachal Pradesh. Isolates showed considerable variation in cultural and pathogenic behaviour, but no correlation was found between linear growth, size of sporangia, chlamydospores, oospores and pathogenic behaviour of these isolates (Shyam and Gupta, 1998). Although the isolates did not exhibit the existence of assemblage(s) of virulence and avirulence; but could be grouped into 18 pathogenic strains on the basis of their aggressiveness patterns and non-differential host-isolate interaction (Gupta and Shyam, 1998).

Etiology

Causal Pathogen

Several species of Phytophthora have been reported to cause fruit rot of tomato (Walker, 1952). *Phytophthora infestans* (Mont.) deBary, incitant of late blight of potato was shown to cause a similar blight in tomato. Buckeye rot has been reported to be caused by *P. maxicana* (Hotson and Hartge, 1923). *P. drechsleri* (Jones *et al.*, 1997), *P. palmivora* (Ramakrishnan and Soumini, 1947; Thomas *et al.*, 1947), *P. capsici* (Kreutzer *et al.*, 1940; Simonds and Kreutzer, 1944). *P. cryptogea* (Williams *et al.*, 1937) and *P. nicotianae* var. *nicotianae* (Weststeijn, 1973; Snowdon, 1991; Jones *et al.*, 1997).

Sherbakoff (1917) first reported *Phytophthora terestria* as the causal agent of buckeye rot of tomato. (Butler,1920), Bewley (1920), Ashby, 1928. Leonian and Geer (1929) and Tucker (1931) agreed that it was synonymous with *P. parasitica*. Later, Richardson (1941) studied the disease in detail and found *P. parasitica* to be responsible for tomato buckeye rot. Waterhouse and Blackwell (1954) in their key for identification of *Phytophthora* species, used temperature as a main criterion for classification and on the basis of this, the fungus was named as *P. parasitica* Dast. The nomenclature was again changed in 1962 (Waterhosue, 1963) and renamed as *Phytophthora nicotianae* Breda de Hann var. *parasitica* (Dastur) Waterhouse (Bose *et al.*, 2002). However, recently the association of *Phytophthora capsici* with this disease from Mexico has also been reported (Fernandez, 2003).

The pathogen belongs to sub-division Mastigomycotina, class Oomycetes, order Peronosporales and family Pythiaceae (Gupta and Paul, 2001). The hyphae of fungus are tough, irregular in width without marked hyphal swellings. The sporangiophores are thinner than the vegetative hyphae, irregularly or sympodially branched. The sporangia are papillate, occasionally have more than one papilla and broadly ovoid, ellipsoid and pyriform to spherical. The apex is not noticeably narrowed. The average and maximum size of sporangia are 38 x 40 and 50 x 40 µm, respectively. They are deciduous with very short, pedicel usually 2 µm long. Antheridia are amphigynous, spherical or oval and 10 x 12 to 16 µm in size. Oogonia are usually produced in single cultures though often very sparsely or not until after some weeks. They are readily produced when opposite strains are grown together in culture. Oogonia measure less than 28 µm (24 to 26), rarely 31 µm. On maturity oogonia become rough, thick walled, yellow brown. Oospores are aplerotic, 18-20 µm in diameter with 2 µm thick wall.

The fungus produces chlamydospores abundantly. They are mostly intercalary, yellowish brown and measure up to 60 µm in diameter. It is intracellular in the affected fruit and both inter and intra-cellular in endosperm and ovule.

Symptomatology

The buckeye rot pathogen has been reported to cause damage to different parts of tomato plant other than the fruits. The disease manifests different phases at various phenological stages of crop *viz.*, damping- off, collar rot, blossom blight and fruit rot (Sharma, 1971).

Damping-off of tomato due to *Phytophthora nicotianae* var. *parasitica* has been reported by Bewley (1921), Taylor (1924), Goidanich (1936), Singh and Srivastava (1966). Infection at seedling stage result in post-emergence damping- off (Gupta and Paul, 2001). Root infection on plants include water soaked lesions, which gradually dry and turn dark brown (Jones *et al.*, 1997). Infection progress upwards on seedling along the vascular bundles and into the developing pith tissue. Growth is stunted and the foliage shows a dull dark-green colour. Before the seedlings develop a complete cylinder or lignified xylem, tissue desiccation ensues thereby causing the stem base to shrivel and the plants girdle and die quickly (Jones *et al.*, 1997). Root and stem infection may also take place when mature plants get infected (Raicu and Satan, 1973).

Infection may spread directly from the soil to leaves which are in direct contact with the ground. Water-splashes can helps in dissemination of pathogen. In either case wilting and foliar discolouration with small necrotic spots result (Gupta and Paul, 2001). Leaves may be attacked but leaf blight is not a common symptom (Jones *et al.*, 1997; Singh, 1999).

On infected petiole and stem blackish green lesions develop leading to their collapse. Petiole infection generally originates from infected leaf blades and may eventually lead to the invasion of stem, but can also develop from fruit stalks (Raicu and Satan, 1973). Severe stem rotting of tomato has been observed under water logged conditions (Clark and Ann, 1979). The infection to fruit has been reported to occur at different stages of maturity. Injury to fruit help in quick establishment of the infection, however, green stage is most susceptible (Sharma *et al.*, 1975). Usually rot symptoms appear at blossom end of the immature green fruits touching or nearly touching the soil (Singh, 1999; Gupta and Paul, 2001). In warm humid weather, the spots enlarge rapidly and generally cover half or even more of the first fruit surface forming concentric zones of narrow dark brown and wider light brown bands (Bose *et al.*, 2002). The diseased portion of the fruit retains its natural shape and consistency. However, under high humid conditions it is usually covered with a white downy growth of the fungus. Initially fruit remains firm, there is no rotting of the skin but eventually decay progresses rapidly and internal flesh may be discoloured up to the core. The fruits shrink and become mummified. Due to invasion of secondary rot causing organism, older fruits soon decompose (Singh, 1999; Gupta and Paul, 2001).

Mucilage inside the fruit turns dark brown to black and decomposes to a watery consistency in advanced stages (Sohi and Sharma, 1989). Decay continues even after harvest and spreads into neighbouring healthy fruits (Snowdon, 1991).

Predisposition

The disease initiation and its further development mostly depend on weather (Sharma *et al.*, 1976a). The fruit rot has been reported from all parts of the world where warm wet weather, high relative humidity and abundant soil moisture favour its development (Singh, 1999). The disease incidence is directly influence by various environmental and soil conditions (Sharma *et al.*, 1978; Bhardwaj *et al.*, 1985; Saha, 2002). Though infection takes place between 18 and 30 °C, maximum infection occurs at 25 °C (Bose *et al.*, 2002). A relative humidity of 75.6 per cent and above is suitable for the initiation of infection and disease development, with optimum disease at 98.5 to 100 per cent (Sokhi, 1994). A humid atmosphere permit the development of an off-white mould bearing numerous sporangia (Snowdon, 1991).

The infection of fruit take place at temperature between 13-35° C (Singh, 1999) but 22-25°C has been found to be optimum (Sharma *et al.*, 1978). The disease generally does not occur if weekly temperature remain at or below 20° C, though at 22.5° C or above even slight rainfall (10 mm) would manifest disease appearance after 4 days (Gupta and Paul, 2001). Contrarily, Singh (1999) found temperatures slightly above 10° C with precipitation in the preceding week responsible for high disease incidence. The highest number of diseased tomato plants (66 per cent) occurred at 25 to 30°C (Singh, 1999). On the other hand, Thareja *et al.* (1989) observed maximum disease incidence at temperatures of 20 to 25°C with average relative humidity of more than 60 per cent. Similarly, in experiments conducted in Cuba with tomato cultivar 'Campbell 28' average temperature of 22-26°C and high relative humidity was found conducive to disease (Perez *et al.*, 1989). At lower temperatures, the sporangia germinate by producing biflagellate planaspores while at temperatures above 25°C, these produce germ tubes directly, restricting the disease level (Singh, 1999; Gupta and Paul, 2001). Zoospores retained motility up to 20 hours at 20° C but at higher temperature motility was short-lived (Jones *et al.*, 1997).

Large amount of rainfall within a short period of time (Wilson, 1956) or a single heavy irrigation (Thompkins and Tucker, 1941) may result in sudden appearance of buckeye rot. Rosenbaum (1920) demonstrated a high incidence of buckeye rot when tomato fruits were placed in soil that was saturated either by watering in pots or by irrigation in the field. The significant effects of rainfall and irrigation on disease dessimination has been well documented (Bowers and Mitchell, 1990; Café-Filho and Duniway, 1995). The effect of water stress imposed either before or after inoculation of tomatoes on the severity of *P. nicotianae* var. *parasitica* was evaluated. Pre-inoculation water stress increase disease severity as compared to post-inoculation stress (Ristaino and Duniway, 1989).

Phytophthora induced water stress during critical stages of crop development and had major impact on plant growth, phenology and yield in tomato. Disease symptoms developed more rapidly and was observed earlier in plants given prolonged or normal irrigation which caused significant reductions in total plant, leaf and fruit. Less frequent irrigation of infested plots caused a delay in disease onset and reduced the impact of disease (Ristaino *et al.*, 1989). The role of soil moisture in the release of zoospores and subsequent infection of the host has been emphasised by various workers. Thomson and Hine (1972) reported that zoospores of *P. parasitica* were present in water placed on field. Higher soil moisture and a soil temperature of 18-20° C have been found favourable for the disease incidence. A soil moisture range of 32 to 35 per cent showed 17.8 per cent disease incidence. Whereas at 44 to 47 per cent soil moisture, it was 45.4 per cent indicating thereby that high soil moisture contents and is essential for development of epiphytotics (Sharma, 1971).

The increase in incidence of disease might be due to an increase in the soil inoculum by continuous cropping with tomato (Kapoor, 1988). Infection occur in association with soil residue where mud containing inoculum has been splashed onto the fruit (Jones *et al.*, 1997). Application of light irrigation and ensuring optimum crop water use were important in reducing buckeye rot (Bhardwaj and Masand, 1995). Phytophthora root rot is most damaging in compact soils or soils with poor drainage (Fulton and Fulton, 1951).

Fruits touching the ground or at lower heights get infected first and show higher disease incidence. The height from the soil at which fruits are attached on plants appear to play an important role on severity of the disease. Thareja *et al.* (1989) recorded maximum number of infected fruits when in direct contact with soil or at 5 cm height. However, infection up to the height of 45 cm has also been reported by various workers (Sharma and Thapa, 1976). The infection on lower fruits mostly takes place directly from soil through mycelium or zoospores of the fungus present in soil, whereas on upper fruits it may occur due to zoospores released from sporangia produced on diseased fruits and disseminated by surface water, spattering rains, rain water or wind or by mutual contact of fruits (Springer and Johnston, 1982; Jones *et al.*, 1997).

Host Nutrition

Nutrition plays a primary role in disease development as well as in its control. Therefore, striking a balance is essential to effectively manage disease. Nutrient manipulation through amendment or modification of soil environment has been an important control for plant diseases (Engelhard, 1990). The role of nitrogen, phosphorus and potassium fertilizers in disease development depends upon the nature of pathogen and host interaction on one hand, and quality and quantity of fertilizers on the other (Singh, 2001).

Nitrogen is an important constituent of major bio-molecules such as amino acids, proteins, growth hormones, enzymes, phytoalexins and phenols and a determinant of cell wall characteristics, is known to exhibit significant influence on disease development caused by soil-borne fungi (Kaufman and Williams, 1964). The disease severity in general increases as the rate of nitrogen application is increased.

High N has been reported to increase the diseases by *Phytophthora* spp. (Awan and Struchtemeyer, 1957; Gallegly and Niederhauser, 1959). Moderate to high fertility has been reported to increase the severity of diseases caused by *P. parasitica* on tomato (Weststeijn, 1973). Host nutrition appreciably affects buckeye rot severity. Sharma and Sohi (1982) observed that the disease was more at 150 and 200 kg N/ha applied in the form of urea, than at 100 kg N/ha.

Sharma and Sohi (1983) evaluated different levels of fertilizers and reported that higher doses of nitrogen improved yield of healthy fruits, with a corresponding upward curve in buckeye rot of tomato. The effect of *P. nicotianae* var. *parasitica* on tomato was studied with different concentrations of N, P and K. The lowest disease rates were registered in plants which had received adequate nutrients, while the plants deficient in some nutrients were the most affected. The greatest correlations were found between P and N doses and disease incidence (Perez *et al.*, 1989).

In a greenhouse experiment on the incidence of early blight of tomato the highest disease index (85.18 per cent) was obtained at increasing levels of N while the lowest disease index (31.11 per cent) was obtained in the control where no N was applied (Singh *et al.*, 2001).

Phosphorus, the second most important element applied to soil as fertilizers, is essentially required for meristematic tissue and root proliferation. The vigorous root growth promoted by adequate phosphorus applications help plants bear the damage caused by root pathogens (Huber, 1980). Mean disease index significantly decreased with increasing levels of P (Singh *et al.*, 2001).

The effects of P on Phytophthora disease have been inconsistent. The results with diseases caused by *P. parasitica* have ranged from less disease (Thyagarajan *et al.*, 1972) to no response (Dukes and Apple, 1968) to increased disease (Olunloyo and Adeniji, 1976). At higher P levels there were more healthy fruits and less disease particularly up to 90 kg P_2O_5/ha (Sharma and Sohi, 1983). Higher doses of phosphorus (60 and 90 kg P_2O_5/ha) reduced the disease incidence (Sharma and Sohi, 1983).

Potassium, third major nutrient supplement in soil is essential for plant growth. Besides, regulating enzyme activity in most cellular functions such as photosynthesis, phosphorylation, protein synthesis and reproduction. It induces formation of thicker cuticular and epidermal cell wall and modifies disease reactions in plants both directly or indirectly (Huber, 1980). The direct effects involve formation of adequate structural barriers and to reduce penetration, multiplication, survivability and aggressiveness of invading pathogen; while as indirectly, the delayed initiation of senescence by potassium avoids infection by facultative parasites (Kiraly, 1976). An increase in potassium from deficiency level causes increased crop growth and yield with simultaneous reduction in disease susceptibility. The host resistance to pathogen, however, remains unchanged if K_2O level is enhanced beyond certain optimum dose. The excess K_2O level is known to inhibit uptake of other cations like Ca^{++}, Mg^{++} and Mo^{+++} so essential for plant growth (Balaji and Vaitheeswaran, 1988). It is believed that the synthesis of high molecular weight compounds like proteins, starch and cellulose, in K-deficient plants gets impaired and low molecular weight compounds accumulated which increase susceptibility to pathogen infection (Huber, 1980).

High doses of K are thought to decrease disease severity. Although, response to K depend on the rate of N used (Pal and Grewal, 1976) yet different levels of potassium showed no significant effect (Sharma and Sohi, 1983).

Several workers have evaluated the role of major elements for early blight incidence in tomato (Bedi and Dhiman, 1983). The application of N, P and K @ 100, 125 and 150 kg ha^{-1}, respectively and FYM @ 30 t ha^{-1} have been found to reduce the incidence of *Alternaria solani* (Ramakrishan and Jeyarajan, 1986).

Disease Management

Soil Amendments

Soil amendments in the form of plant debris, green manures, farmyard manure (FYM), compost, oil cakes and fertilizers have been known to improve crop productivity to improving nutrient status and soil tilth, besides increasing microbial activity in the rhizosphere to suppress certain soil-borne diseases (Sivaprakasham, 1991).The degradation of organic waste materials results in the formation of compost by diverse microbial population (Hadar and Mandelbaum, 1992). The compost is of high value to crops due to its chemical and physical properties that enhance soil fertility, structure, water-holding capacity and overall plant health (Hadar and Mandelbaum, 1992; Stofella and Khan, 2000).

Composts have been used in agriculture with beneficial effects for years (Kelman and Cook, 1977). During the past few decades, several reports have documented the suppressive effects of composts on a variety of soil-borne plant pathogens (Baker and Cook, 1974; Singh, 1983). Moreover, composted swine waste (CSW) potting medium enhanced suppression of *Phytophthora parasitica* var *nicotianae* (Hoitink *et al.*, 1991; Fichtner *et al.*, 2004). Examples of difference in efficacy among compost types against different diseases abound. Composts prepared from tree leaves or rice hulls were effective against club root of Chinese cabbage, whereas compost prepared from saw dust was much less effective (Tamura and Taketani, 1977). Chicken manure (poultry manure) incorporation results in reduction of disease in tomato due to reduced pathogen population and increased microbial activity and anti-fungal substances (Homma *et al.*, 1979). Corrales *et al.* (1990) recorded 30 and 65 per cent reduction in Phytophthora blight of capsicum by amending soil with chick manure and compost. Soil amendments with biogas slurry also proved superior to other forms of manures evaluated in reducing the lentil wilt (Kaushal and Sharma, 1995). Phytophthora rot of pepper were also controlled by soil amendment with composted sewage sludge (Lumesden *et al.*, 1983).

Amendments with dry natural plant materials of soybean and oat @ 1 per cent w/w along with ammonium nitrate fertilizer reduced root rot (*Rhizoctonia solani*) of beans (Davey and Papavizas, 1960). Ramaskrishnana and Jeyarajan (1986) reported reduced cotton root rot (*R. solani*) by soil amendment with farmyard manure (FYM) due to increased soil microbial population. The amendments with FYM was found effective against root rot wilt complex of peas (Kadian and Kaur, 1999). The presence of composts is believed to provide a proper substrate and pH range of 6.35-7.75 for the growth and sporulation of antagonists (Dhiman, 1999). Compost amendments at different doses also reduced safflower root rot (Lukode and Rane, 1989), chilli root rot (Dickerson, 1996) and tomato wilt (Padmodaya and Reddy, 1999).

Vermicompost

Vermicompost significantly inhibited the infection of tomato by *F. oxysporum* f. sp. *lycopersici* as it proved suppressive to the pathogen. The protective effective increased in proportion to the rate of application of vermicompost (Szczech, 1999). However, the incidence of chilli die back was higher in treatments comprising organic manures like vermicompost superimposed with organic pesticides than in those combined with inorganic fertilizers and chemical pesticides (Fugo, 2000). In other experiment with tomatoes the addition of vermicompost suppressed pathogen like *F. oxysporum* and had a positive effect on yields (Szczech and Brzeski, 1994).

Farm Yard Manure

The composts in general, enhance microbial activity in amended soils resulting in a reduction in inoculum density, better plant growth and disease control *Fusairum oxysporium* f. sp. *lycopersici*) causing wilts in tomato (Harender *et al.*, 1997).

Potting mixture of well decomposed FYM with soil (1: 2) has been recommended to minimize the blight in Paulownia caused by *Phytophthora nicotianae* (Mehrotra, 1997). Arulmozyhiyan *et al.* (2002) and Roy *et al.* (2002) while conducting experiments on the incidence of fruit and leaf rot, caused by *P. nicotianae* var. *parasitica* revealed that FYM alone recorded the lowest incidence of fruit rot (13.18 per cent), leaf rot (10.65 per cent) and leaf spot (13.31 per cent). The population of *P. nicotianae* var. *parasitica* in field crop significantly decreased with the use of FYM along with some bio-and chemical agents (Narayanaswamy *et al.*, 2004).

The effects of soil solarization with transparent polyethylene mulch (25 µm) in combination with FYM increased temperature at different soil depths by 3.9-10.5 compared to non-solarized beds and increased seed germination by 2.9-10.2 per cent, decreased the incidence of post emergence damping-off by 2.2 10.2 per cent and root length (7.9 to 2.9 per cent) compared with solarized non-amended treatment (Harender, 2004). The efficacy of soil amendments with dry leaves of pongamia along with FYM and neem seed cakes have been also reviewed (Padmodaya, 2003).

Poultry Manure

The effect of organic fertilizer like poultry manure on pathogenic soil microorganisms have been evaluated. The organic fertilizer increased plant vigour with the corresponding disease of *Phytophthora palmivora* responsible for the root rot (Ramirez *et al.*, 1998; Vawdrey *et al.*, 2002). Control of club root disease of cabbages (*Plasmodiophora brassicae*) was investigated using poultry manure. Mean disease index in non-treated plots was 9.9 than as compared to treated plots (6.6) (Velandia *et al.*, 1998). When used in a 1:2 ratio of husk and poultry manure decreased Phytophthora root rot of coorg mandarin and increased growth (Sawant *et al.*, 1995; Miyasaka, *et al.*, 2001). The composted poultry manure significantly reduced pathogen (*P. cinnamomi*) survival in soil and the development of symptoms on *Lipinus albus* seedlings (Aryantha *et al.*, 2000). Low rates (5 per cent) of poultry manure compost produced more disease control than equivalent or higher rates (10 and 20 per cent) of dairy manure compost in relation to soil-borne disease of strawberry caused by *P. fragariae* (Millner *et al.*, 2004). The treatments fosetyl-Al, poultry manure and transparent polyethylene in combination than alone showed the control of blight in pepper, caused by *P. capsici* (Chavez-Alfaro *et al.*, 1995; Rajan and Sharma, 2000).

Cultural Method

Managing plant diseases of soil-borne nature by manipulating the cultural practices such as crop rotation, organic amendments following, field sanitation, deep ploughing, time and method of planting, irrigation frequency and soil pH adjustment have been attempted successfully by many researchers (Jimenez *et al.*, 1990; Cafe-Filho and Duniway, 1995). Other cultural practices for the management of fruit rots include use of soil amendments (Smith *et al.*, 1979), straw mulches (Welch, 1949; Davis *et al.*, 1970), paper or polyethylene coated paper mulches (Everett, 1971). These practices are sustainable, though some are labour intensive (Thurston, 1990).

Ahmad *et al.* (1989) observed that the prevalence and progress of Phytophthora blight of chilli was influenced indirectly by wet season and directly by rainfall. Periodic flooding also increased mortality (*P. capsici*) of Early California wonder pepper (CWP) by 20-100 per cent depending upon flooding period (Bowers and Mitchell, 1990). Soil moisture levels have been found to influence the severity of other fruit rots. Hot weather conditions of 30° C and soil moisture of 40 per cent water holding capacity also proved conducive for the disease development (Nelson, 1981).

Crop rotation with non-host crops for at least three years proved encouraging in diminishing the inoculum load of *Phytophthora nicotianae* var. *parasitica* as the pathogen remain viable in soil for at least two years (Raicu and Satan, 1976; Gupta and Paul, 2001). Staking of plants has been suggested for the management of fruit rots (Sherbakoff, 1917; Wager, 1935; Sokhi and Sohi, 1974). Staking the plants, stripping of leaves and fruit from the basal 30 cm of stem controls the disease significantly (Sharma *et al.*, 1976b). Sharma and Sohi (1982) recorded maximum yield of healthy fruits and minimum incidence of buckeye rot in staking, mulching along with chemical sprays. Staking with mulching has also been reported effective by early workers like Sherbakoff (1917), Wager (1935) and Welch (1949). Staking and mulching of tomato plants not only reduce the buckeye rot but also minimise the incidence of other soil borne diseases.

Application of polyethylene mulch to obstruct the dispersal of soil borne inoculum has been found effective and economical in the management of this disease (Dodan *et al.*, 1994; Bhardwaj *et al.*, 1995; Gupta and Paul, 2001). It is advantageous to stake the plants so as to prevent contact between fruits and soil (Snowdon, 1991). This may also be achieved by means of a plastic or straw mulch (Springer and Johnston, 1982). Staking and mulching with plastic, newspaper, straw etc help to reduce losses (Sherf and MacNab, 1986). But the disease has been even noticed frequently on staked plants as well (Wilson, 1956; Sharma *et al.*, 1976a).

Spacing

Kuruzwinska and Kuruzwinksa (1991) while studying the effect of planting dates and spacings on *P. infestans* infection in potato, found that planting dates did not have a significant effect on tuber infection while as, planting densities affected both tuber and leaf infection. Fontma *et al.* (1996) reported that despite weekly sprays, all tested varieties during both early and late season crops, proved susceptible to *A. solani* and *P. infestans.*

Spacing has also played a key role in combating the various chilli diseases caused by soil borne pathogens like *Fusarium oxysporum* and *Phytophthora capsici* (Siviero and Centolo, 2001). Wider spacing 4-6 x 1-1. 50 feet has been recommended in un-staked planting to control the disease (Loest, 1948).

Baba (2001) also observed significant differences in the incidence of tomato rots with spacing 45 x 90 cm between plant to plant and row to row than normal (45 x 60 and 30 x 60 cm) and close spacing (30 x 30 cm). The equidistant single row arrangement (75 x 75 cm) and wide twin row arrangement (60 x 90 cm) had reduced foliar and fruit diseases of tomato in a wet year (Warner *et al.*, 2002).

Mulching

The use of dried paddy straw mulch has recorded increase in yield of chilli and tomato with a significant effect on vegetable fruit diseases as well (Zahara *et al.*, 1994; Vinod *et al.*, 2004). Welch (1949), Malla and Sah (2002) also observed a higher tomato marketable (2.4 MT ha^{-1}) and unmarketable (0.98 MT ha^{-1}) yield due to mulching than without (1.38 and 0.39 MT ha^{-1}), respectively. Soil temperature was significantly influenced by organic mulches (paddy straw) as compared to soil moisture which was higher than the control (Ravindar *et al.*, 1997). The paddy straw mulch also increased the net short wave radiation, dry time air temperature, 2-6 days earlier maturity of tomato fruits, protection against frost injury and market diseases (Lal and Singh, 2002). Paddy straw mulch increased water infiltration under low intensity rainfall but increased runoff under high intensity rainfall (Sutrisno *et al.*, 1995).

In field experiments soil-solarization through mulching reduced salinity by 30-50 per cent and gave good control of *P. nicotianae* var. *parasitica* (Satour *et al.*, 1991). Kreutzer and Brayant (1944) while working with tomato fruit rot (*P. capsici*) observed that mulching with straw staking and ridging resulted in more healthy fruits. Considerable control of the disease has been reported in mulches and sprayed tomato crop (Welch, 1949). The interaction of 6.51 fruits of good quality were also obtained in tomato plots treated by grass mulching, no significant yield increase was noticeable (Issiki, 1994). However, in one more experiment paddy straw mulch proved, as compared to plastic or paper mulch, had little effect on yields, proved less efficacious and erratic in disease control of tomato (Jones and McCarter, 1984; Ravindar *et al.*, 1998).

Soil solarization through the use of plastic mulch has been considered as most advanced field technology to control soil borne pathogens and was developed for the first time in Israel to control many plant pathogenic pests, fungi, bacteria, nematodes and weeds (Barbercheck *et al.*, 1986; Katan, 1995). Solarization is a soil disinfestations method of disease control through hydro/thermal soil heating accomplished by covering moist soil with polyethylene (PE) sheets as mulch (Stapleton and Devay, 1982; Patel, 2001) resulting in successful control of soil-borne fungal diseases (Ashworth, 1979; Pullman *et al.*, 1981). Soil solarization is based on trapping solar irradiation by tightly covering the soil, usually with transparent polyethylene sheets. This results in a significant increase above 10-15° C of soil temperatures up to the permit where most pathogens are vulnerable to heat effects (Tjamos, 1996).

Soil heating by PE mulching has proved effective and inexpensive method for controlling soil-borne pathogens like *Phytophthora* spp besides *Fusurium* spp. *R. solani* and *M. phaseoli* (Hasan, 1989). The pathogens could not be detected up to 14 months at any depth after 4 to 8 weeks of solarization, suggesting a long term effect (Lopez-Herrera *et al.*, 1997). It has been observed that soil solarization by mulching with transparent polyethylene sheets control soil borne diseases (Chen and Katan, 1980; Elad *et al.*, 1980; Katan *et al.*, 1980). The long term effectiveness of black agriplastic method solarized soil integrated with chemical fungicide and organic amendment (chicken litter) reduced southern blight of tomato from 71 to 100 per cent (Stevens, 2003).

Phytophthora capsici in tomatoes was effectively controlled by soil solarization (Cartia, 1989). However, there was no effect on established infections of *P. cambivora* but growth in solarized areas was significantly greater than that of plants in untreated or metalaxyl-treated areas (Wides, 1988). Studies on efficacy of solarization or control of *Phytophthora nicotianae* revealed that soil temperature with a clear (30 µm) film reached a maximum of 47°C at 10 cm depth (Coelho *et al.*, 1999).

Mulching of the moistened soil with 30 µm transparent polyethylene film for 8 weeks reduced the incidence of *Pyrenochaeta lycopersici* in tomato. In another field experiment soil solarization by plastic mulching gave good control of *Phytophthora nicotianae* var. *parasitica*, *Pythium* spp and *Rhizoctonia solani* on tomatoes (Satour *et al.*, 1991).

The leaf litter mulch provides a special ecological environment for microbes, and change some physical and chemical characteristics of soil. The quantities of ammonifers and nitrifiers in the nitrogen physiological group increase significantly (Zheng *et al.*, 2004). Leaf litter mulch reduced runoff from 13 per cent of incident rainfall in control to 3 per cent (Sutrisno *et al.*, 1995). It promoted root biomass production, collar diameter and above ground biomass production (Singh *et al.*, 1994). It has been indicated that leaves with a fast rate of decomposition increase conservation of soil moisture and also improve synchronization between N release from the mulch and its demand by crops (Seneviratne *et al.*, 1998). Some more results suggest that leaf litter mulch reduce the survival of *P. palmivora* by accelerating substrate decomposition and by stimulating the activity of antagonistic and hyper-parasitic microbes (Konam and Guest, 2002). The efficacy of pine needle mulch alone or in combination with some fungicides in controlling the buckeye rot of tomato has been well established by Gupta and Shyam (1998).

Chemical Method

Healthy seed is essential to avoid introduction of pathogen into disease free nursery soil. Seed treatment has been reported to manage the buckeye rot (Singh, 1999). Seed should be treated with captan (4 g/kg seed) for eradication of externally seed-borne infection (Sharma, 1974). Sterilization of seeds or seed dressing with fungicides such as captan or metalaxyl have been advocated for management of *Phytophthora capsici* (Kannaiyan and Nene, 1979; Nedumaran and Vidhyasekaran, 1981, 1982; Ferreyra *et al.*, 1984; Chaudhry *et al.*, 1985; Baris *et al.*, 1986; Camino *et al.*, 1987; Kim, 1989; Jimenez *et al.*, 1990). The fumigants basamid (500 kg/ha), solasan (1000 1/ha) and vapam (1500 1/ha) almost completely destroyed *P. nicotianae* var. *parasitica* (Engelhard, 1974; Sokhi and Sohi, 1974). Basamid, solasan, vapam and mancozeb when applied to soil, completely destroyed soil-borne inoculum and reduced disease significantly (Raicu and Satan, 1973; Raicu *et al.*, 1974). Soil drenched with Bordeaux mixture (4: 4: 50) eliminate soil-borne infection of *P. nicotianae* var. *parasitica* (Sharma, 1974). Tomato root rot causd by *P. nicotianae* var. *nicotianae* was controlled by dimethomorph @ 0.05 g a.i. lit^{-1} as soil drench (Washington and McGee, 2000; Washington *et al.*, 2001).

Sridhar *et al.* (1975) screened 19 fungicides *in vitro* against the pathogen and reported brassicol, captan, thiram, difolatan, blitox and miltox highly effective in inhibiting the mycelial growth. Metalaxyl at low concentration inhibited mycelial growth and sporangium formation of *P. nicotianae* var. *parasitica* under *in vitro* conditions. The ethazol (Etridazol) was, however, more toxic. The anionic surfactant Nacconol 90 F (a.i. sodium dedecylbenzene sulphonate) was fungistatic *in vitro* to *P. nicotianae* var. *parasitica* at 75 µg a.i. ml^{-1} for mycelial growth and 60 µg ml^{-1} for spore germination, respectively (Hoy and Ogawa, 1984).

Fungicidal sprays are very important in the effective management of buckeye rot of tomato. Spraying of crop and soil with copper oxychloride, tribasic copper sulphate, zineb and Bordeaux mixture has been reported to be effective by various workers in reducing buckeye rot (Wager, 1935; Fulton, 1954). Earlier attempts to control the disease were made by spray application of Bordeaux mixture (Ramakrishnan and Soumini, 1947), triabasic copper sulphate (Fulton, 1954), Dithane M-45, Basamid, Solasan, Vapam and Ditrapex (Raicu and Satan, 1973; Raicu *et al.*, 1974; Sharma *et al.*, 1976b), Difolatan and copper oxychloride (Sohi and Sokhi, 1972; Sokhi and Sohi, 1974). Good disease control against *P. nicotianae* was obtained with ETMT, captan, chloroneb and Doweo 269 and with Dexon 70 per cent WP and turzole at 40-80 ppm, respectively (Robertson, 1973; Engelhard, 1974).

The control of various Phytophthora diseases were achieved by aluminium tris ethyl phosphonate. A new fungicide 2-cyano-N-ehylamino carbomyl-2-methoxy-iminoacetamide was found suitable for the control of Phytophthora blight in tomato (Smith *et al.*, 1979). Preventive and curative foliar applications of fosetyl @ 800 mg 1^{-1} or 160 µg ml^{-1} were effective for the control of *P. nicotianae* on tomato grown in hydroponic culture (Fenn and Coffey, 1989; Grote and Buesi, 1998). Multiple prophylactic foliar sprays of fosetyl-Al (12 g a.i litre^{-1}) were more effective than a single application in reducing rot of potted tomato plants inoculated with *P. nicotianae* var. *parasitica* (Davis, 1989). Previcar N (Propamocarb hydrochloride) gave good control of *P. nicotianae* var. *nicotianae* causing foot and stalk rot of tomato (Rapp and Richter, 1982).

Chlorothalonil, maneb and mancozeb have been found most effective when applied as protective sprays against fruit rot (Bruck *et al.*, 1981; Paulus *et al.*, 1983). Sequential sprays of captafol, mancozeb and copper oxychloride 40, 55 and 70 days after transplanting (DAT) increased yield by 50.5 per cent by reducing incidence of buckeye rot and defoliation diseases (Bhardwaj, 1991). He also found captafol, mancozeb and copper oxychloride effective against Alternaria rot (*A. solani*) and Phytophthora rot (*P. nicotianae* var. *parasitica*).Three sprays at 15 days interval starting 45 days after transplanting gave best yield of healthy fruits (Bhardwaj, 1991). Captafol (Foltaf 80 WP) at 0.25 per cent gave maximum control. The maneb applied for foliage diseases also provided control of fruit rot fungi (Batson, 1973). Fungicides found effective against Phytophthora rot have been found also effective in reducing the severity of phoma rot.

Cohen *et al.* (1997) reported that fungicidal products of systemic with non-systemic fungicides have proved best chemical control of fruit rots, specifically against the rots initiated by Phytophthora spp. Verma *et al.* (1994) recommended Ridomil-MZ (Metalaxyl + Mancozeb) for effective management of Phytophthroa rot (*P. nicotianae* var. *parasitica*). Khalid *et al.* (1995) also achieved best control of Phytophthora disease with the same fungicide besides Sandofan-M (Mancozeb + Oxidixyl) as compared to captan 50 WP, chlorothalonil and mancozeb. Shyam and Gupta (1994) while evaluating fungicidal sprays and working out their economics recommended a schedule consisting of single spray of Ridomil MZ-72 followed by two sprays of mancozeb, captan or copper oxychloride for the economic and effective management of fruit rots (*Phytophthora nicotianae* var. *parasitica*) and Alternaria rot (*A. alternata*).

The best results were obtained with Ridomil MZ, Dithane M-45 copper oxyhcloride @ 0.2 per cent and Difoltan (Sharma *et al.*, 1976c; Sokhi and Sohi, 1982; Verma *et al.*, 1994; Bhardwaj *et al.*, 1995; Gupta and Paul, 2001). Ridomil MZ + Curzate (0.03 per cent) + Indofil M-45 provided good control of the disease when applied before infection (Shyam and Gupta, 1996a). Metalaxyl and cymoxanil were used against *P. nicotianae* var. *parasitca* in tomato wherein the former was found more effective than the latter (Vanachter *et al.*, 1983). Eteridazole, fenaminosulf and metalaxyl gave good control of *P. nicotianae* var. *nicotianae* on tomato but only etridazole was not phytotoxic. However, treatments using metalazyl + chlorothalonil (80 + 400 g/kg) and metalaxyl + mancozeb (80 + 640 g/kg) were more effective in controlling Phytophthora blight and increasing tomato yields (Vanitha and Ramachandran, 1999; Rodrigues *et al.*, 2000). The application of Ridomil (metalaxyl) significantly reduced buckeye rot blight severity (11.7 per cent) with a marked increase in marketable yield (Harris and Nelson, 1999; Malla and Sah, 2002; Kashina, 2004).

Screening of Germplasm

Despite the continuous efforts in evaluation and breeding for disease resistance against buckeye rot of tomato, nothing concrete has been found so far, though a few cultivars/lines having tolerance to the disease have been identified. Incidence of buckeye rot on 12 tomato varieties under field conditions were recorded by Wilson (1956) wherein he reported minimum incidence (4.6 per cent) in variety Sioux and maximum (15.1 per cent) in Pennhart. Sharma *et al.* (1974) screened 290 tomato varieties by creating artificial epiphytotics and found 10 varieties *viz.* Chamba Exh. 64., Early Market Selection, Flat Large Round, Kopiah PQ x 45727 Molokai, Money Maker 56063, Red Cherry, San Marzan, Tatinter and 11-149 highly resistant. Fifty nine varieties were moderately resistant and 204 moderately susceptible to buckeye rot under Indian conditions. The remaining 17 varieties were found highly susceptible with a disease incidence greater than 60 per cent. The high yielding variety like money maker reported to be resistant to buckeye rot (Sharma *et al.*, 1974) has lost its resistance probably due to appearance of more aggressive strains of the pathogen. (Dodan *et al.*, 1995). Sharma and Bhardwaj (1977) analysed the polyphenol oxidase activity in 4 resistant varieties, namely V 230 Bison, Margalobe, Q-2 and Skopsk Skopskii Marrii and 6 susceptible varieties namely, Rutgers, Earliana, K-945, Kopiah

PQ x 45727, Red Jacket and Antibes, and concluded enzyme activity in resistant varieties ranged from 7.6 to 8.7 per cent as compared to 22.2 to 29.5 per cent in susceptible varieties, respectively.

Higher ascorbic acid content in variety EC-54725 has been reported to be responsible for resistance (Rattan and Saini, 1979). The role of various factors including height of plant, total acidity, nitrogen content of fruits and cuticular thickness of host fruit towards resistance has been investigated (Sharma, 1971). Some researchers linked association of ascorbic acid with the resistance (Rattan and Saini, 1969). When crosses were made between EC-54725 (*Lycopersicon pimpinellifolium*), a small fruit type, resistant to fruit rot and four highly susceptible commercial varieties (Gola, Sioux, S-12 and Lalmani) and (F1, F2 and back cross populations were studied. It was noticed that EC-54725 carries a dominant gene imparting resistance to fruit rot (Rattan and Saini, 1979).

Felix (1948) reported nine tomato varieties namely T 2440-0-1-4, T 244-5-3, T 244-10-3, T 244-2-10-2, T7-013-3-5-52, T 510-51, T 994-51 and T6-02-M6 resistant to Phytophthora fruit rot caused by *P. nicotianae* var. *parasitica* while Sherbakoff 57 Oxhart was most promising with respect to fruit size and resistance. Alexander and Hoover (1995) observed AC 205004, 205011, 205016, 205022, 205026, 205028 and 200051 to be resistant. Among many exotic and indigenous lines tested for resistance under a natural epiphytotic condition, UHF 10, EC 174023 and EC 174041 were found resistant (Dhaliwal and Rattan 1990; Kohli *et al.*, 1996; Rughoo *et al.*, 1997). Out of 21 varieties grown by Berksdale (1968) none was found resistant although the incidence ranged from 0.2 to 50.2 per cent.

Several varieties/hybrids like hybrid N-64, Rossal, EC 12917/P1-2 Ec122-62-1, EC 128-969, EC 129-16, EC 729-01, Hybrid-10, LA 1428, LA 2205 were found resistant from different parts of the world (Obrero and Aragaki, 1965; Perez *et al.*, 1986; Thareja and Srivastava, 1989). Under Indian conditions H-11, Leader, Fournaise, Fire Ball, Early Market Selection, Flat Large Red, Tatinter 11-149 exhibited resistance to buckeye rot (Sohi and Sharma, 1989). Dodan *et al.* (1995) screened 110 cultivars/lines against buckeye rot and reported *Lycopersicon esculentum* var. *cerasiforme* to be resistant under natural epiphytotic conditions. This genotype has the enough potential to be exploited to develop resistant varieties (Gupta and Paul, 2001). Boukema (1983) tested 500 accessions of *Lycoparsicon esculentum*, 20 of *L. pimpinelli folium*, 9 of *L. esculentum X L. pempinellifolium* hybrids, 14 of *L. hirsutum*, 7 of *L. esculantum X L. hirsutum* hybrids, 15 of *L. peruvianum* and *L. gladulosum* and one of *L. chilense*. The majority of accessions had no resistance or vary poor resistance. Commercial hybrids of tomato grown in India were found susceptible to buckeye rot, except Solan Shagun which was somewhat less susceptible (Shyam and Gupta, 1998). Similarly, in Punjab all commercial varieties have been found susceptible except TH 2312 released by PAU which suffer less damage. Some varieties like New Hamsphire, Sune Cup, New Yorker, Nova and Virginia known to be resistant (Yang, 1979) behaved as susceptible host under Indian conditions (Kaur *et al.*, 1988). Out of 14 tomato varieties tested LA 1428 and LA 2205 were resistant to fruit rot caused by *P. nicotianae* var. *parasitica* and LA 1308 and LA 2631, were moderately resistant (Kumar *et al.*, 1997) while as, LA 1312 was found resistant by Kozik *et al.* (1991). However, following inoculation with zoospores of *P. nicotianae* var. *parasitica* under controlled conditions CX8303 and 4 accessions of *L. esculentum* var. *cerasiforme* were rated resistant (Hewitt *et al.*, 1987; Blaker and Hewitt, 1987; Fageria *et al.*, 1998). Although cultivar UC 82B was vulnerable to infection by *P. nicotianae* var. *parasitica* (Snapp and Shenann, 1994). Among many exotic and indigenous lines tested for resistance under natural epiphytotic conditions UHF10 EC 174023, EC 174041 and Sirius showed resistance (Kohli *et al.*, 1996; Rughoo *et al.*, 1997).

Hybrid 1-7 X Patriot and EC 177401 X Syevestra were reported resistant to tomato fruit rot and early blight diseases (Cheema *et al.*, 1992). Moreover, FI-814, Hossiline, New Yorker, Nova, Ottawa-30, Ottawa-31, Pierline, PI 270407 and W-VA-700 behaved as resistant (Sokhi *et al.*, 1993). The evaluation of 40 genotypes of tomato for resistance to buckeye rot revealed that all the commercial cultivars were highly susceptible to disease. However, genotypes such as EC 174076 and SGP_8M_2 mutant had field resistance for buckeye rot (Fageria *et al.*, 1998). After screening 106 tomato cultivars/breeding lines against buckeye rot KT-10, KT-15, tomato 93, Triumph, EC-11958, Florida, Hawaii and C-32 besides Tengeru 97 and Cal-J were found resistant (Bijaya *et al.*, 2002; Kashina, 2004). The line EC, 1296020 recorded the lowest (6.48 per cent) buckeye fruit rot incidence after evaluating 80 lines of tomato (Joshi *et al.*, 2004).

References

Ahmad, S., Ansar, M. and Azmat, I. (1989). Root and collar rot of chillies caused by *Phytophthora capsici*. *Journal of Agricultural Research* 27: 155-156.

Alexander, L.J. and Hoover, M.M. (1995). Disease resistance in wild species of tomato. *Bulletin of Ohio Agriculture Experiment Station* 752: 76.

Arulmozhiyan, Manuel, R.W.W. and Velmurugan, S. (2002). Effect of organic vs. inorganics on betelvine cv. Vellaikodi in open system cultivation. *South Indian Horticulture* 50: 1-3.

Aryantha, I.P., Cross, R. and Guest, D.I. (2000). Suppression of *Phytophthora cinnamoni* in potting mixes amended with uncomposted and composted animal manures. *Phytopathology* 90: 775-782.

Ashby, S.F. (1928). The oospores of *Phytophthora nicotianae* Br. de Hann. notes on the taxonomy of *Phytophthora parasitica* Dast. *Transactions of British Mycological Society* 13: 86-95.

Ashworth, L.J. (1979). Polyethylene tarping of soil in a pistachio nut gome for control of *Verticillium dahliae*. *Phytopathology* 69: 913.

Awan, A.B. and Struchtemeyer, R.A. (1957). The effect of fertilization on the susceptibility of potatoes to late blight. *American Potato Journal* 34: 315-319.

Baba, R.A. (2000). Etiology and management of prevalent field fungal fruit rots of tomato (*Lycopersicon esculutum* Mills) in Kashmir. M.Sc thesis submitted to Sher-e-Kashmir University of Agricultural Sciences and Technology of Kashmir, Srinagar, Kashmir, p. 68.

Baker, K.F. and Cook, R.J. (1974). *Biological Control of Plant Pathogens*. W.H. Freeman and Company, San Francisco, USA, p. 433.

Baker, R.E.D. (1939). Notes on the diseases and fruit rots of tomatoes in the British West Indies. *Tropical Agriculture* 16: 252-257.

Balaji, A. and Vaitheeswaran, M. (1988). Effect of potassium stress on *Hibiscus cannabinus* infected by the root knot nematode, *Meloidogyne incognita*. *Indian Journal of Nematology* 18: 216-218.

Barbercheck, M.E., Broembsen and Ven, S.I. (1986). Effect of soil solarization on plant parasitic nematodes and *Phytophthora cinnamoni* in south Africa. *Plant Disease* 70: 945-950.

Baris, M., Gulsoy, E., Guneu, M., Maden, S., Sagir, A., Senyureu, M., Ulnkus, I., Yalcin, O. and Zengin, H. (1986). Investigations on sources of primary inoculum of cumin blight (*P. capsici*) of capsicum and control measures against the disease. *Bicki Koruma Bullteni* 26: 59-95 [cf: *Review of Plant Pathology* 69: 262].

Batson, W.E. (1973). Characterization and control of tomato fruit rots. *Plant Disease Reporter* 57: 453-456.

Bedi, P.S. and Dhiman, J.S. (1983). Influence of nitrogen, phosphorus and potassium on the development of early blight of tomato. *Indian Phytopathology* 36: 546-548.

Behar, D.S., Kaur, S. and Singh, M. (1993). Prevalence and incidence of pre-harvest fungal rots of tomato fruits. *Plant Disease Research* 8: 179.

Berksdale, T.H. (1968). Rhizoctonia rot and buckeye rot of tomatoes: observational differences in susceptibility. *Plant Disease Reporter* 52: 284-286.

Bewley, W.F. (1920). Damping off and foot rot of tomato seedlings. *Annuals of Applied Biology* 7: 156-172.

Bhardwaj, C.L. (1991). Efficacy and economics of fungicide spray schedules for the management of foliar and fruit disease in tomato. *Indian Journal of Mycology and Plant Pathology* 21: 56-59.

Bhardwaj, C.L. and Masand, S.S. (1995). Effect of mulching and irrigation schedules on buckeye rot (*Phytophthora nicotianae* var. *nicotianae*) of tomato (*Lycopersicon esculentum*). *Indian Journal of Agricultural Sciences* 65: 223-225.

Bhardwaj, S.S., Sharma, S.L. and Dohroo, N.P. (1985). Role of meteorological factors in development of bell pepper fruit rot (*Phytophthora nicotianae* var. *nicotianae* Waterhouse). *Indian Journal of Mycology and Plant Pathology* 15: 3238-329.

Bijaya, A.K., Tewari, R.N. and Verma, T.S. (2002). Screening of tomato varieties/breeding lines against buckeye rot (*Phytophthora nicotianae* var. *parasitica* Dast.). *Indian Journal of Hill Farming* 39-43.

Blaker, N.S. and Hewitt, J.D. (1987). Comparison of seedling and mature plant resistance to *Phytophthora parasitica* in tomato. *Hort. Science* 22: 103-105.

Bose, T.K., Kabir, J., Maity, T.K., Parthasarathy, V.A. and Som, M.G. (2002). *Vegetable Crops*. Vol. I. Naya Prakash, Calcutta, India, pp 1-153.

Boukema, I.W. (1983). Inheritance of resistance to foot and root rot caused by *Phytophthora nicotianae* var. *Breda* de Hann var. *nicotianae* in tomato (*Lycopersicon esculentum* Mill.). *Euphytica* 32: 103-109.

Bowers, J.H. and Mitchell, D.J. (1990). Effect of soil water matric potential and periodic flooding on mortality of pepper caused by *Phytophthora capsici*. *Phytopathology* 80: 1447-1450.

Bruck, R.I., Fry, W.E., Apple, A.E. and Mundt, C.C. (1981). Effect of protectant fungicides on the developmental stages of *Phytophthora infestans* in potato foliage. *Phytopathology* 71: 164-166.

Butler, E.J. (1920). Report of the Agricultural Research Institute and College, Pusa, 1919-1920.

Byrne, J.M., Hausbeck, M.S. and Latin, R.X. (1997). Efficacy and economics of management strategies to control anthracnose fruit rot in processing tomatoes in the Midwest. *Plant Disease* 81: 1167-1172.

Cafe-Filho, A.C. and Duniway, J.M. (1995). Effects of furrow irrigation schedules and host genotypes on *Phytophthora* root rot of pepper. *Plant Disease* 79: 39-43.

Cameron, H.R. (1962). Effect of hydrogen ion concentration on growth rates of *Phytophthora* sp. *Phytopathology* 52: 727.

Camino, V., Despestre, T. and Espinosa, J. (1987). Search for *Capsicum annuum* susceptibility to *Fusarium*. *Capsicum Newsletter* 6: 70.

Chang, H.S. (1983). Crop diseases incited by Phytophthora fungi in Taiwan. *Plant Protection Bulletin* 25: 231-237.

Chaudhry, M.N.A., Akhtar, A.S. and Khan, R.A.A. (1985). Phytophthora problem in chillies and its control. *Capsicum and Eggplant Newsletter* 14: 62-64.

Chavez-Alfaro, J.J., Zavalta, M.E. and Teliz, O.D. (1995). Integrated control of blight in pepper (*Capsicum annuum* L.) caused by the fungus *Phytophthora capsici* Leo. *Fitopatologia* 30: 47-55 [cf: *Review of Plant Pathology* 85: 610].

Cheema, D.S., Dhiman, J.S. and Singh, S. (1992). Multiple disease resistance in hot set tomato genotypes in relation to their economic performance. *Symposium on Integrated Disease Management and Plant Health*, September 25-26, 1992. Dr. Y.S. Parmar University of Horticulture and Forestry, Solan, India.

Chen, Y. and Katan, J. (1980). Effect of solar heating of soils by transparent polyethylene mulching on their chemical properties. *Soil Science* 130: 271-277.

Clark, W.S. and Ann, D.M. (1979). Tomato stem lesion caused by *Phytophthora nicotianae* var. *nicotianae. Plant Pathology* 28: 98.

Cohen, Y., Farkash, S., Reshit, Z. and Baider, A. (1997). Oospore production of *Phytophthora infestans* in potato and tomato leaves. *Phytopathology* 87: 191-196.

Conners, I.L. (1938). Seventeenth Annual Report. *Canadian Plant Disease Survey* 11: 87.

Corrales, C., Vargas, E. and Morura, M.A. (1990). The effect of organic matter on control of foot rot in sweet pepper (*Capsicum annuum*) caused by *Phytopthora capsici. Agronomia Costarricdenxe* 14(1): 9-13 [cf: *Review Plant Pathology* 71: 347].

Critopoulopes, P.D. (1954). Symptoms on tomato fruits incited by three *Phytophthora* species. *Phytopathology* 44: 551.

Cvjetkovic, B. and Jurjevic, Z. (1993). Occurrence of *Phytophthora* species in Goatia. *Fragmenta Phytomedica et hertbologica* 21: 45-56.

Dar, G.M., Dar, G.H., Baba, R.A. and Munshi, N.A. (2004). Causes and management of tomato fruit rots in Kashmir. *Applied Biological Research* 6: 22-26.

Davis, D.R., Kincard, R.R. and Rhodes, F.M. (1970). Mulches reduce soil temperature under tomato and tobacco plants in Florida State. *Horticultural Science* 83: 117-119.

Davis, R.M. (1989). Effectiveness of fosetyl-Al against *Phytophthora parasitica* on tomato. *Plant Disease* 73: 215-217.

Deepak, K. Gopal, K., Sengpta, A., Sahay, R.N., Kumar, D. and Krishma, G. (2003). Effective of IPM module for tomato in the farmers field of Ranchi District, 2003. *Journal of Research*, Birsa Agricultural University 15 125-128.

Dhaliwal, M.S. and Rattan, R.S. (1990). Genetic variability for buckeye rot resistance and horticultural trait in tomato. *Annuals of Biology* 6: 111-115.

Dhiman, J.S. (1999). *Bio-mass enhancement of vegetables seedlings through using biocontrol agents.* Paper presented at the National Symposium on 'Plant Disease Scenario' under Changing Agro-ecosystems' held on 29-30 October at HPKVV, Palampur, India.

Dickerson, G.W. (1996). Compost dressing helps chilli pepper. *Biocycle*, 37: 80-82.

Dodan, D.S. and Shyam, K.R. (1994). Pathogenic variation within *Phytophthora nicotianae* var. *parasitica* isolates from tomato. *Indian Journal of Mycology and Plant Pathology* 24: 196-201.

Dodan, D.S., Shyam, K.R. and Gupta, S.K. (1994). Integrated management of buckeye rot of tomato. *Plant Disease Research* 9: 141-144.

Dodan, D.S., Shyam, K.R. and Bhardwaj, S.S. (1995). Relative response of tomato cultivars/lines against buckeye rot under field conditions. *Plant Disease Research* 10: 133-136.

Dodd, J. (1979). Effects of pests and diseases on the yield and quality of tomatoes in Papua New Guinea. *Papua New Guinea Agricultural Journal* 30: 53-59.

Dukes, P.D. and Apple, J.L. (1968). Inoculum potential of *Phytophthora parasitica* var. *nicotianae* as related to factors of the soil. *Tobacco Science* 12: 200-207.

Elad, Y., Katan, J. and Chet, I. (1980). Physical, biological and chemical control integrated for soil borne diseases in potatoes. *Phytopathology* 70: 418-422.

Elenkov, E. (1977). *Phytophthora capsici* on peppers in greenhouses. *Acta Horticulturae* 58: 401-404.

Engelhard, E.W. (1974). A serious new rot initiated by *Phytophthora parasitica. Plant Disease Reporter* 57: 669-672.

Engelhard, E.W. and Pluetz, R.C. (1979). *Phytophthora* rot, an important disease of poinsetlia (*Euphorbia pulcherrima*). In: *Proceedings of the Florida State Horticultural Society* 92: 348-350.

Engelhard, E.W. (1990). *Soil Borne Plant Pathogens-Management of Diseases with Macro and Micro Elements.* Scientific Publishers, Jodhpur, Rajasthan, India, p. 217.

Ersek, T., Schoelz, J.E. and English, J.T. (1994). PCR amplification of species DNA sequences can distinguish among Phytophthora species. *Applied and Environmental Microbiology* 60: 2616-2621.

Erwin, D.C., Zentmeyer, G.A., Galindo, J. and Niederhauser, J.S. (1983). Variation in the genus Phytophthora. *Annual Review of Phytopathology* 1: 375-396.

The American Phytopathological Society, St. Paul, Minnesota, USA, pp 562.

Everett, P.H. (1971). Evaluation of paper and polyethylene coated paper mulches and fertilizer rates for tomatoes. *Proceedings of Florida State Horticultural Society* 84: 124-128.

Fageria, M.S. (1994). Studies on developing tomato hybrids with multiple disease resistance. Ph.D Thesis submitted to Dr. Y.S. Parmar University of Horticulture and Forestry, Solan, H.P., India, p. 99.

Fageria, M.S., Dhaka, R.S. and Jat, R.G. (1998). Field evaluation of tomato genotypes for resistance to buckeye rot and Alternaria blight. *Journal of Mycology and Plant Pathology* 28: 52-53.

Felix, E.L. (1948). Buckeye rot resistance of tomatoes. *Phytopathology* 38: 569.

Fenn, M.E. and Coffey, M.D. (1989). Quantification of phosphonate and ethyl phosphonate in tobacco and tomato tissues. *Phytopathology* 79: 76-82.

Fernandez, M.C. (2003). *Phytophthora capsici* causal agent of wilt of *Capsicum annuum* in Chile. *Agricultura Teenica* 43: 91-93 [cf: *Review of Plant Pathology* 63: 2-3].

Ferreira, J.F., Nande, S.P. and Thompson, A.H. (1991). First report of *Phytophthora nicotianae* var *nicotianae* on tomatoes in South Africa. *Phytophylactica* 23: 233-234.

Ferreyra, E.R., Tosso, T.J. and Fernandaz, M.F. (1984). Effect of irrigation water management on *Phytophthora capsici* causal agent of *Capsicum annuum* blight. *Agricultura Technica* 44: 319-324.

Fichtner, E.J., Benson, D.M., Diab, H.G. and Shen, H.D. (2004). Abiotic and biological suppression of *Phytophthora parasitica* in a horticultural containing composted swine wate. *Phytopathology* 94: 780-788.

Flett, S.P. (1986). *Phytophpthora nicotianae* var. *nicotianae* causing root and crown rot of direct seeded tomatoes in Victoria. *Australian Plant Pathology* 15: 11-13.

Fontma, D.A., Nono, W.R., Opena, R.J. and Gumedrae, D. (1996). Impact of early and late blight infections on tomato yield. *IVIS Newsletter* 1: 7-8.

Fugo, P.A. (2000). Role of organic pesticides and manures in management of some chilli diseases. *Journal of Mycology and Plant Pathology* 30: 96-97.

Fulton, J.P. (1954). Disease control increases tomato yield. *Arhans Fm. Research* 3: 6.

Fulton, J.P. and Fulton, N.D. (1951). *Phytophthora* root rot of tomatoes. *Phytopathology* 41: 99-103.

Gallegly, M.E. and Niederhauser, J.S. (1959). Genetic controls of host parasite interactions in the phytophthora late blight disease, pp 168-182. In: *Plant Pathology: Problems and Progress* (Eds. C.S. Holton, G.W., Fisher, R.W., Fulton, H. and Hart, S.E.A). Academic Press New York USA, p. 350.

Goidanich, G. (1936). Richerche sulle phytophthorae dee pomodero I. La *Phytophthora parasitica* Dast. Sul Pomoderol. *Bulletin Staz. Pat. Veg.* Rome, W.S. 16: 115-138 [cf: CAB Abstract 1990-92].

Grote, D. and Buesi, C. (1998). Chemical cultural of *Phytophthora nicotianae* on tomato with fosetyl-aluminium in hydroponics. *Mycologia* 63: 78-87.

Gupta, S.K. and Shyam, K.R. (1998). Role of non-foliage application of fungicides in the management of buckeye rot of tomato. *Journal of Mycology and Plant Pathology* 28: 67-68.

Gupta, V.K. and Paul, Y.S. (2001). Diseases of Vegetable Crops. Kalyani Publishers, Ludhiana, India, p.276.

Hadar, Y. and Mandelbaum, R. (1992). Suppressive compost for biocontrol of soilborne plant pathogens. *Phytoparasitica* 20: 113-116.

Hall, G. (1993). An integrated approach to the analysis of variation in *Phytophthora nicotianae* and a redescription of the species. *Mycological Research* 97: 559-574.

Harender, R. (2004). Effect of nelarization of farmyard manure amended soil for management of damping off. *Indian Journal of Agricultural Sciences* 74: 425-429.

Harender, R., Kapoor, I.J. and Raj, H. (1997). Possible management of *Fusarium* wilt of tomato by soil amendments with composts. *Indian Phytopathology* 50: 87-395.

Harris, A.R. and Nelson, S. (1999). Progress towards integrated control of damping of disease. *Microbiological Research* 154: 123-130.

Hasan, M.S. (1989). Soil solarization by nelar heating in Iraq. *Arab Journal of Plant Protection* 7: 122-125.

Hewitt, J.D., Blaker, N.S., Damon, S.E. and Bennett, A.B. (1987). The UCD processing tomato breeding programme. *Acta Horticulturae* 200: 83-89.

Hoitink, H.A.J., Inbar, Y. and Buchm, M.J. (1991). Status of compost amended–posting mixes naturally suppressive to soilborne diseases of floricultural crops. *Plant Disease* 75: 869-873.

Homma, Y., Kubo, C., Ishii, M. and Ohata, K.I. (1979). Mechanism of suppression of tomato Fusarium wilt by amendments of chick manure to soil. *Bulletino Shikoku Agriculture Experimentation Station* 34: 103-121 [cf: *Review of Plant Pathology* 60: 1078].

Hotson, J.W. and Hartge, L. (1923). A disease of tomatoes caused by *Phytophthora mexicana*. *Phytopathology* 13: 520-531.

Hoy, M.W. and Ogawa, J.M. (1984). Toxicity of the surfactant Nacconol for decay–causing fungi of fresh market tomatoes. *Plant Disease* 68: 699-703.

Huber, D.M. (1980). The role of nutrition in defense, pp 381-406. In: *Plant Pathology–An Advanced Treatise*, Vol. 5 (Eds. J.G. Horsfall and E.B. Cowling). Academic Press, NY, USA.

Issiki, M. (1994). Control of bacterial spot disease by plastic rain–shelter in Paraguay. *Japanese Journal of Tropical Agriculture* 38: 232-238.

Jain, S.S., Sharma, S.L. and Juneja, S.L. (1961). Studies on buckeye rot of tomato–new to Himachal Pradesh. *Proceedings of Indian Science Congress Association* 50: 351-352.

Jimenez, J.M., Bustamante, E., Bermudez, W. and Gamboa, A. (1990). Identification and evaluation of sweet pepper lines resistant to fungal blight in Costa Rica. *Turrialba* 40: 228-234 [cf: *Review of Applied Pathology* 71: 210].

Jones, C.W. and McCarter, S.M. (1974). Etiology of tomato fruit rots and evaluation of cultural and chemical treatments for their control. *Phytopathology* 64: 1204-1208.

Jones, J.B., Jones, J.P., Stall, R.E. and Zitter, T.A. (1997). *Compendium of Tomato Diseases*. The American Phytopathological Society, St. Paul, Minnesota, USA, p. 1.

Joshi, A., Gupta, S.K. and Kohli, U.K. (2004). Resistance to septoria leaf spot (*Septoria lycopersici*) and buckeye fruit rot (*Lycopersicon esculentum*). *Indian Journal of Agricultural Sciences* 74: 194-199.

Kadian, O.P. and Kaur, P. (1999). Non-chemical management of pea wilt root rot complex–a case study paper presented at the National Symposium on 'Plant Disease Scenario under Changing Agroecosystems' held on 29-30 October at HPKVV, Palampur, India.

Kakar, K.L., Nath, A. and Dogra, G.S. (1980). Control of tomato fruit borer, *Heliothis armigera* Hubner under mid hill conditions. *Pesticides* 14: 11-13.

Kannaiyan, J. and Nene, Y.L. (1979). Effect of crop rotation and planting time on the incidence of wilt in lentil. *Indian Journal of Plant Protection* 7: 114-118.

Kapoor, J.I. (1988). Fungi involved in tomato wilt syndrome in Delhi, Maharashtra and Tamil Nadu. *Indian Phytopathology* 41: 208-213.

Kashina, B.D. (2004). Evaluation of tomato (*Lycopersicon esculentum* Mill.) cultivars for resistance to late blight disease of tomato caused by *Phytophthora infestans* (Mont.) deBary. *Tests of Agrochemicals and Cultivars* 25: 16-17.

Katan, J., Greenberger, A., Alen, H. and Grinstein, A. (1976). Solar heating by polyethylene mulching for the control of diseases caused by soil borne pathogens. *Phytopathology* 66: 683-688.

Katan, J. (1980). Solar heating of the soil for the control of pink root and other soil borne diseases in onions. *Phytoparasitica* 8: 39-50.

Katan, J.9 1995). Soil solarization–a non-chemical tool in plant protection. *Indian Journal of Mycology and Plant Pathology* 25: 46-47.

Kaufman, D.D. and Williams, L.E. 1964. Efect of mineral fertilization and soil reaction on soil fungi. *Phytopathology* 54: 134-139.

Kaur, S., Singh, S., Kanwar, J.S. and Cheema, D.S. (1988). Variability in tomato for resistance to late blight under field conditions. *Indian Phytopathology* 41: 486-487.

Kaushal, R.P. and Sharma, C. (1995). Effect of organic amendments in the management of lentil wilt. In: National Symposium on *"Recent Trends in the Management of Biotic and Abiotic Stresses in Plants"* held on 2-3 November at HPKVV, Palampur, India.

Kelman, A. and Cook, R.J. (1977). Plant pathology in the People's Republic of China. *Annual Review of Phytopathology* 17: 409-429.

Khalid, M., Shawkat, A., Mohammad, A., Ahmad, S., Majid, K., Ali, S., Aslam, M. and Saleem, A. (1995). Studies on late blight of tomato caused by *Phytophthora infestans* (Mont.) de Bary. *Pakistan Journal of Phytopathology* 7: 128-131.

Kim, C.H. (1989). Soil borne diseases. Phytophthora blight and other diseases of red pepper in Korea. *Republic Extension Bulletin* 302: 10-17 [cf: *Review of Plant Pathology* 70: 385].

Kim, J.Y., Ya, Y.K., Song, Y.H., Kim, J.Y. and Yi, Y.K. (1998). Plant diseases on greenhouse crops in Kyeongbuk areas. *Korean Journal of Plant Pathology* 14: 41-45.

Kiraly, Z. (1976). Plant disease resistance as influenced by biochemical effect of nutrient fertilizers. *Phytopathology* 34: 33-46.

Kohli, V.K., Dev, H. and Thakur, M.C. (1996). Classification of tomato germplasm for resistance to buckeye rot (*Phytophthora nicotianae* var. *parasitica*) and fruit size. *Annals of Agricultural Research* 17: 244-248.

Konam, J.K. and Guest, D.I. (2002). Leaf litter mulch reduces the survival of *Phytophthora palmivora* under cocoa trees in papua New Guinea. *Autralian Plant Pathology* 31: 381-383.

Kozik, E., Foolad, M.R. and Jones, R.A. (1991). Genetic analysis of *Phytophthora* rot in tomato (*Lycopersicon esculentum* Mill.). *Plant Breeding* 106: 27-32.

Kreutzer, W.A., Bodine, E.W. and Durrell, L.W. (1940). Curcurbit disease and rot of tomato fruits caused by *Phytophthora capsici*. *Phytopathology* 30: 974.

Kreutzer, W.A. and Brayant, L.R. (1944). A method of producing an epiphytotic of tomato fruit rot in the field. *Phytopathology* 34: 845-847.

Kumar, V., Singh, B.M. and Sugha, S.K. (1997). Reaction of tomato genotypes to factorial wilt and *Phytophthora* root rot. *Plant Disease Research* 12: 90-94.

Kuruzwinska, H. and Kuruzwinska, J. (1991). Effect of some agrotechnical factors on leaf and tuber infection of selected potato cultivars caused by *Phytophthora infestans* (Mout.) de Bary. *Journal of Ogronicto* 19: 81-90 [cf: *Review of Plant Pathology* 73: 1026].

Lal, S.D. and Singh, P.K. (2002). To evaluate the influence and modified micro-climate on growth, development and yield of tomato fruits. *Indian Journal of Dryland Agricultural and Development* 17: 115-119.

Leonian, L.H. and Geer, H.L (1929). Comparative value of the size of *Phytophthora sporagia* under standard conditions. *Journal of Agricultural Research* 39: 293-311.

Loest, F.C. (1948). Plant diseases and cultural operations. *Farm Magazine of South Africa* 23: 304-307.

Lopez, H.C.J., Perez, J.R.M. and Meleno, V.J.M. (1997). Effect of soil solarization on the control of *Phytophthora* root rot in Avocado. *Plant Pathology* 46: 329-340.

Lukode, G.M. and Rane, M.S. (1989). Effect of organic and inorganic soil amendments on pre- and post-emergence rot and yield of safflower. *Indian Phytopathology* 42: 301.

Lumesden, R.D., Lewis, J.A. and Milliner, P.D. (1983). Effect of composted sewage sludge on several soil borne pathogens and diseases. *Phytopathology* 73: 1543-1548.

Malla, S. and Sah, D.N. (2002). Use of cultural and chemical techniques in managing septoria leaf spot and buckeye rot disease in tomato. *Working Paper Lumle Agricultural Research Centre* 4: 1-8.

Matheron, M.E. and Matejka, J.C. (1990). Differential virulence of *P. parasitica* recovered from citrus and other plants to rough lemon and tomato. *Plant Disease* 74: 138-140.

Matsuzaki, M., Kan, M. and Kiso, A. (1980). Phytophthora rot in Kyushu Japan. *Annals of the Phytopathological Society of Japan* 46: 179-184.

Mehrotra, M.D. (1997). *Diseases of Paulowmia and their management*. *Indian Forester* 123: 66-72.

Millner, P.D., Ringer, C.E. and Maas, J.L. (2004). Suppression of strawberry root disease with animal manure composts. *Compost Science and Utilization* 12: 298-307.

Miyasaka, S.C., Hollyer, J.R. and Kodani, L.S. (2001). Mulch and compost effects on yield and corm rots of taro. *Field Crops Research* 71: 101-112.

Narayanaswamy, H., Pradeep, S., Chanrashekhar, S.Y. and Basavaraja, M.K. (2004). Integrated management of black shank disease of tobacco in field crop. *Plant Disease Research* 9: 66-67.

Nedumaran, S. and Vidhyasekaran, P. (1981). Control of *Fusarium semitectum* infection in tomato seed. *Seed Research* 9: 28-31.

Nedumaran, S. and Vidhyasekaran, P. (1982). Damage caused by *Fusarium semitectum* in tomato. *Indian Phytopahtology* 35: 322-323.

Nelson, P.E. (1981). Life cycle and epidemiology of *Fusarium oxysporum*. In: *Fungal Wilt diseases of Plants*, pp. 51-78 (Eds. M.E. Mace, A.A. Bell and C.H. Bechma). Academic Press, New York, USA.

Nema, S. and Sharma, N.D. (1997). Phytophthora blight of Chrysanthemum. *Journal of Mycology and Plant Pathology* 27: 345.

Nema, S. and Sharma, N.D. (1999). *Phytophthora nicotianae*: incitant of blight of bougainvillea. *Journal of Mycology and Plant Pathology* 29: 262.

Obrero, F.C. and Aragaki, M. (1965). Some factors influencing infection and disease development of *Phytophthora parasitica* Dast. on tomato. *Plant Disease Reporter* 49: 327-331.

Olunloyo, O.A. and Adeniji, M.O. (1976). The influence of NPK fertilizers on the root and stem rot disease of Rosella caused by *Phytophthora parasitica* var. *nicotianae*. *Nigerian Journal of Plant Protection* 2: 84-87.

Padmodaya, B. and Reddy, H.R. (1999). Effect of organic amendments on seedling disease of tomato caused by *Fusarium oxysporum* f. sp. *lycopersici*. *Indian Journal of Mycology and Plant Pathology*, 29: 38-41.

Padmodaya, B. (2003). Effect of soil amendments at different concentrations and incubation periods of decomposition against *Fusarium* wilt of tomato. *Journal of Mycology and Plant Pathology* 33: 317-319.

Pal, M. and Grewal, J.S. (1976). Effect of NPK fertilizers on *Phytophthora* blight of pigeon pea. *Indian Journal of Agricultural Sciences*, 46: 32-35.

Pane, A., Agosteo, G.E. and Cacciola, S.O. (2000). *Phytophthora* species of tomato in southern Italy. *Bulletin–DePP* 30: 251-255.

Patel, B.N., Patel, K.D. and Patel, H.R. (1996). Integrated management of damping–off in fidi tobacco nursery. *Tobacco Research* 22: 14-17.

Patel, D.J. (2001). Soil solarization for management of soil borne plant diseases. *Journal of Mycology and Plant Pathology* 31: 1-8.

Paulus, A.O., Otto, W.H. and Koboyaski, R. (1983). Fungicide for late blight in tomato. *California Agriculture* 37: 8-9.

Perez, P.M.C., Hemandez, M.A. Pulidol, L. and Mayca, S.S. (1989). Epidemiology of the soil fungus *Phytophthora nicotianae* var. *parasitica* in tomato cultivation. *Centro Agricola* 16: 53-59.

Perez, P.M.C., Torres, F., Malina, F. and Verona, G. (1986). Resistance to buckeye rot (*Phytophthora parasitica* Dast.) in varieties of tomato (*Lycopersicon esculentum* Mill.) *Centro Agricula*, 13: 3-16.

Pullman, G.S., Deray, J.E., Garber, R.H. and Weinhold, A.R. (1981). Soil solarization on *Verticillum. Phytopathology*, 61: 954-959.

Raicu, C. and Satan, G. (1973). *Phytophthora parasitica* Dast. Its symtomatology and culture. *Revista de Horticltura si viticulture* 22: 65-70.

Raicu, C., Satan, G. and Costache, M. (1974). Control of *Phytophthora parasitica* Dast. *An. Inst. Care.Legumicult, Si. Floricut* 3: 349-356.

Raicu, C. and Satan, G. (1976). Technology of integrated control of the fungus *Phytophthora parasitica* Dast. on tomato. *Horticultura* 25: 46-50.

Rajan, P.P and Sharma, Y.R. (2000). Effect of organic soil amendments and chemical fertilizers on root rot pathogen (*Phytophthora capsici*) of spices and aromatic plants challenges, and opportunities in the new century, pp. 249-253. Centennial conference on spices and aromatic plants, Calicut, Kerala, India, 20-23 Sept. 2002. In: Indian Society of Spices, Calicut, Indian.

Ramakrishan, G. and Jeyarajan, R. (1986). Biological control of seedling disease of cotton caused by *Rhizoctonia solani*. In: *Management of Soil borne Fungal Diseases of Crop Plants*. Department of Plant Pathology, CPP's TNAU of Coimbatore, India, pp. 45-48.

Ramakrishnan, T.S. and Soumini, C.K. (1947). Fruit rot of tomatoes caused by *Phytophthora palmivora, Proceedings of Indian Academy of Science* 25: 39-42.

Ramirez, L., Duran, A. and Mora, D. (1998). Integrated control of root rot of papaya fruit (*Phytophthora* sp.) under nursery conditions, *Agronomia mesoamericanhe* 9: 72-80 [cf: CAB Abstracts, 2000-2001].

Rao, R.N.M. (1984). Growth and sporangial production in *Phytophthora nicotianae* var. *parasitica. Indian Phytopathology* 37: 633-636.

Rapp, L. and Richter, J. (1982). Effect of propamocarb-hydrochloride (previeur N) on several isolates of some Phytophthora species *in vitro* and *in vivo*. *Zeitschrift-fur-Pflanzenkrank-Leiten* 89: 487-497.

Rattan, R.S and Saini, S.S. (1979). Production and management of tomato. *Vegetable Science* 6: 21-26.

Ravindra, K., Srivastava, B.K. and Kumar, R. (1997). Effect of different mulch materials on the soil temperature and moisture in winter tomato *Crop Research* 14: 137-141.

Ravindar, K., Srivastava, B.K. and Kumar, R. (1998). Influence of different mulch materials on yield and quality of winter tomato. *Biological Sciences* 68: 279-282.

Richardson, L.T. (1941). A Phytophthora tomato disease new to Ontaria *Canadian Journal of Research* 19: 446-483.

Ristaino, J.B. and Duniway, J.M. (1989). Effect of preinoculation and postinoculation water stress on the serenity of *Phytophthora* root rot in processing tomatoes. *Plant Disease* 73: 349-352.

Ristaino, J.B., Duniway, J.M. and Marois, J.J. (1989). Phytophthora root rot and irrigation schedule influence growth and phenology of processing tomatoes. *Journal of the American Society or Horticultural Science* 114: 556-561.

Robertson, G.I. (1973). Fungicides for control of damping–off and root rot of seedlings caused by *Phytophthora. Seed and Nursery Trader* 71: 1-9.

Rodrigues, C., Ribeiro, L.G., Lopes, J.C., Fruitas, F.S. de and Azenedo, L.A.S. de. (2000). Efficacy of metalatyl to control tomato late blight *Horticultura Brasileina* 18: 65-67.

Rosenbaum, J. (1920). Infection experiments on tomato with *Phytophthora terrestria* and a hot water treatment of the fruit. *Pytopathology* 10: 101-125

Roy, H.S.G. (1997). Screening of fruits of different varieties of brinjal for varietal resistance and reaction to other fruit vegetables against *Phytophthora nicotianae* var. nicotianae *Environment and Ecology* 15: 70-73.

Roy, J.K., Sengupta, D.K., Dasgupta, B., Sen, C. (2002). Effect of organic and inorganic fertilizers on yield and disease incidence of major diseases of betelvine. *Horticultural Journal* 15: 59-69.

Rughoo, M., Govinder, N. and Gowrea, B. (1997). Yield and characteristics of cooking tomato var. Sirius. *Revue Agricole et Sucrire de Ille Maurice*, 76: 38-43 [cf: CAB Abs. UD 981016].

Saha, L.R. (2002). *Handbook of Plant Protection*. Kalyani Publishers, Ludhiana, India, p. 928.

Satour, M.M., El-Sherif, E.M., El-Ghareeb, L., El-Haddad, S.A. and El-Wakil, H.R. (1991). Achievements of soil solarization in Egypt. *FAO Plant Production and Protection Paper* 109: 200-212.

Sawant, I.S., Sawant, S.D and Nanaya, K.A. (1995). Biological control of *Phytophthora* root-rot of coorg mandarun (Citrus reticulata) by Trichoderma species grown on coffee waste. *Indian Journal of Agricultural Science* 65: 842-846.

Sehgal, S.P. and Prasad, N. (1966). Variation in the pathogenicity of single zoospores isolates of sesamum *Phytophthora. Indian Phytopathology* 19: 154-158.

Seneviratne, G., Holm, L.H.J. van and Kulasooniya, S.A. (19980. Quality of different mulch materials and their decomposition and N release under low moisture regimes. *Biology and Fertility of Soils* 26: 136-140.

Sharma, A.K. (1992). Integrated management of buckeye rot of summer tomatoes in the hills. *Indian Journal of Mycology and Plant Pathology* 22: 287-288.

Sharma, R.C. and Sharma, S.L. (1982). Cultural variability within isolates of *Phytophthora nicotianae* var *parasitica* from different hosts. *Indian Journal of Mycology and Plant Pathology* 12: 57-59.

Sharma, R.C. and Sharma, S.L. (1987). Variability in *Phytophthora micotianae* var. *parasitica. Journal of Research Punjab Agricultural University* 24: 49-452.

Sharma, R.C., Hari, S.S. and Singh, H. (1992). Occurrence of *Phytophthora nicotianae* var. *parasitica* on hybrid tomato in Punjab. *Plant Disease Research* 7: 90-91.

Sharma, S.L. (1971). *Buckeye rot of tomato caused by Phytophthora parasitica* Dast. Ph.D thesis, Punjab University, Chandigarh, India p. 168.

Sharma, S.L. (1974). Buckeye rot-a destructive disease to tomato in Himachal pradesh. *Pesticides* 8: 31-32.

Sharma, S.L., Sohi, H.S. Chowfla, S.C. and Peshin, S.N. (1974). Note on the reaction of tomato varieties in field to the causal organism of buckeye rot disease. *Indian Journal of Agricultural Sciences* 44: 323-324.

Sharma, S.L., Chowfla, S.C., Sohi, H.S. and Sharma, M.M. (1975). Factors affecting resistance of tomato varieties to buckeye rot (*P. parasitica* Dast.). *Indian Journal of Experimental Biology* 13: 323-325.

Sharma, S.L. and Thapa, C.D. (1976). Effect of pre-harvest fungicidal sprays on storage rots of tomato. *Indian Journal of Mycology and Plant Pathology* 6: 78-79.

Sharma, S.L., Chowfla, S.C. and Sohi, H.S. (1976a). Production of oospores in *Phytophthora nicotianae* var. *parasitica. Kavaca* 4: 61-63.

Sharma, S.L., Chowfla, S.C. and Sohi, H.S. (1976b). Control of buckeye rot of tomato by cultural practicer. *Mycology and Indian Journal of Plant Pathology* 6: 51-54.

Sharma, S.L., Chowfla, S.C. and Sohi, H.S. (1976c). Chemical control of buckeye rot of tomato. *Indian Journal of Mycology and Pathology* 6: 130-134.

Sharma, S.L. and Bhardwaj, S.S. (1977). Relation of polyphenol oxidase activity in tomato to buckeye rot. *Indian Journal of Experimental Biology* 15: 688-689.

Sharma, S.L., Chowfla, S.C. and Sohi, H.S. (1978). Epidemiology and forecasting of tomato buckeye rot. *Indian Phytopathology* 31: 429.

Sharma, S.R. and Sohi, H.S. (1982). Control of common diseases in tomato. *Indian Journal of Horticulture* 39: 146-148.

Sharma, S.R. and Sohi, H.S. (1983). Buckeye rot of tomato as influenced by different levels of NP and K fertilizers, *Journal of Turkish Phytopathology* 12: 19-26. [cf. *Review of Plant Pathology* 63: 2100].

Sherbakoff, C.D. (1917). Buckeye rot of tomato fruit. *Phytopathology* 7: 119-129.

Sherf, A.F. and McNab, A.A. (1986). *Vegetable Diseases and their control.* John Wiley and Sons, New York, USA. p. 728.

Shyam, K.R. and Gupta, S.R. (1994). Management of buckeye rot and *Alternaria* leaf spot of tomato through, fungicide application. *Plant Disease Research* 9: 233-234.

Shyam, K.R. and Gupta, S.K. (1996a). Fungicidal control of buckeye fruit rot (*Phytophthora nicotianae* var. *parasitica*) and *Alternaria* leaf spots of tomato. *Indian Journal of Agricultural Sciences* 66: 309-312.

Shyam, K.R. and Gupta, S.K. (1996b). Management of buckeye fruit rot of tomato through non-foliage application of fungicides. *Plant Disease Research* 11: 202.

Shyam, K.R. and Gupta, S.K. (1998). *Buckeye rot of tomato.* Directorate of Research Dr. Y.S. Parmar University of Horticulture and Forestry, Solan–H.P. India, pp. 20

Simonds, A.O. and Kreutzer, W.A. (1944). Infection phenomenon in tomato fruit rot caused by *Phytophthora capsici Phytopathology,* 34: 813-817.

Singh, A.K., Meena, S.L., Williams, A.J., Banerjee, S.K. and Gupta, B.N. (1994). Effect of mulches and watering frequency in combination with Jalshakti on growth of *Albina procera* in skeletal soil. *Environment and Ecology* 12: 104-109.

Singh, B. and Srivastava, H.C. (1966). Damping–off of tomato seedlings. *Journal of Indian Botanical Society* 32: 1-16.

Singh, R. (2001). Effects of nutrient levels on the incidence of buckeye rot (*Phytophthora nicotianae* var. *parasitica*) in tomato hybrid. *Himachal Journal of Agricultural Research* 27: 42-45.

Singh, R.P., Singh, S. and Rana, M.K. (2001). Effect of host nutrition on early blight of tomato. *Journal of Mycology and Plant Pathology* 31: 248-250.

Singh, R.S. (1983). Organic amendments for root disease control through amendment of soil microbiota and the host. *Indian Journal of Mycology and Plant Pathology*, 13: 1-16.

Singh, R.S. (1999). *Disease of Vegetable Crops*. Oxford and IBH Publishing Co. Pvt. Ltd. New Delhi India p. 406.

Siradhana, B.S., Elleth, C.W. and Schmitthenner, A.F. (1968). Pathogenic and cultural variation in *Phytopthora nicotianae* var. *parasitaca* for green house plants.*Phytopathology* 58: 718-719.

Sivaprakashan, K. (1991). Soil amendments for crop disease management In: *Basic Research for Crop Disease Management* (Ed. P. Vidyasekharan), Daya Publishing House, New Delhi, India, p. 409.

Siviero, P. and Centolo, A. (2001). Cultivation technique for chilli pepper. *Information Agraria* 57: 33-38. (cf. CAB Abs. UD.)

Skadow, K., Schaffrath, J., Gohler, F., Drews, M. and Lanckow, H.J. (1984). *Phytophthora nicotianae* and *Pythium aphanidermatum* as pathogens in NFT culture. *Nachrichtenblate fur den pflanzenschutz in der DDR* 38: 169-172 [cf: Agaricola 98627].

Smith, P.M. Beach, B.G.W., Smith, J.M., Tomluysion, J.A., Faithful, E.M., Chalandon,A., Crisinel, P., Harriere, D., King, J.M., Gallinelle, G. and Absi, M. (1979). Control of phycomycetes. *Proceedings of British Crop Protection Conference Pests and Disease*, Brighton, England. Vol. 1 and 2. British Crop Protection Council, London, UK

Snapp, S.S. and Shenann, C. (1994). Salinity effects on root growth and senescence in tomato and the consequences in tomato and the consequences for security of *Phytophthura* rot infection. *Journal of the American Society for Horticulture Science* 119: 458-463.

Snowdon, A.L. (1991). *Post Harvest Diseases and Disorders of Fruits and Vegetable*. Welfare Scientific Ltd. Aylesburg, UK. 2: 76-77.

Sohi, H.S. and Sharma, S.R. (1989). Buckeye rot of tomato. pp 288-299. In: *Perspectives in Phytopathology* (Eds. V.P Agnihotri, N. Singh, H.S. Chause, U.S. Singh and T.S. Divivedi). Today and Tomorrow Printers and Publishers, New Delhi, India.

Sohi, H.S. and Sokhi, S.S. (1972). Combat disease of tomato. *Indian Horticulture* 17: 17-19.

Sokhi, S.S. and Sohi, H.S. (1974). Chemical control of buckeye rot of tomato. *Indian Phytopathology* 27: 444-445.

Sokhi, S.S. and Sohi, H.S. (1982). Assessment of losses in tomato caused by buckeye rot *Phytophthora nicotianae* var. *parasitica*. *Indian Phytopathology* 45: 675-676.

Sokhi, S.S., Thind, T.S. and Dhillon, H.S. (1993). *Late Blight of Potato and Tomato*. Punjab Agricultural University, Ludhiana, Punjab, pp 19.

Sokhi, S.S. (1994). Integrated approaches in management of vegetable diseases in India. *Indian Phytopathology* 47: 371-376

Springer, J.K. and Johnston, S.A. (1982). Black polyethylene mulch and *Phytophthora* blight of pepper. *Plant Disease* 66: 281.

Sridhar, T.S., Ullasa, B.A. and Sohi, H.S. (1975). Occurrence of a new disease on grape seedlings caused by *Phytophthora nicotianae* var. *parasitica* (Dastur) Waterhouse from India. *Current Science* 44: 406.

Srivastava, M.P. (1981). Buckeye rot in tomato. *Haryana Agricultural University Journal of Research* 11: 563-564.

Stapleton, J.J. and Devay, J.E. (1982). Effect of soil solarization on population of selected soil borne microorganisms on growth of deciduous fruit tree seedlings. *Phytopathology* 72: 223-326.

Stevens, C., Khan, V.A., Rodrigulz, K.R., Ploper, L.D. and Igwegbe, E.C.K. (2003). Integration of soil solarization with chemical, biological and cultural control for the management of soilborne diseases of vegetable. *Plant and Soil* 253: 493-506.

Stofella, P.J. and Khan, B.A. (2000). *Compost Utilization in Horticultural Cropping Systems*. Lewis Publishers, NY, USA.

Sutrisno, A.Y.A., Till, A.R. and Nursasongko (1995). The role of mulches and terking in crop production and water soil and nutrient management in Indiana. *Soil Organic Matter Management for Sustainable Agriculture*. A workshop held in Ubon, Thailand, 24-26 1994.

Szczech, M.M. and Brzeski, M.W. (1994). Vermicompost–fertilizer or biopesticide? *Journal of Phytopathology* 41: 77-83.

Szczech, M.M. (1999). Suppressiness of vermicompost against *Fusarium* wilt of tomato. *Journal of Phytopathology* 147: 155-161.

Tamura, M. and Taketani, K. (1977). Biology and control of club root of Chinese cabbage. *Ishikawa Prefecture Agriculture Experimentation Station Bulletin* 9: 1-26 [cf: *Annual Review of Phytopathology* 24: 94-114].

Taylor, W.H. (1924). Tomato culture–some further notes on disease control. *New Zealand Journal of Agriculture* 29: 40-43.

Thapa, C.D. and Sharma, S.L. (1976). Incidence of fruit rot diseases of tomato in Solan area. *Indian Journal of Mycology and Plant Pathology* 6: 83-84.

Thareja, R.K., Srivastava, M.P. and Singh, S. (1989). Epidemiology of buckeye rot of tomato caused by *Phytophthora nicotianae* var. *parasitica*. *Indian Phytopathology* 42: 58.

Thomas, K.M., Soumini, C.K. and Balakrishnan, M.S. (1947). Studies in the genus Phytophthora. *Proceedings of Indian Academy of Science* 27: 55-73.

Thomson, S.V. and Hine, R.B. (19720. A typical sporangium like structure of *Phytophthora parasitica*. *Mycologia* 64: 457-460.

Thurston, H.D. (1990). Plant disease management practices of traditional farmers. *Plant Disease* 74: 96-102.

Thyagarajan, P., Rao, A.V., Varadarajan, S. and Sundarajan, R. (1972). Studies on betelovine wilt disease. *Madras Agricultural Journal* 59: 187-189.

Tjamos, E.C. (1996). Chemical treatment and soil solarization for management of soil borne diseases. pp 262-281. In: *Management of Soil Borne Diseases* (Eds. R.S. Utkhede and V.K. Gupta). Kalyani Publishers, Ludhiana, India

Tompkin, C.M. and Tucker, C.M. (1941). Buckeye rot of tomato. *California Journal of Agricultural Research* 62: 467-474.

Tucker, C.M. (1931). Taxonomy of the genus *Phytophthora* de Bary. Research Bulletin, Missouri Agricultural Experimentation Station, No. 153.

Vanachter, A., Wambeke, E.V., Droogenbroeck, J.V. and Assche, C.V. (1983). Disease control on tomatoes grown in recirculating nutrient solution. *Modedelingen Van de Faculteit Landbou wwetenschappen Rijksuniversiteit Gent* 48(3): 617-623 [cf: CAB Abstracts 1984, 89642].

Vanitha, S. and Ramachandram, K. (1999). Management of late blight disease of tomato with selected fungicides and plant products. *South Indian Horticulture,* 47: 306-307.

Vawdrey, L.L., Martin, T.M. and Fareri, J-de. (2002). The potential of organic and inorganic soil amendments and biological control agent (*Trichoderma* sp) for the management of *Phytophthora* root rot of pawpaw in far northern Queensland. *Australian Plant Pathology* 31: 391-399.

Velandia, J., Galindo, R.P. and Morino, C.A. (1998). Poultry manure evaluation in the control of *Plasmodisphora brassicae* in cabbage. *Agronomia Columbia* 15: 1-6 [cf: Hort. CD. 85362].

Verma, T.S., Ramesh, C., Lakhanpal, K.D., Sharma, S.C. and Chand, R. (1994). Use of fungicides in controlling fungal fruit rots and increasing fruit and seed yield in tomato. *Himachal Journal of Agricultural Research* 20: 44-48.

Vinod, K., Dharmatti, P.R. and Bhagavanthgoudar, M.H. (2004). Economics of weed control in tomato production. *Scientific Horticulture* 9: 93-96.

Voorst, G.V., Os, E.A. and Zadoks, J.C. (1987). Dispersal of *Phytophthora nicotianae* on green tomatoes by nutrient film technique in a green house. *Netherlands Journal of Plant Pathology* 93: 195-199.

Wager, V.A. (1935). Brown rot of tomato fruits due to *Phytophthora parasitica* Dast. *South African Journal of Science* 32: 235-237.

Walker, J.C. (1952). *Disease of Vegetable Crops*. McGraw Hill Book Company, New York, USA, pp 489-491.

Warner, J., Haq, X. and Zhang, T.Q. (2002). Effects of row arrangement and plant density on yield and quality of early, small vined processing tomatoes. *Canadian Journal of Plant Science* 82: 765-770.

Washington, W.S. and McGee, P. (2000). Dimethomorph soil and seed treatment of potted tomatoes for control of damping off and root rot caused by *Phytophthora nicotianae* var. *nicotianae. Australian Plant Pathology* 29: 46-51.

Washington, W.S., McGee, P., Flett, S.P., Jerie, P.H. and Asheroft, W.J. (2001). Cultivars and fungicides affect Phytophthora rot in processing tomatoes. *Australian Plant Pathology* 30: 309-315.

Waterhouse, G.M. and Blackwell, E.M. (1954). *Key to the species of Phytophthora recorded in the British Isles*. Mycology Paper No. 57. CMI, Kew, Survey, England, p. 30.

Waterhouse, G.M. (1963). Key to the species of *Phytophthora* de Bar. Mycological Paper No. 92, CMI, Kew, Survey, UK, pp 32.

Weimer, J.L. (1920). The distribution of buckeye rot of tomato. *Phytopathology* 10: 172.

Wei, C.T. and Cheo, P.C. (1944). Diseases of tomato in the vicinity of Chengtu. *China Journal of Agricultural Sciences* 1: 288-291.

Welch, E.J. (1949). Year's record of research in Mississippi farm problems. *Mississippi Farm Research* 12: 3-8.

Weststeijn, G. (1973). *Phytophthora nicotianae* var. *parasitica* in tomato. *Netherlands Journal of Plant Pathology* 79: 1-86.

Williams, P.H., Orchard, O.B., White, H.L., Oyler, E., Ainsworth, G.C. and Read, W.H. (1937). *Plant Diseases Report of Experimental Research Station*, Chestn 1836: 42-58.

Wilson, T.J. (1956). Comparative control of buckeye rot of tomato by various fungicides. *Phytopathology* 46: 511-512.

Yang, C.Y. (1979). Bacterial and fungal diseases of tomato. In: *Proceedings of First International Symposium on Tropical Tomato*. 23-27 October, 1979. Shandhu Taiwan, Republic.of China, pp. 111-113.

Zahara, A., Vimla, P., Ismail, E.A.E., Hasjadi, S.S. (1994). Management of bris sandy soil for chilli and tomato production in Malayasia. *Acta Horticulturae* 369: 299-305.

Sustainable Disease Management of Agricultural Crops (2011) *Pages* **118–132**
Editors: **S.K. Biswas and S.R. Singh**
Published by: **DAYA PUBLISHING HOUSE, NEW DELHI**

Chapter 7

Brown Spot Disease of Rice: Present Status and Management Options

☆ ***Chinmay Biswas[1], S.K. Biswas[2] and S.S.L. Srivastava[2]***
[1]*Central Research Institute for Jute and Allied Fibres, Barrackpore, 24-Parganas (N), Kolkata – 700 120*
[2]*Department of Plant Pathology, C.S. Azad University of Agriculture and Technology, Kanpur – 208 002, U.P.*

Rice (*Oryza sativa* L.) is the third most important food crop in the world after wheat and maize. It is cultivated in 114 countries and is the staple food for over half the world's population. Rice farms cover about 11 per cent of the world's arable area and world rice production is about 600 million tonnes. Although its cultivation is mainly concentrated in Asia where more than 90 per cent of the world's rice is grown and consumed, rice is also an essential staple food and a source of income for millions of others in Africa and South America. The top ten rice growing countries are China, India, Indonesia, Bangladesh, Vietnam, Thailand, Myanmar, Japan, Philippines and Brazil (Anonymous, 2006). In India, the slogan 'Rice is life' is most appropriate as this crop plays a vital role in our national food security. India has the largest area (43.5 mha) under rice cultivation with a total production of 91.0 million tonnes (2005-06) (Anonymous, 2007).

The rice crop is attacked by a number of diseases among which Brown spot caused by *Drechslera oryzae* Subramanian and Jain (*Heliminthosporium oryzae* Breda de Hann) is an important one occurring in almost all the rice growing areas of the world. In India, it occurs more or less every year in mild or severe form, occasionally taking epiphytotic form. This disease was one of the principal causes of great Bengal famine of 1942-43 which caused death of two million people (Padmanabhan, 1973).

Occurrence and Importance

Brown spot of paddy has a world-wide distribution and it has been reported in all rice-growing countries in Asia, America and Africa (Ou, 1972). The disease is of great importance in several countries including USA, Japan, Italy, Philippines, India, Sri Lanka and Turkey (Ocfemia, 1923; Chaiappelli, 1929; Surek, 1995; Rangaswami and Mahadevan, 1999). In India, the disease is widely distributed throughout the country specially in West Bengal, Orissa, Andhra Pradesh, Assam, Uttar Pradesh, Punjab, Bihar, Tamil Nadu, Karnataka and Haryana (Rangaswami and Mahadevan, 1999).

The pathogen *Drechslera oryzae* can infect paddy in all stages of crop growth viz, seedling, tillering, panicle initiation and grain filling. However, the occurrence of the disease is more when the crop approaches maturity (Padmanabhan and Ganguly, 1954). There are three phases of damage caused by this disease: poor germination of infected seeds, leaf infection resulting in reduction of effective leaf surface and shrivelling and poor setting of the seed (Singh, 2005). Normally the damage caused by the pathogen is not much serious, but under certain conditions enormous losses are caused when it can assume epiphytotic proportions.

Pedrozo *et al.* (1994) reported that the disease severity varied from 25-75 per cent in the western and eastern coasts of Zamboanga, Philippines. Primarily, brown spot of paddy is a seed borne disease and causes blight of seedlings grown from heavily infected seeds. Ocfemia (1924a) reported 10-58 per cent seedling mortality in the Philippines. Varietal trials in Puerto Rico recorded death of seedlings from infected seeds upto 15 per cent (Tucker, 1927). Nisikado and Miyake (1922) reported the disease incidence of about 90 per cent at seedling stage in the seed bed. Padmanabhan *et al.* (1948) also reported seedling blight leading to severe losses as a result of epiphytotics in India.

Zulkifli *et al.* (1991) reported that the increasing disease severity of the parent plant decreased seed germination by up to 40 per cent and reduced seedling height by about 3-20 per cent with substantial reduction in yield.

Yield losses from leaf infection or leaf spots are generally not very serious. But when the pathogen attacks the panicles including the grains, heavy economic losses are caused. The reduction in yield can be about 45 per cent with high infection and 12 per cent in moderate infection. However, there is no loss in yield in mild infection (Anonymous, 2003). During the brown spot epidemic in Bengal in 1942 the yield losses were to the tune of 50-90 per cent (Ghose *et al.*, 1960).

When the grains are infected, it lowers grain quality and weight. Bedi and Gill (1960) reported from Punjab that the disease caused 4.58-29.1 per cent loss in weight of rice grains. Yield loss due to this disease has been estimated to be around 0.15 million tonnes annually in Eastern India (Variar, 1996).

El-Kazzas and Osman (1989) reported that artificial inoculation of rice plants with the brown spot pathogen at flowering and milky stage reduced maximum yield, but almost no yield reduction was recorded when inoculation was made at maturity.

Taxonomy and Nomenclature of the Pathogen

In 1809, Link established the genus *Helminthosporium* with the type species *H. velutinum* Link ex Gray. The genus *Helminthosporium* was divided into two subgenera namely *Cylindro-Helminthosporium* and *Eu-Helminthosporium* on the basis of shape and germination of spores by Nisikado (1928). Later, Ito (1930) changed the subgenus *Cylindro- Helminthosporium* to *Drechslera*.

Luttrel (1951) recognized three major groups *viz.*, *H. sigmoideum* group with conidia borne on sterigmata, *H. gramineum* group having sessile cylindrical conidia with slightly tapering ends which germinate from any or all cells and *H. sativum* group with ellipsoidal to fusoid conidia which germinate by polar germ tube. Shoemaker (1959) differentiated the graminicolous species into two categories. The species that germinated from any or all the cells were included under *Drechslera* and those with fusoid phragmospore and polar germination were designated as *Bipolaris* with *B. maydis* as lactotype.

But Subramanian and Jain (1966) suggested that all the graminicolous *Helminthosporium* should be included in a single genus *Drechslera.*

The present fungus under study, causing brown spot disease of paddy was first described by Breda de Haan (1900) and named as *Helminthosporium oryzae.* Hori (1901) later described it in Japan and also used the name *H. oryzae.* In India, the disease was first reported by Sundaraman in 1922 with the causal organism as *H. oryzae.* Subramanian and Jain (1966) transferred *H. oryzae* to *Drechslera* as *D. oryzae.*

Ito and Kuribayashi (1927) found the perfect stage of the fungus in culture and gave it the name *Ophiobolus miyabeanus.* Drechsler (1934) considered the fungus to belong to his new genus *Cochliobolus* and Dastur (1942) formally transferred it to that genus. The name *Cochliobolus miyabeanus* (Ito and Kuribayashi) Drechsler ex Dastur is used at present.

Symptomatology and Host Range

The symptoms of the disease occur on the coleoptile, the leaves, the leaf sheaths and also the glumes. However, the most conspicuous symptoms appear on the leaves and the glumes.

Seedling blight occurs in the nurseries. Small brown spots develop on the coleoptile or leaf blade or sheath of older seedlings. When several spots are formed the leaves dry up and the seedlings get killed. On the coleoptile the spots are brown, small and circular to oval. They rarely take the form of long streaks. On the leaves the spots vary in size and shape from minute dots to circular, eye-shaped or oval spots measuring 1-14 x 0.5-3 mm (Singh, 2005). On susceptible varieties the spots are larger and reach 1 cm or more in length. Some times, numerous spots occur and as a result the leaf withers. Concentric lines or zones on the spot have been observed occasionally (Goto, 1954; Sato, 1965).

The younger spots are dark brown or purplish-brown. The larger spots are dark brown at the edge but towards the centre they may be pale yellow or grey. Sometimes the spots are surrounded by a yellowish halo and may coalesce with each other thus becoming irregular in shape. In case of severe infections the entire leaf may turn brown and dry out. The symptoms on the leaf sheath are similar to those on the leaf. El-Kazzas and Osman (1989) reported brown oval spots with grey centres on leaves and glumes as the most conspicuous symptoms of brown spot disease in paddy.

Dark brown spots appear on the glumes and in severe cases the greater portion or the entire surface of some of the glumes may be covered. Under favourable climatic conditions, dark brown conidiophores and conidia develop on the spots to give a velvety appearance. In such cases, the fungus may penetrate the glumes and leave blackish spots on the endosperm. The seeds sometimes become shrivelled and discoloured. Sachan and Agarwal (1995) reported seed-borne inoculum of *H. oryzae* causing light to dark brown dot like spots in the seed coat and endosperm of infected seeds. The presence of mycelium in infected rice seeds was demonstrated by Fazli and Schroeder (1966).

The pathogen has a wide host range which is evident from its presence in various cereals, grasses and other hosts. Nisikado and Miyake (1922) reported that some wild grasses, including *Cynodon dactylon* and *Digitaria sanguinalis,* were infected by the fungus upon artificial inoculation. Ocfemia (1923) also obtained infection of the pathogen on many species of grasses in 23 genera. Some strains of the fungus were reported to infect maize and naked barley, but wheat, common barley and oats were comparatively resistant (Tochinai and Sakamoto, 1937).

Thomas (1940) reported that the fungus readily attacked wheat and *Setaria italica,* slight infection was obtained on *Eleusine coracana,* oats and maize but none occurred on sorghum. Shaw (1921) found that a *Helminthosporium* sp. from wheat attacked rice and one from rice attacked maize, sorghum, oats, barley and sugarcane. The various species of grasses infected by artificial inoculation have seldom or never been found to be attacked under natural conditions. But the fungus has been successfully inoculated on *Setaria glauca, S. italica, Zea mays, Echinocloa frumentacea, Sorghum vulgare, Panicum miliaceum, P. milare* and *Euchlaena mexicana* and re-inoculated on rice (Padmanabhan, 1974).

Natural occurrence of *H. oryzae* on *Leersia hexandra* has been reported by Chattopadhyay and Chakrabarti (1953). Su (1936) noticed a *Helminthosporium* sp. similar to *H. oryzae* on *Panicum colonum* in close proximity to rice fields. Premlatha Dath and Chakrabarti, 1973 also reported that wild rice species are alternate hosts of the fungus.

Perpetuation and Spread

The pathogen is soil and seed-borne. The mycelium and conidia are carried over from season to season on the seed and diseased crop debris left in the field. However, the pathogen does not survive for long in natural soil (Hiremath and Hegde, 1985). Earlier, in 1953 Padmanabhan *et al.*, had studied the viability of *D. oryzae* in seeds, leaves, nodes and internodes, in stubbles left in the field after harvest and in the soil till the time of next sowing (December to July). It was found that the pathogen remains viable only in the seed during this period. Wani *et al.* (2004) isolated the fungus from rice seeds.

Primary infection through diseased seed is probably most common, although diseased seeds do not necessarily always give rise to infected seedlings. The coleoptiles are often infected from diseased seeds but the subsequently developed leaves may not show lesions due to rapid growth of the leaves under normal conditions. Leaf spots mostly arise from secondary infection (Thomas, 1940; Ganguly, 1946).

Padmanabhan *et al.* (1953) made aeroscope studies in India and reported that conidia of the fungus remains over rice fields during the sowing season or earlier (April to July). This air borne inoculum from some external source appears to cause primary infection. The sources may be perennial grass hosts and early sown or winter rice crops. There are reports that this disease occurs naturally on as many as 20 different wild species of *Oryzae.* A few collateral hosts like *Digitaria sanguinalis, Leersia hexandra, Echinochloa colona, Pennisetum typhoides, Setaria italica* and *Cynodon dactylon* on which the fungus is recorded through artificial inoculation, may serve as the sources of primary inoculum (Rangaswami and Mahadevan, 1999). Naimuddin *et al.* (1994) reported airborne inoculum of the pathogen during and after the threshing of paddy at the end of each crop season.

Although the pathogen overwinters on infected seeds, volunteer crops, infected plant debris and several weeds, the main source of infection is seed and air borne inoculum. Secondary infection is caused by the first formed spores on seedlings, which become wind borne and spread from one plant to another in the field. The number of these air borne conidia over rice fields are highest during October to January in India (Sreeramulu and Seshavataram, 1962).

Predisposing Factors

The disease development is favoured by certain predisposing factors which may be soil related, crop related or weather related.

Soil Related Factors

Nutritional deficiencies, primarily those of nitrogen and potash predispose rice to infection of *D. oryzae* (Misawa, 1955; Padmanabhan *et al.*, 1962). Chattopadhyay and Dickson (1960) reported that deficiency and excess of nitrogen both increased brown spot development. Vidhyasekaran *et al.* (1983) confirmed that excess nitrogen too increases the susceptibility to brown spot.

Kaur *et al.* (1984) demonstrated a non-significant positive correlation between nutrient status of N, P and Fe/Mn ratio and the disease expression. A positive significant correlation between Fe status and disease expression and a negative significant correlation between K, K/N ratio and Mg status and disease expression was also reported. Kaur *et al.* (1986) further reported a negative correlation between Ca and disease severity in clay and loam soils.

Presence of hydrogen sulphide in the root zone is reported to be associated with high disease intensity (Kaur *et al.*, 1979). Herrera Isla (1991) reported that low organic matter and reduced cationic exchange increased susceptibility to brown spot.

Leaching of nutrients from soil, low potassium status, upsetting of Fe/Mn ratio, low as well as excess of nitrogen, excessive soil moisture as well as dry condition are the important predisposing factors for disease development (Hemmi and Suzuki, 1931; Misawa, 1955; Baba, 1958; Chattopadhyay and Dickson, 1960; Chakrabarti and Padmanabhan, 1962; Padmanabhan *et al.*, 1962).

Crop Related Factors

The intensity of brown spot infection is more in transplanted rice than in direct seeded rice (Chattopadhyay and Chakrabarti, 1957). Youngest top most leaves of rice are most resistant and second leaves from the top are most susceptible to infection (Das Gupta and Chattopadhyay, 1975)

Susceptibility to *D. oryzae* increases with age of rice plants. The plants become most susceptible at the time of flowering (Padmanabhan and Ganguly, 1954).

Weather Related Factors

The optimum temperature for the mycelial growth of the fungus is 27-30°C and for conidial germination is 25-30°C. The formation of conidia takes place at temperatures between 5°C to 38°C and at pH 4-10 (Nisikado, 1928). Ocfemia (1924b) reported the cardinal temperatures for the growth of *D. oryzae* to be 16°C, 28°C and 40°C.

Chattopadhyay and Das Gupta (1965) reported that although the conidia formation can take place at temperatures ranging from 5°C-38°C, the optimum temperature is between 21°C-26°C. They also reported that a relative humidity of 92.5 per cent or above favours conidial production. It was reported that no infection of rice seedlings by *H. oryzae* occurred at relative humidity below 85 per cent (Katsura, 1937).

Studies on aerial dispersal of conidia revealed that maximum flight of conidia takes place at a wind velocity of 4.0-8.8 km/hr. Minimum temperature and relative humidity are also important but their effect is masked by the effect of wind velocity. Minimum temperature of 27°C-28.5°C, relative humidity 70-99 per cent, wind velocity of 4.0-8.8 km/hr and low to moderate rainfall of 0.4 to 14.4 mm were found most conducive factors for dispersal of conidia in atmosphere (Sreeramulu and Sheshavataram, 1962; Sreeramulu and Ramlingam, 1966; Gangopadhyay and Chattopadhyay, 1974).

El-Kazzas and Osman (1989) reported that the relative humidity, dew period and number of trapped spores showed positive correlation with disease development. Minnatullah and Abdus-Sattar (2002) reported that both temperature and humidity had significant effects on the severity of the disease. Minimum temperature of 7.5-12.3°C, coupled with high humidity (88-89 per cent) favoured high severity of brown spot. High relative humidity (86-100 per cent), optimum temperature between

16-30°C and leaf wetness of 8/24 hrs were found favourable factors for disease development (Anonymous, 2003).

Salive (1994) reported that factors which predispose rice plants to grain damage include high relative humidity during rainy season, alternate periods of humidity and dryness during flowering, prolonged rain during ripening, nutritional imbalance caused by deficiencies of K, P or an excess of N.

Management Options

The conventional management options include cultural practices, chemical control, biological control and use of resistant varieties.

Cultural Practices

Removal and destruction of diseased stubbles and plant remains after harvest is a very effective measure to control brown spot of rice because the pathogen survives in the diseased crop debris left in the field. Eradication of collateral hosts *viz., Leersia hexandra, Echinochloa colonum* etc. may be helpful in controlling the disease (Das Gupta, 1988).

Grummer and Roy (1966) reported that intervarietal mixture, *i.e.* planting alternate rows of varieties of varying reaction to different strains of the fungus reduced the spread of the pathogen.

As the disease is primarily seed borne, hot water treatment (53-54°C for 10-15 min) of the seeds could be an effective method to kill he pathogen. Nisikado (1928) started the treatment of seeds by hot water and many others experimented with the same methods.

Leaching of more mobile elements like K, Fe and Mn makes the rice crop more susceptible of brown spot. Soil amendments or foliar sprays which restore the balance and make the deficient ions available to the crop is a possible approach to avoid serious losses (Padmanabhan, 1963). To provide proper nutrition is the most important factor in controlling brown spot (Ou, 1972). Vidhyasekaran *et al.* (1983) reported that application of 87-110 kg N/ha through coated urea reduced the disease incidence. Gangopadhyay (1968) also observed earlier that disease intensity was low at moderate levels of nitrogen.

Tembhurnikar (1989) reported that green manuring with water hyacinth and *Ipomoea carnea* reduced the disease incidence. Varughese and Padma Kumari (1993) advocated organic farming for judicious management of the disease.

Chemical Control

Nisikado (1928) tried seed treatment with copper sulphate to control brown spot of paddy. Since then many workers have suggested seed treatment with various chemicals including fungicides and antibiotics to control this disease. Thirumalachar (1967) suggested a mixture of the antibiotic aureofungin (20 ppm) and copper sulphate (20 ppm) for rice seed treatment. Dharam Vir *et al.* (1970) recommended seed treatment with 0.3 per cent Dithane M-45.

Seed treatment with chemicals like agrimycin (0.01 per cent), Potassium chloride (1 per cent), di-ammonium phosphate (1.5 per cent) etc. significantly reduced brown spot disease of rice (Prasad *et al.*, 1978). Esuruoso and Joaquin (1981) reported that seed treatment with Agron SW (Toly/mercury acetate), Benlate (Benomyl), Dithane M-45, Furnasan 75 W (Thiram + BHC) etc. reduced seed infection of *D. oryzae*.

Seed treatment with Pyroquilon or Tricyclazole (0.2 per cent) can be recommended in areas where the disease is severe. Treatment with Carbendazim (0.2 per cent) or Mancozeb (0.3 per cent) would also be effective (Geetha and Sivaprakasam, 1993). Sachan and Agarwal (1995) reported that seed treatment with Bavistin+ Dithane M-45 and Bavistin + Thiram were superior to other treatments. Seed treatments with Captan, Thiram, Chitosan, Carbendazim, Mancozeb etc. have been found effective against seedling infection (Anonymous, 2003).

Viswanathan and Narayanasamy (1990) reported that seed treatment with Tricyclazole (0.4 per cent) followed by spraying with Mancozeb (0.25 per cent) + Tricyclazole (0.8 per cent) during tillering and late boot leaf stage provided better control of the disease. However, the efficacy was reduced when the seed treatment was omitted. Serghat *et al.* (2002) reported that seed treatment with Tricyclazole, Mancozeb and Carboxin + Thiram reduced the diseased severity.

Spraying of Carbendazim @ 0.1 per cent at 30, 50 and 70 days after transplanting reduced the disease severity (Alagarsamy *et al.*, 1986). Foliar spray of Mancozeb (0.25 per cent) at mid tillering, maximum tillering, late boot and early grain filling stage has been suggested for the control of the disease (Viswanathan and Narayanasamy, 1990). Chhetry (1993) reported that spraying of Dithane M-45 + Tricyclazole was effective to reduce the disease.

Muthusamy *et al.* (1988) reported that combinations of soil applications and foliar sprays of Copper sulphate (1 kg/ha to soil + 0.5 per cent foliar spray), Ferrous sulphate (25 kg/ha to soil + 0.5 per cent foliar spray) and Manganese sulphate were effective against the disease and the disease incidence was least with Ferrous sulphate treatment.

Biological Control

Butler observed the inhibitory effect of *Trichoderma* sp. on the growth of *Cochliobolus sativus*, perfect stage of *Drechslera sorokiniana* as early as in 1953. *Trichoderma viride* was reported to be antagonistic to *D. sorokiniana*, causal agent of spot blotch of wheat (Prasad *et al.*, 1978). Fernandez (1992) reported that *T. harzianum* reduced the incidence of *Cochliobolus sativus* in wheat and barley. *Chaetomium globosum* was also reported to inhibit the growth of *D. sorokiniana* (Mandal *et al.*, 1999) *Bacillus subtilis* was found to inhibit the condial germination of *Helminthosporium oryzae* and some other fungi.

Kamala Nayar and Vidhyasekaran (1998) reported that *Pseudomonas fluorescens* strain P1 inhibited the growth of brown spot pathogen *H. oryzae in vitro.* The talcum powder based formulation of the bacterial strain was also effective in controlling the disease when applied as seed treatment @ 4 g/kg (1.4 x 10^9 c.f.u.) seedling root dipping in 0.5 per cent (1.8 x 10^9 c.f.u./l) of the product and foliar spray @ 0.1 per cent (3.6 x 10^9 c.f.u./l).

Kumawat (2006) reported that *Trichoderma viride* and *T. harzianum* both inhibited the growth of *Drechslera oryzae in vitro.* Both the bioagents were effective in reducing the severity of brown spot in pot culture. However, *T. harzianum* performed better than *T. viride.*

Control by Artifungal Plant Products

Leaf extracts *Azadirachta indica, Datura metel, Ocimum sanctum, Vinca rosea, Lantana camara, Solanum melongena, Poliyalthia longifolia* and *Agel marmelos* showed antifungal activity against the brown spot pathogen, *H. oryzae* (Ganguly, 1994). Alice and Rao (1987) reported that leaf extract of *Mentha pipenita,* extract of *Allium sativum* and seed extract of *Piper nigrum* reduced the mycelial growth of the pathogen and were effective in reducing seed infection.

Turmeric in aqueous $NaHCO_3$ (10 g/l) (Gangopadhyay, 1998) as well as the rhizome extract of turmeric (Harish *et al.*, 2004) were effective against the development of brown spot disease. Rajappan *et al.* (2001) reported that neem oil formulations inhibited mycelial growth of the pathogen and controlled rice grain discoloration.

Spraying of *Allium sativum* extract (@ 50 per cent concentration) was reported to reduce the disease severity significantly (Raju *et al.*, 2004). It was also reported that the products of *Allium sativum, Bougainvillea spectabilis, Lantana camara, Azadirachta indica* and *Peosopis cineraria* inhibited the conidial germination of *H. oryzae*.

Use of Resistant Varieties

Before the introduction of dwarf rice varieties over large areas many varieties were tested for their reaction to brown spot disease. The varieties Ch 13, Ch 45, T 141, T 298-2A, Co 20, BAM 10, T-998 m, T-2112, T-2118 and T-960 were reported resistant to brown spot (Ganguly and Padmanabhan, 1959; Padmanabhan *et al.*, 1966).

Varieties, Padma and IR-24 are relatively resistant to the disease (Singh and Sharma, 1975). Earlier, Bedi and Gill (1961) reported varieties Russia 894, 5, 287, China 47, 63, 988, 996, 1007 and HR-47 resistant against *Helminthosporium* spot.

Vivekanandan *et al.* (1995) while testing some scented rice varieties against brown spot noted that hybrids were generally susceptible indicating that susceptibility was dominant over resistance. The most resistant parents were ADT 39 and Pusa Basmati-1. Least disease incidence was recorded on ADT 39 x Kasturi, IWP x Pusa Basmati-1 and IWP x Kasturi.

Singh (1988) tested as many as 118 indigenous rice varieties in Eastern Uttar Pradesh for their reaction to brown spot. Indigenous varieties *viz.*, Badam Farm, Badshah Pasand, Rambhog, Pankhali, Ghaghari, Bajarbang, Chinnaur, Hansraj, etc. were found moderately resistant.

Induced Resistance

Induced resistance refers to heightened resistance in a plant towards pathogens as a result of a previous treatment with a pathogen, an attenuated pathogen or a chemical that is not itself a pesticide (Deverall and Dann, 1995). There is enhancement in resistance in response to an extrinsic stimulus without a known alteration of the genome. The protection is based on the stimulation of defense mechanisms by metabolic changes that enable the plants to defend themselves more efficiently (Hammerschmidt and Kuc, 1995). The inducing agents could be either biotic or abiotic.

Resistance Induced by Bioagents

Pre-inoculation with Pathogen/Non Host Pathogens

Ganguly and Padmanabhan (1962) reported that rice plants grown from seeds soaked in culture extracts of *Helminthosporium oryzae* suffered much less damage from the same pathogen than the plants grown from untreated seeds.

Pre-inoculation with Avirulent/Mildly Virulent Isolates

There are many instances where inoculation with avirulent or mildly virulent isolate of a pathogen confers protection against the concerned virulent race. Sinha and Das (1972) reported that inoculation with spore suspension of a mildly virulent race of *Helminthosporium oryzae* induced considerable resistance in rice cultivars *viz.*, Dular and Latishal to a challenge inoculation with its virulent race. Prior inoculation with the mild race reduced both the number and size of brown spot lesions caused by the virulent race. This effect gradually increased with an increase in the concentration of mild race inoculum up to a certain level (c. 5×10^5 spores / ml of suspension), but rapidly declined thereafter. It had also been possible to induce considerable resistance in susceptible rice plants (cv. Dular and Dharial) to a virulent race of *H. oryzae* by previously inoculating them with an avirulent race (Sinha and Trivedi, 1972).

Use of Biocontrol Agents

Antagonistic micro-organisms are generally used in direct control of plant pathogens by virtue of their capacity of hyperparasitism, inhibition, antibiosis, etc, and the phenomenon is called biological control. However, some biocontrol agents are also known to induce resistance upon pre-inoculation. Kumawat (2006) reported induction of resistance against brown spot (*Drechslera oryzae*) by preinoculation with *Trichoderma harzianum* and *T. viride* with increased levels of protein and phenolic content. The senior author has also noticed about 68-70 per cent reduction in disease severity due to pretreatment with *Trichoderma* spp.

Resistance Induced by Abiotic Agents

Sinha *et al.* (1993) reported that seed treatment with low doses (0.002-0.02 per cent) of chitosan,an oligomer of fungal origin induced resistance to many fungal pathogens. Soil application of calcium chloride(15ppm) and dipotassium hydrogen phosphate(25mM) reduced the brown spot severity in rice by 67 and 64 per cent respectively (Biswas, 2008).

Although both biotic and abiotic agents have been reported to induce resistance in rice, chemicals would be more preferred because they are easy to formulate and handle (Manandhar *et al.,* 1998a). However, much research is still needed before any commercialization of the inducers may take place and several issues need to be addressed. For the chemical inducers it is important to thoroughly study their toxic effects if any, on the environment. The costs of using the inducers should also be considered. Experiments are required to further determine the optimal crop growth stages for application of the inducers in order to reduce consumption of the active ingredients. The use of induced resistance in a crop like rice has certain advantages, *i.e.* seeds or seedlings of paddy can be treated in several ways before transplanting to the field (Manandhar *et al.,* 1998a). Chemical agents may be more applicable as inducers than biotic agents, provided that they are inexpensive and not hazardous.

References

Alagarsamy, G., Alice, D. and Reddy, S.P. (1986). Chemical control of brown spot (BS) and sheath rot (ShR) in Tamil Nadu. *Intern. Rice Res. Newsletter* 11 (5): 18.

Alice, D. and Rao, A.V. (1987). Antifungal effects of plant extracts on *Drechslera oryzae* in rice. *Intern. Rice Res. Newsletter* 12 (2): 28.

Anonymous (2003). Rice doctor, *Intern. Rice Res. Inst.* Los Banos, Philippines. 10-20 pp.

Anonymous (2006). *Rice India* 16 (1): 91-92.

Anonymous (2007). *Economic Survey*, 2006-2007, Government of India.

Baba, I. (1958). Nutritional studies on the occurrence of *Helminthosporium* leaf spot and 'akiochi' of the rice plant. *Bull. Natn. Inst. Agric. Sci.* Tokyo, ser. D, 7: 1-157.

Bedi, K.S. and Gill, H.S. (1960). Losses caused by brown spot disease of rice in Punjab. *Indian Phytopath.* 13 (2): 161-164.

Bedi, K.S. and Gill, H.S. (1961). Relative reaction of different varieties of rice to the brown leaf spot disease in Punjab. *Indian Phytopath.* 14: 42-47.

Biswas,C. (2008).Studies on induced resistance in paddy against brown spot caused by *Drechslera oryzae* Ph.D Thesis, CSA Univ. of Agric. and Tech. Kanpur,101pp.

Breda de Haan, J. (1900). Vorlaufige Beschreibung von Pilzen, bei tropischen kulturpflanzen beobachtet. *Bull. Inst. Bot. Buitenzorg* 6: 11-13.

Butler, F.C. (1953). Saprophytic behaviour of some cereal root rot fungi II. Factors influencing saprophytic colonization of wheat straw. *Ann. Appl. Biol.* 40: 298-304.

Chaiappelli, R. (1929). Rice field invaded by *Helminthosporium oryzae. Grorn Dirislcolt.* 19 (10): 155-156.

Chakrabarti, N.K. and Padmanabhan, S.Y. (1962). Studies on *Helminthosporium* disease of rice. VI. Nutritional factors on disease expression. *Proc. 49th Indian Sci. Congr.* III: 479 (Abstr.)

Chattopadhyay, S.B. and Chakrabarti, N.K. (1953). Occurrence in nature of an alternate host (*Leersia hexandra*) of *Helminthosporium oryzae. Nature* 172: 550.

Chattopadhyay, S.B. and Chakrabarti, N.K. (1957). Effect of method of cultivation on the incidence of brown spot of "aus" paddy. *Indian Agric.* 1: 45-52.

Chattopadhyay, S.B. and Das Gupta, C. (1965). Factors affecting conidial production of *Helminthosporium oryzae. Indian Phytopath.* 18: 160-167.

Chattopadhyay, S.B. and Dickson, J.G. (1960). Relation of nitrogen to disease development in rice seedlings infected with *Helminthosporium oryzae. Phytopathology* 50: 434-438.

Chhetry, G.K.N. (1993). Effect of some fungicides on the brown leaf spot in five paddy varieties. *Ann. Plant Protec. Sci.* 1 (2): 135-136.

Das Gupta, M.K. (1988). *Principles of Plant pathology.* Allied Publishers Limited, New Delhi, 1040 pp.

Das Gupta, M.K. and Chattopadhyay, S.B. (1975). Variation in susceptibility of rice leaves to the brown spot. *Indian Phytopath.* 28: 17-21.

Dastur, J.F. (1942). Notes on some fungi isolated from 'black point' affected wheat kernels in the central provinces. *Indian J. Agric. Sci.* 12: 731-742.

Deverall, B.J. and Dann, E.K. (1995). Induced resistance in legume. In: *Induced Resistance to disease in plants.* (Eds. Hammerschmidt, R. and Kuc, J.) Kluwer Academic Publishers, Dordrecht, Netherlands, 182 pp.

Dharam Vir, Mathur, S.B. and Paul Neergaard (1970). Control of seed borne infection of *Drechslera spp.* on barley, rice and oats with Dithane M-45. *Indian Phtopath.* 23: 570-572.

Drechsler, C. (1934). Phytopathological and taxonomic aspects of *Ophiobolus, Pyrenophora, Helminthosporium* and a new genus, *Cochliobolus. Phytopathology* 24: 953-985.

El-Kazzas, M.K. and Osman, Z.H. (1989). Brown spot disease of rice in Egypt. In: Rice farming systems new directions. *Intern. Rice Res. Inst.* Los Banos, Laguna (Philippines), 347-348 pp.

Esuruoso, O.F. and Joaquim, M.A. (1981). Control of some important seed borne fungi of rice (*Oryza sativa* L.) in Nigeria. *African J.* 2: 29-42.

Fazli, S.F.I. and Schroeder, H.W. (1966). Effect of kernel infection of rice by *Helminthosporium oryzae* on yield and quality. *Phytopathology* 56:1003-1005.

Fernandez, M.R. (1992). The effectiveness of *Trichoderma harzianum* on fungal pathogens infecting wheat and black oat straw. *Soil Biol. Biochem.* 24: 1031-1034.

Gangopadhyay, S. (1968). Studies on the nitrogen relationship and epidemiology development of brown spot disease of rice incited by *Helminthosporium oryzae* (Breda de Haan.) Ph. D. Thesis, Kalyani Univ., India.

Gangopadhyay, S. (1998). Turmeric (*Curcuma longa*), an ecofriendly ingredient to control seed borne rice diseases. *Curr. Sci.* 75 (1): 16-17

Gangopadhyay, S. and Chattopadhyay, S.B. (1974). Weather conditions and air borne spread of *Helminthosporium oryzae* in West Bengal. *Indian J. Mycol. Plant Pathol.* 4: 206-208.

Ganguly, D. (1946). *Helminthosporium* disease of paddy in Bengal. *Sci. Cult.* 12: 220-223.

Ganguly, D. and Padmanabhan, S.Y. (1959). *Helminthosporium* disease of rice. III. Breeding resistant varieties; selection of resistant varieties from genetic stock. *Indian Phytopath.*12: 99-108.

Ganguly, D. and Padmanabhan, S.Y. (1962). *Helminthosporium* disease of rice. IV. Effect of cultural extract of the pathogen on the reaction of rice varieties to the disease. *Indian Phytopath.* 15: 135-140.

Ganguly, L.K. (1994). Fungitoxic effects of certain plant extracts against rice blast and brown spot pathogen. *Environ. Ecol.* 12 (3): 731-733.

Geetha, D. and Sivaprakasam, K. (1993). Treating rice seeds with fungicides and antagonists to control seed borne diseases. *Intern. Rice Res. Notes* 18 (3): 30-31.

Ghose, R.L.M., Ghatge, M.B. and Subrahmanyan, V. (1960). *Rice in India* (Revised edition). Indian Council of Agricultural Research, New Delhi, 474 pp.

Goto, I. (1954). Studies on the leaf spot of the rice plant caused by *Ophiobolus miyabeanus.* Ito et Kurib. V. On the concentric line of lesion. *J. Yamagata Agric. For. Soc.* 6: 7-8.

Grummer, G. and Roy, S.K. (1966). Intervarietal mixture of rice and incidence of brown spot disease (*H. oryzae* Breda de Haan). *Nature* 209: 1265-1267

Hammerschmidt, R. and Kuc, J. (1995). *Induced resistance to disease in plants.* Kluwer Academic Publishers, Dordrecht, Netherlands.

Harish, S., Saravanan, T., Radjacommare, R., Ebenezar, E.G. and Seetharaman, K. (2004). Mycotoxic effects of seed extracts against *Helminthosporium oryzae* Breda de Hann, the incitant of rice brown spot. *J. Biol. Sci.* 4 (3): 366-369.

Hemmi, T. and Suzuki, H. (1931). On the relation of soil moisture to the development of *Helminthosporium* disease of rice seedlings. *Forschn Geb. Pflkrankh.* 1: 90-98.

Herrera-Isla, L. (1991). Determination of the intensity of the brown spot disease caused by *Helminthosporium oryzae* in Cuba. *Centro Agricola* 18 (2): 3-7.

Hiremath, P.C. and Hegde, R.K. (1985). Saprophytic activity of *Drechslera oryzae* in soil. *J. Soil Biol. Ecol.* 5: 1-6.

Hori, S. (1901). Ine no hagare-byo (Leaf blight of rice plant.). *Bull Cent. Agric. Exp. Stn.* Nishigahara, Tokyo, 18: 67-84

Ito, S. (1930). On some new ascigerous stages of the species of *Helminthosporium* parasitic on cereals. *Imper. Acad.* Tokyo, 6: 352-355.

Ito and Kuribayashi, K. (1927). Production of the ascigerous stage in culture of *Helminthosporium oryzae. Ann. Phytopath. Soc.* Japan, 2: 1-8.

Kamala Nayar and Vidhyasekaran, P. (1998). Evaluation of a new biopesticide formulation for the management of some foliar diseases of rice. In: Ecological agriculture and sustainable development: Vol.2. *Proc. Intern. Conf. on Ecological Agriculture: Towards Sustainable Development*, Chandigarh, India, 15-17 November, 1997. 382-388 pp.

Katsura, K. (1937). On the relation of atmospheric humidity to the infection of the rice plant by *Ophiobolus miyabeanus* Ito and kuribayashi and to the germination of its conidia. *Ann. Phytopath. Soc.* Japan 7: 105-124.

Kaur, P., Kaur S. and Padmanabhan, S.Y. (1979). Effect of manganese and iron on incidence of brown spot disease of rice. *Indian Phytopath.* 32: 287-288.

Kaur, P., Kaur, S. and Padmanabhan, S.Y. (1984). Relationship of host nutrient status and brown spot disease expression in rice. *Indian Phytopath.* 37: 156-158.

Kaur, P., Kaur, S. and Padmanabhan, S.Y. (1986). Effect of calcium on the development of brown spot disease of rice. *Indian Phytopath.* 39: 57-61

Kumawat, G.L. (2006). Effect of *Trichoderma* on *Drechslera oryzae* and its role in defense against brown leaf spot of paddy. M.Sc. Thesis. *C.S.A. Univ. Agric. and Tech., Kanpur,* 78 pp.

Link, H.F. (1809). Observations in ordines plantarum naturales. I. Andhrarum Ordines Epiphytes, Mucedines Gastramyces at Fungus. *Gasell. Nat. Freund Berlin Magazine* 3: 3-42.

Luttrell, E.S. (1951). A key to the species of *Helminthosporium* reported on grasses in the United States. *Plant Dis. Reptr. Suppl.* 201: 59-67.

Manandhar, H.K., Jorgensen, H.J.L. Mathur, S.B. and Smedegaard-Petersen, V. (1998). Resistance to rice blast induced by ferric chloride, dipotassium hydrogen phosphate and salicylic acid. *Crop Protec.* 17 (4): 323-329.

Mandal, S., Srivastava, K.D., Agarwal, R. and Singh, D.V. (1999). Mycoparasitic action of some fungi on spot blotch pathogen (*Drechslera sorokiniana*) of wheat. *Indian Phytopath.* 52: 39-43.

Minnatullah, M. and Abdus-Sattar (2002). Brown spot development in Boro rice as influenced by weather conditions. *J. Appl. Biol.* 12: 71-73.

Misawa, T. (1955). Studies on the *Helminthosporium* leaf spot of rice plant. *Jubilee Publ. Commem. 60th Birthday of Profs. Y.Tochinai and T. Fukushi*: 65-73. Sapporo. Japan.

Muthusamy, S., Mariappan, V., Narasimhan, M., Muthusamy, M. and Eswaramoorthy, S. (1988). Effect of micronutrient on rice brown spot (BS) incidence. *Intern. Rice Res. Newsletter* 13 (6): 32-33.

Naimuddin, R. Chakrabarty, U.N. and Chakrabarty, R. (1994). Airborne fungal load in agricultural environment during threshing operations. *Mycopathologia* 127(3):145-149.

Nisikado, Y. (1928). Studies on the *Helminthosporium* disease of gramineae in Japan. *Ohera Inst. Agric. Res. Spec. Rept.* 4: 391.

Nisikado, Y. and Miyake, C. (1922). Studies on the helminthosporiose of the rice plant. *Ber. Ohara. Inst. Landw. Forsch.* 2: 133-194.

Ocfemia, G.O. (1923). The *Helminthosporium* disease of rice. *Phytopathology* 13: 53 (Abs.)

Ocfemia, G.O. (1924a). The *Helminthosporium* disease of rice occurring in the Southern United States and in the Philippines. *Am. J. Bot.* 11: 385-408.

Ocfemia, G.O. (1924b). The relation of soil temperature to germination of certain philippine upland and lowland varieties of rice and infection by the *Helminthosporium* disease. *Am. J. Bot.* 11: 437-460.

Ou, S.H. (1972). *Rice diseases.* Commonwealth Mycological Institute, Kew, Surrey, England. 368 pp

Padmanabhan, S.Y. (1963). The role of therapeutic treatments in plant disease control with special reference to rice diseases. *Indian Phytopathol. Soc. Bull.* 1: 79-84.

Padmanabhan, S.Y. (1973). Great Bengal Famine. *Ann. Rev. Phytopath.* 11: 11-24.

Padmanabhan, S.Y. (1974). Control of rice diseases in India. *Indian Phytopath.* 27:1-28.

Padmanabhan, S.Y. and Ganguly, D. (1954). Relation between age of rice plant and its susceptibility to *Helminthosporium* and blast diseases. *Proc. Indian Acad. Sci.* 39B: 43-50.

Padmanabhan, S.Y. Ganguly, D. and Chandwani, G.H. (1966). *Helminthosporium* disease of rice. VII. Breeding resistant varieties: relection of resistant varieties of early duration from genetic stock. *Indian Phytopath.* 19: 72-75.

Padmanabhan, S.Y., Abhichandani, G.T., Chakrabarti, N.K. and Patnaik, S. (1962). Studies on *Helminthosporium* disease of rice. 2. Effect of potassium. *Proc. 49th Indian Sci. Congr.* III: 479-480.

Padmanabhan, S.Y., Chowdhury, K.R.R. and Ganguly, D. (1948). Nature and extent of damage caused by *Helminthosporium* disease of rice. *Indian Phytopath.* 1: 34-47.

Padmanabhan, S.Y., Ganguly, D. and Balakrishnan, M.S. (1953). *Helmintho sporium* disease of rice. II. Source and development of seedling infection. *Indian Phytopath.* 6: 96-105.

Pedrozo, J.P., Yumol, R.D. and Ducanes, A.A. (1994). Incidence and severity of brown spot (*Helminthosporium oryzae*) and other major diseases of rice in Zamboanga city (Philippines). In: *Integrated Pest Management*: *Learning from experience.* Pest Management Council of the Philippines. Laguna (Philippines). 73 pp.

Prasad, K.S.K., Kulkarni, S. and Siddaramaiah, A.L. (1978). Antagonistic action of *Trichoderma* spp. and *Streptomyces* sp. on *Drechslera sativum* (*Helminthosporium sativum* Pam. King and Bakke). Subram. and Jain. *Curr. Res.* 12: 202-213.

Premlatha Dath, A. and Chakrabarti, N.K. (1973). Wild rice species, the alternate hosts of *Helminthosporium oryzae. Sci. Cult.* 39: 394-397.

Rajappan, K., Vshamalini, C., Subramanian, N., Narasimhan, V. and Abdul Kareem, A. (2001). Management of grain discoloration of rice with solvent free EC formulations of neem and pungam oils. *Phytoparasitica* 29 (2): 171-174

Raju, K. Manian, S. and Anbuganapathi, G. (2004). Effects of certain herbal extracts on the rice brown leaf spot pathogen. *J. Ecotoxicol. Environ. Monitoring* 14 (1): 31-37.

Rangaswami, G. and Mahadevan, A. (1999). *Diseases of Crop Plants in India.* (4th Ed.). Prentice Hall of India Pvt. Ltd., New Delhi, 165-169 pp.

Sachan, I.P. and Agarwal, V.K. (1995). Seed discolouration of rice: location of inoculum and influence on nutritional value. *Indian Phytopath.* 48 (1): 14-20.

Salive, A. (1994). Various factors influencing darkening of the grain. *Arroz* 43: 30-35.

Sato, K. (1965). Studies on the blight diseases of rice plant. *Bull. Inst. Agric. Res. Tohoku Univ.* 16: 1-54.

Serghat, S., Mouria, A., Tauhami, A.O. and Douira, A. (2002). *In vivo* effect of some fungicides on the development of *Pyricularia grisea* and *Helminthosporium oryzae*. *Phytopathologia Mediterranea* 41 (3): 235-245.

Shaw, F.J.F. (1921). Report of the Imperial Mycologist. *Sci. Rep. Agric. Res. Inst.* Pusa, 1920-21:34-40.

Shoemaker, R.A. (1959). Nomenclature of *Drechslera* and *Bipolaris* grass parasites segregated from *Helminthosporium*. *Can. J. Bot.* 35: 879-887.

Singh, R.A. and Sharma, V.V. (1975). Evaluation of rice germplasm and varieties for resistance of brown spot. *IL RISO* 24: 383-386.

Singh, R.N. (1988). Evaluation of indigenous rice varieties for resistance to brown spot. *Indian Phytopath.* 41 (4): 635-637.

Singh, R.S. (2005). *Plant diseases.* (8th Ed.) Oxford and IBH Publishing Co. Pvt. Ltd., New Delhi. 439-444 pp.

Sinha, A.K. and Das, N.C. (1972). Induced resistance in rice plants to *Helminthosporium*. *Physiol. Plant Pathol.* 2: 401-410.

Sinha, A.K. and Trivedi, N. (1972). Resistance induced in rice plants against *Helminthosporium* infection. *Phytopath. Z* 74: 182-191.

Sinha, A.K., Chaudhary, A.K. and Das, A.R. (1993). Chitosan induces resistance in crop plants against their fungal pathogens. *Indian Phytopath.* 46 (4): 411-14.

Sreeramulu, T. and Ramalingam, A. (1966). A two year study of the air spora of a paddy field near Visakhapatnam. *Indian J. Agric. Sci.* 36: 111-132.

Sreeramulu, T. and Seshavataram, V. (1962). Spore content of air over paddy fields. I. Changes in a field near Pentapadu from 21st September to 31st December, 1957. *Indian Phytopath.* 15: 61-74.

Su, M.T. (1936). Report of the Mycologist, Burma, Mandalay, for the year ending 31st March, 1936. *Rep. Oper. Dep. Agric.* Burma, 1935-36:35-39.

Subramanian, C.V. and Jain, B.L. (1966). A revision of some graminicolous *Helminthosporia*. *Curr. Sci.* 35: 352-355.

Sundaraman, S. (1922). *Helminthosporium* disease of rice. *Bull. Agric. Res. Inst.* Pusa, 128: 1-7.

Surek, H. (1995). Diseases of rice in Turkey. Edirne (Turkey). *Thrace Agric. Res. Inst.* 4 pp.

Tembhurnikar, S.T. (1989). Effect of green manuring on the development of bacterial leaf blight and brown spot. *Oryza.* 26: 110.

Thirumalachar, M.J. (1967). Aureofungin in the control of seed borne *Helminthosporium oryzae* infection and seedling blight. *Indian Phytopath.* 20: 277-279.

Thomas, K.M. (1940). Detailed administration report of the government mycologist, Madras, 1939-40, 18pp.

Tochinai, Y. and Sakamoto, M. (1937). Studies on physiological specialization of *Ophiobolus miyabeanus* Ito and kuribayashi. *J. Fac. Agric. Hokkaido Univ.* 41:1-96.

Tucker, C.M. (1927). Report of the plant pathologist. *Rep. Puerto Rico Agric. Exp. Stn.* 1925: 24-40 pp.

Variar, M. (1996). Upland Rice Research: Achievements and Perspectives. central Rainfed Upland Rice Research Station. (A unit of Central Rice Research Institute, Cuttack). Hazaribag, India. 28 pp.

Varughese, A. and Padma Kumari, G. (1993). Effect of organic manure and inorganic fertilizers on the disease incidence in rice. *J. Trop. Agric.* 31 (2): 251-253.

Vidhyasekaran, P., Ranganathan, R., Palaniappan, S.P. and Ramasamy, R. (1983). Effect of slow release nitrogen fertilizers on rice brown spot disease. *Intern. Rice Res. Newsletter* 8: 11.

Viswanathan, R. and Narayanasamy, P. (1990). Chemical control of brown leaf spot of rice. *Indian J. Mycol. Plant Pathol.* 20 (2): 139-143.

Vivekanandan, P., Giridharan, S. and Narasimhan, V. (1995). Reaction of some scented rice hybrids to brown spot disease. *Madras Agric. J.* 82 (4): 327.

Wani, M.A., Kurucheve, V. and Gincy Devasia (2004). Distribution of external and internal mycoflora of paddy seeds. *Environ. Ecol.* 22 (4): 310-311.

Zulkifli, E., Klap, J. and Castano, J. (1991). Effect of grain discoloration in upland rice on some yield components. *Intern. Rice Res. Newsletter* 16 (4): 20.

Sustainable Disease Management of Agricultural Crops (2011) *Pages* **133–148**
Editors: **S.K. Biswas and S.R. Singh**
Published by: **DAYA PUBLISHING HOUSE, NEW DELHI**

Chapter 8

Biological Control of Plant Diseases

☆ *S. Kulkarni*[1], *S.R. Singh*[2] *and S.K. Biswas*[3]

[1]*Department of Plant Pathology, UAS, Dharwad, Karnataka*
[2]*Krishi Vigyan Kendra, Belatal, Mahoba – 210 423, U.P.*
[3]*Department of Plant Pathology,*
C.S. Azad University of Agriculture and Technology, Kanpur – 208 002, U.P.

Plant diseases have been with mankind since agriculture began in the earth.The population explosion has caused a need for more food with increasing amount of land devoted to crop production. Intensification and monocropping have resulted in increasing disease pressure among the crops. Although the application of synthetic fertilizers and pesticides has given higher yields, environmental balances have been disrupted and considerable crop damage by insects and pathogens still; occur. Plant diseases contribute 13-20 per cent of losses in crop production worldwide. Control of plant diseases by chemicals can be spectacular but this is relatively a short term measure and more over, the accumulation of harmful chemical residues some times causes serious ecological problems. In recent years, the increasing use of potential hazardous pesticides and fungicides in agricultural has been the result of growing concern of both environmentalist and public health authorities. Moreover, use of such chemicals entails a substantial cost to the nation and developing country like India cannot afford it.

Today we live in constant threat of man created irreversible phenomena. This statement is applicable to pesticides for in the case of DDT, the time between its first use and the publics general awareness of the danger it presents to the environment has been in the order of a quarter of a century.Even with proper use, chemicals may be hazardous indirectly in various ways namly toxicity of residuals, injury to non target crops and endangered species, development of resistance and insurgence at more serious levels. The continued use of chlorinated pesticides has been responsible for the presence of residues in grain, milk, vegetables as well as in human tissues like the placenta, cord blood. As a result of outcries against the tragedies mentioned the discovery and development of alternatives to chemical control is an essential agenda for agricultural scientists. So, Biological control have provided an effective and ecofriendly management of plant diseases.These are specific to the targeted pathogens and do not lead to residue problems and accumulation of toxic pollutants in the soil or underground water.

One of the major challenges facing organic producers is disease management.The losses due to diseases can be significant and in some cases can devastate entire crops. Cultural methods of disease control are commonly used on organic farms. The application of organic chemicals is often a last resort and regulated while biological control is still not readily available. Biological control is the reduction of the amount of inoculums density or disease producing activity of a pathogen accomplished by one or more organisms other than man.

Components of Biological Control

The Host

The host plant is a main component in biological control of plant diseases, aimed at suppression of the disease producing activities of the pathogen and regulation of the amount of pathogen inoculum. The host plant also indirectly involved in biological control by producing the site for action of one or more antagonists of the target pathogen.

The host also in suppression or termination of pathogenesis or reproduction of the pathogen by one or more mechanism of the host resistance. Plants also have general resistance to pathogens that cannot yet defined genetically which is controlled by the collective action of multiple physiological factors associated with active host metabolism and growth.

Antagonist

In biological control of plant pathogens antagonists are biological agents with the potential to interfere in the life processes of plant pathogens. Antagonists includes all classes of organisms *viz,* fungi, bacteria, nematodes, viruses, viroids, protozoa and seed plants. Mild strain of plant virus may be classified as antagonists when introduced in to a plant, they cross protect that plant against severe strain and thereby biological control. The antagonist need not be a complete virus. A small satellite self replicating RNA molecule when introduced into cucumber mosaic virus causes increased virulence of CMV on tomato but reduces symptoms caused by CMV in squash, tobacco, sweet corn referred to as biological control. The bioagent referred to as CARNA 5 regulates disease expression. When disease expression is suppressed CARNA5 is acting as biological agent an antagonist of CMV.

Pathogen

Pathogens include pathogenic fungi, bacteria, nematodes, seed plants, algae, viruses and viroids.The physical relationship of a pathogen to its host during pathogenesis may be either as an epiphyte, or endophyte or both. Most fungal pathogens have both epiphytic and entophytic phases. The vascular wilt fungi have a very brief epiphytic existence *ie,* The period of pre penetration growth between germination of the propagule and penetration in the host root, but there after they exist entirely as endophytes within the root cortex and vascular tissue. In general the more internal the pathogen during the host pathogen interaction the less vulnerable the pathogen to control by antagonist. The pathogen may be subjected to antagonism at any time during their life cycle whether dormant or active during their saprophytic or parasitic existence. In some cases the secondary colonists add to the plant damage and also have the potential to retard the spread of pathogens in the host. The longer the time between infection and sporulation the more vulnerable the pathogen displacement by non pathogen(secondaries).

Following points presently supporting management of plant diseases through bio-agents:

1. It avoids environmental pollution of soil, air and water as being experienced in chemical control.
2. It avoids the residue toxicity or crop products when consumed but it is very much experienced in fungicide control.
3. It avoids adverse effect on the beneficial microorganisms including antagonists in the soil whereas chemicals are lethal.
4. It is less expensive compared to chemical control, as they are expensive.
5. Continuous use of bio-agents avoids the development of resistant strains of pathogens whereas the continuous use of systemic and specific fungicides has resulted in development of new chemical resistant strains thus making chemical ineffective.
6. Bio-agents application is usually once and does not need repeated application whereas fungicides need to be repeatedly applied as they loose effectiveness after sometimes.
7. Bio-agents are more effective especially for soil borne diseases, as fungicides may not reach pathogen site.
8. Also the quantity of fungicide used in several cases is more whereas in bio-agents it is very less.
9. In the absence of satisfactory control measures in virus diseases biological control is presently satisfactory.
10. It aims at risk free management of disease whereas in chemical control risks are many including phytotoxicity.
11. It becomes part of modern large-scale agriculture and helps in increasing crop production within existing resources maintaining biological balance.

The microorganism used in the biological control of the plant diseases are termed as "antagonists", an antagonist is a microorganism that adversely affects another growing organism in association with it. To be efficient the antagonist should possess the following desirable characters:

1. It should survive and grow in the rhizosphere or spermosphere (to prevent infection) or in the hyposphere near the pathogens resting structures or in the soil mass (to reduce survival).
2. It should be growing on available substrates producing broad-spectrum antibiotics effective against array of pathogen strains.
3. The antibiotics produced by one antagonist should not be inhibitory to other associated antagonists.
4. The antibiotic produced should not cause damage to host plants.
5. The antagonist should be adaptable to large-scale commercial production and handling.
6. Spore germination should be quick and prolific but returning to dormancy stage should be slow.
7. It should be more adaptable to varied environmental extremes than pathogen.

The antagonist belongs to fungi, bacteria, actinomycetes, viruses and nematodes. Antagonism is the balance wheel of nature and operates through various mechanisms, which ultimately reduces the plant diseases.

Table 8.1

Biocontrol Agents	*Target Pathogen*	*Diseases/Crop*	*References*
Aspergillus niger AN 27	*Fusarium* spp.	Seed rot, damping off, vascular wilt of vegetables, fruits, ornamentals and fibre crops.	Sen (1999)
	Pythium sp.	Damping off	
	Macrophomina phaseolina	Charcoal rot of field and vegetables crops	
	Rhizoctonia solani	Root and stem rot, canker, sheaths blight and ear rot in monocoat and dicoat.	
	Sclerotinia sclerotiorum	Vegetables and ornamental crops	
Gliocladium roseum	*Bipolaris sorokiniana* *Fusarium moniliforme*	Barley Maize	Knudsen *et al.* (1995) Vakili (1992)
G. Virens	*M. phaseolina* *Phythium ultimum*	Mungbean, sunflower, cotton	Hussain *at el.* (1990) Howell *et al.* (1993)
	F. solani f.spp. *phaseoli*	Navy bean	Tu (1991)
G. virens GL-3	Sclerotium rolfsii	Snapbean	Papanvizas and Collins (1990)
G. virens GL-21	*R. solani*	Carrot	Ristaino *et al.* (1994)
G. virens G6	*P. ultimum*	Cucumber	Wolffhechet and Funck Jensen (1992)
	R. solani	Cotton	Howell and Stipanovic, (1995)
Penicillium oxicilum	*Pythium* spp.	Chickpea	Trapero-casas *et al.* (1990)
Pythium oligandrum	*Aphanomyces cochlioides*	Sugar beet	Mc Quilken and Whipps (1995)
Trichoderma harzianum	*M. Phaseolina*	Mungbean Sunflower	Hussain *et al.* (1990)
T. harzianum 1295-22	*P. ultimum*	Cucurbeet cucumber	Taylor *et al.* (1991)
T. hazianum	*Glomerella glycines* *M. phaseolina*		Fernandez (1992)
	P. ultimum	Lettuce	Lynch *et al.* (1991)
T. harzianum MTR 35	*F.o.*f.sp. *radicis-lycopersici*	Tomato	Rattink (1993)
T. harzianum	*S. sclerotiorum*		Kundren *et al.* (1991)
Trichoderma hamatum	*R. solani*	Radish	Chung and Hoitink (1990)
T. koningii	*S. rolfsii*	Tomato	Latunde-Dada (1993)
T. viride	*F.o.*f. sp. *sesame* *R. solani*	Sesamum	Chung and Choi (1990)
Chaetomium globosum	*Sclerotium cepivorum*	Onion	Kay and Stwart (1994)

Contd...

Table 8.1–Contd...

Biocontrol Agents	*Target Pathogen*	*Diseases/Crop*	*References*
C. globosum (Cg-13)	*P. ultimum*	Sugar beet	Di Pietro *et al.* (1992)
C. globosum	*Drechslera sorokinina*	Wheat	Biswas (1997)
Coniothyrium minitans	*S. sclerotiorum*	Sunflower Carrot	Mc Laren *et al.* (1994) Mc. Quilken *et al.* (1995)
F. oxysporum F_0 47 (non pathogenic)	*F.o.*f.sp. *lycopersici* *F. solani*	Tomato Pea	Alabouvette *et al.* (1993) Oyarzun *et al.* (1994)
F. oxysporum (non pathogenic)	*F.o.*f.sp. *dianthi*	*Carnation*	Garibaldi and Gullino, (1990)
*F.o.*f.sp. *nivium* (non pathogenic)	*F.o.*f.sp. *nivium*	Water melon	Martyn *et al.* (1991)

Tabl 8.2: Bacteria as Biocontrol Agents Against Soil Borne Pathogens

Biocontrol Agents	*Target Pathogen*	*Diseases/Crop*	*References*
B.cereus UW-85	*Phytophthora sojae*	Soyabean	Osbura *et al.* (1995)
Bacillus cereus UW 85	*Phytophthora megasperma* f. sp. *medicaginis*	Alfalfa	Handelsman *et al.* (1990)
B. subtilis PAP-183	R. solani	Common bean	Andrade *et al.* (1994)
B. subtilis GB-03	*Fusarium oxysporum* f.sp. *vasinfectum* *R. solani*	Cotton Cotton	Backman *et al.* (1994) Brannen and Backman, (1994)
B. subtilis 2005	*R. solani*	Oil seed	Fiddman and Rossall (1995)
P. aureoginosa 7 NSK-2	*Pythium* sp.	*Tomato*	Buysens *et al.* (1993)
P. cepacia AMMD	*Aphanomyces euteiches* f.sp. *pisi*	Pea	Parke *et al.* (1991)
P. cepacia N 24	*S. rolfsii*	Sunflower	Hebbar *et al.* (1991) Hebbar *et al.* (1992)
P. cepaci	*P. ultimum* *R. solani* *S. rolfsii*	Cucumber Cotton Bean	Fridlender *et al.* (1993)
P. cepacia 89 G-120	*R. solani*	Cotton	Press and Kloepper, (1994)
P. fluorescens	*P. ultimum* *R. solani* *S. sclerotiorum*	Cucumber Cucumber Kanda Sunflower	Park and Yeom (1994) Reddy *et al.* (1994)
P. fluorescens AB 254	*P. ultimum*	Maize	Callan *et al.* (1990)
P. *fluorescens* pf-5	*P. ultimum*	Cucumber	Kraus and Loper (1992)
P. fluorescens PRA–25	*Aphanomyces euteiches* f. sp. *pisi*	Pea	Park *et al.* (1991)
P. fluorescens Q 2-87	*Gaeumannomyces graminis* var *tritici*	Wheat	Mazzola *et al.* (1995)

Contd...

Table 8.2–Contd...

Biocontrol Agents	*Target Pathogen*	*Diseases/Crop*	*References*
P. fluorescens Q 2-92 -80	*Pythium* sp.	Chickpea	Trapero (1990)
P. fluorescens CH-33	*P. aphanidermatum*	Cucumber Cucumber	Ranking and Paulitz (1994) Moulin *et al.* (1994)
P. putida	*S. sclerotiorum*	Sun flower	Expert and Digat (1995)
P. putida NIR	*Pythium ultimum*	*Cucumber*	Paulitz and Loper (1991)
P. putida R 104	*R. solani*	Wheat	De Freitas and Germida (1991)
Streptomyces sp.	*M. phaseolina*	Mungbean sunflower	Hussain *et al.* (1990)
S. griseoviridis	Damping off	Pepper	Landenpera *et al.* (1991)
Enterobacter agglomerans	*R. solani*	Cotton	Chernin *et al.* (1995)
P. aureofaciens PA-157-2	*Phytophthora megasperma*	Asparagus	Carruthers *et al.* (1995)

Identification and Development of Effective Strains

- ☆ Screening and selection of local strains.
- ☆ Improvement of strains-biotechnology, mutation
- ☆ Genes which code for biocontrol should be added
- ☆ Genes which code for pathogenecity should be deleted.

Biocontrol is not totally Successful

- ☆ Usage of technology itself
- ☆ *In- vitro* studies- successful.
- ☆ *In- vivo* studies-failure–so, need for suitable strains for field application.
- ☆ Bio agents works well in some geographical area.
- ☆ Mass production of antagonist.
- ☆ Large scale demonstration to convince the farmer's about the time and method of application
- ☆ Slow action of Bioagents not impressed by farmer's.
- ☆ Quality control of Biofungicides- producers exploiting the situation commercially.
- ☆ Avirulent strain may become virulent
- ☆ Antagonist altered genetically–fear of their safety for plant and human beings i.e, antibiotics produced by antagonist.
- ☆ Expression of Biocontrol mechanisms in trangenic plant.

Types of Formulations

1. Powder formulation
2. Encapsulations in organic polymer like sodium alginate.
3. Pelleting biomass and bran with sodium alginate.

4. Wheat bran saw dust water for soil application.
5. Molasses enriched (Kaolin) clay granules.
6. Liquid coating formulations bioprotectant as powder on which suspension of aqueous binder is sprayed on seeds for 0.1 mm. thick layer.
7. As spray from emulsifiable concentrate with 10 spores/1. cfu needed is 1-10 cfu/ml.

Methods of Application

1. Broadcast (125-150 kg/ha)
2. Furrow application (130-160 kg/ha)
3. Root zone application (Formulation mixed in soil @ 1 kg/plant)
4. Seed treatment (4 g/kg of seeds)
5. Wound application
6. Spraying (10^6–10^8 cfu/ml)

Trichoderma spp.

Characteristics of *Trichoderma*

Trichoderma has become a model biocontrol agent in the management of plant disease because of the following characters they are-

1. *Trichoderma* is ubiquitous in nature.
2. Easy to isolate and culture.
3. Grows rapidly on many substrates.
4. Affects a wide range of pathogens.
5. Non-pathogenic to higher plants.
6. Acts as mycoparasite, competes well for food and site and produces antibiotics such as trichodermin, dermin and viridin etc.
7. Primarily reported as antifungal as well as antibacterial and also anti nematodes.
8. Suppresses pathogens by all three mechanisms *i.e.*, antibiosis, mycoparasitism and competition.
9. Its formulations are seed inoculants and soil. The applications are easy to handle and apply.
10. It is not producing any adverse effects during handling.
11. Easily handled to produce efficient strains through or radiation, biotechnological techniques such as protoplast fusion techniques etc.

Development of Formulation

Biopesticide formulations are required to have many desirable characters such as ease in preparation and application, stability during transport and storage, good shelf life, sustainability of efficacy, acceptable cost and adequate market potential. Generally, the formulation from solid-state fermentation does not require sophisticated procedures. These products are dried, ground and added with the carriers for use. However, such preparations are bulky, take longer time and are therefore prone to contamination. The liquid fermentation is devoid of such problems. The large quantity of propagules can be produced within few days and the biomass can easily be formulated into dust, granule, pellet, wettable powder or emulsion. Various carriers used in *Trichoderma* formulations are celatom, vermiculite, talc, diatomaceous earth, molasses, gypsum, lignite, kaolin, peat, wheat bran, clay, perlite, sodium alginate, calcium chloride, bentonite, rice flour, gluten and pyrex (Kelly, 1976;Khan *et al.*,2001;Moody and Gindrant,1977; Papavizas *et al.*,1984;Papavizas and Lewis,1989). For bacterial bioagents, diatomacious earth, talc, calcium alginate, lignite, peat, FYM, cow dung cake and wood charcoal have been used by different workers. A brief review on the formulation of fungal and bacterial biopesticies has been given by Khan and Sinha(2005).

Key to Success of Biological Control

1. Selection of virulent strain of antagonists.
2. There should be sufficient growth and sporulation on the mass culture media in case of facultative before applying.
3. Advance application to provide enough time for interaction between antagonist and target pathogens.
4. The soil temperature and moisture should be optimum for establishment and growth of the antagonist.
5. Sufficient organic matter (food base) should be present in the soil for the multiplication, survival and activity of the antagonists.
6. Monitoring the population of the target pathogen and antagonist from time to time on selective medium.
7. Handling, production, storage should be easy and also product should be cheap and available easily.
8. Integration of biocontrol with tolerant resistant varieties and or chemical control may be more feasible.

Mycoparasitism

Trichoderma spp. parasitize a range of other fungi. The events leading to mycoparasitism are complex, and take place as follows: first, *Trichoderma* strains detect other fungi and grow tropically towards them; remote sensing is at least partially due to the sequential expression of cell-wall-degrading enzymes (BOX 4). Different strains can follow different patterns of induction, but the fungi apparently always produce low levels of an extracellular exochitinase. Diffusion of this enzyme catalyses the release of cell-wall oligomers from target fungi, and this in turn induces the expression of fungitoxic endochitinases, which also diffuse and begin the attack on the target fungus before contact is actually made. Once the fungi come into contact, *Trihoderma* spp. attach to the host and can coil around it and form appressoria on the host surface. Attachment is mediated by the binding of carbohydrates in the *Trichoderma* cell wall to lectins on the target fungus. Once in contact, the *Trichoderma* produce several fungitoxic cell-wall-degrading enzymes and probably also peptaibol antibiotics. The combined activities of these compounds result in parasitism of the target fungus and dissolution of the cell walls. At the sites of the appressoria, holes can be produced in the target fungus, and direct entry of *Trichoderma* hyphae into the lumen of the target fungus occurs. There are at least 20–30 known genes, proteins and other metabolites that are directly involved in this interaction, which is typical of the complex systems that are used by these fungi in their interactions with other organisms. In the Figure

8.1, 'R' indicates hyphae of the plant pathogen *Rhizoctonia solani*, 'T' indicates hyphae of *Trichoderma* spp., and 'A' indicates appressoria-like structures. In a, the *Trichoderma* strain is in the process of parasitizing a hypha of *R. solani*. The coiling reaction is typical of mycoparasitic interactions. 'b' shows a higher magnification of the *Trichoderma–R. solani* interaction, showing the appressoria-like structures, and c shows an *R. solani* hypha from which the *Trichoderma* has been removed. The scale bar in 'b' represents 10 µm; *R. solani* hyphae are typically 5–6 µm in diameter, and mycoparasitic hyphae of *Trichoderma* spp. are ~3 µm in diameter. Photomicrograph in b reproduced with permission from REF. 92 © (1987) British Mycological Society. Photomicrograph in 'c' reproduced with permission from REF. 93 © (1983) American Phytopathological Society.

Antibiosis

Antibiotsis or secondary metabolites are produced by a number of genera of fungi, bacteria and yeasts been *in- vitro*. These chemicals play an important role in disease suppression. Some examples of antibiotics or secondary metabolites produced by fungi, bacteria which may be involved in pathogenic invasion are as follows:

Table 8.4: Effect of Antibiotics/Secondary Metabolites on Pathogens Produced by Bioagent Bacteria

Antagonist	*Antibiotics*	*Pathogen*	*References*
A. radiobacter K-84	Agrocin 84	*A. tumifaciens* on several woody plants	Jones *et al.* (1988)
A. radiobacter 434	Agrocin 434	*A tumifaciens*	Donner *et al.* (1993)
Bacillus cereus UW 85	Ammonia	*Phytophthora cactorum*	Gilbert *et al.* (1990)
S. hygrascopius var *geldanus*	Geldanamycin	*R. solani*	
P. putida M-17	Unidentified	*Erwinia* sp.	
P. fluorescens pf.–5	Pyoluteorin Pyrrolnitrin	*P. ultimum* *R. solani*	Howell and Stipanovic (1980)
P. fluorescens	HCN	*Thielaviopsis basicola* (tobacco)	Voisard *et al.* (1989)
P. fluorscens HV 37a	Oomycin	*P. ultimum* (cotton)	Gutterbon (1990)
B. cereus	Zwittermycin A	*Phytophthora medicaginis*	Whipps (1997)
P. aureofaciens PA 147-2	Unknown	*P. megasperma*	Carruthers *et al.* (1995)
Botrytis cinerea	Botrycine, oxalic acid	Yeast	

Indirect Method

Cross Protection

Cross protection can be defined as the inhibition of disease symptoms resulting from the prior or simultaneous inoculation of the host with a closely related species of the pathogen. The phenomenon of cross protection certainly stabilized for viruses been defined as when a mild stain of viruses is injected into a plant, such plant become resistance against relative viruses as described by Mckinney (1929).

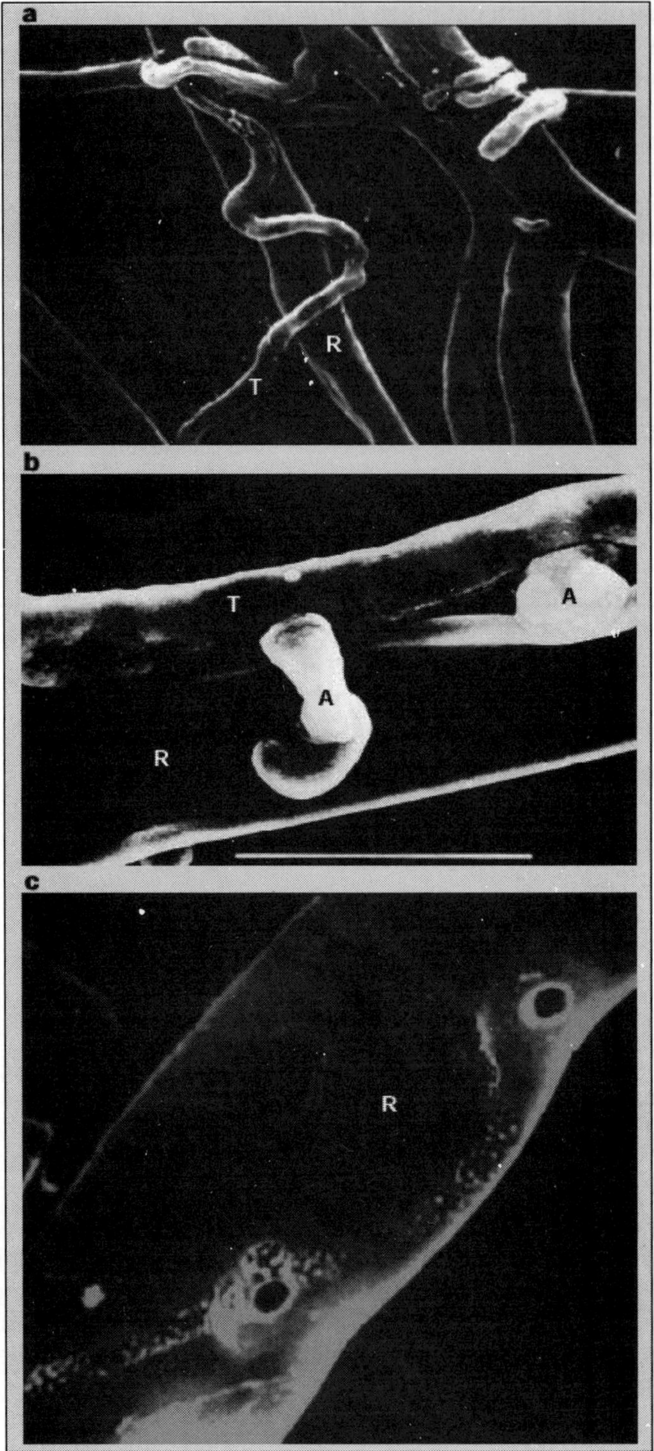

Figure 8.1

Induced Resistance

It is the phenomenon in which a plant once appropriately stimulated exivites and enhance resistance upon challenge inoculation with a pathogen. To a certain degree, susceptible plants can be altered in their resistance against pathogen without genetic manipulation. This enhancement of resistance in response to exogenous stimulates without a non alteration of genomes is called induced resistance. Pre inoculation with certain strain of nonpathogenic, plant growth promoting rhizobacteria bioagent and chemicals which is not consider as fungicide can induce resistance in plant against several pathogens some examples as follows (Table 8.5).

Table 8.5: Biocontrol Agent Mediated Induced Resistance in Plants

Diseases	*Pathogens*	*Biocontrol*	*Name of the Plant*
Blossom blight of apple	*Enwinia amylovora*	*Pseudomonas fluorescens*	Apple
Grey mold of bean	*Botrytis cineria*	*T. harzianum* T 39 *Pseudomonas fluorescens*	Bean
Grey mold of cotton	*Botrytis cineria*	*T. harzianum* T 39	Cotton
Anthracnose of cucumber	*Colletotrichum orbiculare*	*T. harzianum* T 203 *P. putida* 89B-27	Cucumber

Growth Promotion

Growth promotion in plants is another indirect activity of biocontrol agents for management of disease. Their have been several reports on growth promotion following treatment of seed root, cutting and soil with bacterial and fungal biocontrol agent. Some examples of given in the Table 8.6.

Table 8.6: Effect of Bio-agents on Growth Parameter of Plants

Antagonist	*Plant*	*Growth promotion effect on*	*References*
A. niger AN-27	Cauliflower Brinjal	Seedling growth, and yield Seedling and yield shoot	Sen *et al.* (1999)
Trichoderma sp.	Lettuce Marigold	Shoot and dry weight Shoot and dry weight	Ousley *et al.* (1994)
T. harzianum BR105	Tomato	Dry weight of root and shoot	Besnard and Davet (1993)
T. viride	Lettuce	Fresh shoot weight	Colly-smith *et al.* (1991)
T. harzianum T 203	Bean Radish Tomato Pepper	Germination Height, leaf area dry weight Germination Germination	Kleifeld and Chet (1992) Chang *et al.* (1986) Inber *et al.* (1994)
C. globosum	Wheat	Shoot and root length and leaf area	Biswas *et al.* (2001)
T. harzianum	Rice	Seedling growth root shoot length leaf area	Pandey (2004)
T. harzianum T 95	Chrysanthemum	Fresh and dry weight	Chang *et al.* (1986)

Source: Prasad (2005).

References

Alabouvette, C., Lemancean, P. and steinberg, C. (1993). Recent advances in the biological control of fusarium wilts. *Pesticide Science*, 37: 365-373.

Andrade, O.A., Mathre, D.E. and Sands, D.C. (1994). Suppression of *Gaeumannomyces graminis* var. *tritici* in Montara soils and its transferability between soils. *Soil Biology and Biochemistry*, 26:397-402.

Backman, P.A., Brannen, P.M. and Mahaffee, W.F. (1994). Plant response and disease control following seed inoculation with Bacillus subtitis. In: "*Improving plant Productivity with Rhizosphere Bacteria*" (M.H. Ryder, P.M. Stephens and G.D. Bowen, ends), PP 3-8 CSIRO, Division of soils, Adelaide, Australia.

Besnard, O. and Davet, P. (1993). Mise en evidence de souches de *Trichoderma* spp. a la fois antagonistes *Pythium ultimum* et stimulatricer de la croissance des plantes. *Agronomie*, 13: 413-421.

Biswas, S.K. (1997). *Biocontrol potential of Chaetomium globosum against Drechslera sorokiniana*, M.Sc. Thesis, IARI,New Delhi pp 63.

Biswas, S.K., Srivastava, K.D., Aggarwal, R., Prem Duraja and Singh, D.V. (2000). Antogonism of *Chaetomium globosum* to *Drechslera sorokiniana* the spot blotch pathogen of wheat, *Indian Phytopath.*, 53(4):436-440.

Brannen, P.M. and Backman, P.A. (1994). Suppression of Fusarium wilt of cotton with *Bacillus subtilis* hopper box formulaltions. In: "*Improving Plant Productivity with Rhizosphere Bacteria*" (M.H. Ryder, P.M. Stephens and G.D. Bowen, ends), PP. 83-85. CSIRO, division of soils, Adelaide, Australia.

Buysens, S., Huybrechts, M. and Hotte, M. (1993b). Role of Pyochelin in the antagonistic effect of *Pseudomonas aeruginosa* 7NSK2 against *Pythium* sp. In: "*Fourth International Symposium on Pseudomonas: Biotechnology and Molecular Biology*". Abstract P. 213. Vancouver, BC, Canada.

Callan, N.W., Mathre, D.E. and Miller, J.B. (1990). Bio-priming seed treatment for biological control of *Phythium ultimum* pre-erergence damping-off in Sh2 sweet corn. *Plant Disease*, 74: 368-372.

Carruthers, F.L., Shum-Thomsa, T., Conner, A.J. and Mahanty, H.K. (1995). The significance of antibiotic production by *Preudomonas aureofaciencs* PA 147-2 for biological control of *Phytophthora megasperma* root rot of asparagus. *Plant and soil*, 170: 339-344.

Chang, Y.C., Chang Y.C., Baker, R., Kleifeld, O. Chet (1986). Increased growth of plants in present of the biological control agent *Trichoderma harzianum*. *Plant Disease*, 70, 145-148.

Chernin, L., Ismailov, Z, Haran, S. and Chet, I. (1995). Chitinolytic *Enterobacter agglomerans* antagonistic to fungal plant pathogens. Appl. Envir. Microbiol., 61: 1720-1726.

Chung, H.S. and Choi, W.B. (1990). Biological control of sesame damping off in the field by coating seed with antagonistic *Trichoderma viride*. Seed Sci. and Tech., 18: 451-452.

Chung, Y.S. and Hoitink, H.A.J. (1990). Interactions between thermophilic fungi and *Trichoderma hamatum* in suppression of Rhizoctoria damping off in a bark compost-amended container medium. *Phytopathology*, 80: 73-77.

Coley-Smith, J.R., Ridout, C.J., Mitchell, C.M. and Lynch, J.M. (1991). Control of bottom rot disease of lettuce (*Rhizoctoria solani*) using preparations of *Trichoderma viride*, *T. harzianum* or tolclotos-methyl. *Plant Pathology*, 40: 359-366.

de Freitas, J.R. and Germida, J.J. (1991). *Pseudomonas cepacia* and *Pseudomonas putida* as winter wheat inoculants for biocontrol of *Rhizoctonia solani. Can. J. Microbiol.*, 37: 780-784.

Di Pietro, A. Gut-Rella, M., Packlatko, J.P. and Schwinn, F.J. (1992). Role of antibiotics produced by *Chaetomium globosum* in biocontrol of *Phythium ultimum*, a causal agent at damping-off. *Phytopathology*, 82: 131-135.

Donner, S.C., Jones, D.A., Mcclure, N.C., Rose warne, G.M., Tate, M.E., Kerr, A., Fajardo, N.N. and Clare, B.G. (1993). Agrocin 434, a new plasmid encoded agrocin from the biocontrol *Agrobacterium* strains K84 and K1026, which inhibits biovar two agrobacteria. *Physiol. Mol. Pl. Path.*, 42: 185-194.

Expert, J.M. and Digat, B. (I1995). Biocontrol of Sclerotinia wilt of Sunflower by *Pseudomonas fluorescens* and *Pseudomonas putida. Can. J. Microbiol.*, 41: 585-691.

Fernandez, M.R. (1992). The effect of *Trichoderma harzianum* on fungal pathogens infesting soybean residues. *Soil Biology and Biochemistry*, 10: 1027-1029.

Fiddaman, P.J. and Rossall, S. (1995). Selection of bacterial antagonists for the biological control of *Rhizoctonia solani* in oilseed rape (*Brassica vapus*). *Plant Pathology*, 44: 695-703.

Fridlender, M., Inbar, J. and Chet, I. (1993). Biological control of soil-borne pathogens by a β-1, 3 glucancase-producing *Pseudomonas cepacia. Soil Biology and Biochemistry*, 25: 1211-1221.

Garibaldi, A. and Gullino, M.L. (1990). Biological control of Fusarium wilt of carnation. In: "*Proceedings of the Brighton crop Protection Conference-Pests and Diseases*". Vol. 3A-1, pp. 89-95. British Crop Protection Council, Thornton Heath.

Gilbert, G.S., Hande Isman, J. and Parke, J.L. (1990). Role of ammonia and calcium in lysis of zoorpores of *Phytophthora cactorum* by *Bacillus cereus* strain UW85. *Experimental Mycology*, 14: 1-8.

Gutterson, N. (1990). *Microbial fungicides: recent approaches to elucidating mechanisms. Critical Reviews in Biotechnology*, 10: 69-91.

Handelsman, J., Raffel, S., Mester, E.H., Wunderlich, L. and Grau, C.R. (1990). Biological control of damping-off alfalfa seedlings with *Bacillus cereus* UW85. *Appl. Environ. Microbiol.*, 56: 713-718.

Hasegawa, S., Kodama, F. and Kondo, N. (1991a). Suppression of Fusarium wilt of adzukibean by rhizosphere microorganisms In: "*Natural Occurring Pest Bioregulators*" (P.A. Hedin, ed.), pp. 408-416. American Chemical Society, Washington, D.C.

Hebbar, K.P., Davey, A.G., Merrin, J., Mc Loughlin, I.J. and Dart P.J. (1992). *Pseudomonas cepacia*, a potential suppressor of maize soil-borne diseases-seed inoculation and maize root colonization. *Soil Biology and Biochemistry*, 24: 999-1007.

Hebbar, P., Berge, O., Heulin, T. and Singh, S.P. (1991). Bacterial antagonists of sunflower (*Helianthus annus* L.) fungal pathogens. *Plant and soil*, 133: 131-140.

Howell, C.R. and Stipanovic, R.D. (1980). Suppression of *Pythium ultimum* induced damping-off of cotton seeddlings by *Preudomonas fluorescens* and its antibiotic pyoluteorin. *Phytopathology*, 70: 712-715.

Howell, C.R. and Stipanovic, R.D. (1995). Mechanisms in the biocontrol of *Rhizoctonia solani*-induced cotton seedling disease by *Gliodadium virens*: antibiosis. *Phytopathology*, 85: 469-472.

Howell, C.R., Stipanovic, R.D. and lumsden, R.D. (1993). Antibiotic production by strains of *Gliocladium virens* and its relation to the biocontrol of cotton seedling diseases. *Biocontrol Science and Technology*, 3: 435-441.

Hussain. S., Ghaffar, A. and Aslam, M. (1990). Biological control of *Macrophomina phaseolina* charcoal rot of sunflower and mung bean. *Journal of Phytopathology*, 130: 157-160.

Inbar, J., Abramsky, M., Cohen, D. and Chet, I. (1994). Plant growth enhancement and disease control by *Trichoderma harzianum* in vegetable seedlings grown under commercial conditions. *European J. Pl. Path.*, 100: 337-346.

Jones, D.S., Ryder, M.H., Clare, B.G., Forrand, S.K., Kerr, A. (1988). Construction of a Tra deletion mutant of p Ag K84 to safeguard the biological control of crown gall. *Mol. Gen. and Gen.*, 212: 207-214.

Kay, S.J. and Stewart, A. (1994). Evalution of fungal antogonist of control of onion white rot in soil box trails. *Plant Pathology*, 43: 371-377.

Kleifeld, O. and Chet, I. (1992). *Trichoderma harzianum*-interaction with plants and effect on growth response. *Plant and soil*, 144: 267-272.

Knudsen, G.R., Eschen, D.J., Dandurent, L.M. and Bin, L. (1991). Potential for biocontrol of *Sclerotina sclerotiorum* through colonization of sclerotia by *Trichoderma harzianum*. *Plant Disease*, 75: 466-470.

Kraus, J. and Loper, J.E. (1992). Characterization of a genomic region required for production of antibiotic pyoluteorin by the biological control of pythium damping off of cucumber. *Phytopathology*, 82: 264-271.

Landenpera, M.L., Simon, E. and Uoti, J. (1991). Mycostop-a novel biofungicide based on streptomyces bacteria. In: "*Biotic Intractions and soil-borne Diseases*" (A.B.R. Beemster, G.J. Bollen, M. Gerlagh, M.A. Ruissen, B. Schippers and A. Tempel, eds), PP. 258-263. Elsevier, Amsterdom.

Latunda-Dada, A.O. (1993). Biological control of southern blight disease of tomato caused by *Sclerotium rolfsii* with simplified mycelial formulation of *Trichoderma koningii Plant Pathology*, 42: 522-529.

Lyuch, J.M., Lumsden, R.D., Alkey, P.T. and Cusley, M.A. (1991). Prospects for control of Pythium damping-off of lettuce with *Trichoderma, Gliocladium*, and *Enterobacter* spp. *Biology* and *Fertility of soil*, 12: 95-99.

Martyn, R.D., Biles, C.L. and Pillard, E.A. (1991). Induced resistance to Furarium wilt of water melon under simulated field conditions. *Plant Disease*, 75: 874-877.

Mazzola, M., Fujimoto, D.K., Thomashow, L.S. and Cook, R.J. (1995). Variation in sensitivity of *Gaeumannomyces graminis* to antibiotics production by Fluorescent *pseudomonas* spp. and effect on biological control of take all of wheat. *Applied and Environmental Microbiology*, 61: 2554-2559.

Mc Laren, D.L., Huang, H.C., Kozub, G.C. and Rimmer, S.R. (1994). Biological control of Sclerotinia wilt of sunflower with *Talaromyces flavus* and *Coniothyrium minitans*. *Can. J. Microbiol.*, 78: 231-235.

Mc Quilken, M.P. and Whipps, J.M. (1995). Production, survival and evaluation of solid-substracte inocula of *Coniothyrium minitans* against *Sclerotinia sclerotiorum*. *European J. Pl. Path.*, 101: 101-110.

Moulin, G. Lemanceau, P. and Alabouvette, C. (1994). Control by Fluorescent pseudomonads of *Pythium appanidermatum* root rot, responsible for yield reduction in soillers culture of cucumber. In: "*Improving plant productivity with Rhizosphere Bacteria* "(M.H. Ryder, P.M. Stephens and A.D. Rovira, eds).

Osburn, R.N., Milner, J.L., Oplinger, E.S., Smith, R.S. and Handelsman, J. (1995). Effect of *Bacillus cereus* UW 85 on the yield of soybean at two field sites in Wisconsin. *Plant Disease,* 79: 551-556.

Ousley, M.A., Lynch, J.M. and Whipps, J.M. (1994). The effects of addition of *Trichoderma* inocula on flowering and shoot growth of bedding plants. *Scientia Horticulture,* 59: 147-155.

Oyarzum, P.J., Portma, J., Luttikholt, A.J.G. and Hoogdang, A.E. (1994). Biological control of foot and root in pea caused by *Fusarium solani* with non pathogenic *Fusarium oxysporum* isolates. *Can. J. Botany,* 72: 843-852.

Papavizas, G.S. and Collins, D.J. (1990). Influence of *Gliocladium virens* on germination and infectivity of sclerotia of *Sclerotium rolfsii. Phytopathology,* 80: 627-623.

Park, C.S. and Yeom, J.R. (1994). Biological control of cucumber damping off and enhancement of seedling growth by low temperature-tolerent *Pseudomonas fluorescens* M45 and MCO7. In: *Improving plant productivity with Rhizosphers Bacteria* (M.H. Rgder, P.M. Stephens and G.D. Bowen eds), pp. 54-56. C SIRO, Division of soils, Adelaide, Australia.

Parke, J.I., Rand, R.I., Joy, A.E. and King, E.B. (1991). Biological control of Pythium damping off and Aphanomyces root rot of peas by application of *Pseudomonas cepacia* or *P. fluorescens* to seed. *Plant Diseases,* 75: 987-992.

Paulitz, J.C. and Loper, J.E. (1991). Lack of a role for fluorescent siderophores production in the biological control of Pythium damping off of cucumber by strain of *Pseudomonas putida, Phytopathology,* 81: 930-935.

Paulitz, T., Windham, M. and Baker, R. (1986). Effect of peat vermiculite mixes containing *Trichoderma harzianum* on increased growth response of radish. *American Soc. Hort. Sci.,* 111: 810-814.

Press, C.M. and Kloepper, J.W. (1994). Growth chamber and field screening procedures for evaluation of bacterial inoculants for control of Rhizoctoria solani on cotton. In: *Improving Plant Productivity with Rhizosphere Bacteria"* (M.H. Ryder, P.M. Stephens and G.D. Bowen, eds), pp. 61-63. CSIRO, Division of soils, Adelaide, Australia.

Rankin, L. and Paulitz, T.C. (1994). Evaluation of rhizosphere bacteria for biological control of Phythium root rot of greenhouse cucumbers in hydroponic culture. *Phytopathology,* 78: 447-451.

Rattink, H. (1993). Biological control of Fusarium crown and root rot of tomato on a recirculation substrate system. *Mededelingen Faculteit Landbouwwet enschappen Rijksuniversitcit Gent,* 58 (3b), 1329-1337.

Reddy, M.S., Covert, D.C., Craig, K.A., Rennie, R.J. Mckenzie, D. Verma, P.R. and Mortensen, K. (1994). Suppression of pre emergence damping off of canola, caused by *Rhizoctonia solani* by rhizobacteria. In: *"Improving Plant Productivity with Rhizosphere Bacteria".* M.H. Ryder, P.M. Stephens and G.P. Bowen, eds). pp 80-82. CSIRO, Division of Soils, Adelaide, Australia.

Ristaino, J.B., Lewis, J.A. and Lumsden, R.D. (1994). Influence of isolates of *Gliocladium virens* and delivery systems on biological control of southern blight on carrot and tomato in the field. *Plant Disease,* 78: 153-156.

Pandey, S.K. (2004). *Effect of some antagonists on Drechslera oryzae, causal agent of brown leaf spot of Paddy.* M.Sc. Thesis, C.S.A.U.A. and T. pp 49.

Sen, B. (1999). Kalisena-an ecofriendly biopesticide-cum-biofertilizer from *Aspergillus nigar* AN-27. Division of Plant Pathology, IARI, New Delhi pp 10

Taylor, A.G., Min, T.G., Harman, G.E. and Jin, X. (1991). Liquid coating formulation for the application on of biological treatments of *Trichoderma harzianum*. *Biological control*, 1: 16-22.

Trapero-casas, A., Kaiser, W.J. and Ingram, D.M. (1990). Control of Phythium seed rot and pre emergence damping-off of chickpea in the US Pacific Northwest and spain. *Plant Disease*, 74: 563-569.

Tu, J.C. (1991). Comparison of the efficacy of *Glicladium virens* and *Bacillus subtilis* in the control of seed rots and root rots of navy beans. *Mededelingen Faculteit Landbouwwetenschappen Rijksuniversiteit Gent* 56(2a): 229-235.

Vakili, N.G. (1992) Biological seed treatment of corn with mycopathogenic fungi. *Journal of Phytopathology*, 134: 313-323.

Voisard, C., Keel, C. Haas, D. and Defago, G. (1989). Cyanide production by *Pseudomonas fluorescens* helps suppress black root rot of tobacco under gnotobiotic conditions. *European Moolecular Biology Organisation Journal*, 8: 351-358.

Whipps, J.M. (1997). Developments in the Biological Control of soil-borne plant pathogens. Academic Press Ltd., pp. 133.

Wolffhechet, H. and Funck Jenson, D. (1992). Use of *Trichoderma harzianum* and *Gliocladium virens* for the biological control of Post-emergence damping-off and root rot of cucumbers caused by *Phythium ultimum*. *Journal of Phytopathology*, 136: 221-230.

Sustainable Disease Management of Agricultural Crops (2011) *Pages* **149–167**
Editors: **S.K. Biswas and S.R. Singh**
Published by: **DAYA PUBLISHING HOUSE, NEW DELHI**

Chapter 9

Integrated Disease Management Strategies in Pulses

☆ *R.K. Prajapati[1], R.G. Chaudhary[2], S.R. Singh[3] and R.K. Gangwar[4]*
[1]Krishi Vigyan Kendra (Baghpat), S.V. Patel University of Agriculture and Technology, Modipuram, Meerut, U.P.
[2]Division of Crop Protection, Indian Institute of Pulses Research, Kanpur – 208 024, U.P.
[3]Krishi Vigyan Kendra, Belatal, Mahoba – 210 423, U.P.
[4]Krishi Vigyan Kendra, Chomu (Tankarda), Jaipur, Rajasthan

Introduction

Pulses are an important component of farming systems of India both ecologically and in terms of human and animal nutrition. Pulses have been an integral part of Indian agriculture since time immemorial. It is gaining importance globally, as they are the cheap source of vegetable protein. Awareness about the utility of pulses in human food, animal feed and soil health in increasing world production of pulses now stands at 58.0 million tonnes. India has the distinction of being the top producer of pulses in the world accounting 25 per cent of the global output, the expected production of pulses for 2003-04 being 15.2 million tonnes in the country. Inspite of this, the country has experienced progressive decline in per capital availability of pulses from 69g in 1961 to 37 g at present. The present productivity of pulses is only 640 kg ha-1, which is quite low as compared with other countries. The present Indian population of about 1015 million is expected to rise to 1350 million by 2020 AD. Keeping in view the directly requirements of protein, a minimum of 27 million tons of pulses is required by 2020 AD.

The common pulses are chickpea (*Cicer arietinum* L.) pigeonpea [*Cajanus cajan*(L.) Millsp.], mungbean (*Vigna radiata* (L.) Wilczek], urdbean [*V. mungo* (L.) Hepper], lentil (*Lens culinaris* Medik.) and fieldpea (*Pisum sativum* L.) are the major pulse crops grown under varied agro-climatic situations *viz.*, North West Plain Zone (west Uttar Pradesh., Delhi, Punjab, Haryana and west Rajasthan), North East Plain Zone (eastern Uttar Pradesh, Bihar, West Bengal and Assam), Central Zone (Madhya Pradesh, Maharashtra, Gujarat and Rajasthan), Southern Zone (Tamil Nadu, Andhra Pradesh and

Karnataka. The major diseases of pulses are wilt, root rot, mould, ascochyta blight, phytopthora blight, alternaria blight, leaf spot, mildews, rust, anthracanose, sterility mosaic and common bean mosaic known to limit yields of these crops. Management of losses caused by diseases are therefore, an important component of the crop production technology. Integrated Disease Management (IDM) is the only answer to combat the huge losses. IDM encourages the most compatible and ecologically sound combinations of available disease suppression techniques, to keep the incidence/intensity of diseases below economically damaging levels.

Disease Problems

Most of the pulses are cultivated on marginal and rainfed area and are prone to attack by a large number of plant pathogens. Among these, fungi, bacteria, nematodaes and virus are very important and cause diseases which are of economic significance. Chickpea and pigeonpea, the two most important crops are attacked by over 170 and 210 pathogens, respectively (Nene *et al.*, 1972). Similarly, mungbean, urdbean, lentil, pea and rajmash are also infected with an array of pathogens. Several workers have time to time reviewed the diseases of different pulse crops. Some of the extensive and recent ones are being mentioned here on pulses (Gurha *et al.*, 2003; Singh and Naimuddin, 2004; Gurha, 2004; Chaudhary *et al.*, 2004; Vishwa Dhar and Chaudhary, 2004;Gurha and Vishwa Dhar, 2004 and Vishwa Dhar *et al.*, 2004) detailed information. The major diseases of pulse crops are economically important in India are listed in Table 9.1. In this chapter we attempt to review the available management options that could be integrated to improve the productivity and reduce the present gap between potential and realized yields.

Table 9.1: Major Disease Problems of Pulse Crops in India

Crop	*Disease*	*Pathogen*	*Problem Areas*
Chickpea	Wilt/root rot	*Fusarium oxysporum* f.sp.*ciceri*, *Rrhizoctonia* spp., *Fusarium* spp.	NEPZ, CZ, SZ
	Grey mould	*Botrytis cinerea*	Tarai, UP, Bihar
	Ascochyta blight	*Ascochyta rabiei*	NWPZ
Pigeonpea	Wilt	*Fusarium udum*	NEPZ, CZ, SZ
	Sterility mosaic		NEPZ, SZ
	Phytophthora blight	*Phytophthora drechsleri* var .*cajani*	NEPZ, NWPZ
	Alternaria blight	*Alternaria alternata*	NEPZ (post rainy)
Mungbean and urdbean	Yellow mosaic	Mungbean yellow mosaic virus	NEPZ, NWPZ
	Cercospora leaf spot	*Cercospora cruenta* and *C. canescens*	SZ (rabi), NEPZ (kharif)
	Powdery mildew	*Erysiphe polygoni*	CZ(kharif), SZ (rabi)
Fieldpea	Powdery mildew	*Erysiphe pisi*	All zones
	Rust	*Uromyces viciae fabae* syn. *U. fabae*	NEPZ
Lentil	Rust	*Uromyces viciae fabae* syn. *U. fabae*	NEPZ
	Wilt/root rot	*F. oxysporum* f.sp. *lentis*, *Rhizoctonia* spp.	NEPZ, CZ
Rajma	Bean common mosaic	Bean common mosaic virus	NEPZ (rabi)
	Root rots	*Rhizoctonia* spp.	

Management Strategies

Although management practices for major diseases have been developed in the past, these have not integrated well with crop production packages. Development of means of disease management such as cultural practices and use of bio-agents in combination with host plant resistance is essentially required to develop integrated disease management packages for major diseases. IDM modules for wilt, phytophthora blight in pigeonpea wilt/root rot of chickpea have been developed. Crop rotations, wider spacing, limited use of fungicides and occasional cropping of disease tolerant varieties are among the measures being practiced by traditional farmers. Though some progress in recent past has been made towards the development of cultivars with resistance to few diseases, but their effective utilization in management of these diseases has been very limited and needs to be looked into. There seems to be great scope for exploitation of other means of disease management such as cultural practices and use of bioagents in combination with host plant resistance. These options may be evaluated in farmers' fields for adoption and further refinement.

Pigeonpea

Wilt

Fusarium wilt (*Fusarium udum* Butler) is widely distributed in almost all pigeonpea growing areas is *i.e.* India, Kenya, Malawi, Nepal, Tanzania and Uganda and is considered as the major yield reducing factor. Mean prevalence of the disease has been reported to vary from 0.1 per cent in Rajasthan to 22.6 per cent in Maharashtra and the annual crop losses in India have been estimated as 97000 tons (Kannaiyan *et al.,* 1984). Butler, 1910, has described the pathogen. Rai and Upadhyay 1979 have reported *Gibberella indica* as the perfect state of *F. udum*. Other workers have indicated existence of strains in *F. udum*. Gupta *et al.* (1988) reported seven strains while as many as 11 strains have been recorded by Gaur and Sharma (1989). Reddy *et al.* (1996) suggested prevalence of five pathotypes in different parts of the country. Mishra and Vishwa Dhar (2003) reported prevalence of three variants from Kanpur, which indicate occurrence of possibly a greater variability in the pathogen than thought earlier.

Cultural practices like removal of wilted plants and crop residues, summer ploughing, soil solarization using polythene sheets and green manuring reduce disease incidence (Chauhan *et al.,* 1988; Chaudhary, 1998). Early sowing, well distributed rains, weed free, high soil fertility, long duration cultivars, low plant density and higher biomass increase crop vulnerability to wilt (Reddy *et al.,* 1994). Intercropping with cereals,tobacco and few other crops has been reported as most effective. For example Sorghum and pigeonpea in the ratio of 1:1 reduced the wilt disease. (Natarajan *et al.,* 1985). Seed treatment with Benlate,Thiram,Carbendazim and Benomyl + Thiram and Carbendazim + Thiram recommended for reducing the disease incidence (Vishwa Dhar and Chaudhary,1998). *Trichoderma harzianum, T. viride, T. koningi,Bacillus subtilis* and *Gliocladium virens* has been evaluated and found most effective (Chaudhary,1998). Soil amendments like zinc, boron (Subramaniam,1963), organic amendments, oil cake, neem cake, saw dust (Dasgupta and Sengupta,1989), green manuring with *Sesbania* and *Crotolaria* is also effective in reducing *F. udum* population and reducing the disease incidence (Dasgupta and Sengupta,1989). Wilt resistant varieties have listed resistant source of *F. udum* (Vishwa Dhar *et al.*2000) and (Vishwa Dhar and Chaudhary, 2001).

Sterility Mosaic

Sterility mosaic is of great economic importance in high endemic areas of Uttar Pradesh, Bihar, Gujarat, Karnataka and Tamil Nadu. The disease is now confirmed to be caused by a virus of asymmetric morphology, which has been named as " pigeonpea sterility mosaic virus" (Lava Kumar *et al.*, 2000) under field conditions. The disease is transmitted by an eriophyid mite (*Aceria cajani*) which is specific to pigeonpea (Seth, 1962).

Intercropping with non-host crops has not been found to influence the disease. Breaking the pigeonpea cycle by removal of off-season pigeonpea plants and discouranging perennial crops in the disease endemic areas may reduce the off season survival of vector and pigeonpea inoculum resulting in disease escape during the subsequent main crop season. Seed dressing with 10 per cent aldicarb @ 3 g kg/ha seed has been found to protect the crop up to 45 days after sowing (Reddy *et al.*, 1990). Spraying of acaricides like Kelthane, metasystox or dimethoate at 0.1 per cent effectively controls the vector and reduces disease incidence (De *et al.*, 1995). A large number of resistant sources, which have been listed by different workers (Nene *et al.*, 1989; Reddy and Vishwa Dhar, 1998; Vishwa Dhar and Chaudhary, 2001).

Phytophthora Blight

Phytophthora blight is considered primarily a seedling disease causing heavy mortality in low laying areas of the field and in water paths as it is promoted by higher humidity. It is of north and north eastern plains of the country affecting predominately the stem and occasionally the leaves. The disease is of more economic importance in the short duration genotypes as compared to long duration varieties. The losses due to the disease can be very high even in long duration varieties in season of high rainfall and water logging. Avoid pigeonpea cultivation in field of water logging areas of the field which are prone to water logging. Sowing the crops on rides or raised beds than on flat beds as disease incidence can minimize. Seed dressing with Metalaxyl (6 g/Kg seed) is very useful in minimizing losses caused by the disease (Kannaiyan and Nene, 1984). Intercropping of pigeon pea with crops like help in minimizing the disease. These intercrops provide quick ground cover and present splashing of inoculum from soil to the foliage of pigeonpea under low level of disease pressure, these individual components may help in managing the disease. But under high disease pressure, irrigation of these components was found necessary to keep the disease under control. Foilar spraying with metalaxyl (0.1 per cent) or metalaxyl +mancozeb (0.3 per cent) was also found necessary conditions favourable for the disease. Varieties tolerant/moderate resistant have been reported by Vishwa Dhar and Chaudhary, 2001).

Chickpea

Wilt

Wilt (*Fusarium oxysporum* f. sp. *ciceri*) is estimated up to 10 per cent yield losses through India. It is considered as the most serious disease. It is a typical vascular disease causing xylem necrosis. The disease is systemic in nature and plants may be infected at any stage. The pathogen is both soil as well as seed borne. Four pathogenic race (1,2,3,4) has been reported in India.

The avoiding of chickpea in a heavily infested soil can be reduced the yield losses cultivation of susceptible genotypes should be avoided for a longer period in endemic areas. Deep ploughing in hot summer months and removal of host plant debris from the field helps in reducing the pathogen population in the soil. Transparent polythene sheet for 6-8 weeks during April-May effectively reduced pathogen population in the soil (Chauhan *et al.*, 1988). As the polythene sheet may be expensive, deep ploughing during hot summer months may adopted (Singh and Singh 1984, Dahiya *et al.*, 1988). Intercropping or crop rotation with wheat, barley, linseed and mustard may help in management of wilt (Gurha and Srivastava, 2002). Seed treatment with Benlate T(30 per cent benomyl +30 per cent

Thiram) @ 1.5 g kg/seed or Bavistin @ 2.5 g/kg seed (Haware *et al.*, 1978). Carbendazim @ 2.0 g/kg seed, thiram + carbendazim @ (2+1) g/kg of seed of or *T. viride* + vitavax (4+1) g/kg seed may reduce the wilt incidence in the field (Gupta,1995). Judicious use of fungicides as seed treatment before sowing may ensure better plant stand in the field of growth from infection of different soil borne pathogens.

The use of bioagents like *Trichoderma* spp., *Gliocladium virens, Bacillus subtilis, Pseudomonas fluorescencs*, etc., has been advocated as an integral part of plant disease management (Mukhopadhyay,1988 and 1989, Mukhopadhyay *et al.*, 1992). A large number of chickpea genotypes having moderate to high level of resistance against wilt have been identified through multilocation testing in wilt sick plots under the All India Coordinated Pulses Improvement Project since 1985. The management of dry root rot caused by *Rhizoctonia bataticola* may be adopted similar strategies. Kranti, Bharti, JG 130, Sadbhawana have fairly good tolerance against dry root rot disease in chickpea.

Ascochyta Blight

Ascochyta blight (*Ascochyta rabiei*) is of great economic importance as it can be cause 100 per cent yield loss. Cool wet weather favours the disease development. The pathogen is higher variable and occurrence of six races from Syria, 12 from India and 7 from Pakistan have been reported (Singh, 1990). The disease is both externally as well as internally seed borne and the pathogen can survive up to 5 months on seed surface spread at 25-30 °C and in the crop residues. The secondary spread of the disease is through conidia and ascospores of the fungus.

Agronomic manipulation likes late sowing, wider row and plant spacing and intercropping with wheat, barley, etc., reduces the disease incidence in the field. Destruction of diseased plant debris after the harvest of the crop and deep ploughing which will hasten the decomposition of infested straw. Seed should be produced in disease free areas to prevent the introduction of disease in new areas or more virulent races into already infested areas. Crop rotation with wheat, barley,rapeseed and mustard particularly in the areas where the pathogen produces perfect stage and destruction of weed hosts of the pathogen are also helpful in reducing the disease incidence.Thiram, benomyl, calixin-M, thiobendazole, bavistin+thiram (1: 2), hexacarp, captan, captafol @3 g/kg seed and Rovral @ 2.5 g/kg seed are recommended as seed treatment.Secondary spread of disease may be managed effectively by 2-3 sprays of hexacarp,captaf, Indofil M-45 or Kavach @ 350 g/acre in 100 litres of water. During February and March if weather remains cloudy with intermitted rains, one protective spray should be applied immediately. High level of resistance is available in wild species. There is need to use biotechnology tools for transferring resistance from these wilt to cultivated types.

Botrytis Gray Mold

Botrytis gray mold (*Botrytis cinerea*) appears regularly every year in moderate to severe form depending upon the environmental condition. Epidemic of this disease was recorded during 1982-83 in India at North western states causing complete damage to the crop. The pathogen *B. cinerea* can survive through infected seed, crop debris and other host-plants either parasitically or as internally seed borne. In soil it can survive as mycelium and sclerotia both. Eight plant species and 21 weeds are reported to be host of this fungus.The crop may be sown late in the first fort night of November particularly in irrigated areas, lower seed rate should be used, wide row and plant spacing should be adopted (Kumar,1995). Tall, erect and compact genotypes may be grown which attract less incidence of this disease. Chemical seed treatment with a combination of bavistin +thiram (1:2), dithane M-45, bavistin @ 3 g kg/seed, thiobendazole @ 2 g kg/seed, Rovral and Ronilan @4 g/kg seed (Grewal and

Laha, 1983;Singh and Kaur,1990). For the control of secondary infection, dithane M-45 @ 350 g, thiobendazole or bayleton @200g per acre in 100 litres of water should be sprayed (Haware *et al.*, 1997). Spray can be repeated depending on the weather conditions. Some of the tolerant varieties identified are Pusa 209, SAKI 9516 and Sweta, which should be grown in combination with other management.

Mungbean and Urdbean

MYMV (Mung bean Yellow Mosaic Virus)

The disease is caused by mungbean yellow mosaic virus is the most destructive and widely distributed disease of mungbean and urdbean. MYMV is transmitted by a white fly bigeminivirus. It occurs in almost all the states of India wherever these crops are grown but attains serious proportions in Uttar Pradesh, Uttaranchal, Madhya Pradesh, Chhattisgarh, Bihar, Jharkhand, Orissa, Punjab, Haryana, Delhi, Rajasthan and Andhra Pradesh. The yield losses from 10 to 100 per cent, caused by this disease, depending upon the crop stage at which the plants get infected. The major setback to pod bearing and yield in urdbean occurred at 25 per cent disease intensity and MYMV infection delayed pod maturity (Singh *et al.*, 1982). Thakur and Agrawal (1991) recorded 57.6,44.3 and 36.5 per cent reduction in grain yield in plants manifesting symptoms at 30,40 and 50 days after sowing, respectively.

Cultivation of resistant varieties is the most economical method, which should be supported by control of vector by using chemicals or adopting cultural practices. Since MYMV develops more severe symptoms and causes heavy yield losses in early infected plants. It is suggested to avoid the early infections by cultural means or vector control. A large number of mungbean and urdbean varieties/ lines were screened under the AICRP on MULLaRP at different locations broad based resistance against the disease (Amin, 1985;Vishwa Dhar, 1996; Singh *et al.*, 2000; 2002). IPU 94-1 (Uttara and DPU 88-3) highly resistant to MYMV. These could be utilized for breeding MYMV resistant varieties (Singh *et al.*, 2000; 2002).

Systemic insecticides, aldicarb and disyston were found effective in reducing the disease incidence in mungbean when applied in the soil at sowing time @ 1 kg a.i./ha. Soil application of disulfoton @ 2 kg a. i./ha reduced 84 per cent whitefly population,21 per cent yellow mosaic incidence and increased yield by 42 per cent whereas, phorate (2 kg a.i./ha) resulted in minimizing 73 per cent disease incidence and yield increase in mungbean. Higher per cent increase in yield is not only due to reduced MYMV incidence but also as a result of control of insect pests affecting the crop (Singh *et al.*, 1980;Varma *et al.*,1992). Foliar applications of metasystox at weekly intervals starting from 15 days old crop has been found effective in vector control reducing the MYMV incidence. Malathian, ambithion and baythion @ 0.1 per cent and monocrotophos (0.05 per cent) were found highly effective in quickly killing the white fly populations. Mineral oil sprays (2 per cent) are also found effective in killing the vector within 1-3 minutes in glass house tests. A fungus *Paecilomyces farinosus* has been found parasitic on the vector, *Bemisia tabaci* but the practical utility of the fungus to control the white fly is yet to be explored.

Cercosporal Leaf Spots

Cercospora leaf spot (CLS) (*Cercospora cruenta, C. canescens*) causes leaf spotting and defoliation of mungbean and urdbean in warm and humid climates. Severe damage from CLS has been reported from Indonesia, India, Bangladesh, Taiwan, Philippines, Thailand and West Malaysia. In India, the disease occurs in most of the mungbean and urdbean growing states. Singh *et al.*2000 reported yield losses to the tune of 50 per cent in several disease field.

Disease free seed may help in reducing the seedborne inoculum of the pathogen. Field sanitation, crop rotation, destruction of infected crop debries and avoiding collateral hosts in the vicinity of the crop may help in reducing the disease incidence. Adjustment of date of sowing in mungbean so that the crop may escape heavy rains has been recommended to control the disease. Delayed sowing of mungbean and urdbean reduces the disease severity. The early sown crop of mungbean in second week of July at 20 cm row spacing control more disease as compared to delayed sowing with wider spacing of 40 cm. Mulching reduces primary inoculum of the pathogen thereby decreasing disease incidence in urdbean. Mancozeb,carbendazim, copper oxychloride and benomyl are reported to reduce the disease incidence considerably. However, two sprays of carbendazim (300 g/ha) at 30 and 45 days after sowing increased the yield in mungbean. Four sprays of carbendazim at 10 days interval applied 30 DAS effectively controlled, however, highest cost: benefit ratio of 1:5.5 with one sprays of carbendazim (0.025 per cent) at 30 DAS with yield increase by 250 kg/ha was observed (Singh and Gurha, 1996). Thiophenate methyl (0.025 per cent) gave the maximum disease control closely followed by carbendazim (Singh *et al.*, 2001). However, carbendazim delayed crop maturity by about a week.

Powdery Mildew

Powdery mildew (*Erysiphe polygoni*) is wider spread and occurs throughout the Indian subcontinental and south East Asia countries. The disease causes serious damage to mungbean and urdbean especially in Maharashtra, Andhra Pradesh, Tamil Nadu, Madhya Pradesh, Rajasthan, Himachal Pradesh and Uttar Pradesh. In India, the losses due to powdery mildew in winter sown urdbean and mungbean are more as compared to rainy season crop (Tiwari, 1977). Delayed sowing of mungbean and urdbean with wider spacing has been found to considerably reduce the disease severity. However at Coimbatore, early sown crop of urdbean in July showed lower incidence of the disease than the late planted crop in August- January (Raguchandan and Rajappan, 1995). Thiovit,elosal,Karathane,Calixin, Bavistin,Benlate, Topsin–M, Microsulf,Cosam, wet sulf,sulphur dust and Indofil Z-78 have been found effective to control the disease under field conditions. A fungus *Ampelomyces quisqualisces* indicated hyper parasitism on powdery mildew fungus. The possibility of its use as biological agent for powdery mildew control on wide scale may be explored.

Pea

Pea is an important winter pulse crop in Uttar Pradesh, Bihar,Rajasthan,Madhya Pradesh and West Bengal. Among many biotic stresses diseases like seed and root rots (*Rhizoctonia solani,Sclerotium rolfsii* and *Fusarium* sp., rust (*Uromyces viciae-fabae*) and powdery mildew (*Erysiphe pisi*) are the key factors,which causes heavly yield losses. While seed and root rots reduce plant stand in the field,rust and powdery mildew affect the aerial plant parts during podding and maturity stages. The latter tow diseases causes forced drying of the crop during Feb-March resulting in heavy yield losses in pea crop.

Seed and Root Rot/Wilt Diseases

Soil is the main harbour of primary inoculum of all important seedling and root diseases either in state or as resting structures in the crop debrics. Therefore, crop rotation,intercropping,destruction of crop debris and clean cultivation do help mainly in the management of seed rots, damping off, collar rots,root rot and wilt. Crop rotations with sugarbeet and fodder maize after every 5-6 years intervals help in wilt management(Yainkov,1989). Soil application with phosphorus,potassium and calcium and optimum moisture during germination also facilitates uniform plant stand by checking seed and seedling rots and damping off diseases. Seed treatment with carbendazim, thiram, carbendazim+thiram, molybdenum, boron, thiram+graphite and metalaxyl was found effective for control the disease. Use of zinc salt of 2,4,5-T@2-4g/kg seed found effective against pre-emergence mortality. (Meuli *et al.*, 1947). Paul and Rathour (1947) recommended 2-3 sprays of bavistin (1.5 per cent) or Indofil M-45(0.2 per cent) at flowering for the elimination of the seed borne pathogens causing seedling and root rots diseases.

Antagonists found effective against root rot are *Pseudomonas fluorescence* (Park, 1991), *Bacillus subtilis* and *Gliocladium virens* (Hwang and Chakravarty, 1992) has been found effective. Utikar and Sulaiman, 1997 and Puzio-Tdzkowaska, (1990) reported many varieties/cultivars has combined resistance against wilt and ascochyta foot rot.

Aerial Parts Diseases

Powdery mildew (*Erysiphe pisi*), rust (*Uromyces viciae-fabae*), white rot (*Sclerotinia sclerotiorum*), downy mildew (*Peronospora viciae*), stem rot (*Sclerotium rolfsii*), blight (*Ascochyta pinodes*), leaf spots (*Alternaria* spp., *Cercospora pisi* var. *sativae, Collectrichum, Phullosticta pisi,* grey mold (*Botrytis cinerea*), pod rots (*Phytophthora parasitica*). Among these powdery mildew and rust are economically caused huge losses.

Powdery mildew and rust are the most important diseases of pea in India sub-continent. Certain cultural management is do help in minimizing the infectious of aerial diseases. Sowing in October helps in escaping the damage caused by rust and powdery mildew (Sangar and Singh, 1994) in plains and to some extent ascochyta blight in the hills. Similarly, growing early maturing short duration varieties, avoiding over crowding of plants, excessive irrigation before podding help in reduce downy mildew, rust and white rot disease. Pea intercropping with mustard, linseed, wheat, gram and barley delays the appearance and buildup of powdery mildew. Among systemaic fungicides, benomyl, chlorothalonil and carbendazim are quite effective.Control of white rot can also be achieved through carbendazim,captafal or triadiomefor sprays (Srivastava *et al.*, 1993). Singh and Dhar (1997) found seed treatment with Mancozeb and two sprays each of Mancozeb and Dinocop starting at 75 DAS as most effective against powdery mildew,rust, blights on variety T163 and Rachna. Singh *et al.* (1992) suggested combinations of seed treatment with carbendazim or thiram and two foliar sprays each of Mancozeb and Dinocap followed by one sprays of Endosulfan for powdery mildew,rust and other leaf diseases. Singh (1991) reported *Penicillium* spp.,*Paecilomyces lilacinus,Aspergillus* spp., *Trichoderma harzianum* against soil inoculum. Spraying of ginger extract @20000 ppm or neemazole (neem product) was found highly effective against pea powdery mildew (Singh *et al.*, 1991 and Singh, 1997).

Lentil

Lentil (*Lens culinaris* Medik.) is an important pulse crop grown in countries of India subcontinent, Mediterranean, Latin America and parts of East Africa and all part of the world. The total world area under lentil cultivation is about 3.2 million hectare of which approximately 31 per cent is in India. Lentil suffers due to a large number of diseases caused by fungi, viruses and nematodes. Important key diseases including vascular wilt (*Fusarium oxysporum* f. sp. *lentis*), collar rot (*Sclerotium rolfsii*), rust (*Uromyces viciae fabae*), ascochyta blight (*Ascochyta lentis*), Downy mildew (*Peronospora lentis*), powdery mildew (*Erysiphe polygoni*), gray mould (*Botrytis cinerea*), and pea seed borne mosaic virus (PSbMV).

Root and Stem Diseases

Vascular Wilt

The first published record of vascular wilt of lentil was made from India (Vasudeva *et al.*, 1952). In Madhya Pradesh, as high as 50-78 per cent incidence of this diseases has been recorded (Agrawal *et al.*,1993 and Khare *et al.*, 1979). The plants affected by wilt mostly die but in few cases seeds may also form, as such losses may not be exactly equal to the incidence of the diseases. It has been observed that every 10 per cent increase in wilt incidence causes 8.77 per cent losses in yield (Erskine *et al.*, 1989). Symptoms of the disease are different at three stages. The germinating seed may rot, post emergence stage and browning of radicle with stunting and finally collapse of seedling is observed. True vascular wilt is seen at adult stage of the plant.

The wilt is caused by *Fusarium oxysporum* var. *lentis*.The pathogen is highly host specific and having great variability in its various isolates does exit on the basis of the difference in virulence. The pathogen grows in xylem vessels and initiates lysis formation. It produces thermostable non-specific metabolites capable of producing wilt symptoms and hydrolytic and pectolytic enzymes. It is seed and soil borne pathogen. The disease is favoured in sandy loam soils with sufficient aeration, 25 per cent moisture, 7.5-8.0 pH and 20-30°C soil temperature (Sharma *et al.*, 1971). The varieties viz, L 80, Pusa 3, Pant 234, Pant L 4 and Pant L 406 found resistant to wilt and ascochyta blight. The fungicidal seed treatment protects the seed from lentil wilt pathogen. Some of the seed dressers like benomyl or thiram + carbendazim (2:1) 205 to 3 g/kg of seed may reduce the wilt incidence also. Soil application of non-targeted chemicals like rogors 5G, diptrex 5G or solvirex 5G @ 20kg/ha control the wilt at seedling stages. Pre soaking of seed with 80 ppm. Mn and Zn have also been found effective against lentil wilt. High doses of phosphorus reduce the population of *Fusarium* in soil as well as incidence of lentil wilt. Soil amendment with zinc 10 kg/ha is also effective. Delayed sowing, crop rotation with cereals control the disease. The best diseases controlled by seed treatment with carboxin @ 05 g/kg and soil application of mass culture of *Gliocladium virens* @ 200 kg/ha.

Collar Rot

The collar rot of lentil caused by *Sclerotium rolfsii* Sacc, occurs mainly in India and several other tropical and sub-tropical countries. The pathogen attacks young seedlings at the collar portion exhibiting damping off symptoms. The plants infected at advanced stages gradually turn pale, droop and dry. White fungal growth with mustard seed like sclerotia develops on the affected part of the plant. The fungus usually survives in the soil on organic matter but it is detected from seed also. The pathogen has very wide host range but paddy, sorghum and maize are not infected. The disease is favoured by high soil moisture, high humidity, and good sunshine with 25-30°C temperature which prevail in India generally in September. Hence, early sown crop suffers more due to this disease.

High doses of potassium and phosphorus reduce the disease considerably. (Agrawal *et al.*1976; Kannaiyan and Nene, 1978; Khare *et al.*, 1979). Pusa 2, Pant 234, JL 80, Pusa 1, LP 18 and pant 638 have been identified as resistant varieties against collar rot (Abu-Mohammad and Umesh Kumar, 1986; Agrawal *et al.*1993 and Kannaiyan and Nene, 1976). Seed treatment with thiram + mancozeb or vitavax is reported to reduce the early stage death of the plant due to collar rot. The disease is controlled greatly by manipulating sowing dates in such a way so that seedling stage of the crop does not maturing with high soil residues helps in reducing the inoculum potential of the pathogen. Seed or soil treatment with *Trichoderma harzianum, Bacillus subtilis* or *Gliocldium virens* controls the disease to a greater extent (Shreshta and Mukhopadhyay, 1992; Agrawal *et al.*,1975)

Foliar Diseases

Rust

The earliest record of lentil is from India (Butler, 1918). The disease may assume epiphytotic form under favourable weather conditions. It is estimated that every 1 per cent increase in the disease index reduces the lentil yield by 11.5 kg/ha (Singh *et al.*, 1986). Yellowish white aecia appear singly or arranged in small groups in circular manner on pods and lower surface of the leaves. The uredia develop on both the surface of leaves and are brown, powdery, and oval to circular and may coalesce with each other.

In India, the disease is caused by *Uromyces viciae fabae* (Pers.) Schroet. Six pathotypes of the pathogen have been identified (Singh *et al.*, 1980). Infection by the rust pathogen increases electrolyte leakage of cells and free amino acids in plants which is more in susceptible varieties. In India, lentil rust appears mostly in January-February as pycnia and aecia. High humidity, cloudy and drizzling weather with 20-22°C temperature favour the disease development. Broad bean, lathyrus and some species of *vicia* are the collateral hosts and may play important role in the epiphytotics of the disease (Couner and Bernier, 1982;Kapooria and Sinha, 1966)

Several varieties like Pant L406, Pant L4 in India; Takoa, Laird and Centrenela in Chile;INIA 406 in Equador; NEL 358 in Ethiopia and ILL 4605 in Morocco have been found resistant against rust. The rust inoculum from seed may be eliminated by its proper cleaning and treatment with Diclobutazole. In field, the disease is controlled effectively and economically by spraying with dithane M-45@ 2500 ppm, zineb + 0.5 per cent nickel or calixin M. Spray of crop with triadimefon 05 kg+ propineb 2 kg/ha at pre-bloom,full bloom and early pod maturity stage controls the lentil rust very effectively. Use of early varieties, irrigation and destruction of remains after harvest may help in reducing the inoculum of rust pathogen.

Ascochyta Blight

This disease was first recorded from India Sattar (1934). The disease may cause up to 40 per cent losses in yield. Small, tan to brown coloured spots with dark margins appears on leaves, stem and pods on which dark brown to black pin head size pycnidia develop in concentric rings. The lesions may girdle the stem causing rapid death of plant parts above the girdle. *A. lentis* is named as lentil isolate. Variability in some characters has been also observed in different isolates of pathogen from lentil itself (Kaiser and Hannan, 1982). The disease is transmitted through infected crop residues.Infected seed is the only source of inoculum while presence of infected crop stubbles causes an early and severe epiphytotic. Seed treatment with benomyl, carbendazim, carbathin, iprodion, thiobendazole and metalaxyl @ 1 per cent has been reported to disinfect and protect the lentil seed from *A. lentis.* The disease can be controlled with 3 sprays of benoymyl 0.2 per cent coupled with yield increase of 81 per cent. Sun drying, burning of diseased crop debris, crop rotation, early sowing to escape moist weather at the harvest and use of disease free seed can minimize the disease incidence.

Mildews

Downy mildew (*Peronospora lentis* Gaumann) is considered as a minor disease and occurs in India. Symptoms of the disease include pale green yellow spots on upper surface with grayish downy fungal growth on opposite side. Leaves may become chlorotic and defoliation occurs especially in lower branches leaving stunted plants with bunch of similar leaves on the top. Powdery mildew of lentil caused by *Erysiphe polygoni* has been reported from Argentina, Bangladesh, India, Sudan, Tanganyika and former USSR. The disease may appear at any stage of plant but more severely at flowering. White powdery spots appear on leaves, which later cover most of the foliage. The plants get defoliated and killed. The entries J 441, J 500, J 809, J 814 and J 1609 were found free from the downy mildew. While Pant L 639, JPL 970, DL315 and LG 224 have been found resistant against powdery mildew at India.

Gray Mould

This disease occurs mainly in humid cool regions and can cause serious losses to lentil under favourable conditions. The disease is caused by *Botrytis cinerea* Pers. Ex Fr. The symptoms include development of dark brown lesions on the foliage along with the hair like erect conidiophores of the fungus. The pathogen is transmitted through seed as well as infected remains of plants. The seedling stage is more prone to infection, which is further favoured by moist cool weather and thick plant canopy. Seed treatment and spraying of young plants with 1 per cent thermal control the disease effectively.

Table 9.2: Cultural Control of Major Diseases of Pulse Crops

Crop	*Disease*	*Method*
Chickpea	Wilt/root rot	Sanitation – debris
		Deep ploughing – summer
		Soil solarization polythene sheet
		Late sowing
		Deep sowing (10-15 cm)
		Soil amendments –oilcake
	Ascochyta blight (AB)	Sanitation-debris
		Deep sowing (10-15 cm)
	Botrytis gray mold (BGM)	Intercropping-wheat, barley, mustard
		Late sowing
		Wider row and plant spacing
Pigeonpea	Wilt	Deep ploughing-summer
		Sanitation-debris
		Crop rotation-sorghum
		Intercropping-sorghum
	Sterility mosaic(SM)	Removal-stray plants
	Phytophthora blight (PSB)	No ratoon/perrenial crop
		No water stagnation
		Ridge sowing
		Intercropping with cowpea, mungbean, urdbean
Fieldpea	Powdery mildew (PM) and Rust	Early sowing
		Timely sowing
Mungbean and urdbean	Yellow mosaic virus (YMV)	Early sowing during spring
Lentil	Wilt	Deep ploughing – summer
		Sanitation-debris
		Crop rotation

Table 9.3: Disease Resistant Varieties of Pulses Released in India

Crop	*Disease*	*Resistant Varieties*
Pigeonpea	Wilt	NP (WR) 15, BDN 1, BDN 2, Sharda, C 11, TT 6, Maruthi, Asha, Jawahar, BSMR 736,853,MA3, Amar, NA-1, MAL13
	SM	Bahar, Hy 3C, Sharad, Pusa 9, NA-1, BSMR 175, BSMR 736, ICPL 87051,MA3, 6
	Alternaria blight+SM	Sharad, Pusa 9
	Wilt+SM	Asha, BSMR 736, DA 11 (Wilt)
	Phytophthora blight	Bahar, NA-1, Amar, Azad, MA6, MAL13, Sharad, Pusa 9
Chickpea	Wilt	Pusa 212, JG 315, ICCC 32, Avrodhi, Phule G 5, Vishwas, Vijay, Vishal, GF 89-36, Pusa 372, KWR 108, KG 1168, Haryana channa 1, PBG 1, GNG 663, JG 74, GPF 21, ICCV 10
	Wilt+root rot	ICCV 10, H 355
	Ascochyta blight (tolerant)	Gaurav, GNG 146, GNG 469, C 235, BG 261, PBG 1
Mungbean	YMV	ML 267, Pant M 2, Pant M 3, PDM 11, PDM 54, MUM 2, NM 1, Pant M 4, ML 613, SML 134, Pusa 105, Asha, BM 4, MH 85-111
	YMV+CLS	ML 131
	PM	TARM 1, Pusa 105 (t)
Urdbean	YMV	PDU 1, Pant U 19, Pant U 30, UG 218, PDU 88-31, WBU 108, Mash 338, Narendra urid 1
	PM	LBG 17 (rabi)
Lentil	Rust	Pant I 406, Pant L 639, DPL 15, DPL 62, L 4076, L 4147, Pant 4, LH 84-8
		Pant L 77-2, DPL 15, DPL 62 (tolerant), Pant L 4
	Wilt	DPL 15, DPL 62, Pant L 4
Fieldpea	PM	DMR 11, HUP 2, HFP 4, Pant P 5, Rachna, DMR 7, KFP 103, HFP 8909, J 885
	PM+Rust	Malviya Matar 15 (HUDP 15)

Approaches for Integrated Disease Management

Use of chemicals, resistant varieties, and biological agents with modified cultural practices may help in controlling the diseases to some extent. However, considering the diversity in pathogen and varied agro-climatic conditions influencing the diseases, only single practice may not achieve the desired goal. Therefore, it would be worthwhile to integrate available compatible control practices to develop viable and economic integrated management strategies. Efforts to develop integrated management of major diseases have been initiated and in this pursuit modules for management of wilt, Phytophthora blight and sterility mosaic of pigeonpea have been developed (Gurha and VishavaDhar, 2005). Similar module for pea powdery, mildew and rust has recently been advocated by Chaudhary and Naimuddin, 2004. These may be evaluated in farmers field for adoption and further refinement. Development of similar modules for other pulse crops is required.

Table 9.4: Effective Chemicals for Control of Major Diseases of Pulse Crops

Crop	*Disease*	*Chemicals*	*Mode of Application*
Chickpea	Wilt/root rot	Carbendazim, Benomyl, Thiram, Benlte, Carbendazim+ Thiram	ST
	Ascochyta blight	Tridemorph M, Thiram, Benomyl, Carbendazim+ Thiram Captan, Tridemorph M, Chlorothalonil, Dithianon, Mancozeb	ST
	Botrytis gray mold	Venclozolin, Carbendazim, Thiram, Iprodione, Carbendazim+Thiram, Tridemefon, Carbendazim, Thiabendazole Mancozeb, Rovral, Bayleton	FS ST FS
Pigeonpea	Wilt	Carbendazim, Benomyl, Thiram, Benlate T, Carbendazim+Thiram	ST
	Phytophthora blight	Formulations of Metalaxyl	ST+FS
	Sterility mosaic	Acaricides (Kelthane, Morestan)	FS
	Alternaria	Mancozeb	FS
Mungbean/ urdbean	Cercospora leaf spot	Carbendazim	FS
	Yellow mosaic virus	Phorate, Carbofuran granules Dimethoate, Monocrotophos	ST
	Powdery mildew	Carbendazim, Tridemorph, Dinocap, Wett. Sulphur	ST
	Root rot/wilt	Carbendazim, Benomyl, Thiram, Benlate T, Carbendazim+Thiram	ST
Fieldpea/ lentil	Powdery mildew	Wettable Sulphuir, Carbendazim, Trideomorph, Dinocap	FS
	Rust	Mancozeb, wettable sulphur	FS
	Root rot/wilt	Carbendazim, Benomyl, Thiram, Benlate T, Carbendazim+ Thiram	ST

ST: Seed treatment; FS: Foliar spray.

Table 9.5: Integrated Approach of Disease Management of Pulses Crop

Crop	*Disease*	*Management Options*	
Pigeonpea	Wilt	Field	Deep summer ploughing,removal of stubbles
		Varieties	MA-13,NA-1,Amar
		Seed	Carbendazin+ Thiram(1:2 g/kg seed) or Carbendazim (2g kg seed),*Trichoderma harzianum* or *T. viride* (@10 g/ka seed)
		Intercropping	Sorghum (1:1 ratio)
		Sowing time	Timely sowing
		Spacing	Normal
	Phytophthora blight	Field	Removal of perennial or rationed pigeonpea in the vicinity
		Varieties	Bahar, NA-1, Azad, MA-6, MAL13, Sharad,Pusa 9
		Seed	Aldicarp (10 per cent)@ 3g/kg seed
		Crop rotation	3 years with cereals
		Sowing time	Normal

Contd...

Table 9.5–Contd...

Crop	*Disease*	*Management Options*	
		Spacing	Normal
		Foliar spray	Spray of Acaricide like kelthane,metasystox or dimethoate @0.01 per cent at first notice of disease
Chickpea	Wilt	Field	Deep summer ploughing and field sanitation
		Varieties	JG315, Avrodhi,DCP92-3,JG74,BG372,KWR108
		Seed	4 g *T. viride*,+ 1 g carboxin or carbendazim + thiram (1+ 2 g/kg seed)
		Sowing time	Normal
		Intercropping	With linseed (2:1) or mixed cropping with linseed (1:1)
	Ascochyta blight	Seed	Use of healthy seed and treatment with any recommended fungicides
		Varieties	Resistant/tolerant varieties
			If cloudy weather provides one protection spray of the fungicides should be applied
		Secondary spared	For the management of secondary spred of the disease, 2-3 sprays of any recommended fungicide at 10-15 days interval may be applied
		Spacing	Wide row and plant spacing and growing chickpea with wheat, barley or mustard is recommended
Mungbean and urdbean		MYMV	Resistant and tolerant varieties
		Intercropping	Inter/mixed cropping of mungbean and urdbean with non-host crops like sorghum, pearl millet and maize
		Rouging	Destruction of affected plants as soon as noticed in the field
		Sowing time	Adjustment of sowing dates to avoid the disease
		White flies maggots	Application of phorate, disulfoton granules @ 1-2 kg a.i./ha in furrow during sowing
		White flies	To control white flies spray of metasystox, melathion (0.1 per cent), dimethoate (0.02 per cent), monocrotophos (0.05 per cent) at the initiation of disease and subsequently at 10-15 days interval as per disease pressure is recommended

Conclusion and Future Thrusts

Pulse crops are predominantly grown in the developing countries, and to cut down the ravages caused by diseases, the farmers with limited resources need effective and low cost protection technology, which is only possible through reliable integrated disease management systems. The informations generated on the management practices for individual diseases, need to be brought together and tested to develop IDM modules for specific locations to encounter multiple diseases.

For a systematic approach to develop long term and sustainable IDM systems, the foremost pre-requisite will be mapping of major diseases in respect of their occurrence and severity in the different agro-climatic regions. Detailed disease distribution maps based on precise surveys need to be prepared and updated at regular intervals. This will enable to develop need-based strategy for resistance breeding, exploiting the available infrastructure facilities and resources in the country.

Recent informations on the identification of sources of multiple disease resistance are encouraging. The ongoing programme should be further vigorously persued to identify stable and broad based multiple disease resistance sources and their subsequent utilization in developing high yielding cultivars according to the requirements of growing situations.

The potential of genetic engineering in understanding the host-pathogen interactions, knowledge of the structure and function of disease resistance genes and in developing stable; non race specific resistance against disease and multi-adversities, (pathogens, insects, nematodes, weeds etc.) needs to be exploited and explored at a faster speed in order to design sustainable IDM systems. A joint venture and teamwork of Breeders, Pathologists, Agronomists and Biotechnologists can only accomplish this. Biotechnology may also help in evolving super strains of antagonists with high survival, root colonization potential and wide spectrum antifungal compound production for enhancing field application of biocontrol practices in IDM systems.

Active involvement of farmers at all levels during the evaluation of newly designed IDM systems for its components is perhaps the most important factor that will ultimately stimulate the acquisition and use of technological informations.

References

Agrawal, S.C. and Nema, S. (1988). Resistance in mung bean and urd bean against powdery mildew. *Indian Journal of Pulses Research*, IIPR, Kanpur 2: 76-77.

Amin, K.S. 1985. *Plant Pathology: Consolidated Report on Kharif Pulses*, (1984-85). All India Coordinated Pulses Improvement Project, DPR, Kanpur.132. pp.

Butler, E.J. (1910). The wilt disease of pigeonpea and the parasitism of *Necosmospora vasinfecta* Smith. Memories of the Department of Agriculture in India, *Botanical series* 2: 1-64.

Chaudhary, R.G. (1998). Epidemiology and management of wilt of pigeonpea (*Cajanus cajan* L.) caused by *Fusarium udum* Butler. Ph. D. thesis, C.S. Azad Univ. of Agric. and Tech., Kanpur.

Chaudhary, R.G. (2004).Biological control of soil borne diseases of pulse crops. *In: Pulses in New Perspective* (Eds Ali *et al*). ISPRD, IIPR, Kanpur, India pp. 345-361.

Chaudhary, R.G., Naimuddin, De, R.K. and Rai, A.B. (2004). Masoor, Matar Evam Rajma Ke Rog aur Unka Prabandhan, Bharatiya Dalhan Anusandhan Sansthan, Kanpur, India pp.237-267.

Chauhan, Y.S., Nene, Y.L., Johansen, C., Haware, M.P., Saxena, N.P., Sarder Singh, Sharma, S.B., Sahrawat, K.L., Burford, J.R., Rupela, O.P., Kumar Rao, J.V.D.K. and Sitanantham, S. (1988).Effect of soil solarization on pigeonpea and chickpea: *Research Bulletin No.* 11, ICRISAT, Patancheru, Andhra Pradesh, India.16 pp.

Dahiya, S.S., Sharma, S. and Faroda, A.S. (1988). Effect of date and depth of sowing on the incidence of chickpea wilt in Haryana. India. *International Chickpea Newsletter* 18: 24-25.

Dasgupta, A. and Sengupta, P.K. (1989). Effect of different soil amendments on wilt of pigeonpea (*Cajanus cajan* (L.) Millsp) caused by *Fusarium udum* Butler. *Beirage Zur Tropischen Landwirt Schrat and Veterindermedizin* 27 (3): 341-345.

De, R.K., Vishwa Dhar and Rathore, Y.S. (1995). Control of sterility mosaic of pigeonpea with insecticides and acaricides. *Indian Journal of Pulses Research* 9: 87-89.

Gaur, V.K. and Sharma, L.C. (1989). Prevalence of Fusarium wilt of pigeonpea in Rajasthan.*Indian Phytopathology* 42 (3): 468.

Grewal, J.S. and Laha, S.K. (1983). Chemical control of *Botrytis* blight of chickpea. *Indian Phytopathology* 36:516-520.

Gupta, O., Kotasthane, S.R. and Khare, M.N. (1988). Strain variation in *Fusarium udum in* Madhya Pradesh, India. *International Pigeonpea Newsletter* 7: 22-25.

Gupta, S.K. (1995). Studies on the management of chickpea wilt complex. M.Sc. Thesis, Rajendra Agriculture University Pusa, Bihar. 70 pp.

Gurha, S.N. (2004). Chana Mein Rog Prabhandhan. *In: Dalhan* (Eds Masood Ali *et al.,)* IIPR, Kanpur, pp 201-212.

Gurha, S.N. and Srivastava, R. (2002). Effect of root exudates of wheat (*Triticum aestivum*) and Indian mustard (*Brassica junca*) on growth of *Fusarium oxysporum* f. sp. *ciceri* isolates. *Indian Journal of Agricultural Sciences* 72(12): 747-748.

Gurha, S.N. and Vishwa Dhar. (2005). Chickpea diseases and their management. *Chickpea Production and Productivity Constraints* (Eds.Amerika Singh *et al.*) NCIPM, IARI, New Delhi, India, pp.99-110.

Gurha, S.N., Gurdip Singh and Sharma, Y.R. (2003). Disease of chickpea and their management. *Chickpea Research*.Kanpur, India, pp 195-227.

Haware, M.P., Nene, Y.L.and Rajeshwari, R. (1978). Eradication of *Fusarium oxysporum* f. sp. *ciceri* transmitted in chickpea seed. *Phytophatology* 68:1364-1367.

Haware, M.P., Tripathi, H.S., Rathi, Y.P.S. Lenne, J.M. and Jayanthi, S. (1997). Integrated management of Botrytis gray mould of chickpea, cultural, chemical, biological and resistance options. Pages 9-12 *In*: *Botrytis Gray Mold of Chickpea* (Eds.,M.P.Haware, D.G.FarisandC.L.LGowda), ICRISAT, Patancheru, and Andhra Pradesh, India.

Hwang, S.F. and Chakravarty, P. (1992). Potential for the integrated control of Rhizoctonia root rot of *Pisum sativum* using *Bacillus subtilis* and a fungicide. *Zeitschrift fur Pflanzenkrankheiten und Pflanzenschutz* 99: 626-636.

Kannaiyan, J., Nene, Y.L., Reddy, M, V., Ryan, J.G. and Raju, T.N. (1984). Prevalence of pigeonpea diseases and associated crop losses in Asia, Africa and the Americas. *Tropical Pest Management* 30 (1): 62-71.

Kumar, B. (1995). Integrated Management of *Botrytis* gray mold of chickpea.M.Sc. Thesis, Punjab Agricultural University, Ludhiana.

Lava Kumar, P., Jones, A.T., Sreenivasulu, P. and Reddy, D.V.R. (2000). Breakthrough in the identification of the causal virus of pigeonpea sterility mosaic disease. *Jounal of Mycology and Plant Pathology* 30 (2): 249.

Mahendra Pal. (1989). Multilocational testing of pigeonpea for broad based resistance to Fusarium wilt in India. *Indian Phytopathology* 42 (3): 449-453.

Meuli, L.J., Twigs, B.J. and Fymn, G.K. (1947). The zinc salt of 2,4,5-Trichlorophenol as a seed fungicide. *Phytopathology* 37: 474-480.

Mishra, S. and Vishwa Dhar. (2003). Comparative conidial morphology and virulence of *F. udum* from different pigeonpea varieties. *Farm Science Journal* 12 (2): 132-134.

Mukhopadhyay, A.N. (1988). Biocontrol efficacy of *Trichoderma* spp. in controlling damping off in tomato and egg plants wilt complex of lentil and chickpea. *In*: *Proceeding of the 5th International Congress of Plant Pathology*, Kyoto, Japan.

Mukhopadhyay, A.N. (1989). Microbioal control of soil borne pathogens by *Trichoderma* and *Gliocladium.In: National Seminar and 7th Workshop of AICRP on Biological Control*, 23-25 October1989, Lucknow.

Mukhopadhyay, A.N., Shrestha, S.N. and Mukerjee, P.K. (1992). Biological seed treatment for control of soil borne plant pathogens. *FAO Plant Protection Bulletin* 40(1-2): 21-30.

Natrajan, M., Kannaiyan,J., Willey,R.W. and Nene,Y.L. (1985). Studies on the effect of cropping system on Fusarium wilt of pigeonpea. *Field Crop Research* 10(4):336-346.

Nene, Y.L., Reddy, M.V., Beniwal, S.P.S., Mahmood, M., Zote, K.K., Singh, R.N. and Sivaprakakam, K. (1989). Multilocational testing of pigeonpea for broad based resistance to sterility mosaic in India. *Indian Phytopathology* 42 (3): 444-448.

Parke, J.L.1991. Efficacy of *Pseudomonas cepacia* AMMD and *Pseudomonas fluorescence* PRA 25 in biocontrol of *Pythium* damping off and *Aphanomyces* root rot of pea. *Bulletin* SROP 14: 30-33

Paul, Y.S. and Rathoure, Rajeev. (1947). *Integrated management of pea diseases in Himachal Pradesh.*IPS-Golden Jubilee International Conference on Integrated Disease Management for Sustainable Agriculture. 10-15th Nov. 1997. New Delhi. Pp 301 (abstract).

Puzio-Jdzkowska, M. (1990). Evaluation of the resistance of hybrids of garden and field pea to *Fusarium* wilt). *Biuletyn Institute Hodowli I Aklimatyzaeji Roslin* 173-174: 113-115.

Raguchander, T. and Rajappan, K. (1995). Relationship between powdery mildew and meteorological factors in urdbean. *Plant Disease Research* 10: 5-9.

Rai, B. and Upadhyay, R.S. (1979). Discovery of perfect state of *F. udum* Butler *In: Proceeding of the Annual Conference of the Society for Advanced Botany* 5,25(Abstr.)

Reddy, M.V. and Vishwa Dhar. (1998). Disease resistance in major pulse crops. Pages 281-300. *In: Recent Advances in Pulses Research* (Eds.,A.N.Asthana and M. Ali),ISPRD,IIPR,Kanpur,India.

Reddy, M.V. Vishwa Dhar, Lenne, J.M. and Raju, T.N. (Eds.) (1996). *Proceeding of the Asian Pigeonpea Pathologists Group Meeting and Monitoring Tour, 20-25* November. 1995. International Crop Research Institute for the Semi-Arid Tropics, Patancheru, Andhra Pradesh, India 32 pp.

Reddy, M.V., Sharma, S.B. and Nene, Y.L. (1990). Pigeonpea: diseases Management. Pages 303-347. *In: The Pigeonpea* (Eds., Y.L.Nene, S.D.Hall and V.K. Sheila), CAB International/ICRISAT, and Wallingford, U.K.

Sangar, R.B.S. and Singh, V.K. (1994). Effect of sowing dates and pea varieties on the severity of rust, powdery mildew and yield. *Indian Journal of Pulses Research* 7:88-89.

Seth, M.L. (1962). Transmission of pigeonpea sterility by an eriophyid mite. *Indian Phytopathology* 15: 225-227.

Singh, D.P. (1991). Genetics and Breeding of Pulse Crops, Kalyani Publishers, New Delhi and Ludhiana, 371 pp.

Singh, G. and Kaur, L. (1990). Chemical control of gray mold of chickpea. *Plant Disease Research* 5: 132-137.

Singh, G. (1990). Identification and designation of physiologic races of *Ascochyta rabiei* in India. *Indian Phytopathology* 43: 48-52.

Singh, R.A. and Grurha, S.N. (1996). Economics of carbendazim spray for control of cercospora leaf spots in mungbean. *Indian Journal of Pulses Research* 9: 52-54.

Singh, R.A. and Naimuddin(.2000). Mung Evam Urd Mein Rog Prabhandhan. *In: Dalhan* (Eds Masood Ali *et al.*) IIPR, Kanpur, pp 225-235.

Singh, R.A., De, R.K., Gurha, S.N. and Ghosh, A. (2000). Yellow mosaic disease of mungbean and urdbean. Pages 337-348. *In: Advances in Plant Disease Management* (Eds., U. Narain,K. Kumar and M. Srivastava). Advance Publishing Concept, New Delhi, India.

Singh, R.A., De, R.K., Gurha, S.N. and Ghosh, A. (2001). Management of cercospora leaf spot of mungbean. Page 146 *In: National Symposium on Pulses for Sustainable Agriculture and National Security*, 17-19 April 2001,New Delhi, ISPRD, IIPR, India.

Singh, R.A., De, R.K., Gurha, S.N. and Ghosh, A. (2002). Yellow mosaic virus of mungbean and urdbean. Pages 395-408. *In: IPM System in Agriculture Key Pathogens and Diseases* Vol. 8 (Eds. R.K. Upadhyay, D.K.Arora and O.P. Dubey), Aditya Books Pvt. Ltd., New Delhi, India.

Singh, S.K. and Dhar, S.L. 1997. Integrated management of major foliar diseases caused by fungi in field pea. IPS-Golden Jubilee International Conference on Integrated Disease Management for Sustainable Agriculture 10-15th Nov.1997, New Delhi. Pp 294(Abstract).

Singh, S.K., Rahman, S.J., Gupta, B.R. and Kalha, C.S. (1992). An integrated approach to the management of major disease and insect pests of peas in India. *Tropical Pest Management*. 38: 265-267.

Singh, U.P. (1997). Powdery mildew of pea: Biological and management. IPS-Golden Jubilee International Conference on Integrated Disease Management for Sustainable Agriculture. 10-15th Nov.1997, New Delhi, pp.72 (Abstract).

Singh, U.P. and Singh, R, B. (1984). Effect of date of sowing on the sclerotinia stem rot and wilt of gram (*Cicer arietinum* L.) Paper presented in All India Rabi Pulses Workshop, Bhubneshwar, Orissa.

Singh, U.P., Srivastava, B.P., Singh, K.P., Mishra, G.D. (1991). Control of powdery mildew of pea by ginger extract. *Indian Phytopathology*, 44: 55-59.

Singh.D. (1991). Biocontrol of *Sclerotinia sclerotiorum* (Lid) de Bary by *Trichoderma harzianum. Tropical Pest Management*, 37: 374-378.

Srivastanva, J.S. and Singh, B. (1993). Control of Sclerotinia stem and pod rot of pea.*In: Current Trends in Life Sciences.* (Vol. 19)- *Recent Trends in Plant Disease Control* (Eds. H.B. Singh, D.N. Upadhyay and L.R.Saha). Today and Tomorrows Printers and Publishes, New Delhi, pp.143-146.

Subramaniam, S. (1963). Fusarium wilt of pigeonpea. I. Symptomatology and infection studies. *Proceedings of the Indian Academy of Sciences*, Section B 57: 134-138.

Tiwari, A.S. (1977). Mungbean varieties requirements in relation to cropping seasons in India. Pages 129-131. *In: First International Mungbean Symposium.* Asian Vegetable Research and Development Centre,Shanhua,Taiwan.

Varma, A., Dhar, A.K. and Mandal, B. (1992). MYMv transmission and control in India. Pages 8-27. *In: Proceedings International Workshop* (Eds, S.K. Green and D.H. Kim), 2-3 July 1991, Bangkok, Thailand, AVRDC, Shanhua, Taiwan.

Vishwa Dhar and Chaudhary, R.G. (2001). Disease resistance in Pulse Crops: Current status, future approaches. Pages 144-157. *In: Role of resistance in intensive agriculture* (Eds. S. Nagarajan)

Vishwa Dhar and Chaudhary, R.G. (2004). Arhar Mein Rog Prabandhan. *In: Dalhan* (Eds Ali *et al.*). ISPRD, IIPR, Kanpur, India, pp 213-223.

Vishwa Dhar, Chaudhary, R.G. and Naimuddin. (2000). Disease management in pulses: Present status and future thrusts. *In: Advances in Plant Disease Management* (Eds., Udit Narain, Kumud Kumar and M. Srivastava), Advance Publishing Concept, New Delhi, India.

Vishwa Dhar, Reddy, M.V. and Chaudhary, R.G. (2005). Major diseases of pigeonpea and their management *In: Advances Pigeonpea Research* (Eds. Ali,M and Shiv Kumar). IIPR, Kanpur, India, pp. 229-261.

Vishwa Dhar (1996). Consolidated on mungbean and urdbean. All India Coordinated Research Project on Improvement of MULLaRP, IIPR, Kanpur, India.

Yainkov, I.I. (1989). (Fusarium wilt of peas) *Zaschehita Rostini* (Moskova). 4: 18.

Sustainable Disease Management of Agricultural Crops (2011) *Pages* **168–175**
Editors: **S.K. Biswas and S.R. Singh**
Published by: **DAYA PUBLISHING HOUSE, NEW DELHI**

Chapter 10

Integrated Management Strategies of Plant Viral Diseases

S.R. Singh[1], S.K. Biswas[2] and R.K. Pandey[3]
[1]*Krishi Vigyan Kendra, Belatal, Mahoba – 210 423, U.P.*
[2]*Department of Plant Pathology, C.S. Azad University of Agriculture and Technology, Kanpur – 208 002*
[3]*Division of ACB and PDM, Amity University, Noida – 201 303, U.P.*

The disease of plants caused by viruses result in enormous economic losses. The losses are mainly in the tropics and semi tropic region which provide ideal environmental conditions for the perpetuation of viruses and their vectors. The estimated loss caused by mungbean yellow mosaic virus is about $ 400 million in the productivity of mungbean, soybean and black gram (Varma *et al.*, 1992). Crop losses are also occur in the temperate region (Hull and Davis, 1992). The crop losses are even more alarming in perennial plants where there is a steady decline as a result of infections by viruses and virus like organisms. During the last fifty years such diseases have resulted in the destruction of over 2500 million citrus, cocoa, sandal trees and coconut in the various parts of the world.

From the last few years there has been increasing pressure to reduce the use of pesticides in agriculture. There are several reasons on this aspect. The pesticides are perceived to be dangerous for human health either through residue in food and soil or direct effect. These chemicals are also harmful for beneficial microorganisms. Thus for high value food products, there is a demand for low value of pesticides residue and as detection instrumentation sensitivity increases, the acceptable level will continue to decrease.

The successful disease management in most cropping systems involves an integrated programme. Often cultural measures are necessary to improve efficacy of other control measures. Indeed cultural methods of disease management were often devised even before; farmers knew the cause of the diseases themselves.

Although there are several types of management strategies have been adopted for management of diseases most of them can be placed exclusion, eradication avoidance of source of infections, control of vectors, and modification in cultural practices and resistance of host plants.

Avoidance of Source of Infection

The main sources of viral diseases infection are planting material, adjoining crops, collateral hosts, volunteers and ground keepers. Occasionally vectors are also source of primary infection.

Sanitation of Field

Sanitation of the field is one of the most important controls major of plant diseases. It eliminates or reduces the pathogen inoculums in the field which prevent the spread of inoculums from infected plant material to healthy plant. It is well know that some viral diseases are carried in weeds and some other viruses are perpetuate in to field. Infected, weed seed can introduce viruses in to field. Therefore, Weed control is useful for limitation of viral infection weeds lost should be pullout and destroyed from the crop field to minimize the viral infection.

Crop Rotation

Monoculture is a practice of growing a single plant species repeatedly on the same land. It is in contrast to inter cropping and crop rotation. Crop rotation is the practice of growing a sequence of plant species on the same land. The benefits of crop rotation are to maintain crop yield and suppress the soil borne diseases. There are numbers of viruses and vectors of other virus survive in soil in absence of host both soil borne viruses and vector can be controlled by crop rotation.

Use of Virus Free Seed

Over 130 viruses are known to be transmitted through the seed. To a varying extent, seed carrying viruses not only result in a poor crop strand but also provide inoculums for the secondary spread of viruses. The extent of secondary viruses spread varies from virus to virus depending on the efficiency of transmission by the vectors. In case of lettuce mosaic virus, 1.6 per cent infection of lettuce seed resulted in 40 per cent incidence whereas, for cowpea banding mosaic viruses 4.5 per cent seed transmission was required for similar final incidence (Varma, 1976)

The viruses are carried through seeds either as surface contaminants for example tomato mosaic virus in tomato, or in testa, as is southern bean mosaic virus in cowpea and beans, or in the embryo, as in black gram mottle virus. The mere presence in seed does not ensure transmission. For example, black gram mottle virus is transmitted through seed only when embryonic axis of black gram seed has a maximum of 60ng virus" (Varma *et al.*, 1992) whereas, SBMV is not transmitted through the seeds of cowpea var. V152 even when present in very large amount in the testa.

Not much attention has been given to curing virus infected seeds. Treating seeds with dry heat, chemicals, γ-irradiation or storage for various period has helped to cure seeds to different viral infection(Varma,1976) but the large scale application of such techniques is difficult. The most practical approach for avoiding this source of infection is to produce seed in virus free areas and provide certified seeds to the growers, as has been recommended for pea seed borne mosaic virus (Varma *et al.*, 1991).

Use of Virus Free Vegetative Material

Primary infection through the planting material is even more serious in vegetatively propagated crop like sweet potato, sugarcane, potato, banana and citrus. Viral diseases in all these crops cause considerable losses in crop production. The use of virus free planting material is imperative in case of crop cultivation with vegetatively propagated materials. It requires careful testing because even propagules obtained from healthy looking plants may not be free from infection, as was found for cassava infected with Indian cassava mosaic virus (Malathi *et al.*, 1989). Determination of freedom from infection by a pathogen is restricted by detection limits. Recently advance development of biotechnological methods, like polymerase chain reaction which is 5000-fold more sensitive than ELISA (Wetzel *et al.*, 1992) have made it possible to detect even a single pathogen in propagating material. Such diagnostic techniques will be very useful in developing programmes for the use of healthy planting materials.

Heat Treatment for Virus Free Seed Material

Heat treatment has been found effective in some viral diseases usually transmitted through bulb, cuttings and grafts. Heat treatment cure planting materials of virus free like, potato leaf roll and sugarcane mosaic viruses because the viruses are inactivated at temperature which is not lethal to plants (Varma,1976). Sugarcane mosaic virus can be successfully controlled by heat treatment of cuttings sets at 52-54°C.

In vegetatively propagated crops, infection of an entire cultivar is common. Virus free plants can be obtained by meristem tip cultiure technique. Successes in eliminating a viral infection is depend of the size of meristmatic tissue culture (Kassanis and Varma 1967). These techniques have been widely used to obtain virus free clones of crops like banana, cassava and sugarcane. In citrus, *in vitro* shoot tip grafting has been very successful in producing virus free plants (Vogal *et al.*, 1988).

Avoiding of Collateral Hosts

Collateral hosts play an important role in the perpetuation of viruses. Virus transmitted through weeds not only help in perpetuation but also in spatial distribution, because these viruses are transmitted through the seeds of their host plants. Therefore, control of weeds, may reduces the losses due to viruses. Weeds are important reservoir of virus belonging to groups like cucumo, poty and gemini viruses. Removal of malvaceous weeds around plants of Bhindi and observing a closed period before sowing the principal crop reduced the incidence of damage caused by Bhindi yellow vein mosaic virus (Varma *et al.*, 1976).

Avoidance of Vectors

This is a good approach for virus disease management. Therefore, to avoid the population of vectors northern India is selected for true potato seed because crop is harvested before development insect population. In north India, mungbean yellow mosaic virus (MYMV) is a major constraint in the production of mungbean during *kharif* which is the main growing season for this crop. Short duration varieties of mungbean well developed which could be fruitfully cultivated during the *zaid* season under irrigation. So that losses due to mungbean yellow virus can be avoided. However, in recent years as a result of changes in cropping pattern the vector has started appearing early and causing heavy incidence of MYMV even in the zaid crop. Resulting in a decline in the aerea under summer mung (Varme *et al.*, 1992 a).

Chemical Control of Vectors

This is a very common approach for the controlling viral disease. It has been tried either at the source of the vectors to prevent the primary spread of the virus or in the crop to minimize both primary and secondary spread. Timely application of chemical is very important for management of vector as well as viral disease U.K. sugar beet yellow disease is managed very effectively by applying insecticides at the initial stage of development of vector population(Varma,1976). Whitefly transmitted viruses are more difficult to control by chemical treatment owing to the ability of the vector and the relatively short acquisition and transmission access periods, although they transmit virus in persistent manner. The best disease control and improvement in economic yields has been obtained by treatment with systemic insecticides like Carbofuran, Disulfotan(Sharma and Verma, 1982 a)

Biological Control

Biological control of vectors has not received the attention it deserves. It was very successful in checking the spread of carrot mottley dwarf virus in Australia and New Zealand through the introduction of the wasp, aphids, which reduced the population of the vector *Cavariella aegopodi*. It however does not work in Europe(Varma,1976).Similarly, the effect of citrus greening disease, caused by a bacteria-like organism was successfully restricted through the biological control of its psyllid vector *Diaphorina citri* and *Trioza erytreae* in the Reunion islands. Within 24 months of the release of 30-50 *eulophid* ectoparasites *Tetrastichus dryi* and *T.radiatus* per 2 km the population if psyllid vector was reduced and spread the greening disease checked. But *T.radiatus* could not able to control the population of *D.citri* in India(Aubert and Quilici 1984).

Botanical

Spraying plants with minerals oils reduced infection caused by virus which are transmitted by aphids in non persistent manner and also those transmitted by whiteflies. It is not, however, very effective against the persistent type of aphid borne viruses (Sharma and Varma, 1982b). This practices are useful approach particularly for vegetable crops, as it avoids the use of toxic chemicals.

Use of Resistant Varieties

This is the most important and practical approach to managing viral diseases. Genetic resistance to plant viruses may operate through resistance to vectors, transmission by vectors/seeds, virus multiplication, symptom development, intracellular movement hypersensitivity or immunity to virus infection. Immunity is the most desirable form of resistance for managing viral diseases because no detectable multiplication of virus occurs in such plants. Two contrasting positive and negative genetic mode have been suggested to explain resistance to viruses in plants (Fraser 1987).The positive model postulates that the resistant plants contain a gene or genes which prevent viral multiplication. The negative model proposes that such plants lack a gene or genes required for susceptibility. Although there is no conclusive evidence in support of either models the non-specificity of the immunity and the specificity of the susceptibility suggest the negative model. This is also supported by the occurrence of viral diseases like cowpea, golden mosaic, cotton leaf curl and okra leaf curl in India, which was earlier free from such diseases although the causal viruses existed in low number in the wild, owing to the inadvertent introduction of susceptibility genes. Both the positive and negative models, however may determine various other forms of resistance (Varma *et al.*, 1991b).

Resistance genes have been identified for nearly 140 host virus combinations, but there is no set pattern for the mechanism of resistance. Of the 57 host varieties with monogenic resistance to one or more of 29 different potyviruses, 47 per cent are dominant and 44 per cent recessive and the remainder incompletely dominant. Similarly, of the three different host identified to have monogenic resistance to cucumber mosaic virus, two are dominant and one is recessive. Thus, mechanism of resistance varies not only from virus to virus, but even for the same virus in different host. For a long time, resistance to viruses was considered to be more durable than that to the other groups of pathogens but in recent years resistance breaking strains of viruses have been reported for over 75 per cent of the resistance host-virus combination tested. The frequency of resistance breaking strains is greater for dominant than recessive or incompletely dominant genes. The rapid development of resistance to makes breeding for resistance to viruses a difficult task (Varma *et al.*, 1991a). Resistance may break down through the development of new biotypes or new strains of viruses. Many high-yielding cultivars of rice resistance to rice tungro virus are widely grown in Phillipines. Resistance to RTV in some of these cultivars has been overcome by the vector leaf hopper becoming able to feed on and colonize these cultivars continuously for several years (Dahal *et al.*, 1990). New strains arise frequently in viruses with divided genomes, like white-fly transmitted gemini viruses. It is estimated that resistance to mung bean yellow mosaic virus in mungbean breaks down in less than six years (Verma *et al.*, 1992) in many crops; sources of resistance to viruses are available within the species. For many others they are found in wild relatives particularly at the centers of origin, owing to co-evolution of the crops and their pathogens. Useful resistance in crops has also been found in secondary centers of diversity (Lenne and Wood 1991). Introduction of new crops and extensive movement of germplasm both essential for agricultural development, bring in pathogens and diseases new to either the crops or the region, cocoa swollen shoot is an example of the former situation (Thresh, 1981).In crops where resistance is not available in desirable cultivars, the identification of rate-reducing resistance would be helpful.

Cross Protection

The exact a mild strain to prevent infection by a severe strain is not clearly understood(Gibbs and Stocknicki1986). It has been successfully used in different countries for protecting tomato, citrus and papaya from infection by severe strains of citrus tristeza, tomato mosaic and papaya ring spot viruses respectively. The most expensive application of cross protection has been in citrus industry. Such protection, however, is not useful for sustained growth and (a) the plant may not be protector against all the severe strains, and (b) the protected plants may be predisposed to other viruses. In Florida, mild isolates of citrus triesteza virus failed to protect sweet orange on sour orange rootstock from infection by severe isolates, leading to decline. In South India, protected lime trees yielded less than the unprotected trees (S.R.Sharma, 1988). Satellite component containing mild strains of cucumber mosaic virus provided protection from the severe strains by 22 to 83 per cent and increased yield by up to 56 per cent in peppers in China (Tien *et al.*, 1987).

Systemic Acquired Resistance (SAR)

Many compounds, such as plant growth hormones, nucleic acid base analogous microbial metabolites,alkaloids,plant extracts a virulent race of pathogens and drugs have been tested for the ability to inhibit infection by viruses(Varma,1976).Treated plants develop nonspecific -systemic acquired resistance which is associated with systemic induction of pathogenesis related proteins like chitinases and gluconases. Similar resistance also develop in the hypersensitive response and to damage caused by feeding insects. Recent evidences suggests that salicylic acids, systemic and ethylene act as endogenous molecular signals in systemic acquired resistance(Enyedi *et al.*,1992).

Resistance in Transgenic Plants

Biotechnological development have opened up the possibility of developing plant resistant to specific virus through transfer of alien genes, particularly of viral origin, using a variety of system. Several strategies like transforming DNA copies of satellite RNAs that reduce symptoms expression, on structural gene sequences, defective interfering molecules, ribosome mediated and antisense RNA mediated have been tried for development of transgenic crop against viruses. (Hull and Davis,1992). The most effective approach has been coat protein mediated resistance(Varma and Sinha,1992).This has been efficient against a large number of viruses including those belonging to important crops like cucumo, potex, poty, tobamo and tobraviruses.The mechanism of resistance is not well understood. It appears to vary in different host virus combinations. In the cases of cucumber mosaic virus, transgenic plant that express larger amount of virus coat protein are more resistant than those expressing a smaller amount(Quemada *et al.*, 1991), whereas, the reverse was also found for some potyviruses(Lindbo and Dougherty, 1992). Nucleocapsid protein mediated resistance has been very effective in developing resistance to not only homologous but also heterologous strains of tomato spotted wilt virus(Pang *et al.*, 1992) which is emerging as the most serious pest in various parts of the world.

Use of Antiviral Phytoproteins in Field Conditions

There are many antiviral proteins used in field condition for management of viral diseases. Antiviral phytoproteins are used in several host plants *i.e.* tobacco, urdbean, bhindi, sunhemp, tomato and tobacco infected by several viruses those producing local lesions is well as systomic symptoms in crops of field condition. Antiviral phytoproteins help to protect the susceptible host during the vulnerable easy stages of development disease. The pre-inoculation sprays of Bouganivilli esculentum leaf extract prtoected crop of cucumismeco against cucumber green mottle mosaic virus, crotdlaria juncea against sunhemp roselt virus and tomato against tomato mosaic virus for 6 days (Verma and Dwivedi, 1983).

Since the duration at resistance conferred by SRI's was only upto 6 days, it was realized that the durability of resistance to be prolonged in field condition for getting better protection against virus infection. If has been shown that protenaceous modifiers like papain enhanced the activity of CA-SRI and prolonged the durability of induced resistance against sunhemp tomato virus in crotolaria juncea upto 12 days (Verma and Versha, 1995). Natural plant compound called NS 83 was demonstrated to reduce and delay the disease symptoms by TMV, PVK and in tobacco and tomato plants under field conditions and the yield in tomato was increased by 23.5 per cent (Xin-yun *et al.*, 1988) with the precipitious withdrawn at some of the toxic plant protectants, it may be profitable to explore natural plant products such as alternatives particularly against virus diseases where all other methods fail. The phytoproteins or their smaller pepetides may prove valuable as lead structure for the development of synthesis compounds. It behaves us to explore more intensely the rich source of antivirals. The value at these antiviral protein is unlimited because they are quite safe, non-toxic even after repeated and prolonged use, enhance substantially the plant growth and yield. Antiviral phytoproteins may be more useful if they are integrated with other strategies of virus disease management.

Management of viral diseases is possible only through integration of various approaches like those described in this chapter. This has been demonstrated in vegetatively propagated crops such as potato and citrus, for which entire stocks and areas have been freed from viral and virus like disease through serious efforts. Given the benefits of technological developments it is now possible to device and implement similar programmes for all crops affected by viral disease for sustainable agricultural production.

References

Aubert, B., Quilici (1984). Biological control of the African and Asian Citrus psyllids (Homoptera: psylloidea) through eulophid and enycrtid parasites (Hymenoptera: Chalcidoidea) in Reunion Island. In: *Proceedings of the 9th conference of the International Organisation of Citrus Virologist*. IOCV, Riverside, CAP 100-108.

Dahal, G., Hibino, H., Gabinagan, R.C., Tianco, E.R., Flores, Z.M. and Aquiero, V.M. (1990). Changes in cultivar reaction to tungro due to changes in virulence of the leaf hopper vector.*Phytopath.*, 80: 659-665.

Enyedi, A.J., Yalpani, N., Silverman, P. and Raskin, I. (1992). Signal molecules in systemic plant resistance to pathogens and pests. *Cell*, 70: 879-886.

Fraser, R.S.S. (1987). Genetics of plant resistance to viruses. In: Plant resistance to viruses. Wiley, chichester (Cibo faind Symp 135) pp 6-22.

Gibbs, A.J. and Skotnicki, A. (1986). Strategic defense initiatives; plant, viruses and genetic engineering. In Verma A, Verma JP(eds.)Vistas in plant pathology, Malhotra Publishing House, New Delhi, P-467-480.

Hull, R. and Davies, J.W. (1992). Approaches to non conventional control of plant virus diseases. *Crit. Rev. Pl. Sci.*, 11: 17-33.

Kassanis, B. and Varma, A. (1967). The production of virus free clones of some British potato varieties.*Ann Appl Biol*.59: 750-754.

Lenne, J.M. and Wood, D. (1991). Plant diseases and the use of wild germplasm. *Annu. Rev. Phytopathol.* 29: 35-63.

Lindbo, J.A. and Dougherty, W.G. (1992). Pathogen-derived resistance to a potyvirus: immune and resistant phenotypes in transgenic tobacco expressing altered forms of a poty virus coat protein nucleotide sequence.*Mol.Plant Microbe Interact* 5: 144-153.

Malathi, V.G., Varma, A. and Nambisan, B. (1989). Detection of Indian cassava mosaic virus by ELISA. *Curr. Sci.* 58: 149-150.

Pang, S.Z., Nagpala, P., Wang, M., Slighton, J.L. and Gonsalves, D. (1992). Resistance to heterologous isolates of tomato spotted wilt virus in transgenic tobacco expressing its nucleocapsid protein.*Phytopath*.82: 1223-1229.

Quemada, H.D., Gonsalves, D. and Slighton, J.L. (1991). Expression of coat protein gene from cucumber mosaic virus strain C in tobacco: protection against infection by CMV strains transmitted mechanically or by aphids. *Phytopath*. 81: 794-802.

Sharma, S.R. and Varma, A. (1982a). Control of yellow mosaic of mungbean through insecticides and oils. *J. Entomol. Res.*, 6: 130-136.

Sharma, S.R. and Varma, A. (1982b). Control of virus diseases by oil sprays. *Zbl. Mikrobiol.*, 137: 329-347.

Thresh, J.M. (1981). *Pest pathogens and vegetations*. Pitman London

Tien, P., Zhang, X., Qiu, B., Qin, B. and Wu, G. (1987). Satellite RNA for the control of plant diseases caused by cucumber mosaic virus. *Ann. Appl. Biol.* 111: 143-152.

Varma, A. (1976). Recent trends in control of plant virus diseases. Proc. Natl. Acad. Sci., India, Sect B *Biol. Sci.*, 46: 192-206.

Varma, A., Dhar, A.K. and Mandal, B. (1992a). Mung bean yellow mosaic virus, its vector and their control in India. In: Green, J. (ed) *Mung bean yellow mosaic virus*. Asian vegetable research and Development Center, Taipei, Taiwan, p-8-27.

Varma, A., Kheterpal, R.K. and More, T.A. (1991b). Genetical resistance in plants to viruses. In: programme of the golden Jubilee symposium on genetic research and education. *Current Trends and the Next Fifty Years.* Indian society of Genetics and Plant Breeding, IARI, New Delhi p. 127-128(abstr.)

Varma, A., Kheterpal, R.K. and Vishwanath, S.M. (1991a). Detection of pea seed-borne mosaic virus in commercial seeds of pea and germplasm of pea and lentil. *Indian Phytopath.*, 44: 107-111.

Varma, A., Krishnareddy, M. and Malathi, V.G. (1992b). Influence of the amount of blackgram mottle virus in different tissues on transmission through the seeds of *Vigna mungo. Pl. Pathol.*, 41: 274-281.

Vogal, R., Bove, J.M. and Nicoli, M. (1988). Le programme francais de selection sanitaire des drgrumes. *Fruits*, 43: 709-720.

Wetzel, T., Candresse, T., Macquaire, G., Ravetanandro, M. and Denez, J. (1992). A highly sensitive immunocapture polymerase chain reaction method for plum pox potyvirus detection *J. Virol. Methods*, 39: 27-37.

Sustainable Disease Management of Agricultural Crops (2011) *Pages* **176–185**
Editors: **S.K. Biswas and S.R. Singh**
Published by: **DAYA PUBLISHING HOUSE, NEW DELHI**

Chapter 11

An Overview of *Pseudomonas fluorescens* and its Role in Disease Management

☆ *S.P. Singh*[1], *A. Kirankumar Singh*[1], *Rupankar Bhagawati*[1] *and S.R. Singh*[2]

[1]*ICAR Research Complex for NEH Region, Arunachal Pradesh Centre, Basar – 791 101*
[2]*Krishi Vigyan Kendra, Belatal, Mahoba – 210 423, U.P.*

Pseudomonas are most widely used biocontrol agents since they are reported to have antifungal, antinematode, plant growth promoting, and defence activities and competitive activity for nutrients space and time. Other mechanism include production of hydrogen cyanide (Defago *et al.*, 1990) and degradation of pathogenic toxin (Borowitz *et al.*, 1992). Moreover *P. fluorescens* produce plant growth promoting substances such as auxins and gibberellins which enhance plant growth and yield (Dubeikosky *et al.*, 1993, Gade *et al.*, 2008).

Siderophores, the unique compounded meant for uptake for iron, are produced virtually by all bacteria under iron limiting conditions. These are reported to be involved in suppression of soil borne fungal pathogens (Padghan and Gade, 2006).

Although there are several environment factors which can modulate the production of antifungal compounds, these produces can not be differentiated based on biochemical test (Kumar *et al.*, 2002). Molecular analysis using PCR based RAPD method is thus useful to differentiate *Pseudomonas* strains at interspecific levels. RAPD analysis helps us to determine the genetic diversity of the organism possessing similar functions. A small group of genetically related *Pseudomonas* spp. to be identified and organism having similar properties may share some genetic relationship among themselves irrespective of geographical location.

Biological control of plant diseases using antagonistic bacteria (*Pseudomonas fluorescens*) may be considered as a promising alternative to the use of some hazardous chemical fungicides.

Pseudomonas fluorescens have good prospectus in future as they give very high cost benefit ratio. In view of this first assumption is to isolate the *P. fluorescens* bacteria from the various field crops rhizosphere with maximum antagonistic activity against soil borne fungal pathogens under Indian environmental condition and determine the ability of selected bacterial isolates to suppress the soil borne fungal pathogens under *in -vitro* condition.

One of the mechanisms of *P. fluorescens* is their ability to produce siderophores for sequestering iron. The secreted siderophores binds to the Fe^{3+}that is available in the rhizosphere and thereby effectively prevent growth of pathogens in that region, this information will be helpful to study siderophore medicated antibiosis of *P. fluorescens* against *Fusarium oxysporum* f. sp. *ciceri, Rhizoctonia bataticola* and *Sclerotium rolfsii.*

Utility of PCR based RAPD analysis using know sequence primers to identify bacterial strains after biochemical and morphological characteristics is needed to differentiate *pseudomonas* strains at interspecific levels.

Use of bacteria for management of plant diseases and yield improvement has increased steadily since the mid 1960's (Baker, 1987) and in this connection *Pseudomonas fluorescent* spp. have emerged as the largest and potentially most promising group of plant growth promoting rhizobacteria (PGPR) involved in the biocontrol of plant diseases (Kloepper *et al.*, 1980).

Characteristics

Pseudomanas spp.are gram negative, aerobic, rod shape bacteria. It secretes large amount of a fluorescent yellow green siderophore under iron limited condition. The enhanced plant growth resulting from the introduction of specific root colonizing *Fluorescent Pseudomonads* into the rhizosphere is thought to be due to their production of siderophores under iron limited conditions and to the consequent competition with deleterious rhizosphere.

The genus *Pseudomonas* was first named by Migula in 1894. *Pseudomonas fluorescens* and *P. putida* were identified as important organisms with ability of plant growth promotion and effective in disease management (Tiwary and Thrimurthy, 2007). *P. fluorescens* is strictly aerobic, Gram negative, straight or curved rod shaped, polarly flagellated bacterium with size 1.50-0.45 × 0.51-0.03µ. It has an ability to grow at 4 °C and hydrolyze gelatin. Optimum temperature requirement for growth is 20-25°C. It produces circular, tabular, colonies with greenish centre on gelatin media.

Selective Isolation and Study of Population in Soil

Randomly selected and counted number of plants are gently uprooted using trowel etc. Utmost care should be taken to avoid any damage to the root system while uprooting. The shoot portion of the plants is cut off and the roots with rhizosphere soil are brought to the laboratory in polythene bags for isolation of rhizobacteria. Loosely adhering soil practices are removed by gently tapping and thereafter, closely associated with roots (rhizosphere) is collected and one gram of the soil from each sample is added to 9 ml of sterilized distilled water.

The soil suspended in the tube shaken gently but thoroughly to mix soil particles and then uniformly dispersed on solidified medium. For isolation of rhizoplane bacteria, roots (after removing the rhizosphere soil) are cut into pieces and then one gram of such roots are transferred to 9 ml of sterilized distilled water. One ml of the suspension (from rhizosphere soil as well as rhizoplane) from the first dilution (10^{-1}) are aseptically transferred to another tube (10^{-2}). This procedure is repeated till the original sample is diluted upto 10^{-6}. For isolation of *Pseudomonads fluorescent*, 0.1 ml of soil suspensions from the dilutions (10^{-3} to 10^{-6}) is transferred in sterilized plates. Twenty ml of King's B medium (KBM) (King *et al.*, 1954) is poured in each plate and gently rotated to ensure uniform distribution of the soil/root suspension. After solidification, the plates are incubated at 28°C for 24 hours. They are examined under UV light and colonies with yellow-green colour pigmentation are marked and recorded. Individual fluorescent colony may be picked up with the help of sterilized loop and inoculated on solidified KBM by zigzag streaking. The plates should be incubated at 28°C for 24 hours. The colony growing at the last tip of the zigzag line is transferred to KBM slants and kept in refrigerator at 4°C for further studies.

Characterization of the Isolates (Buchanan and Gibbons, 1974)

Morphological Tests

Cell Shape and Arrangement

Young cultures are Gram stained by smearing loopful of the cells on grease free clean micro slides, stained accordingly and examined microscopically for shape, arrangement and Gram reaction.

Colony Morphology

Pure cultures are streaked on King's B medium and visual observations are made after 48 hours of incubation at incubated at 28°C.

Motility Test

A loopful of freshly grown broth culture is placed at the centre of a "hanging drop slide", vaseline should be applied around the depression of the drop and the slide inverted taking care not to let the drop fall or touch the bottom. Primarily, the drop was located under low power and then shifted to oil immersion to observe length and mobility.

Biochemical Characterization of *Psedomonas fluorescens*

The confirmation of the *Psedomonas fluorescens* isolates is to be performed with the following biochemical tests.

Starch Hydrolysis

Starch is a complete carbohydrate of the polysaccharide type, hydrolysed by bacterium. The positive test indicates by the presence of amylase enzyme.

Medium

Starch Agar–nutrient agar + 0.2 per cent soluble starch

Test Reagent–Gram' Iodine

Inoculate bacterial culture on starch agar plates and incubate for 7 days. After incubation, the plates flooded with Lugol's iodine solution. Presence of starch hydrolysis indicates by the appearance of clear zone. Reddish zone indicate the starch is partially hydrolyse to dextrin. Negative indicates absence of amylase enzyme.

H_2S Production

The activity of bacterium on sulpher containing amino acids frequently results in production and liberation of H_2S gas.

Medium

Peptone water = peptone 1 per cent Cystine 0.01 per cent

Lead acetate paper was prepared by moistening the filter paper in saturated solution of lead acetate.

Lead acetate paper was inserted in the tubes containing peptone water inoculated with bacterium and hold by the plugs above the culture, without touching the medium. If filter paper strips turns black, it indicated the positive test for H_2S production.

Gram's Reaction

Identification was made by gram staining and by studying the morphological characters of the isolates.

Procedure

1. First a smear is to be prepared of bacterial cells by holding a clean slide by grapping at edges.
2. A loopful of bacterial suspension be transferred in the center of slide, with the help of wire loop.
3. The drop be smeared over slide and air dried.
4. Then dried smear is to befixed by passing the slide 3-4 times rapidly over the flame.
5. The smear needs to be flooded with crystal violet for 30 seconds, washed in the tap water.
6. Immerge smear in potassium iodide/Lugol' iodine solution for 30 seconds washed in tap water then decolorized with 95 per cent alcohol and rinsed with water.
7. Counterstained with saffranin for 10 second, again washed with tap water and air dry.
8. Drop of cedar wood oil was placed on the side and examined the smear under oil immersion lense.

IAA Production

Indole is nitrogen containing compound formed degradation of amino acids tryptophan by various bacteria. Pure tryptophan is not used in the test medium. Use tryptone digestion product of certain protein as substrate since it contains considerable amounts of tryptophan.

Medium

1 per cent aqueous solution of tryptone

Trypton broth (as a control–trypton broth of 1 per cent glucose)

Test Reagent

Kovac's regent (P-di-methylamino benzaladehyde 50 g. amyl alcohol 750 ml, HCL 250 ml).

Procedure

The medium was distributed in test tubes and autoclaved. The bacteria was inoculated and incubated for 24 hours. Kovac's regent 1ml was added in incubated test tube. The alcohol layer was separated from aqueous layer upon standing and a rendering of alcohol layer from tryptone broth within a few minutes taken as a positive test for indole.

Arginin Test

This test is used to differentiate gram negative enteric bacilli on the basis of their ability to decarboxylate amino acids.

Media–Prepare according to Fay and Berry (1972)

Positive–purple colour

Negative–yellow colour or no no colour change.

Table 11.1: Some Identifying Characteristics for Differentiation of *Pseudomonas* species

Characteristics	*P. fluorescens*	*P. putida*	*P. aeruginosa*
Pyocyanin	–	–	–
Fluorescin	+	+	+
Oxidase test	+	+	+
Nitrate reduction	–	–	+
Gelatin liquefaction	+	–	+
Growth at 4 ºC	Yes	Yes	No
Growth at 35 ºC	Yes	Yes	Yes

Test Reagent (Nessler's Reagent)

Add a loopful of test culture to Nessler's reagent on a slide. The development of an orange to brown colour indicates the presence of ammonia. The sterile tube tested at the same time show pale-yellow or no colour reaction.

Gelatin Hydrolysis

The gelatin medium is sterilized by autoclaving. Stab tubes are inoculated with test isolates and incubated at 28°C for 48 hours. After incubation, the tubes are chilled in ice bath. Control tube without any inoculums solidified and the tubes where gelatin was hydrolysed remained in liquid state which suggests a positive reaction. Care should be taken to agitate the tubes while testing.

Oxygen Requirement

Yeast extract tryptone agar is prepared, melted and held at 100°C in tubes to expel, and rolled for uniform distribution. The agar is allowed to solidify and the tubes incubate at 28°C for 48 hours.

Surface growth of the organism indicates them to be aerobic, homogenous growth throughout the tube indicates the organism to be anaerobic in nature.

Oxidase Test

Test cultures are streaked on nutrient agar plates and incubated for 48 hours at 28°C. Thereafter, the plates are flooded with p-amino dimethyl aniline oxalate solution (1g/100 ml of *p*-amino dimethyl aniline oxalate is dissolved, heated, filtered and stored at 4°C).

Positive test is shown by the pink colour of the colonies. On standing (exposed to light), colonies turn to dark pink or black colour.

Ammonification

Sterile tubes containing 4 per cent peptone solution are inoculated with test isolates and incubated at 28°C for 48 hours. A drop of Nessler's reagent is placed on spot plate along with a test isolate sample, change of colour to yellow is the indication of presence of ammonia.

Levan Formation

King's B medium is supplemented with 4 per cent sucrose and plates inoculated with test bacteria are incubated at 28°C for 48 hours. After incubation, the isolates which formed slimy colonies with excess levan are taken as positive test for levan formation.

Cellulose Production

Cultures are inoculated in the Czepek mineral salt agar medium and incubated for 48 hours at 28°C. After incubation, the plates are flooded with 1 per cent (w/v) aqueous solution of hexadecyl trimethyl ammonium bromide. A clear zone around the colonies indicates positive test.

In vitro Screening of the *Fluorescent Pseudomonas* Isolates Siderophore Production

By Chrome Azurol 'S' Assay

Petri plates are poured with Chrome Azurol S agar medium and allowed to solidify. Desired numbers of *Pseudomonads fluorescent* isolates are spotted at equidistance and incubated for 30 hours at 27°C. Each treatment should be replicated thrice. Isolates producing an orange halo may be considered as positive for siderophore production. The isolates that produce siderophore should be scored out and reinoculated on fresh seeded plates with one isolate per plate and incubated. After incubation, the diameter of the orange zones formed is measured. Those strains which produce similar results in all the three replicates are considered as positive siderophore producers (Schwyn and Neilands, 1987).

HCN Production

HCN production is tested as the method described by (Castic and Castric (1983). Prepare solution of 0.05 per cent Picric acid (1 per cent sodium carbonate in sterilized flasks). Inoculate one ml culture of each test isolate on succinate agar separately. The control plates will not receive the inoculum. A disc of Whatman filter paper No.1 of the diameter equal to Petri plate. Impregnated with alkaline picric acid solution and placed in upper lid of inoculated Petri plates under aseptic condition. The plates incubate up side up at 28 ±1°C for 24 hours. Indicates changes in colour from yellow to light brown moderate brown or strong reddish brown of HCN production.

Phosphate Solubilization

Phosphate solubilization test to be performed by spot inoculation of test organism on Pikovaskay's medium. Incubate the plates at 28 ±1°C for 4-5 days. Consider formation of a clear inhibition zone around the colony positive for phosphate solubilization.

Plant Growth Response

In order to assess the plant growth promoting activity and inhibitory effects of the bacterial isolates, seed bacterisation and root dipping methods are generally employed.

Seed Bacterisation

Among several methods developed for application of biocontrol agents (BCA), the seed bacterisation (treating/coating seeds with BCA's inoculums) is the most common one.

The rhizobacterial isolates are grown at 28°C for 48 hours on King's B medium. The bacterial cells are scrapped from the plates by adding 10 ml of sterilized distilled water per plate and prepared a slurry by suspending in 1 per cent (w/v) carboxy methyl cellulose (CMC). Similarly formulated products may be suspended in water to make slurry. A known number of seeds are surface disinfested with 1 per cent sodium hypochlorite (NaOCl) for 3 minutes followed by 4-5 washings with sterilized distilled water. These seeds are then dipped into bacterial cell suspension + CMC slurry and mix thoroughly to ensure uniform coating. These coated seeds are spread on plastic sheets and air dried for 24 hours. Treated seeds may either be placed at equal distance on sterilized moist towel/paper/

plastic pots having sterilized soil. Observations may be recorded with respect to shoot length, root length, increased/decreased germination percentage, fresh and dry shoot and root weight.

Root Dipping

Root dipping method is mainly used for the crops grown by seedlings *e.g.* rice, tomato, etc. The roots are dipped in bacterial suspension (10^8 cfu/ml) and subsequently placed in plastic pots having sterilized soil. Observations with respect to growth are recorded as described above.

Biocontrol Potential

The isolates of rhizobacteria are tested for their biocontrol potential on King's B medium/PDA.

Antifungal Activity on Solid Media

Each isolate is inoculated in 30 ml Nutrient Broth Yeast Extract medium or in King's B broth. Inoculated flasks are incubated in an incubator shaker at 28°C for 24 hours. A 5 mm sterilized paper disc impregnated/dipped in bacterial suspension of individual isolate is placed at a distance of 5 mm from the periphery of plate. A plug (5 mm diameter) of the test fungus taken from the leading edge of the culture grown on PDA is placed in the opposite direction. Plates are then incubated for 5 days and/or until the leading edge of the fungus reached the edge of the plate. Finally the zone of inhibition that developed is measured (distance between the leading edge of the bacterial colony and the nearest edge of the fungal colony around each bacterial isolate).

Viability of Fungal Mycelium

The plates of test fungus with good inhibition zones of rhizobacterial isolates are selected. The 5 mm mycelia disc is aseptically cut in the vicinity/periphery of inhibition zone. While removing, it is ensured that half of the disc should be taken from the growing zone while the remaining half from the inhibition zone. Similarly, a 5 mm disc from different places with mycelia growth should be cut out and placed over sterilized PDA plates impregnated with 1000 µg/ml streptopenicillin. The incubated plates are examined after 48-72 hours for the growth of the fungus.

Antifungal Activity in Broth

Measured quantity of King's B broth and potato dextrose broth (PDB) are placed in 100 ml conical flasks and autoclaved. A 5 mm mycelia disc of test fungus is picked up aseptically from actively growing colony is placed in each flask. These flasks are then individually inoculated with 0.1 ml of bacterial suspension prepared from 48 hours old culture of bacterial isolate grown on King's B medium. The flasks are inoculated with fungal mycelium alone serve as control. The flasks are incubated at 28°C for 10 days. In case of sclerotial pathogen, observations may be recorded on sclerotia formation. The cultures are filtered through Whatman filter paper No. 40. The mycelia mats are collected and dried for 24 hours at 50°C in hot air oven. The dry weight of mycelium is recorded. Per cent inhibition should be calculated using control as standard.

Viability of Sclerotia

Sclerotial pathogens *viz. Rhizoctonia solani,S. rolfsii* etc are grown on PDA and after formation and maturation of sclerotia, these are mechanically harvested and surface sterilized with sodium hypochlorite solution, repeatedly washed with sterile water and dried by placing them between sterilized filter paper layers. These sclerotia are dipped and incubated in cell suspension of *Pseudomonas* for different time periods, taken out and transferred to sterilized PDA plates impregnated with 1000 μg/ml streptopenicillin. The incubated plates are examined after 72 hours for the growth of the fungus.

Biocontrol Activity Mediated by the Synthesis of Allelochemicals

Offensive PGPR colonization and defensive retention of rhizosphere niches are enabled by production of bacterial allelochemicals, including iron–chelating siderophores, antibiotics, biocidal volatiles, lytic enzymes, and detoxification enzymes. Competition for iron and the role siderophores. Iron is an essential growth element for all living organisms. The scarcity of bioavailable iron in soil habitats and on plant surfaces from a furious competition.Under iron-limiting conditions PGPR produce low molecular weight compounds called siderophores to competitively acquire ferric ion (Schippers *et al.*, 1987). Although various bacterial siderophores differ in their abilities to sequester iron, in general, they deprive pathogenic fungi of this essential element since the fungual siderophores have lower affinity. Some PGPR strains go one step further and draw iron from heterologous siderophores produced by cohabiting microorganisms (Raaiyakers *et al.*, 1995). Siderophores biosynthesis is generally tightly regulated by iron sensitive fure proteins, the global regulators GacS and GacA, the sigma factors RpoS, PvdS, and Fpvl, quorum-sensing autoinducers such as N-acyl homoserine lactone and site-specific recombinases. However, some data demonstrate that none of these global regulators is involved in siderophores production. Neither GacS nor RpoS significantly affected the level of siderophores synthesized by Entrobacter cloacace CAL2 and UW4 (Saleh and Glick, 2001). RpoS is not ivolved in the regulation of siderophores production by *Pseudomonas putidu* strain WCS358.

In addition, GrrA/GrrS, but not GacS/GacA, are involved in siderophores synthesis regulation in *Serratia plymuthica* strain 1C1270, suggesting that gene evoluation occurred in the siderophore producing bacteria (Ovadis *et al.*, 2004). A myriad of environmental factors can also modulate siderophores synthesis, including pH, the level of ion and form of iron irons, the presence of other elements, and an adequate supply of carbon, nitrogen, and phosphorus (Duffy and Defago, 1999). The strongest factor related to the action of Pseudomonas stimulated growth response is the production of siderophores. Growth promotion of potato plants in response to seed bacterization in soil cropped frequently to potato was due to siderophore–mediated suppression of microbial cyanide production. The hypothetical interactions are believed to take place.The uptake of iron by siderophore of DRB(S) from the soil enhances their HCN production from the root exudates component "X". HCN inhibitors the energy metabolism of the root cell and thus decreases uptake of nitrogen, phosphorus and potassium (NPK), competition for iron by the siderophore of a *Pseudomonas* inhibits the HCN production by DRB, thereby increasing the available energy of the root cell for uptake of NPK. The DRB inhibit the potato root cell function by producing HCN. Cynide production is dependent on availability of Fe^{++}, and competition for iron by siderophores of pseudomans inhibits HCN production by DRB. The siderophores of PGPR having higher affinity for Fe^{++}, and ability to use the siderophores of DRB, starve the DRB of iron nutrition. This leads to non-production of HCN and consequent increase in the availability of energy to root cells for the uptake of P, K and N. The smooth explanation of the interaction runs into trouble when it is noted that there are Fluorescent pseudomonas which are DRB and which can use the FE 3^{+} siderphore complex of a particular PGPR. They can nullify the initial beneficial effects of *P. fluorescence*. Such a process may be involved in cases of *P. fluorescence* failures in fields (Dube, 2001).

Rhizosphere Competence

The competitive exclusion of deleterious rhizosphere organisms is directly linked to its ability to successfully colonise on root surface. All disease suppressive mechanisms exhibited by fluorescent pseudomonas are essentially of no real value unless these bacteria could successfully establish themselves in the root environment. It is well known that different pseudomonads have different abilities to colonise a particular root-niche.

A desired number of seeds are surface disinfested with 1 per cent sodium hypochlorite for 3 minutes and air dried. Seeds are then treated with fluorescent pseudomonads + 1 per cent (w/v) carboxy methyl cellulose (CMC) slurry mixed thoroughly to ensure uniform coating and air dried. Seeds treated with 1 per cent (w/v) CMC and sterilized distilled water serve as control. The treated seeds are placed in aluminium foil cylinders having 1000 g of autoclaved sandy loam soil. In each cylinder, a desired number of seeds are sown. At the end of 7th and 10th day of incubation, the cylinders are cut into five different segments (0-1,1-2,2-3,3-4 and 4-5 cm). The cut root segments from different zones are collected separately along with adhered rhizosphere soil. Population dynamics along roots is recorded and the cfu per cm root or per g rhizosphere soil is calculated. Two methods are generally employed to study the colonization of the roots by individual rhizobacterial isolates.

Dilution Plate Technique

The cut root segments from different zones are collected separately with adhered rhizosphere soil. One g of such samples from each zone are taken to assay the population density of the bacterial isolates. Serial dilutions followed by plating of the suspension are done on King's B medium. Plates are incubated at 28°C and cfu recorded after 24 hours.

Impression Method

Roots are cut from 7 day old seedlings and placed over the King's B agar plates and press with the help of sterilized forcep and needle so as to get complete impression of root system of seedlings. The plates are then incubating at 28°C for 48 hours and observations are recorded after 24 hours.

Mass Multiplication and Formulation

King's B broth medium is inoculated with selected rhizobacterial isolates and incubated at 28°C and 120 rpm in a rotator shaker for 48 hours. Different types of carriers are used for making the formulations of rhizobacterial isolates *e.g.* farmyard manure, talc powder, kaolinite, lignite, vermiculite, peat, charcoal, etc. Generally talc powder is preferred. Talc powder ($CaCO_3$) mixed with 1 per cent carboxy methyl cellulose (CMC) is autoclaved twice and mixed with 48 hours old *Pseudomonas* broth @ 400 ml broth per 1000 g of talc-CMC powder. Formulation was further diluted with talc-CMC in order to obtain 10^8 cfu per g formulation.

References

Baker, K. F. (1987). Evolving concepts of biological control of plant pathogens. *Ann. Rev. Phytopathol.* 25: 67-85.

Borowitz, J.J., M. Stanki- Dicz, T. Lewicka and Z. Zukowsaka, (1992). Inhibition of fungal cellulose peptinose, xylanase activity of plant growth promoting *fluoresecent Pseudomonas*. Bulletin OILB/ SROP. 15: 103-106.

Buchanan, R. E. and Gibbons, N.E. Eds. (1974). *Bergey's Manual of Determinative Bacteriology*. 8th ed. Baltimore, Williams and Wilkins.

Castric, K.F. and P.A. castric, (1983). Method for rapid detection of cyanogenic bacteria. *Appl. Environ. Microbiol.* 45: 700-702.

Chao, W.L., Nelson, E.B.; Harman, G.E. and Hoch, H.C. (1986). Colonisation of rhizosphere by biological control agents applied to seeds. *Phytopathology*, 76:60-65.

Defago, G., C.H. Berling, U. Burger, D. Hass. G. Kahr, C. Keel, C. Voisard, P. Wirthner and B. Wuthrich. (1990). Suppression of black rot of tobacco and other root diseases by strains of Iapseudomonas *fluorescens*. Potential application and mechanism. In: D. Honby (Ed) *Biological control of soil borne plant pathogens*. CAB International, Wellingford. Oxon UK: 93-108.

Dube, H.C. (2001). Rhizobacteria in biological control and plant growth promotion. *J. Myco. Pl. Patholpgy*. 31 (1): 9-21.

King, E.O.; Ward, M. K. and Raney, D. E. (1954). Two simple media for the demonstration of pyocyanin and fluorescein. *Journal of Laboratory and Clinical Medicine*, 44: 301-307.

Kloepper, J. W.; Leong, J.; Teintze, M. and Schroth, M.N. (1980). Enhanced plant growth by siderophores produced by plant growth promoting rhizobacteria. *Nature* (London), 285: 885-886.

Palleroni, N. J.; Kunisawa, R.; Contopoulou, R. and Dondoroff, M. (1973). Nucleic acid homologies in genus Pseudomonas. *Int. J. Sys. Bacteriol.*,23: 333-339.

Schwyn, B. and Neilands, J.B. (1987). Universal chemical assay for detection and determination of siderophores. *Annal. Biochem.*,160:47-56.

Sustainable Disease Management of Agricultural Crops (2011) *Pages* **186–193**
Editors: **S.K. Biswas and S.R. Singh**
Published by: **DAYA PUBLISHING HOUSE, NEW DELHI**

Chapter 12

Management of Diseases of Horticultural Crops through Organics

☆ ***Shripad Kulkarni and V.I. Benagi***
Department of Plant Pathology, UAS, Dharwad, Karnataka

Shift of Indian agriculture from a state of food deficiency to food sufficiency through introduction of fast growing, high yielding hybrid varieties, usage of high dose of chemical fertilizers, pesticides and weedicides. No doubt green revolution was yielded rich dividends but at the same time gifted serious pest problems and degradation of environment followed by deleterious effects on environment.

World over it has been estimated that more than 67,000 different pest attack crop. To protect them, many chemical pesticides are used which have unsafe environment impact, and hence there is a pressure for decreased reliance on such agents and greater regulatory control of their use. Besides, many of the pathogens developed many fold of resistance to fungicide. (Chandra *et al.*, 2005).

Most of horticultural crops are eaten fresh or used for health care; hence any contamination in the form of pesticide/chemical residue may lead to health hazards; therefore, organic horticulture offers a better possibility of producing healthy food. Plant diseases caused by different groups of organisms belonging to fungi, bacteria, viruses, rickettsia, spiroplasma, nematodes and few others have remained important in causing significant losses in different crops indicating the urgent need of their integrated management. The continuous and indiscriminate use of chemical pesticides has posed several serious problems such as pesticide residue, development of resistant strains, environmental pollution and adverse effect on beneficial microorganisms and created a greater concern over global food safety and security.

Organic farming relies on crop protection, crop residues, animal manures, legumes, green manures, off farm organic wastes, cultural practices, mineral bearing rocks and aspects of biological pest control to maintain soil productivity and to supply plant nutrients and to control diseases, insects, weeds and other pests.

Cultural Methods

By practicing the following methods we can regulate/modify the pest and disease incidence effectively.

Selection of Widely Adopted and Resistant Varieties

By choosing varieties which are well adopted to the local environmental conditions (such as temperature, nutrient supply, pests and disease resistance) by which crop is allowed to grow healthier and stronger against attack of pests and pathogens.

Cropping System

In a particular agricultural ecosystem plays major role in plant disease management. By adopting a suitable cropping system in right time, we can avoid most of the harmful pests and diseases. Some of the practical practices are as follows.

1. Activity in the soil and enhance the presence of beneficial organism. For instance plants colonized by arbuscular mycorrhizal may increase pests also. So careful selection of proper green manure is essential.
2. Crop rotation: Cauliflower–paddy–cauliflower rotation is highly effective in controlling stalk rot of cauliflower disease caused by *Sclerotinia sclerotiorum* and it reduces infection by >60 per cent and > 161 per cent increase in seed yield has been observed.
3. By adopting mixed cropping system pest and disease incidence can be minimized since pest has less host plants to feed on.
4. By following crop rotation practices we can increase the soil fertility and reduce the chances of soil borne diseases.
5. By cultivating green manure cover crops like Horse gram, cow pea, sunhemp, sesbania, dhaincha, glyrisidia, we can increase the biological activity in the soil and enhance the presence of beneficial organism. For instance plants colonized by arbuscular mycorrhizal may increase pests also. So careful selection of proper green manure is essential.

Selection of Clean Seed and Planting Materials

Seeds and planting materials are the primary sources of diseases. Hence, selection and use of disease free seeds after inspection for pathogens and weeds is very much essential. Further, it is advised to get seeds and planting materials from the reliable sources only. Use of healthy seed and it's hot water treatment at 50°C for 25 minutes or for sodium phosphate 90 gram/Lit. for 20 minute is effective for tomato mosaic which is most dangerous disease of tomato.

Selection of Optimum Planting/Sowing Time and Spacing

Most of the pests or diseases attack the crops only in a certain life stage. By adopting sufficient spacing between plants we can reduce the spread of a disease as well as allows good sunlight to the plants which facilitates less moisture on the leaves leading to hinder and of pathogen development and infection. In the same way more sunlight allows plants to do more photosynthesis. This practice not only avoids disease and pests in cropping system but also increase the crop productivity.

Balance Organic Nutrition

Gradual and steady growth makes plants less vulnerable to infection. So this steady growth could be achieved by applying organic fertilizers timely and moderately because, excess and

Rajapuri (Highly susceptible)

Sakkarebale (Highly resistant)

Figure 12.1: Banana Variety Against Sigatoka Disease

indiscriminate use of fertilizers often results in damaging the roots of plant. This damage facilitate to secondary infection. To overcome this problem we can adopt integrated nutrient management system with organic manures like FYM, compost, nutrients slowly when the plant needs. Further, by using liquid bio–fertilizers like potash mobilizers namely *Frateuria aurentia* along with organic manures provides balanced potassium and contributes to the prevention of fungi and bacterial infections.

Addition of More and More Organic Matter

Organic content of the soil is directly related to density and activities of microorganisms in the soil there by pathogenic and soil borne fungal population can be reduced. Besides this, organic matter provides.

- ✰ All the nutrients that are required by the plants.
- ✰ Corrects C:N ratio in the soil.
- ✰ Good physical chemical and biological support to soil
- ✰ More water holding capacity to the soil.
- ✰ Cover from evaporation losses of the moisture from the soil

Ultimately, organic matter supplies substances which strengthen plants with their own protection mechanisms.

Soil Amendments

The decomposition of organic matter helps in alteration of the physical, chemical and biological conditions of the soil and also reduce the inoculum potential of a soil borne pathogens. In addition, the practice improves soil structures, which promotes root growth of the host. Various biochemical substances like antibiotics and phenols are released during decomposition, which in turn induce resistance in the root system. Soil amendments like sunflower, rape seed cakes, mustard cake, gypsum and bean straw can be used.

Rhizoctonia solani causal organism of damping off in nursery stage and wire stem, bottom rot and head rot after transplanting plants in crucifiers. Soil amended with neem cake 1 kg/m^2 and solarized by covering with white polythene sheet for two weeks and treatment of seeds with *Trichoderma harzianum* proved to be effective in the management off damping off of vegetables crops. Soil amended with cellulose powder was also found effective in reducing the disease incidence, which may be due to increase in C:N ratio in soil. This increase results in the decrease of fungal population but actinomycetes and bacterial populations increased.

Organic amendment is used here to mean organic material incorporation into the soil that comes from external sources such as processing residues or industrial waste products. Organic material added as fresh crop residue and grown in the field in rotation–break, cover, trap, antagonistic or green manure crops- are discussed below. Incorporation into the soil of large amounts of any organic material will reduce nematode densities. Oil cakes, coffee husks, paper wastes, crustacean skeletons, sawdust and chicken manure, amongst others, have been used with some success. Nematode control may be due to any one or more of the following mechanisms:

- ✰ Toxic and non-toxic compounds present in the organic material;
- ✰ Toxic metabolites produced during microbial degradation; or
- ✰ Enhancement of the soil antagonistic potential.

Chitin amendments have received much interest in the past as an organic amendment in that they stimulate the antagonistic potential in soil towards nematodes. Organic amendments have also been combined with various biocontrol agents with reports of enhanced levels of control. The use of organic amendments is often limited by availability and, in some cases, by the large quantities needed. In addition to their effects on nematode density, organic amendments also improve soil structure and water-holding capacity, reduce diseases and limit weed growth, all of which ultimately lead to a stronger plant and improved tolerance to nematode attack.

Biofumigation

This term normally refers to suppression of soil-borne pests and pathogens by biocidal compounds, principally isothiocyanates, released in soil when glucosinolates in cruciferous crop residues are hydrolysed. Soil amended with fresh or dried cruciferous residues at 38°C day and 27°C night temeperatures reduced *Meloidogyne incognita* gall by 95-100 per cent after 7 days' incubation in controlled environmental condition. It should noted here, that many cruciferous plants are good hosts of some important species of *Meloidogyne*.

The term biofumigation is now used more freely whenever volatile substances are produced through microbial degradation of organic amendments that result in significant toxic activity toward a nematode or disease causing pathogens. The release of toxic compounds already present in antagonistic plants are neem, marigold and castor, or the production of toxic compounds due to microbial fermentation of nutrient-rich organic amendments, *e.g.* velvet bean, sunhemp or elephant grass, lead to significant levels of nematode control.

Biofumigation under these circumstances is greatest when there is an optimum combination of organic matter, high soil temperature and adequate moisture to promote microbial activity leading to toxin production. In tropical and subtropical production systems, plastic mulch and drip irrigation improve effectiveness of biofumigation. Transporting organic amendments to the field or incorporating cover crops that produce large amounts of biomass into the soil, together with plastic mulch and drip irrigation, should significantly increase the level of control attained.

Biofumigation using fresh marigold as an amendment is used effectively in root knot management in protected cultivation in Morocco. Tagetes is grown in the raised beds prior to the planting of susceptible horticultural crops. The crop is then incorporated into the soil after 2-3 months. The beds are fitted with drip irrigation and covered with plastic mulch. The soil in the bed is then biofumigated under conditions of the high temperature and optimum soil moisture.

Control due to any form of biofumigation is probably the result of multifaceted mechanisms including:

1. Non-host or trap cropping depending on the host status of the plant used.
2. Lethal temperature due to solarization.
3. Nematicidal action of toxic by-products produced during organic matter degradation.
4. Stimulation of antagonists in the soil after biofumigation.

Water Management

By practicing good water management water logging in the field and stress on plants can be avoided otherwise pathogens take chance and infect the crop. Further, sprinkling water on foliage shall be avoided as it increase diseases by giving chance to pathogenic fungal spores to germinate.

Use of Proper Sanitation Measures

"Pull and Burn" is the best method to control disease and removal of infected plant parts leaves, fruits) from the ground to prevent the disease from spreading. Eliminate residues of infected plants after harvesting.

Soil Solarization

Soil solarization for 4 weeks during summer months coupled with application of neem cake @ 400 g/m^2 proved effective against damping-off (76.9 per cent) in nursery and resulted in significant, higher number of healthy transplants.

Figure 12.2: Soil Solarization Method

Figure 12.3: Healthy Turmeric Crop in Soil Solarized Plot

Mechanical Methods

1. By removing and burning clipping of lower leaves upto 20 cm and weading to reduce the alternaria blight in tomato

2. This is the best method, before reaching the loss beyond economic level. To save crop from fungal and bacterial diseases pull and burn method is most effective.

Botanicals

Plants during their long evolution, have synthesized a diverse array of chemicals to prevent the colonization by pathogens. They produce secondary metobolites like terpenoids alkaloids, flavonoids, phenolic compounds. These secondary metabolites are having disease suppressing properties. In India several plants such as nicotinoids, natural pyrethrins, rotenoids, neem products have been used in the past for suppression of diseases. Among the botanical pesticides neem occupies very important place in the pest and disease control. Different parts of neem tree can affect more than 200 insects and diseases, some of them are effective against nematodes, fungi, bacteria and viruses.

Biological Control

Management of soil borne pathogens is difficult because of non-availability of desired level of resistance against major soil borne diseases caused by species of *Fusarium, Sclerotium, Macrophomina, Pythium, Phytophthora* and few others forced to search or resorting to new approaches to manage the diseases. Therefore, biocontrol agents or antagonists are means of plant disease control has gained importance in the recent years. The biocontrol agents multiply in soil and remain near root zone of the plants and offer protection even at later stages of crop growth.

Table 12.1: Successful Control of Plant Diseases Using *Trichoderma* Species

Chick pea wilt	Carbendazim seed treatment (ST) + *T. harzianum* or neem cake + *T. harzianum*
Redgram wilt	ICP8863 + ST with *T. harzianum*
Foot rot of paper	Neem cake + *T. harzianum* (soil solarization and application of Metalaxyl MZ (two spray)
Cotton root rot	*T. harzianum* + Benomyl
Sclerotium wilt of sugarbeet	Captan + *T. harzianum*
Damping off of tobacco, tomato, egg plant	Metalaxyl MZ + *T. harzianum*
Sclerotium rolfsii in tomato, beans, groundnut, sugar beet and chick pea	Furrow application of 130-160 kg/ha with 4 g/kg of seed treatment of (*T. harzianum T. viride, T. hamatum*)
Bean root rot *Macrophomina phaseolina*	Furrow application of *T. harzianum* (130-160 kg/ha)
Fusarium oxysporum in bean, chrysanthemum, cotton, melon and redgram	Soil application of *T. harzianum* at 130-160 kg/ha
Different wood rotting fungi in wood trees	Dusting *T. viride* on wounds.
Grey mold, *Botrytis cinerea* in grapes	Aerial spray with *T. viride* (106–108 cfu/ml)
Stalk rot of rabi sorghum	*T. harzianum* @ 4g kg/seed)

Mechanisms of Biocontrol

- ✰ Competition
- ✰ Mycoparasitism
- ✰ Antibiosis and Enzymes
- ✰ Cross protection
- ✰ Induced systemic resistance (ISR)

Methods of Application

1. Broad cast application (125-250 kg/ha)
2. Furrow application (130-160 kg/ha)
3. Root zone application (Formulation mixed in soil @ 1 kg/plant)
4. Seed treatment (4g/kg seeds)
5. Wound application (Applied to wounds in peach, plum etc)
6. Spraying (106–108 cfu/ml)

Advantages of Biopesticides

1. Avoids environmental pollution (soil, air and water)
2. Avoids adverse an effect on beneficial organisms *i.e.* maintains healthy biological control balance
3. Less expensive thou pesticides and avoids problems of resistance
4. Biopesticides are self maintaining in simple application and fungicide needs repeated application
5. Biopesticides are very effective for soil borne pathogen where fungicidal approach is not feasible.
6. Biopesticides eco-friendly, durable and long lasting
7. Very high control potential by integrating fungicide resistant antagonist
8. Biopesticides help in induced system resistance among the crop species.

Use of Bleaching Powder

12 kg/ha found to reduce bacterial blight by 80 per cent when applied in furrows at the time of planting.

References

Chandra, Krishna, Greep, S. and Srivatha, R.S. H. (2005). *Biocontrol Agents and Biopesticides.* Published by Regional Center of Organic Farming, Hebbal, Bangalore.

Gupta.S.K. and Thind T.S. (2006). *Disease problems in vegetables production.* Scientific publishers (India), Jodhpur.

Ray.A.B., Sharma, B.K and Singh.U.P. (2004). Medicinal properties of plants-Antifungal, Antibacterial and Antiviral Activities. International Book Distributing Co- Lucknow- 226004 (U.P).

Singh R.S. (2005). *Plant Diseases* (Eight Edition) Oxford and IBH publishing Co. Pvt. Ltd. New Delhi.

Sustainable Disease Management of Agricultural Crops (2011) *Pages* **194–202**
Editors: **S.K. Biswas and S.R. Ṣingh**
Published by: **DAYA PUBLISHING HOUSE, NEW DELHI**

Chapter 13

Development of Bio-formulation Products

P.K. Gupta[1], Yogita Gharde[2] and S.R. Singh[3]
[1]*Division of Plant Pathology,*
Indian Agricultural Research Institute, New Delhi – 110 012
[2]*IASRI, New Delhi – 110 012*
[3]*Krishi Vigyan Kendra, Belatal, Mahoba – 210 423, U.P.*

Importance of Microbial Biopesticides

Studies of natural epizootics of entomopathogenic bacteria and fungi during the latter half of the nineteenth century and the first half of the twentieth century, stimulated man's interest in employing them as microbial pesticides and myco-insecticides to control agricultural pests. Mass production of the selected bacteria and fungi is a necessary prerequisite for any large-scale field application. The revival of interest in microbial insecticides over the last 20 years, has led to large-scale production of *Bacillus thuringiensis* (Berliner). While, the production of bio-insecticides is common worldwide, there is little information available on the biotechnology of entomopathogenic fungi, and their industrial production is still relatively unsophisticated. A microbial toxin can be defined as a biological poison derived from a microorganism, such as a bacterium or fungus. Pathogenesis by microbial entomopathogens occurs by invasion through the integument or gut of the insect, followed by multiplication of the pathogen resulting in the death of the host, *e.g.*, insects. Studies have demonstrated that the pathogens produce insecticidal toxin important in pathogenesis.

In the short term, with our present level of knowledge, it seems likely that commercial production of myco-insecticides is going to be restricted to small-scale production in specific areas. At present there appears to be a situation in which biological control agents including fungi would provide a viable commercial option, where conventional chemical control gives insufficient control or where there is insecticide resistance; where conventional chemicals are too expensive; or where government restricts application of chemicals.

Steps Leading to Commercialization

Once any new system has been identified and characterized in the laboratory, the following steps must be completed before it can be successfully commercialized.

Process of Development

A process must be developed which can be carried out on a large enough scale to ensure that an adequate amount of the material can be made. This process must be sufficiently reliable to provide a product, which is both safe and effective. In addition, the production cost must allow manufacturers to make a profit.

Organism Storage

The first problem in producing a microorganism is storing it in a way that ensures the retention of desirable features. The most obvious feature to be retained is effectively pathogenicity; the second is productivity in terms of yield in the commercial production process. This is typically measured either by total biomass, or by number of infective propagules, *i.e.* spores or fragments of mycelium, produced per litre/hour of fermentation time. Many microorganisms are known to lose desirable features either on storage or after repeated sub-culturing. Among insect pathogens, *Beauveria bassiana* has been reported to reduce virulence after subculturing. Organisms, which lose virulence, may sometimes be restored to their former potency by passing them through their normal host, *i.e.* the target insect; however, such a technique would be cumbersome as a routine part of a production process and always presents the risk of contamination. Therefore, the problem is often avoided by storing a large number of elements of a single spore isolate in a deep frozen or freeze-dried condition. Samples are checked periodically to make certain that virulence and productivity have not diminished with time.

Fermentation Method

The standard method of production of microorganisms is the process of fermentation. There are many types of fermentation; the two most common are "submerged" and "semi-solid".

Submerged or deep-tank fermentation is, as the name implies, a growth of microorganisms in a fully liquid system. There are a number of advantages to fully liquid systems, which include the ability to hold temperature and pH constant, the ability to pump large quantities of air into the system and disperse it by means of stirring impellers, and the ability to generate reasonably homogeneous conditions to maximize the growth of microorganism.

Despite, the many advantages of submerged fermentation some fungi will not yield a satisfactory product by this technique. Semi-solid fermentation offers an alternative in which the fungi grow primarily on the wet surface of a solid material, often some form of processed cereal grain to which nutritional adjuvants have been added, though attempts are made frequently to use "waste" materials or media of low value, such as straw. This allows fungi to grow in conditions more similar to those found in nature. Semi solid fermentations are relatively easy to develop on a small scale. Scaling them up to the sizes necessary for commercial product presents numerous problems; aeration becomes a major difficulty as the volume of a semi-solid mass increases more rapidly than the available surface area. This requires either a very large area of relatively shallow media, *e.g.* on trays, or in a vessel which can agitate or tumble the media. On any scale, trays are very difficult to sterilize and keep sterile. The development of large vessels for semi-solid media fermentation requires the invention of a number of techniques or pieces of equipment for:

- ✰ Keeping the media friable after sterilization. Its tendency is to set solid when it cools, rather like oatmeal.
- ✰ Inoculation with the desired fungus without contamination
- ✰ Aeration and agitation during fermentation
- ✰ Drying the material prior to opening the fermentor in order to avoid contamination, etc.

Medium Development

Whichever, type of fermentation is chosen, nutrients must be provided so that the microorganism can grow. Which nutrients are chosen will markedly affect how fast the organism grows, how much is produced and often, how infective the final product is. Nutrients to be provided include a carbon source, *e.g.* glucose or molasses, nitrogen source, *e.g.* soybean meal or yeast extract, and a "defined", in which case the precise nature and quantity of every nutrient is known, or, alternatively, they can contain ingredients of an indeterminate and occasionally variable nature *e.g.* molasses; most commercial media are latter type.

Laboratory Processing

Cleanliness and Safety

The normal technical procedures designed to keep cultures sterile or pure and prevent cross-infection of laboratory materials contribute considerably to the safety of the individuals as most laboratory workers observe reasonably high standards of personal hygiene and wash their hands after handling cultures of any kind and before touching the rest of their person, handling their food or smoking.

Safety Cabinets (Laminar Air Flow Bench)

There are cabinets, which are fitted with filters, which remove particles down to 0.01 mm and discharge their effluence into the laboratory air.

Ultra-violet Light

Continuous irradiation of the air above eye level is possible and reduces the bacterial content. Ultra-violet light can be used to disinfect the insides of protective cabinets but too much reliance should not be placed on this as a disinfecting agent. Chemical methods should also be used (75 per cent ethylalcohol). Ultra-violet light does not penetrate surfaces.

Use of the Microscope

One important microbiological technique will be introduced: the use of the oil immersion objective is essential to most microscopic studies of bacterial cells. The best type of microscope is with Phase-contrast.

Preparation of Dilution

Examples of Dilutions

- ✰ 1×10^{-2}or 1: 100; 1.0 ml + 99.0 ml or 1.0 ml: (99 ml) = 1/100
- ✰ 2×10^{-2} or 1: 50; 2.0 ml + 98.0 ml
- ✰ 5×10^{-3} or 1: 200; 0.5 ml + 99.5 ml
- ✰ 2×10^{-4} means 2/10.000 = 1/5.000

Dilution Plate Counting

Preparation of Dilutions: Shake each stock dilution well. Using aseptic techniques transfer 0.1 ml of the 100 stock dilutions to a 9.9 ml dilution blank (1:10,000 dilution). Mix well to obtain even distribution of organisms. With a pure pipette, place 1.0 ml of the 1:10,000 dilution into a sterile Petri plate and with the same pipette transfer 1.0 ml to a 9.0 ml dilution blank (1:100,000). Discard the pipette and mix the dilution to a fresh 9.0 ml dilution blank (1:1,000,000). Mix with a fresh pipette and transfer 1.0 ml of the final dilution to a sterile Petri dish. Pour approximately 15-20 ml of nutrient agar which has been steam-pressure sterilized (autoclaved) and cooled to 45°C in a water bath; then rotate each plate to mix the inoculum and medium. After the agar has completely solidified (15-30 minutes) invert the plates and place in incubator space to dry off excess surface moisture. Time of incubation is 2 days at 25 or 28°C. After incubation count the number of colonies.

Pure Culture Techniques Tube Transfers

Transfers from agar plates to tubes, and from tube culture to tube are a common and simple procedure, but requires careful attention to certain details. Transfer from a colony on an agar plate to an agar slant or tube of broth may be made either a straight wire or a loop.

Pure Culture

Pure cultures of different microbial species and subspecies may be isolated from the highly complex mixed populations in nature by a variety of methods; the most common of which will be used is dilution plating. Each discrete colony, which arises in or on a well-prepared dilution plate, is assumed to be composed of cells, all of which are descendant of a single cell. Sub-cultures from such a colony should produce colonies all of the same kind, any one of which is composed all of the same kind of cells as the original colony.

After culture purity has been established in subcultures, a stock culture of the organism is established. The stock culture may be maintained on an agar slant kept in the refrigerator, but unless it is transferred at suitable intervals to establish a fresh stock, it may loss its viability. Covering the slant with a layer of sterile mineral oil to exclude oxygen and prevent evaporation may prolong viability of agar slant cultures.

Preparation and Testing of Culture Media

Copper or zinc containers must not be used. Large quantities of media can be made in stainless steel buckets. Smaller amounts can be made in glass laboratory flasks. Distilled water should be used. Dextrose, when autoclaved with salts such as phosphate may yield inhibitory substances. It is best to add this and other carbohydrates as sterile solutions after the medium have been sterilized. Excessive heating and remelting may destroy growth factors and gelling capacity and cause darkening or pH drift of culture media.

Adjustment of pH

Pipette 10 ml of the medium into a 152 mm x 16 mm test-tube and measure pH with indicator test-paper., and can be corrected by N/20 HCl or N/20 NaOH. All final readings of pH must be made with the medium at room temperature because hot medium will give a false reaction with some indicators.

Sterilization

The usual objective in heat sterilization is the destruction of all microorganisms with the least possible damage to the material being sterilized. Methods employing moist or dry heat or filtration are used in the sterilization of culture media and of materials and equipment used in the bacteriological laboratory.

Steam Sterilization

Steam under pressure in a closed system (autoclave) is frequently used to sterilize heat stable culture media, heat stable chemicals, dilution blanks, rubber articles, and to sterilized heat stable culture media (antibiotic) must be before sterilization. A temperature of 121°C at 15 lbs pressure per sq. feet for 15 minutes generally is required for moist heat sterilization. The most important point in the operation of the autoclave is the removal of all air from the chamber. The temperature rather than the pressure is the lethal agent. Therefore, it is important that the desired temperature is reached before timing the sterilization cycle. Adequate time must be allowed for heat convection current to penetrate the centre of the material. Preheating large quantities of media is often desirable to reduce the over-exposure to sterilization temperature. Viscosity reduces convection, and therefore, a long period of time is required for sterilization of agar than of broth. Two to five minutes must be added for agar sterilization.

Flaming

Although not achieving red heat, flaming is a method commonly used for decontaminating the mouths of bottles, flasks, culture tubes, glass slides, scalpels etc. by passing them though a Bunsen flame without allowing them to become red-hot.

Hot Air

The process of sterilization by hot air ovens is simple in principle and the materials to be processed can be prepacked in craft paper or sealed in metal containers. The method is used for sterilizing such articles as glass syringe etc. Material such as dry powders, fats, oils and petroleum jelly in small shallow containers can also be sterilized by this method. Air is not a good conductor of heat so that hot air sterilization poses problems of penetration. Special apparatus "fan" is needed to ensure that all parts of the load have been maintained at the required temperature. The hot air must circulate between the packages. The Heating-Up Period is the time taken to reach sterilization temperature; this may take about 1 hour.

The holding periods at different sterilization temperatures recommended are:

- 160°C for 45 min.
- 170°C for 18 min.
- 180°C for 7 min.

The cooling down period is carried out gradually to prevent glassware from cracking as a result of a too rapid fall in temperature. This period may take up to 2 hours. The full sterilizing process from loading the oven to unloading it may therefore take about 3-4 hours, depending on the size of load. Long exposure to a lower dry heat temperature has been a method of sterilization recommended like a dry heat cycle of 135°C for 22 hours.

Elaboration of Processing

The development of a process includes three steps: elaboration on laboratory scale, small scale and pilot large-scale production and finalization of product *e.g.* formulation of microbial pesticides.

Laboratory Scale

The goal of the laboratory research is to develop

1. Preservation of the culture,
2. Culture of the strain on solid media and its propagation,
3. Nutrient medium of a suitable composition,
4. Techniques of laboratory submerged or semi-solid scale culture.

Small Scale Production

The first task in small-scale production is the development of a basic culture procedure affording reproducible results. Absolute production yields are of secondary importance, the main emphasis being put on the expected standard course of the microbial process with the attendant changes in growth, pH, consumption of nutrients, production of metabolites, and in the physiological state of the microorganisms. On attaining reproducible results the search for individual optimum conditions can begin.

The first factors to be explored are usually aeration and stirring. Next to be examined is the effect of changes in the ratio of individual components of the nutrient medium, with special reference to the source of carbon and nitrogen and their mutual proportions. The effect of different types and amounts of antifoam agents is also studied since the physical action of these agents can affect the dispersion of air in the culture and respiration of the microorganisms; some antifoam agents (vegetable oils, lard oil) can be utilized by the microorganisms, thus changing the metabolism of the culture. Fluid from laboratory and small-scale fermentors can be used for tentative isolation of metabolites and preliminary tests of their quality.

Problems of Contamination of Microbial Processes

Microbial contamination of fermentation processes and principles and techniques for its elimination should be appraised from several viewpoints:

1. Type of the metabolite produced;
2. Specific properties of the culture;
3. Machinery and equipment used in the process;
4. Nutrient medium and the raw materials used for its preparation;
5. Technology adopted in the process.

Contaminating microorganisms affect negative the microbial process by:

1. Destroying the cells of the production strain,
2. Inactivating the synthesized metabolites,
3. Producing substances affecting the producer's metabolism and thus decreasing the production of the required metabolite,
4. Exhausting compounds crucial for growth and product synthesis from the medium.

However, the presence of a contaminating microorganism need not always results in a drop in metabolite or spores production; in the absence of this drop, it is often difficult to detect contamination.

In semi-solid fermentation process, contamination need not always lead to failure. The term "protected fermentation" is used for such processes in which foreign contamination is suppressed by the presence of an antimicrobial agent. This agent is added to the medium in suitable form.

Sterility of Microbial Process

Checks of the absence of alien microorganisms are required during all crucial stages of the process, culture, product isolation, and its final formulation. Identification of microbial contamination is done by microscopic and culture methods. Microscopic examination represents a useful tool for immediate detection of massive contamination of cultures with an alien microflora. In most cases viable cells are difficult to distinguish from dead ones. The differentiation is somewhat aided by the so-called vital test which, however, is not always fully reliable. The most reliable method for identification of alien microorganisms is by culturing.

Pilot-plant-Fermenters

Although crucial technological factors may be assessed in laboratory fermenters, it is often necessary to verify the laboratory results in devices more closely resembling industrial fermenters. These devices are pilot-plant fermenters; they serve to complete, modify and verify the data obtained in laboratory fermenters and to test the suitability of construction, regulation and monitoring devices. Since, with the same type of agitator with dimensions proportional to fermenter dimensions, laboratory fermenters require higher impeller speed to achieve the same aeration effect as in large fermenters, and since the aeration effect depends largely on the transfer of mechanical force into liquid via the impeller, the verification of agitation and aeration conditions is of utmost importance. Different types of agitators, their relative size, different types and dimensions of baffles, aeration systems, and the effect of their position inside the fermenter are also studied on pilot-plant scale.

Product Development

Many systems for controlling pests are successful in the laboratory. However they fail when tried in the field. This is one of the many problems, which must be solved by product development. Products must be manufactured and formulated in a way which makes them stable for the longest possible time, as convenient as possible to use, and as immune as possible to "use and abuse" *i.e.* the failure of many people to store or use the product as directed and then to blame the product for poor performance. For novel products, testing is necessary to establish that they are safe to use; in addition, quality control tests must be devised which will ensure that every batch will be safe and effective. Once any testing is completed and quality control protocols developed, the appropriate government authority must be approached for permission to sell the product; at present, this must be done individually for every country in the world.

Once a product has been sold, it is often helpful to visit growers who have used it, either to confirm its success or if it has failed, to investigate and determine the cause. Further development work can than be carried out to find ways of avoiding product failure in the future.

Local Marketing and Sales

Once made, the product must be sold and used if all the effort spent in development is to be worthwhile. There are many obstacles to selling alternative pesticides, especially fungi. Users are unfamiliar with how to use these types of products and how they work. Marketing requires extra efforts both to familiarize growers with the new product and to provide a substantial back-up service for the product.

Example: *Verticillium Lecanii*

Verticillium lecanii is a well-documented entomopathogen of aphids, scale insects and whiteflies in tropical and subtropical regions. Also, *V. lecanii* sometimes hyperparasitizes phytopathogenic fungi, mostly rusts and powdery mildews. Laboratory culture is possible on all conventional mycological media, but the best growth is on Sabouraud's agar. On solid media, conidia are produced and in submerged culture are produced blastospores. Colonial growth rate was optimal at 23°C–24°C. Both germination and growth declined steeply above 25°C and ceased above 30°C. Sporulation responded over slightly narrower temperature range than growth or germination, ceasing at 30°C.

Effect of Humidity

Virtually all fungi require humidity for spore germination, growth and sporulation. Thus, to ensure maximum germination of spores and hence highest possible levels of infection of insects, spore sprays should be synchronized with optimal humidity, which for most crops should occur in the evening as ambient temperature falls. The half-life of conidia in distilled water varies, both at 2°C (110–160 days) and -17°C (60-120 days), blastospores on the whole are even shorter-lived and more variable (100-150 days at 2°C).

Spread of Infection

Fungal spores in slime/heads adhere firmly to the mycelium when dry *V. lecanii* conidia did not become airborne from dried cultures or from *V. lecanii* killed aphids. Presumably, infection of a new aphid population on a new crop originates from soil.

Control of Aphids

Control of aphids and scale insect 0.02 per cent Triton X-100 or Tween 20 in the same concentration are used as wetting agent with concentration (5 x 10^7 blastospores/ml). Concentrations (10^7-10^8 blastospores/ml) died 2-3 times faster (LD_{50} 1.9-2.3 days) than those treated with a lower concentration, 10^5 spores/ml (LD_{50} 4.8–6.2 days). *Aphis gossypii* is controlled within 14 days.

Mode of Action Penetration of the Host Integument

Once the spore has attached to the insect, it must germinate to produce a germ tube, which will then, penetrated the host cuticle. However, highly pathogenic strains germinate quicker and penetrate the epicuticle directly whilst strains of low pathogenicity took longer to germinate and grew extensively over the cuticle surface with only limited penetration. Virulence of *V. lecanii* has been associated with high *extracelluar chitinase* activity.

Production and Storage of *V. lecanii*

The choice of infectious material is between conidia and blastospores. Production of conidia on agar is too expensive and it is also difficult to ensure culture purity. Alternatively, conidia can be produced on a cheap granular solid media such as grain, jack seeds (*Arthocarpus*) pearl millet or potato extract for 5-6 days in aerated vessels.

Small-scale Production

Polyethylene cushions made of large thin walled, polyethylene tubing sealed into sections, which are partially filled (1 cm high layer absolutely horizontal) with submerged culture of *V. lecanii* after 2 + 1 days cultivation and inflated with sterile air. Product is harvested (ca. after 14-16 days) by discarding the medium and retaining the mad. Yield from a 0.8 per cent peptone, 1 per cent sorbitol medium was 1 x 10^{12-13} conidia/1 m^2.

Formulated Product

Harvesting of mud is done by mixing it with Siloxyd (ca. 100-125 g/1 liter of mud and wetting agent (Tween 80) and, after predrying at 30°C one day, the wet cake is crushed using a meat-mincer, and the product if dry to perfection at 30°C can be stored or milled and packaged like water-dispersable powders which is prepared for dilution with water into a final spray. The application dose per hectare or meter square is determined depending of the conidia content.

Standardization

Despite theoretical strain stability, a commercial product must be shown to have constant potency. This should be measured by viable spore count and bioassay. The viability of *V. lecannii* conidia and blastospores can easily be assessed by an agar-slide technique (Hall, 1976). Pathogenicity of production batches of spores should be measured by bioassay in comparison with a standard since variation between consecutive assays is often significant. *V. lecanii* is a clearly promising biological control agent against aphids, scales, thrips under glass, humid tropic and as mycofungicide against rust diseases.

Registration

The absence of records of *V. lecanii* in man and other vertebrates is an impressive evidence for its innocuity. All *V. lecanii* strains so far examined by Hall (1981) cannot grow at 37°C and so the likelihood of infecting warm-blooded vertebrates internally is exceedingly remote. As reaction to the dose of 10^6 conidia injected intravenously, no adverse symptoms were observed and, 28 days later, no gross pathological changes were apparent in the internal organs and no signs of the fungus could be found either in sectioned organs or in agar cultures from these. Safety tests are being carried out preparatory to commercial exploitation.

In the field, critical parameters like temperature and humidity cannot be readily manipulated. Temperate weather is often unpredictable and often unfavourable to fungi, except perhaps where the pest occupies a moist microclimate. However, great potential may exist in tropical and subtropical zones where high humidity is normal for prolonged periods. *Verticillium lecanii* strains were first introduced commercially in the U.K. for the control of aphids "Vertalex" and whitefly "Mycotal" on protected ornamental and vegetable crops.

References

Hall, R.A. (1976) A bioassay of the pathogenicity of *Verticillium lecanii* conidiospores on the aphid *Macrosiphoniella sanborni. J. Inverteb. Pathol.* 27: 41-48.

Hall, R.A. (1981) The fungus *Verticillium lecanii* as a Microbial Insecticide against Aphids and Scales. In: *Microbial Control of Pests and Plant Diseases* 1970-1980. Acad. Press 1981, Ed. Burges H.D., p.482-498.

Sustainable Disease Management of Agricultural Crops (2011) *Pages* **203–209**
Editors: **S.K. Biswas and S.R. Singh**
Published by: **DAYA PUBLISHING HOUSE, NEW DELHI**

Chapter 14

Biopesticides: An Ecofriendly, Sustainable and Cost Effective Approach for Integrated Disease and Pest Management

☆ ***R.K. Pandey[1], S.R. Singh[2], P.K. Gupta[3] and Shashi Mala[4]***
[1]Division of ACB and PDM, Amity University, Noida – 201 303, U.P.
[2]Krishi Vigyan Kendra, Belatal, Mahoba – 210 423, U.P.
[3]Division of Plant Pathology, Indian Agricultural Research Institute, New Delhi – 110 012
[4]Aligarh Muslim University, Aligarh

Biopesticides are a class of environmentally friendly pesticides that are used to control pests (fungi, insects, weeds and diseases) using living organisms. In fact, fungi have long been known to infect and kill another fungal, insects and mites. The Chinese recorded infections of *Cordyceps* in the 7th Century. During the 19th century in Italy, Agostino Bassi recorded the infection of silk worms by the pathogenic fungus *Beauveria bassiana.* Elie Metchnicoff won the Nobel Prize for his work in developing a biocontrol agent, the fungus *Metarhizium anisopliae,* as a method for the control of the cereal cockchafer. However, as stated earlier, fungi can be used to control insects, other fungi, and weeds.

It is hearting to observe the growing awareness among the farmers and policy makers about ecologically sustainable methods of pest management. More and more farmers are coming to realize the short-term benefits and long-term positive effects of the use of bioagents and other ecologically safe methods to tackle pests. The present article 'Biopesticides' is much relevance in the context. 'Biopesticides' are certain types of pesticides derived from such natural materials as animals, plants, bacteria, and certain virus. These include for example; fungi such as *Beauveria sp.*, bacteria such as *Bacillus* sp., neem extract and pheromones. Similarly canola oil and baking soda have pesticides applications and are considered as biopesticides. The use of these materials is widespread with applications to foliage, turf, soil, or other environments of the target insect pests. In a much simpler way we can say that these are pest management tools that are based on beneficial microorganisms (bacteria, viruses, fungi and protozoa), beneficial nematodes or other safe, biologically based active ingredients. Benefits of biopesticides include effective control of insects, plant diseases and weeds as well as human and environmental safety. Biopesticides also play an important role in providing pest management tools in areas where pesticides resistance, niche markets and environmental concerns limit the use of chemical pesticide products.

Biocontrol Agents and their Available Bioformulations

Fungi to Control Other Pathogenic Fungi

Plants are infected by number of fungal pathogens which causes different types of symptoms on plant. As fotroot rots on root, mildew on leaf and stem, blight on twing leaf, in these pathogenic fungi can be controlled by several fungal bioagents like *Aspergillus*, *Trichoderma* spp. have been used to control *Fusarium oxysporum*, *Rhizoctonia solani* and *Pythium* on a variety of plants. (Some of the registered microbial products have been listed in Table 14.1.

Table 14.1: Name of Registered Microbial Products

Acremonium spp.	*Metarhizium anisopliae*
Aspergillus niger	*Metarhizium flavoviride*
Aspergillus terreus	*Nuclear Polyhedrosis Viruses*
Fusarium nivale	*Bacillus Thuringienis*
Trichoderma harzianum	*Bacillus sphaericus*
Trichoderma viride	*Pseudomonas fluorescens*
Trichoderma virens	*Chladosporium oxysporum*
Paecilomyces lilacinus	*Verticillium lecanii*
Beauveria bassiana	VAM (*Glomus fasciculatum*)

Fungi to Control Harmful Insects

Many fungi are capable of infecting and killing insects. Mass inoculation of insect using fungal pests was initiated by Krassilstchik in 1888. However, the advance in fungal insect biopesticides did not progress as quickly as initially thought. This was mainly due to the development of chemical pesticides. However, due to the environmental concerns of applying chemical pesticides, fungal insecticides are being revisited and explored. Some of the most widespread fungi used to control insects are *Beauveria bassiana*, *Metarhizium* spp., *Dermolepida albohirtum*, and *Lagenidium giganteum*. The application of fungal insecticides is usually done using fungal spores. However, fungal mycelia are also used as inoculant to control insect pathogens. Few examples of fungal pesticides, the target to some of the companies which supply the biopesticide (Table 14.2).

Table 14.2: Name of Registered Microbial Products

Acremonium spp.	*Trichogramma japonicum*
Beauveria bassiana	*Trichogramma brasilliensis*
Metarhizium spp.	*Trichogramma achaeae*
Lagenidium giganteum	*Trichogramma pretiosum*
Verticillim lecanii	*Trichogramma exiguum*
Nomuraea rileyi	*Bracon species*
Trichogramma chilonis	

Nematodes to Control Pathogenic Fungi

These are also called as Fungivores or myceliophagous nematodes. This group of nematodes has a spear similar to plant parasitic nematodes, but it is used to feed on fungal hyphae and spores. These nematodes by feeding upon harmful fungi accelerate the decomposition process and help in recycling of minerals and other nutrients from fungi.

The following nematodes are used to control the fungi:

Aphelenchus avenae, Aphelenchoides compositicola, Ditylenchus myceliophagus, Ditylenchus triformis.

Nematodes to Control Harmful Insects

These are also called as entomophilic or entomopathogenic nematodes. Entomopathogenic nematodes belonging to the families Steinernematidae and Heterorhabditidae, are soil dwelling lethal parasites of insects that are used for inundative, augmentative or inoculative biological control. They are environment friendly.

The members of family Steinernematidae and Heterorhabditidae are unique in their ability to carry and release specific bacteria of the genus *Xenorhabdus* and *Photorhabdus* respectively inside the insect haemocoel, where they cause septicaemia and kill the insect host and in turn, provide nourishment for the developing nematodes in the insects cadaver. Till now about 56 species of the genus *Steinernema* and 13 of *Heterorhabditis* have been reported throughout the world. Some of these nematodes are being mass multiplied on artificial media and have been commercialized for the biological control of insects. These nematodes have been found to be the highly effective against the insect pests of various crops.

The following species of these nematodes are being used as biocontrol agents for controlling the insects pests of various crops as well as in world.

- ☆ *Steinernema carpocapsae* (Weiser, 1955) Wouts *et al.*, 1982
- ☆ *Steinernema glaseri* (Steiner, 1929) Wouts *et al.*, 1982
- ☆ *Steinernema feltiae* (Filipjev, 1934) Wouts *et al.*, 1982
- ☆ **Steinernema thermophilum* Ganguly and Singh, 2000
- ☆ **Heterorhabditis indica* Poiner *et al.*, 1992
- ☆ *Heterorhabditis bacteriophora* Poiner, 1976

*(*Described from India)*

Besides, the members of the genus *Steinernema* and *Heterorhabditis* are a few other nematodes known to control the insects pests. These are:

1. *Mermis nigrescens*–a parasite of grasshopper
2. *Romanomermis culcivorax*–a parasite of mosquitoes

Relevance of Biopesticides

25 million cases of acute occupational pesticide poisoning in developing countries are being reported each year (WHO, 1990). 14 per cent of all known occupational injuries and 10 per cent of all fatal injuries are caused by pesticides (ILO, 1996). Obsolete pesticides are being stored in developing countries–20,000 tonnes in Africa alone. Pesticides residues in agricultural commodities are being the issue of major concern besides their harmful effect upon human life, wild life and other flora and

fauna. Equally worrying thing is about development of resistance in pest to pesticides. The only solution of all these is use of 'Biopesticide' that can reduce pesticide risks, as-

1. Biopesticide would sustain the soil health and improve the diversity of beneficial microorganisms without leaving any toxic residues, therefore, the environmental quality and plant vigour improves and soil degradation is arrested.
2. Biopesticides are best alternatives to conventional pesticides and usually inherently less toxic than conventional pesticides.
3. Biopesticides generally affect only the target pest and closely related organisms, in contract to broad spectrum, conventional pesticides that may affect organisms as rent as birds, insects and mammals.
4. Biopesticides often are effective in very small quantities and often decompose quickly, thereby resulting in lower exposures and largely avoiding the pollution problems caused by conventional pesticides.
5. When used as a fundamental component of Integrated Pest Management (IPM) programs, biopesticides can greatly decrease the use of conventional pesticides, while crop yields remain high.
6. Amenable to small-scale, local production in developing countries and products available in small, niche markets that are typically unaddressed by large agrochemical companies.

Concept of Biopesticides Production Technology

The Plant Protectionists have been advocating for an eco-friendly, non-chemical and economic approach for plant disease control since the time most of the "chemical killer pesticides" *viz.*, DBCP, MBr, BHC, DDT etc., where banned. Prior to this, high chemical fertilizer inputs and application of chemical pesticides in agriculture cause some problems which are as follows:

Economic Problems

Affected by availability and affordability of chemical fertilizers through imports and subsidies causing unstable burden on the fiscal system. Moreover, the resource poor farmers of rural India could hardly afford to purchase costly chemical fertilizers.

Crop Yield

Over the years, crop production under high chemical input has resulted in soil degradation and loss of fertility leading to a decline in seed germination and crop yield.

Health Hazards

Extensive input of chemical fertilizers in soil has resulted in accumulation of pollutants such as heavy metals, pesticides etc. in the soil posing as potential hazards of chemical contamination of our food and ground water. But the aims of modern agriculture are:

- ☆ Enhancement of agricultural sustainability
- ☆ Environmental safety
- ☆ Reduction of disease incidence

To attain the above mentioned objectives of getting better and safer product of biopesticide cum bio-fertilizer, as substitutes of chemical fertilizers and chemical pesticides respectively, organic agriculture was initiated which apply methods for improving soil quality in terms of both physico-chemical and biological properties. These systems, minimize the use of purchased inputs by substituting them through use of farm-generated inputs, or by exploiting biological systems such as management by biopesticide cum bio-fertilizer of A-Mycorrhizal fungi (*Glomus* spp.) and oilseed cakes (Karanj-*Pongamia pinnata*) with antagonistic fungal bioagents (*Trichoderma* spp., *Paecilomyces lilacinus* etc.) to enhance nutrient availability, plant health and vigour, reduction of agriculture inputs and also for reduction of diseases incidence.

Nature of Biopesticides

Biopesticides fall into three major classes:

1. Microbial pesticides consist of a naturally occurring or generally controlled microorganism (*e.g.*, a bacterium, fungus, virus or protozoan) as the active ingredient. These pesticides can control many different kinds of pests, although each separate active ingredient is relatively specific for its target pest(s). For example, there are fungi that control certain weeds, and other fungi that kill specific insects.

   They suppress pest by:

   (*a*) Producing a toxin specific to the pest.

   (*b*) Parasitism

   (*c*) Preventing establishment of other microorganisms through competition or

   (*d*) Other modes of action.

   An example of a most widely used microbial pesticide is subspecies and strains of *Bacillus thuringiensis*, or "Bt". It is a naturally occurring soil bacterium that is toxic to the larvae of several species of insects but not toxic to untargeted organisms. Bt can be applied to plant foliage or incorporated into the genetic material of crops and as discovered, it is toxic to the caterpillars (larvae) of moths and butterflies. These also can be used in controlling mosquitoes and black flies. Several strains of Bt have been developed and now strains are available that control fly larvae. While some Bt's control moth larvae found on plants, other Bt's are specific for larvae of flies and mosquitoes. The target insect species are determined by whether the particular Bt produces a protein that can bind to a larval gut receptor, thereby causing the insect larvae to starve.

2. Plant-Incorporated-Protectant (PIPs) is pesticide substances that can produce plants from genetic material. For example, scientists can take the gene for the Bt pesticide protein, and introduce the gene into the plant's own genetic material. Then the plant, instead of the Bt bacterium infection, manufactures the substance that destroys the pest.

3. Biochemical pesticides are naturally occurring substances that control pests by nontoxic mechanisms. Conventional pesticides, by contrast, are generally synthetic materials that directly kill or inactivate the pest. Biochemical pesticides include substances, such as insect sex pheromones, that interfere with mating as well as various scented plant extracts that attract insect pests to traps. Man-made pheromones are used to disrupt insect mating by creating confusion during the search for mates, or can be used to attract male insects to traps. Pheromones are often used to detect or monitor insect populations, or in some cases, to control them. Farmers in their traditional wisdom have identified and used a variety of plant products and extracts for pest control, especially in storage. As many as 2121 plant species are reported to possess pest management properties, 1005 species of plants exhibiting

insecticide properties, 384 with antifeedant properties, 297 with repellant properties, 27 with attractant properties and 31 with growth inhabiting properties have been identified. The most commonly used plants are neem (*Azadirachta indica*), pongamia (*Pongamia glabra*) and mahua (*madhuca indica*). 2-5 per cent neem or mahua seed kernel extract has been found effective against rice cutworm, tobacco caterpillar, rice green leafhopper, and several species of aphids and mites. The efficacy of vegetable oils in preventing infestation of storage product pests such as bruchids, rice and maize weevils has been well documented. Root extracts of Tagetes or Asparagus as nematicide and Chenopodium and Bougaivillea as antivirus have also been reported.

Criteria of Developing Potential Biopesticide

The efficacy of many of the biopesticide can equal that of conventional chemical pesticides. However, the mode of action will be different. With many of the biopesticides, the time from exposure to morbidity and death of the target insect may be 2 to 10 days. With the mission to develop a successful and appropriate bioformulations to ease its application under farmers' field conditions in the present investigation attempt is made. The powder-based formulation was desired to possess the essential characters like long retention of viability ease of store and apply with an excellent dispersible nature. Biological control of diseases and insect pests has been achieved by seed treatment with the *Trichoderma* and *Paecilomyces* in recent years successfully by several workers. However, the success of a biocontrol agent depend on its ability to produce inoculum in excess and to survive, grow and proliferate well on seed coat, spermosphere and rhizosphere of the growing plants. Such types of antagonistic establishment need a substrate during seed treatment which can provide a better coating on seed surface and food base to antagonist for better proliferation. The use of microbial biomass as alginate prius and granules (Lumsden *et al.*, 1995, 1996); starch granules (Lewis *et al.*, 1995), wheat bran alginate pellet (Smith, 1996) pyrax (Lewis and Fravel, 1996), Talc powder (Tewari and Mukhopadhyay) extruded granules (Lewis and Larkin, 1997) and cellulose granules (Lewis *et al.*, 1998) has been successfully used against soil-borne plant pathogens as seed/soil application in various crops.

Important Characteristics of Biopesticides

1. They have a narrow target range and a very specific mode of action.
2. They are slow acting.
3. They have suppress, rather than eliminate, a pest population.
4. They have limited filed persistence and a short shelf life.
5. They are safer to humans and the environment than conventional pesticide.
6. Do not have residue problems.

Understanding the fundamental differences in the mode of action biopesticide vs. traditional pesticides is important. Since the use patterns of a biopesticide may be different from traditional pesticides to control a particular pest species. It is important that care to be taken when using any pesticide, even organic or natural or biopesticide. Even if this product is considered to be organic in origin, it is still a pesticide. Just because a product is though to be organic, or natural, does not mean that it is not toxic. Some organic pesticides are as toxic, or even more toxic, than many synthetic chemical pesticides. Organic pesticides have specific modes of action; just as do synthetic pesticides have specific modes of action, just as do synthetic pesticides. While some organic pesticides may be nontoxic or are only slightly toxic to people, they may be very toxic to other animals. For instance, the organic pesticide ryania is very toxic to fish. Also, some organic pesticides may be toxic to beneficial insects, such as honeybees, if they are combined with other materials, such as combining pyrethrins with rotenone. The use of an Integrated Pest Management (IPM) is important to insure success.

Conclusion

It is fact any transition from the existing norms takes some time and if the proper testing etc. has been done for the product only then the community accepts with cost effective and eco-friendly view. Dissemination of the information should be properly done and the community should also develop faith in the cost effective, sustainable and eco-friendly multifarious potential products once a faith is developed, it is obvious that the community will continue to accelerate the growth of the biopesticides cum bio-fertilizers because protection of crop yield is equally important of production.

References

Cook, R.J. and Baker, K.F. (1983). *The nature and practice of biological control of plant pathogens.* St. Paul. M.W. APS. 539pp.

Fravel D.R. (2005). Commercialization and implementation of Biocontrol. *Annu. Rev. Phytopathology,* 43:337-359.

Jones and Lisansky, S. 1947, *Microbial biopesticides.* In *"Microbial insecticides*: Novelty or Necessity?" British Crop Protection Council Proceedings/Monograph series No. 68, 3-10.

Jumanah, F. 1994. *Pesticide policies in developing countries*: Do they encourage excessive use? World Bank Discussion Papers 238, the World Bank, Washington DC.

Lewis, J.A.; Larkin, R.P. and Rogers, D.L. (1998). A formulation of *Trichoderma* and *Gliocladium* reduce damping off caused by *Rhizoctonia solani* and saprophytic growth of the pathogen in the soil less mix. *Plant Dis.* 82: 501-506.

Lewis, A. a. and Larkin, R.P. (1997). Extruded granular formulation with biomass of biocontrol *Gliocladium virens* and *Trichoderma* spp. to reduce damping off of eggplant caused by *Rhizoctonia solani* and saprophytic growth of the pathogen in soil-less mix. *Biocontrol. Sci. Tech.,* 7: 49-60.

Lewis, J.A. and Papavizas, G.C. (1987). Application of *Trichoderma* and *Gliocladium* in alginate pellets for control of Rhizoctonia damping off. *Plant Pathol.*36: 397-404.

Tewari, A.K. and Mukhopadhyay, A.N. (2001). Testing of different formulation of *Gliocladium virens* against chickpea wilt complex. *Indian Phytopath.* 54: 67-71.

Sen, Bineeta (2000). Biological control: A success story. *Indian Phytopath.* 53: 243-249.

Sustainable Disease Management of Agricultural Crops (2011) *Pages* **210–221**
Editors: **S.K. Biswas and S.R. Singh**
Published by: **DAYA PUBLISHING HOUSE, NEW DELHI**

Chapter 15

Alternaria Leaf Spots of Broccoli and their Ecofriendly Management

☆ ***Gireesh Chand*[1], *Udit Narain*[2], *Dharmendra Kumar*[3] and *K.K. Chandra*[3]**

[1]*Department of Vegetable Science, N.D. University of Agriculture and Technology, Kumarganj, Faizabad – 224 229*

[2]*Department of Plant Pathology, C.S. Azad University of Agriculture and Technology, Kanpur – 208 002*

[3]*Krishi Vigyan Kendra, Chandauli, N.D. University of Agriculture and Technology, Kumarganj, Faizabad – 224 229*

Vegetable production must be increased to meet the demand of our country. Taking into consideration, the rapid growth of population and alarming situation of vegetables and nutritive deficiency in the country, there is a need of coordinated effort for improving broccoli as a Brassicaceous vegetable included under "cole crop", a new source of vegetable.

A critical review of the reasons for low yield reveal that diseases are one of the limiting factors in the successful cultivation of broccoli. Some of the important diseases affecting the broccoli are blight, powdery and downy mildews, alternaria leaf spots and blight, seedling blight, wilt, root rot, bacterial wilt, root knot, mosaic and several post harvest diseases. Majority of the diseases, which occur on the Brassicaceous hosts also infect the broccoli crop. Some of the known diseases and their pathogens are given in Table 15.1.

Many pathogens including fungi, bacteria, nematodes, mycoplasma and viruses have been reported on Brassicaceous vegetables from different parts of the world and of these, only a few diseases have been investigated in greater detail. During the survey of diseases of Brassicaceous vegetables in Uttar Pradesh, the broccoli crop included in this group was found to be moderately to severely affected by Alternaria leaf spots and blight mainly caused by *Alternaria brassicae* (Berk.) Sacc. and also *Alternaria brassicicola* (Schwein.) Wiltshire, *Alternaria raphani* Groves and Skolko and *Alternaria alternata* (Fr.) Keissler.

Table 15.1: Diseases of Brassicaceous Vegetables/Broccoli and their Causal Organisms

Sl.No.	Name of Disease	Causal Organism
(A)	**Fungal disease**	
1.	Alternaria leaf spot/blight	*Alternaria spp.*
2.	Black leg and Phoma root rot	*Phoma lingam*
3.	Club root	*Plasmodiophora brassicae*
4.	Cercospora leaf spot	*Cercospora brassicicola*
5.	Damping off	*Fusarium* spp., *Pythium* spp.
6.	Anthracnose	*Colletotrichum higginsianum*
7.	Black mould rot	*Rhizopus stolonifer*
8.	Black root	*Aphanomyces raphani*
9.	Downy mildew	*Peronospora parasitica*
10.	Gray mold	*Botrytis cinerea*
11.	Phytophthora root rot	*Phytophthora megasperma*
12.	Powdery mildew	*Erysiphe polygoni*
13.	Ring spot	*Mycosphaerella brassicicola*
14.	Sclerotinia stem rot	*Sclerotinia sclerotiorum*
15.	Verticillium wilt	*Verticillium albo-atrum*
16.	White blister/rust	*Albugo candida*
(B)	**Bacterial diseases**	
17.	Bacterial leaf spot	*Pseudomonas syringae*
18.	Bacterial soft rot	*Erwinia carotovora*
19.	Black rot	*Xanthomonas campestris*
20.	Crown gall	*Agrobacterium tumefaciens*
(C)	**Viral diseases**	
21.	Cauliflower mosaic	*Cauliflower mosaic virus*
22.	Radish mosaic	*Radish mosaic virus*
(D)	**Nematode disease**	
23.	Root knot	*Meloidogyne* spp.
(E)	**Nutrient deficiency**	
24.	Whiptail	Molybdenum (Mn)
25.	Bronzing/curling/drying	Potassium (K)

Except the preliminary report of the occurrence of different species of *Alternaria* on broccoli, no information is available on distinct symptomatology, etiology of the causative pathogens and their distinguishing characters for species differentiation and host-range of the pathogen etc. The literature on *Alternaira* species with regard to their morphological study and ecofriendly management of the disease through varietal, cultural and biological methods are reviewed herein.

Geographical Distribution and Severity of Alternaria Disease

The fungus, *Alternaria brassicae* was first described in 1836 under the name of *Macrosporium brassicae* Berk. from England by Berkely. Two decades later, Kuhn (1855, 1856) investigated the attack of this fungus on rape in Germany and its name was later changed to *Alternaria brassicae* (Berk.) Sacc. by Saccardo in 1886. In India, the first report of the fungus was made by Mason (1928) from Pusa (Bihar) on *Sarson* (herbarium material). Since then, it is being constantly reported from different parts of the country. The fungus has also been reported on leaves of turnip, cabbage and radish by Jorstad (1945). The appearance of disease was noticed much earlier in U.P. (Dey, 1948).

During the course of investigations on the taxonomic studies of phytopathogenic fungi, particularly of the genus *Alternaria* associated with broccoli, four species *viz., A. brassicae* (Mason, 1928), *A. brassicicola* (Schimmer, 1953), *A. raphani* (Gorves and Skolko, 1944) and *A. alternata* (Suhag *et al.*, 1985) were found to be parasitic on it, causing the characteristic leaf spot and blight, lesions on stems, branches and petioles, inflorescence and siliquae. So, a comparative study of disease symptomatology and etiology was made with ultimate aim to develop a most suitable and feasible key for ready identification of *Alternaria* spp. associated with broccoli, and their eco-friendly management.

Symptomatology

The disease appears initially from the seedling stage on this vegetable crop. Symptoms caused by *A. brassicae* start from margin of leaves or at the tip and spread inward and downward accordingly. The dark brown, irregular spots are accompanied by characteristic yellow halo appearance and have distinct or indistinct zonations, which gradually increase in size in concentric manner. Two or more such spots may coalesce to form larger area. In well advanced stage, some of the spots are covered with dark olive green mass of the fungal spores with shot-hole appearance.

The leaf spot due to *Alternaria brassicicola* are circular to elliptical and olivaceous in colour which vary from 4-12 mm in size with indistinct or distinct zonations. In severe infection, several lesions coalesce to cover large area and cause death of leaves. The young and actively developing leaves are almost free from the disease or may have the small necrotic areas. The severely infected leaves show burning effect, which ultimately dry up and may fall prematurely.

In case of symptoms of Alternaria leaf spot due to *Alternaria raphani,* the spots are brown to dark brown in patches starting from margin of leaves, developing inwards, in irregular fashion. The spots are circular to irregular in appearance, dark brown with indistinct zonations on both surfaces of leaves. The affected areas soon turn blighted and wither. The severely infected leaves show burning effect, which ultimately dry up and may fall prematurely.

Alternaria alternata initially appears to produce minute, dark brown patches which start from the margin or centre of leaf and at later stage of disease development, a number of spots coalesce to form the irregular dark brown spots.

Breoccoli heads show browning, beginning at the margin of the individual flower and flower cluster. On plants grown for seed, dark brown lesions occur on the main axis, inflorescence, branches and on the siliquae. These lesions coalesce rapidly and cause premature ripening and splitting of siliquae and may result in high level of seed infection. Infected seeds are shrunken, discoloured or covered with fungal growth and have low quality. Cankers may form just below nearly mature broccoli heads and result in premature death of plants.

In general, the symptomatology reveals that the disease appears initially as small brown spots on the leaf lamina or on margin of the leaves which later increase rapidly in size or inward and become mostly circular or irregular in shape and turn light to dark brown in colour with slight distinct zonations and with or without halo (Chupp, 1925 and Chupp and Sherf, 1960). It has been reported that in severe cases, the spots coalesce and give the leaf as blighted appearance with dark brown mass of spores. On the some hosts (cabbage and cauliflower) (Narain, 1986) symptoms develop the concentric rings on the leaf (Kadian and Saharan, 1983). The leaf spots later enlarge, become nearly circular, 4-12 mm in diameter, olivaceous in colour with zonations and profuse sooty sporulation in case the infection is specially caused by *A.brassicicola* (Ellis, 1971). The symptoms caused by different species of *Alternaria* on different Brassicaceous hosts have close resemblance with that described by Chupp (1925), Chupp and Sherf (1960), Valkonen and Koponen (1990), Changsri and Weber (1963), Narain and Saksena (1975), Narain *et al.* (1982), Narain (1986), Gupta and Basuchaudhary (1992), Chand and Narain, (2005) and Ahmad and Narain (2007).

Alternaria brassicicola has been reported to be pathogenic on white cabbage, cabbage, knol-khol and radish (Rao, 1977 and Dillar *et al.*, 1998). Sangwan *et al.* (2002) reported that *A. raphani* isolated from radish was pathogenic to a number of Brassicaceous hosts and parasitized the vegetable crucifers as well as oil crops (Suhag *et al.*, 1985).

Elliot (1917), Changsri and Weber (1963), Narain *et al.* (1982) and Narain (1986) have also reported more or less the same type of symptoms caused by *Alternaria raphani*.

The comparative symptom differentiation of *Alternaria* spp. occurring on broccoli has been presented in Table 15.2.

Table 15.2: Comparative Symptom Differentiation of *Alternaria* spp. Occurring on Broccoli

Symptoms on	*Alternaria brassicae*	*Alternaira brassicicola*	*Alternaria raphani*	*Alternaria alternata*
Leaves	Spots dark brown to black with halo chlorotic tissues, 0.5 to 2.0 cm in diam. and in concentric circle.	Numerous, dark brown to black spots, 4-12mm. in diam., zonate, lesions with sooty black appearance.	Spots dark brown or black, 3-3 mm. in size, raised with distinct or indistinct zonations.	Spots minute to large, dark brown in patches often colescing to form irregular dark brown lesions.
Stems	Dot like linear or elongated lesions.	Spots numerous, dark brown to black.	Lesions dark brown to black, in linear manner.	Spots/lesions dark brown in patches.
Siliquae	Linear, elongated or dot like lesions, without concentric rings.	Appearance of minute dark brown to black spots.	Irregular brown to black lesions.	Lesions coalescing and splitting type.
Heads	Bark brown spots in beginning which increase in size with sporulation	Spots brown to black in colour with sooty appearance.	Spots dark brown to black in colour.	Cankers may form just below on nearly mature heads.

Morphological Characters of *Alternaria* spp.

Alternaria brassicae (Berk.) Sacc. (Michelia, 2: 129, 1880)

Colonies fast growing, amphigenous, usually in beginning pale olive and later turn into dark

150 ´ 5-10 mm in size, bearing one to several small but distinct conidial scars; *Conidia* solitary or occasionally in chains upto 4, straight or slightly curved, obclavate, rostrate with 5-10 transverse and 0-8 longitudinal or oblique septa, olive or grayish olive in colour and 40-150´15-30 mm in size; *Beak* usually pale brown, short, cylindrical, 10-130 mm in length and 3-8 mm in width.

Alternaria brassicicola (Schw.) Wiltshire (*Mycol. pap.*, 20: 8, 1947)

Colonies growing as dark olivaceous to dark blackish brown, velvety; *Mycelium* septate, branched, hyaline at first, later brown or olivaceous brown, inter and intracellular, smooth, 25-75´1.5-7.5 mm in size; *Conidiophores* produced either singly or in groups of 2-12, emerging through stomata, usually simple, erect or curved, occasionally geniculate, more or less cylindrical but often slightly swollen at the base, septate, pale to mild olivaceous brown, smooth, 20-75 mm long and 4-8 mm thick; *Conidia* mostly in chains of upto 20 or more, usually tapering slightly towards apex, obclavate, basal cell round, the apical cell being more or less rectangular or resembling a truncate cone with 2-10 cross and 0-6 longitudinal septa, often constricted at the septa, pale to dark olivacecous brown, smooth or becoming slightly verriculose with age and measuring 20-120 mm length and 7-25 mm width; beak usually non-existent.

Alternaria raphani Groves and Skolko (*Can. J.Res., Sect.C.22:227,* 1944).

Colonies grow faster, centre sometimes raised and light brown to dark brown in colour; *mycelium* septate, branched, hyaline at first later become olive buff in colour, 2-7 mm thick, some times swollen to form chlamydospores in chains or in groups; *Conidiophores* simple or occasionally branched, septate, olivaceous brown, 2.5-8.0´2.5-150 mm in size with short break, some times swollen slightly at tip and usually with a single conidial scar; *Conidia* commonly in chains consisting of 3-4, obclavate or ellipsoidal in shape, dark golden brown to olivaceous brown, smooth or some times minutely verruculose with 3-7 transverse and often a number of longitudinal or oblique septa, constricted at septa and measuring 20-60 mm long and 10-30 mm thick; *Beaks* of conidium smaller than that of *A.brassicae*, but longer to that of *A.alternata*.

Alternaria alternata (Fr.) Keissler (*Beih. Bot. Zbl.*, 29: 434).

Colonies fast growing, usually black or olivaceous black with abundant sporulation; *Mycelium* septate, branched, olive buff in colour, 2.5-9.5mm in width, sometimes swollen ranging from 12.5 to 23.5 mm to form chlamydospores; *Conidiophores* usually arise single or sometimes in groups, mostly simple, septate, straight, brown or variously curved and geniculate, pale to mid olivaceous brown, smooth and measure 24.5-68.5 mm in length and 3.50-7.25 mm in width; *Conidia* formed in long and often branched chains, varying in shape and sizes, dark brown in colour, smooth to verruculose and 8.50–53.50 ´ 5.30–21.50 mm in size, *Beaks* of conidia usually short, conical or cylindrical but never equal in the length of conidium, often one third or half of the spore body, usually lighter in colour, 10–45.5 mm in length and 1.5–6.5 mm in width and 0–3 septate.

All the four species of the genus *Alternaria* observed parasitic on broccoli differ from each other on the basis of their mode of formation of spores (Neergaard, 1945) and conidial morphology (Ellis, 1971; Khalid *et al.*, 2004; Chand and Narain, 2005 and Ahmad and Narain (2007). In *A. alternata* and *A. brassicicola* long chains of conidia are formed but latter is without beak. In *A. brassicae*, conidia are solitary or seldom in chains with comparatively larger conidia and beaks but in *A. raphani* conidia are produced in short chains (3-4) and there is frequent formation of chlamydospores in culture (Narain and Saksena, 1975; Verma and Saharan, 1993). Morphological observations of all the four species of *Alternaria* were made in culture (PDA) and in nature (host) and their comparative differentiation has been presented in Table 15.3 and based on their distinguishing characters; a very simple and feasible key has been framed for ready identification of *Alternaria* spp. associated with broccoli.

Table 15.3: Comparative Differentiation in *Alternaria* spp. Occurring on Broccoli

Morphological Characters	*Alternaria brassicae*	*Alternaria brassicicola*	*Alternaria raphani*	*Alternaria alternata*
Conidiophores	Mild-pale grayish to olive, simply erect, swelling at base, geniuculate, septate	Pale mild olivaceous brown, simple, septate, erect or curved	Olivaceous brown, simple or occasionally branched, septate	Olive brown erect, simple and septate
Length	15-150 mm	25-75mm	25-150 mm	24.5-68.5 mm
Width	5-10 mm	1.5-7.5 mm	3.5–7.5 mm	3.50-7.25 mm
Conidia				
Conidial chain	Solitary or occasionally upto 4	Upto 20 or more	Upto 3-4	Upto 20 or more
Shape	Obclavate rostrate	Obclavate	Obclavate or ellipsoidal	Polymorphic
Colour	Olive or grayish olive	Pale to dark brown olivacious	Golden brown to olivaceous	Dark brown
Cross septa	5-16	2-10	3-7	3-10
Longitudinal septa	0-8	0-6	0-5	0-8
Length	40-150 mm	20-120 mm	20-60 mm	8.5-35.5 mm
Width	15-30 mm	7-25 mm	10-30 mm	5.3–21.5 mm
Beak				
Length	10-130 mm	Non-existent	5-75 mm	10-45.5 mm
Width	3-8 mm	-	2.5-6.0	1.5-6.5 mm
Cross septa	0-5	-	0-3	0-3

Key to *Alternaria* spp. Parasitic on Broccoli

I. Conidia solitary or occasionally in chains up to four, obclavate, rostrate tapering gradually into thick cylindrical beak ....*A. brassicae*

II. Conidia in long chains consisting of twenty or even more in a chain.

(a) Conidia usually cylindrical, basal cell rounded and apical cell more or less rectangular and beak usually almost non existent ....*A. brassicicola*

(b) Conidia usually polymorphic, often with short conical or cylindrical short beaks ....*A. alternata*

III. Conidia 3-4 in chain, straight or curved, obclavate or ellipsoidal, generally with short beak, chlamydospores formed abundantly in culture ....*A. raphani*

Among these four species, *A. brassicae* is more destructive one and occurs more frequently on Crucifers including broccoli (Koike, 1997). One more species, *A. cheiranthi*, has also been reported from India, on *Cheiranthus cheiri* (wallflower), a cruciferous plant (Narain and Singh, 1981) but it does not occur on broccoli. *A. brassicae* and *A. brassicicola* have been reported from every continent on Brassicaceous hosts (Anonymous, 1969), and latter is more serious on seed crops (Ellis, 1971).

Ecofriendly Management

Cultural Measures

Among the common cultural practices, the use of clean, large, healthy seed, seed treatment, long crop rotation, weed control, planting at appropriate recommended times, use of balanced nutrients, proper plant density, proper drainage in the fields, use of tolerant/resistant cultivars and application of chemicals at proper time with adequate foliage coverage have been advocated to manage Alternaria disease of Brassicaceous vegetables (Channon and Maude, 1971; Chupp, 1925; Chupp and Sherf, 1960; Chand *et al.*, 2005; Dixon, 1981; Kolte, 1985; Saharan and Chand, 1988; Singh, 1987; Thomas, 1984.

Seed infection can largely be controlled by water treatment at 50°C for 18 minutes (Schimer, 1953). Aging of Brassica seeds for 5-6 months at temperature above 35°C eliminates seed infection and seed borne inoculum (Chahal, 1981 and Kolte, 1985).

The rate of disease spread in plant population and the disease severity were minimum when broccoli in *Rabi* season of the year 2002-2004. The incidence of the disease was minimum when brocccoli was transplanted as early on 5th October followed on 15th Oct, 25th Oct., 5th Nov. and 15th Nov. (Reyeke *et al.*, 1998, 2000 and Chand *et al.*, 2005.

In order to find out the suitable inter cropping of broccoli with saunf to minimize losses from the disease when broccoli in single row was transplanted with Saunf (double rows), followed by broccoli (single row) + Saunf (single row), Broccoli (double rows) + Saunf (double rows), Broccoli (double rows) + saunf (single row) and Broccoli (sole crop). Maximum disease incidence was recorded in broccoli as sole crop. The intercropping of broccoli with Saunf might be detrimental to the pathogen due to various secondary plant substances secreted by the companion crop affecting the infection and incidence of the pathogen (Kroonen-Backbir and Zwarf-Roodzant, 2000 and Chand *et al.*, 2007).

Application of fertilizers containing humus 45 per cent and 5 per cent mullein, + 4gm ammonium nitrate, 3.4gm potassium chloride and 82 per cent super phosphate added to 110gm organic matter reduced the disease better than the organic manure alone (Kupryanova,1957).

To avoid storage and transit losses, the heads should be handled carefully to avoid bruising and surface moisture allowed to evaporate before storage. The storage house should be kept at 33-34°F and ample ventilation provided to reduce humidity (Walker, 1927).

Seed Treatment

Hot Water Treatment

Excellent control of *Alternaria brassicae* in white cabbage and cauliflower seed has been obtained by 30 minutes immersion in water heated to 45°C for 20 minutes at 50°C or even 30 minutes at 40°C. This treatment improves germination by up to 13 per cent along with elimination of other fungi such as *Penicillium* and *Mucor* spp. (Neilson, 1936). Hot water treatment of seed at 50°C for 30 minutes controls the disease (Walker, 1952).

Chemical Treatment

For controlling seed borne *A. Brassicae, A. Brassicicola, A. raphani* and *A. alternata* infections in brassicaceous vegetables, seed dressing with Benlate, Dithane M-45, Dithane Z-78, Bavistin, captan, Thiram, Vitavax and Zineb have been found to be very effective (Atkinson, 1950; Chahal *et al.*, 1977; Ellis,1968; Gupta and Saxena, 1984; Kumar and Singh, 1986; Maude *et al.*, 1972; Maude and Humpherson-Jones, 1980; Randhawa and Aulakh, 1982; Chand *et al.*, 2006).

Bioagents Treatment

Seed treatment with Mycostop, a powdery formulation of mycelium of *Streptomyces griseoviridis* controls *A. brassicicola* causing damping off in cauliflower and cabbage. Biocontrol of seed borne *A. raphani* and *A. brassicicola* in radish has been obtained through other antagonists (Vannaeci and Harman, 1987).

Host Resistance

Source of resistance in different brassicaceous vegetables have been identified, which can be incorporated through conventional and biotechnological techniques under suitable agronomically sound yield and quality bases to be effective against *Alternaria* spp. However, exploitation of this information to manage Alternaria disease under field conditions need much more emphasis.

A field study was carried out with 14 broccoli cultivars (Shogun, Earl, Fiesta, Lord, marathon, Monterery, Montop, Montvy, Tex968, RZ2502, SK4001, E52022, Shadow and Emerald city) and data were presented on yield and quality parameters, including infection by Alternaria. The best results were obtained with Fiesta, Lord and Marathon followed by Earl, Montop and Monterery (Reycke *et al.*, 1998; Chand *et al.*, 2006 and 2007).

Biological Control

Plant Extracts as Fungitoxicants

Essential oil from the roots of radish (1:2500) inhibits *A. brassicae*. The deproteimized leaf extracts of *Accaia nilotica, Enicostema hyssopifolium, Mimosa homata* and *Vitis vinifera* have shown fungistatic activity against *A. Brassicicola* (Umalkar *et al.*, 1977). The extracts prepared from the leaves of *Lawsonia alba*, roots of Datura stramonium and inflorescence of *Mentha piperita* have fungitoxic activity against *A. brassicae*. Extracts from the ferns, *Adiantum caudatum, Diplazium esculentum* and *Pteris vittata* reduce the growth and germinartion of *A. brassicicola* (Yasmeen and Saxena, 1990; Ahmed and Agnihotri, 1977).

Antagonistic for Biocontrol

Saprophytic phylloplane fungi such as *Aureobasidium purpulans* and *Epicoccum nigrum* are pathogenic to *A. brassicicola* (Pace and Cambell, 1974). The metabolites of *Acremonium roseogrseum* and *Aspergillus terreas* inhibit *A. brassicae* (Rai and Singh, 1980). Treating cauliflower seed with *Trichoderma viride* and *Streptomyces* spp. inhibits or reduces damping off caused by *A. brassicicola* (Tahvonen, 1982 and 1988)

Brassicaceous seeds treated with either *Gliocladium virens*-19, *Trichoderma harzianum*-22, *T. harzianum*-50, *Penicillium carylophilum*-36 and *P. oxalicum*-76 has a significantly higher emergence of healthy seedlings (Wu and, 1984).

In vitro, the efficacy of bioagents elucidated that *Trichderma harzianum, T. viride* and *T. virens* proved to be the most effective bioagent as they completely inhibited the growth of *Alternaria* spp. *Isolated* from the broccoli plants in culture (Chand, 2005).

References

Ahmad, S. and U. Narain (2007). Symptomatology, Etiology and Ecofriendly management of *Alternaria* leaf spots and blight of broccoli. In: *Ecofriendly management of plant diseases*. Daya Publishing House, Delhi. Chapter, 31:461-472pp.

Ahmed, S. R. and J. P. Agnihotri (1977). Antifungal activity of some plant extracts. *Indian J. Mycol. Pl. Pathol.*, 7:180-181.

Anonymous (1969). CMI distribution map of *Alternaria brassicicola.* Commonwealth Mycological Institute, Kew.

Atkinson, R. G. (1950). Studies on the parasitism and variation of *Alternaria raphani. Can. J. Res. Sect. C.*, 28:288-317.

Chahal, A. S. (1981). Seed borne infection of *Alternaria brassicae* in Indian mustard and its elimination during storage. *Curr. Sci.*, 50:621-623.

Chahal, A. S., M. S. Kang and J. S. Chauhan (1977). *Alternaria* blight: a serious disease of rapeseed and mustard. *Progressive Farming*, 12:21-22.

Chand, G. and U. Narain (2005b). Eco-friendly management of *Alternaria* leaf spot of broccoli. Paper presented in 6th National Symposium on Sustainable Plant Protection strategies: Health and Environmental Concerns, held at K.K.V., Ratnagiri M.S., Oct. 15-17, 2005. 31p.

Chand, G. and U. Narain (2005a). Characterization of *Alternaria* spp. and their symptoms differentiation in broccoli. Paper presented in 6th National Symposium on Sustainable Plant Protection strategies: Health and Environmental Concerns, held at K.K.V., Ratnagiri, M.S., Oct. 15-17, 2005. 30p.

Chand, G., U. Narain and M. Kumar (2005a). Taxonomy and parasitism of *Alternaria* spp. occurring on broccoli in India. Paper presented in National Seminar on New Horizons in Life science, held at Govt. P.G. College, Narsingpur, M.P., Sept. 2-3, 2005, 35p.

Chand, G., U. Narain and R. Pandey (2006). Efficacy of different fungicides against *Alternaria* leaf spot of broccoli. *Annals of Plant Protection Sciences*, 14(2): 468-469.

Chand, G., U. Narain, M. Kumar and M. Srivastava (2005b). Integrated management of *Alternaria* leaf spot of broccoli. *Indian Journal of Plant Pathology*, 23: 201-204.

Chand, G. U. Narain, M. Kumar and S. Verma (2007). Cultural management of broccoli. *Annals of Plant Protection Sciences*, 15(2):510-51.

Changsri, W. and G.P. Weber (1963). Three *Alternaria* species pathogenic on certain cultivated crucifers. *Phytopathology*, 53: 643-648.

Channon, A. G. and R. B. Maude (1971). Vegetables. In: disease of crop plants, J. H. Weston (Ed.) Macmillon, London. Chapter, 16:323-363.

Chupp, C. (1925). Alternaria leaf spot and blight of crucifers. In: *Manual of vegetable garden diseases.* The MacMillan Co., New York, 146-150 pp.

Chupp, C. and A.F. Sherf (1960). Crucifer diseases. In: *Vegetable diseases and their control.* Ronald Press Company, New York, Chapter 8: 237-288 pp.

Dey, P. K. (1948). *Plant Pathology.* Adm. Rep. Agric. U. P. 46:439-476.

Dillard, H.R., A.C. Cobb and J.S. Lamboy (1998). Transmission of *Alternaria brassicicola* to cabbage by flea beetles. *Plant Disease,* 82 (2): 153-157.

Dixon, G. R. (1981). Pathogens of crucifers crops. In: Vegetable crop disease. A.V.I. Publ. Co. Inc., Westport, Connecticut: 119-123pp.

Elliott, J.A. (1917). Taxonomic characters of the genera *Alternaria* and *Macrosporium. Amerir J. Bot.*, 4: 439-476.

Ellis, M. B. (1968). *Alternaria brassicae*. Commonwealth Mycological Institute. Descriptions of pathogenic and bacteria. No. 162, CMI, Kew, Survey, England.

Ellis, M.B. (1971). *Dematiaceous Hyphomycetes*. Commonwealth Mycological Institute, Kew, Surrey, England, 464-497 pp.

Groves, J.W. and A.J., Skolko (1944). Notes on seed borne fungi II. *Alternaria. Can. J. Res. Sect. C.*, 12: 217-234.

Gupta, R. C. and A. Saxena (1984). Efficacy of some fungicides on incidence of seed borne fungi of *Eruca sativa* grown in Tarai Region of Nainital. *Madras Agric. J.*, 71:81-88.

Gupta, D.K. and K.C., Basuchaudhary (1992). Occurrence and prevalence of *Alternaria* species in crucifers grown in Sikkim. *Indian J. Hill Farming*, 5:129-131.

Jorstad, I. (1945). Parasitic fungi on cultivated and economic plants in Narway. Medd. Plantepat. Inst. Oslo, 1:142.

Kadian, A.K. and G.S. Saharan, (1983). Symptomatology, host-range and assessment of yield losses due to *Alternaria brassicae* infection in rapeseed and mustard. *Indian J. Mycol. Pl. Pathol.*, 13: 319-323.

Khalid, A., M. Akram, U. Narain and M. Srivastava(2004). Characterization of *Alternaria* spp. associated with Brassicaceous vegetables. *Farm Sci. J.*, 13: 195-196.

Koike, S.T. (1997). Broccoli raab as a host of *Alternaria brassicae* in California. *Plant Disease*, 82 (5): 552.

Kolte, S. J. (1985). Disease management strategies for rapeseed mustard crop in India. *Agric. Rev.*, 6:81-88.

Kronen, B. and Zwart, R. (2000). Integrated cultivation of cabbage is competitive. *PAV Bulletin Akkerbonw*, 5:23-26.

Kuhn, J. (1856). Das Befallen des Rapses durch den Rapsverderber sporidesmium exitiosum Kuhn. *Bot. Zeitung*. 15:89-98.

Kuhn J. (1855) Uber das vevallen des Rapses and die Krankhist der Mohrenblather. *Hedwigia*, 1:86-92.

Kumar, K. and D.P.singh (1986). Control of *Alternaria brassicae* infection in mustard and rapseed. *Pesticides*, 20:22-23.

Kupryanova, M. V. K. (1957). Measures for the control of cabbage diseases. *Plant Prot. Mascow*, 5:28-29.

Maude, R.B. and F/M. Humpherson-Jones (1980). Importance and control of seed borne diseases of oilseed rape. *Aspects Appl. Biol.*, 6:335-241.

Maude, R. B., A. M. Kyle, C. G. Moule and C. L. Dudley (1972). Seed dressing with systemic fungicides for the control of seed-borne fungal pathogens. 22nd Annual report for 1971, National Vegetable Research Station, Mellesbourne Worwick, U. K., p 75.

Mason, E.W. (1928). Annonated account of fungi received at the Imperial Bureau of Mycology. I and II (Fascicles), pp. 34, Kew, Survey.

Narain, U. (1986). Studies on *Alternaria* spp. associated with leaf spots of Crucifers in India. *Adv. Biol. Res.*, 4(1 and 2): 187-191.

Narain, U. and H.K. Saksena (1975). A new leaf spot of turnip. *Indian Phytopath,*. 28: 98-100.

Narain, U. and J. Singh (1981). *Alternaria* leaf spot Cheiranthus from India. *Nat. Acad. Sci. Letters*, 4: 3-4.

Narain, U., J. Singh and A.K. Koul (1982). Leaf spot of candytuft caused by *Alternaria rapahani, Nat. Acad. Sci. Letters*, 5: 13.

Neergaard, P. (1945). *Danish species of Alternaria* and *Stemphylium*: taxonomy, parasitism, economical significance. Humphry Millford, Oxford Univ. Press, London, 560 pp.

Nielson, O (1933). Experiments on the control of silique fungus. *Tidsskèr for Planteanl*, 29:437-452.

Pace, M. A. and Campbell (1974). The effect of saprophytes on infection of leaves of Brassica spp. By *Alternaria brassiccicola. Trans. Br. Mycol. Soc.*, 63:193-196.

Rai, B. and D. B. Singh (1980). Antagonistic activity of some leaf surface micro fungi against *Alternaria brassicae* and *Drechslera graminea. Trans. Br. Mycol. Soc.*, 75:363-369.

Randhawa, H. S. and K. S. Aulakh (1982). Comparative evaluation of performance of sixteen fungi toxicants as seed dressing for the control of seed borne fungi of raya (*Brassica juncea* Coss.). *J. Res. Punjab Agric. Univ.*, 19:343-350.

Rao, B.R. (1977). Species of *Alternaria* on some cruciferae. *Geobios*, 4: 163-166.

Reyeke, L., Rooster, L. and Vanparys, l. (1998). Autumn cultivation of broccoli. *Proeftuinnieuws*, 8:24-25.

Vanparys, L. (2000). Purple broccoli: Violet Queen and RS84090 are usefull. *Proeftuinnieuws*, 9(6):18-19.

Saccardo, G.S. (1886). Hyphomyceteae. IN. Sylloge fungorum Omninum hucusque cegritorum. *Pavia, Italy*. 4:804p.

Saharan, G. S. and J.N. Chand (1988). Disease of rape seed and mustard. In: Diseases of oilseed crops. Directorate of Publication, Haryana Agric. Univ. Press, Hissar India, Chapter 3: 84-91.

Sangwan, M.S., N. Mehtaand S.K. Gandhi (2002). Some pathological studies on *Alternarie raphani* causing leaf and pod blight of radish. *J. Mycol. Pl. Pathology.*, 32:125-126.

Scimmer, F.C. (1953). *Alternaria* brassicicola on summer cauliflower seed. *Plant Pathol.*, 2:16-17.

Singh, R. S. (1987). *Alternaria* leaf spot black spot of crucifers. In.: Diseases of Vegetable crops. Oxford and IBH Publ. Co., New Delhi, India, Chapter 5: 159-163.

Suhag, L. S., R.M, Singh and Y. S. Malik (1985). Epidemiology of pod and leaf blight of raddish caused by *Alternaria aternnata. Indian Phytopath.*, 38: 148-149.

Tahvonen, R. T. (1982). Preliminary experiments into the use of Streptomyces spp. Isolated from peat in the biological control of soil and seed borne diseases in peat culture. *J. Sci. Agric. So. Finland*, 54: 357-369.

Tahvonen, R. T. (1988). Microbial control of plant disease with *Steptomyces spp. Bulletin O.E.P.P.*, 18:55-59.

Thomas, P. (1984). *Alternria* black spot (grey leaf spot). In: Canada Growers manual. Canada council of Canada, Winnipeg, Manitoba, Canada 1056-1058.pp.

Umalkar, G. V. D. S. Mukadam and K.M.A. Nehimiah (1977). Fungistatic properties of some deprotenized leaf extracts. *Sci. and Cult.*, 43:437-439.

Vannacci, G. and G. E. Harmon (1987). Biocontrol of seed borne *Alternaria raphani* and *A. brassicicola*. *Can. J. Microbial.*, 33:850-856.

Volkonen, J.P.T and Koponen (1990). The seed borne fungi of chinese cabbage (*Brassica pekinensis*). Their pathogenicity and control *Plant Pathology*, 39:510-516.

Verma, P.R. and G. S. Sharma (1993). *Alternaria brasscae* (Berk). *A. brassicicola* (Schein.) Wiltish. and *A. Raphani* Groves and Skolko: Introduction. Bibliography and subjects index. *Agraic Canada Res. Stn. Sakatoon, Tech. Bull.*, 81pp.

Walker, J. C. (1927). *Diseases of cabbage and related plants*. U. S. Deptt. of Agric. *Farmers Bulletin*, 1439:30p.

Walker, J. C. (1952). *Diseases of crucifers*. In: Diseases of vegetables crops. Mc Graw Hill Book Co., New York, London, Chapter, 6: 150-152.

Wu, W. S. and lu, J. H. (1984). Seed treatment with antagonists and chemicals to control Alternaria brassicicola. *Seed Sci. and Technol.*, 12:851-862.

Yasmeen and S. K. Saxena (1990). Effect of fern extracts on growth and germination of fungi. *Curr. Sci.*, 15:798-799.

Sustainable Disease Management of Agricultural Crops (2011) *Pages* **222–233**
Editors: **S.K. Biswas and S.R. Singh**
Published by: **DAYA PUBLISHING HOUSE, NEW DELHI**

Chapter 16

Root Rot Complex of Pea

☆ ***T.K. Bag*[1] *and Prabhas Chandra Singh*[2]**

[1]*Head, Central Potato Research Station, Shillong*
[2]*Division of Crop Protection, Indian Institute of Vegetable Research, Post Box No. 1, P.O. Jakhini, Shahanshapur, Varanasi – 221 305, U.P.*

Introduction

Root rots occur in every pea growing areas in the world. It usually starts when the pea plant is in the pre- or post-emergent seedling stage. Death soon follows such early infections, which finally results in a poor crop stand. Initially root decay begins on the fine feeder roots which later progresses gradually to the main taproot. Sometimes, however, the taproot is the first to be attacked. Root-rotting fungi can also reduce the quality of shelled peas. Infected plants produce peas, which are irregular in size and variable in maturity with lowered sugar content. Etiology of root rot of pea is complex, varies from geographical areas in the world, and may be caused by any one or a combination of several common soil fungi. It is often difficult or impossible to differentiate the common types of root rot, especially in an advance stage or late in the season. At this stage, it may be really a difficult task to determine whether root rot or a wilt disease has killed the plants. Pea plants damaged by root rot may be pulled out easily from the crop field as compared to those affected by wilt where roots of the affected plants are still intact. The common root-rotting fungi may attack the same plant at the same time. Species of *Pythium*, as well as lesion, stunt, and other nematodes are all commonly a part of the disease complex and can increase the severity of root rot.

Distribution

Root and foot rots caused by a complex of soil borne pathogens are the most destructive diseases of pea (*Pisum sativum* L.). *Aphanomyces euteiches* Drechs. is reported to be the most important pathogen in the pea root rot complex in several countries, *e.g.*, Sweden (Engquist, (1992 and Olofsson, (1967), Norway (Sundheim and Wiggen, (1972), the U.S. (Papavizas and Ayers,(1974; Hagedorn, (1984), The Netherlands (Oyarzun, and van Loon, (1989), and New Zealand (Chan and Close, (1987). Specialized physiological races of *Fusarium oxysporum* f. sp. *pisi*, that cause wilted plants, are also serious problems (Clarkson, (1978). Other reported pathogens causing foot and root rot are *F. solani* (Mart.) Sacc. f. sp.

pisi (F. R. Jones) W. C. Snyder and H. N. Hans., *Phoma medicaginis* Malbr. and Roum. var. *pinodella* (L. K. Jones) Boerema, *Mycosphaerella pinodes* (Berk. and Bloxam) Vestergr., *Pythium* spp., *Chalara elegans* Nag Raj and Kendrick (synonamorph = *Thielaviopsis basicola* (Berk. and Broome) Ferraris), and *Rhizoctonia solani* Kühn (Hagedorn, (1984; Kraft *et al.* (1981; Tu, (1987). The importance of these pathogens varies according to location, climate, and crop management strategy.

Pea wilt is recognized as the most important disease of peas in USA in (1930 and subsequently it is known to be present throughout the world. Pea root rot complex (PRRC), caused by *Alternaria alternata, Aphanomyces euteiches, Fusarium oxysporum* f. sp. *pisi, F. solani* f. sp. *pisi, Mycosphaerella pinodes, Pythium* spp., *Rhizoctonia solani,* and *Sclerotinia sclerotiorum,* is a major yield-limiting factor for field pea production in Canada (Xue, 2003).

Fusarium root rot is a widespread and destructive disease of field pea crops across the Canadian prairies (Chong *et al.*, 2003; Chang *et al.*, 2004; McLaren *et al.*, 2005 and McLaren, *et al.*, 2006). Surveys in Alberta in 2003-2006 showed that the disease occurred in most fields, producing dark lesions on roots and often causing plants to collapse. *Fusarium* species including *F. avenaceum, F. oxysporum* and *F. solani* f. sp. *pisi* were frequently isolated from diseased roots of pea in central Alberta and Manitoba (Yager *et al.*, 2003; Chong *et al.*, 2003; McLaren *et al.*, 2004 and Chang *et al.*, 2005). The pathogen causes significant yield reduction in pea under favourable environmental conditions.

In India, wilt and root rot complex of pea caused by *Fusarium oxysporum* f. sp. *pisi* and *F. solani* f. sp. *pisi* are the most important root pathogens of pea in northern India (Maheshwari and Gandhi, (1998). The disease has been reported from Himachal Pradesh, Bihar, UP, MP and many other states in India. An account of distribution of root rot disease of pea caused by different soil borne fungi in different countries is given in Table 16.1.

Losses

Common root rot of peas, caused by *Aphanomyces euteiches* Drechsler, has been reported in most pea growing areas of North America, Northern Europe, Australia, Japan and New Zealand and is one of the most destructive diseases of peas worldwide (Pfender (1984). Annual yield losses are estimated to be at 10 per cent worldwide (Pfender (1984); however, under favourable conditions complete crop losses are not uncommon. Tu ((1987) reported that root rot of field pea caused by *Pythium ultimum* and *Rhizoctonia solani* caused a crop loss of 2.8 t/ha or a total loss of approximately 24000 t of peas which has been estimated to be worth US$ 6 million in South Ontario. The disease caused by *Pythium ultimum* has also been known to occur and may cause yield losses in the USA (Parke *et al.* (1991; Bowera and Parke (1993). Persson *et al.* ((1997) reported that *A. euteiches* was the most yield-reducing pathogen of pea in Denmark and southern Sweden. Average annual yield losses of pea due to *A. euteiches* are estimated at 10 per cent in the Midwestern United States (Hagedorn, (1984). Entire fields, however, are frequently destroyed during wet years.

Damping off and root rot diseases on pea are caused by single or combination of soil borne fungi *i.e., Fusarium solani* Mart. Sacc, *F. oxysporium, Rhizoctonia solani* Kuhn, *Sclerotium rolfsii* Sacc, *Pythium spp.* and *Phytophthora cactorum* (Sm.et. Sm.) Leonion. These pathogens attack roots during growing season causing substantial losses in yield (Maheshwari *et al.* (1983; Rauf, 2000). Wilt and root rot complex of pea caused by *Fusarium oxysporum* f. sp. *pisi* and *F. solani* f. sp. *pisi* has been reported to cause upto 60 per cent losses during the years of epiphytotics in Northern India (Kumar, (1993). A survey conducted by Kumar and Dubey (2002) in eight locations in Bihar to assess the incidence of collar rot disease in peas revealed that the disease incidence ranged from 23.6 to 52.7 per cent. Losses

Table 16.1: Root Rot Disease of Pea Caused by Different Fungal Pathogens

Sl.No	*Root rot*	*Causal Organism*	*Distribution*	*Reference*
1.	Pea root rot complex (PRRC)	*Alternaria alternata, Aphanomyces euteiches, Fusarium oxysporum* f. sp. *pisi, F. solani* f. sp. *pisi, Mycosphaerella pinodes, Pythium* spp., *Rhizoctonia solani,* and *Sclerctinia sclerotiorum*	Canada	Xue, 2003
2.	Fusarium root rot	*F. avenaceum, F. oxysporum* and *F. solani* f. sp. *pisi*	Canada, central Alberta and Manitoba; New Zealand;	Xue, 2003, Yager *et al.*, 2003; Chong *et al.*, 2003; McLaren *et al.*, 2004; Chang *et al.*, 2005 and Clarkson, 1978.
3.	Root and foot rots	*Aphanomyces euteiches*	Sweden, Norway, the U.S., The Netherlands, and New Zealand	Engquist, 1992; Olofsson, 1967; Sundheim and Wiggen, 1972; Papavizas and Ayers, 1974; Hagedorn,1984; Oyarzun and van Loon, 1989; Clarkson, 1978; Chan and Close, 1987.
4.	Foot and root rot	*F. solani* f. sp. *pisi, Phoma medicaginis* var. *pinodella, Mycosphaerella pinodes, Pythium* spp., *Chalara elegans* and *Rhizoctonia solani*	Many countries	Hagedorn, 1984; Kraft *et al.*, 1981; Tu, 1987
5.	Pea wilt and root rot complex	*Fusarium oxysporum* f.sp. *pisi* and *F. solani* f.sp. *pisi*	India	Maheshwari and Gandhi,1998
6.	Pea Foot and Root Rot	*Aphanomyces euteiches, Phoma medicaginis* var. *pinodella* and *Fusarium solani; F. avenaceum, F. oxysporum, F. culmorum, Chalara elegans, Pythium irregulare,* and *Mycosphaerella pinodes, Phytophthora irregulare*	Sweden and Denmark	Persson *et al.*, 1997
7.	Pea root rot	*Pythium aphanidermatun* and *P. ultimum*	Taiwan	Lin *et al.*, 2002
8.	Pea root rot	*Aphanomyces euteiches, Fusarium oxysporum* f. sp. *pisi* and *F. solani* f. sp. *pisi* and *Rhizoctonia solani*	Pakistan	Khan *et al.*, 1998
9.	Pea root rot	*Thielaviopsis basicola, Fusarium solani, Aphanomyces euteiches, Pythium* spp., and *Rhizoctonia solani* and *F. solani* f. sp. *pisi*	New York, Auburn, Mount Morris, USA	Blume and Harman, 1979
10.	Pea damping off and root rot	*Fusarium solani, F. oxysporium, Rhizoctonia solani, Sclerotium rolfsii Sacc, Pythium* spp. and *Phytophthora cactorum*	Egypt	Abda *et al.*, 1992; Persson *et al.*, 1997 and Ragab *et al.*, 1999
11.	Common root rot of pea	*Aphanomyces euteiches*	Russia	Kotova, 1969, 1971 and 1976
12.	Common root rot of pea	*Aphanomyces euteiches*	Japan	Fukumishi *et al.*, 1976, Yokosawa *et al.*, 1974
13.	Common root rot of pea	*Aphanomyces euteiches*	France	Labrousse, 1933 and 1934
14.	root rot and wilt of pea	*Fusarium oxysporum* f. sp. *pisi* and *F. solani* f. sp. *pisi*	Taiwan	Lin *et al.*, 1984

incurred by root rot specifically caused by *Aphanomyces, Pythium,* or *Rhizoctonia* individually or in combination are not known in India.

Table 16.2: Estimated Pea Yield Loss in Different Countries

Country	Causal Organism of Root Rot of Pea	Loss	References
Worldwide	*A. euteiches*	10 per cent	Pfender (1984)
South Ontario	*Pythium ultimum* and *Rhizoctonia solani*	2.8 t/ha or a total loss of approximately 24000 t of peas (US$ 6 million)	Tu (1987)
Midwestern United States	*A. euteiches*	Average annual yield losses of 10 per cent	Hagedorn (1984)
Northern India	*Fusarium oxysporum* f. sp. *pisi* and *F. solani* f. sp. *pisi*	upto 60 per cent losses	Kumar (1993)

Symptoms

There are four different types of root rot in pea. The symptoms vary with the causal organism involved. The symptoms of four common types of root rot are as follows:

Aphanomyces Root Rot

In *Aphanomyces* root rot, the fungus can infect plants at any age. The most diagnostic symptom of the disease is that the central part of the taproot *i.e.* vascular tissue can be separated easily as a long, fiber-like string from the outer, softened, water-soaked portion (cortex) of the root when the plant is pulled from the soil. The thin side branches or feeding rootlets are killed. The decay can extend up the stem to slightly above the collar region. The decayed surface may often appear slimy in wet soils turning gray, then yellowish or pink, and finally brownish black. The leaves dry and shrivel progressively upward on the stem. Plants appeared usually stunted, and the leaves progressively turn yellow starting from the bottom of the shoot. Pods may be few with a reduced number of small and deformed seeds. In severe cases, the plants collapse and die without forming any pods. Late infected plants appear almost normal aboveground and often produce normal peas.

Fusarium Root Rot

It affects mainly the taproot starting infection close to points where the seed is attached. The primary and secondary roots shows the development of reddish brown streaks which later merge. The external portion of the stem shows brick red, dark reddish brown, or chocolate colored lesions. Whereas the central part of the taproot is a deep red. Plant growth is stunted, the foliage turns grayish, then yellow, the lower leaves wither, and the plant eventually dies. The lower stem is often girdled, causing the plant to fall over. *Pythium ultimum* is often found in Fusarium-infected roots and vice versa.

Pythium Seed Rot and Damping Off

Pythium commonly causes seed rot as well as pre- and post-emergence damping-off in pea. Root rot may also occur in older plants, and often results in pruning of roots that significantly reduces root length. Damage is most common in wet soils and is characterized by soft rot. Roots infected with *Pythium* are typically light brown in colour, soft, and watery. Infected plants are frequently stunted and pale green to yellow in color. Although it primarily causes a seed rot, damping-off, and seedling root rot, *Pythium ultimum* can cause a watery, soft decay of older plants in wet soils at an optimum temperature of 17° to 23°C.

Rhizoctonia Root Rot

Rhizoctonia solani is a polyphagus pathogen and can attack pea plants at any stage of growth. In seed infection, seeds may turn dark brown and decay. Water-soaked, then reddish brown-to-brown lesions form in the seedling epicotyl and hypocotyl. The growing point may die as it emerges from the soil. Seedlings damp-off may recover to produce a normal plant. On older plants, scurfy, reddish brown, sunken lesions are visible to form on the underground stem and roots. The stem may be girdled causing severe plant stunting and yellowing. The brown, thread-like filaments (mycelium) of the causal fungus may be seen with a hand lens on the surface of the lesion or canker.

Etiology

It was Jones and Drechsler who described and established that pea root rot is caused by *Aphanomyces euteiches* (Haenseler, (1923). The very popular and most common name "root rot"was firmly established in (1936 by wakefield and Moore who specifically coined this term for the diseases caused by Aphanomyces *euteiches.* They used the term "black root rot" for the pathological conditioned caused by *Thielaviopsis basicola* in peas. *Aphanomyces euteiches* Drechs. is now well established fact to be the most important pathogen in the pea root rot complex in several countries, *e.g.*, Sweden (Engquist, (1992 and Olofsson, (1967), Norway (Sundheim and Wiggen, (1972), the U.S. (Papavizas and Ayers,(1974; Hagedorn, (1984), The Netherlands (Oyarzun, and van Loon, (1989), and New Zealand (Chan and Close, (1987). A comprehensive survey and nearly 40 yr field study established that *Fusarium solani* f. sp. *pisi* [Jones] Snyd. and Hans.), *Aphanomyces euteiches* Drechs., and *Pythium ultimum* Trow were the most widespread root pathogens which cause root of pea (*Pisum sativum* L.) in New York state. In (1975, it was also established that *Thielaviopsis basicola* (Berk. and.Br.) has association in pea tissue and is a component of pea root rot complex. Occasionally this organism has been recorded to be a pathogen of pea roots in other states (Burke and Kraft, (1974).

Other reported pathogens causing foot and root rot are *F. solani* (Mart.) Sacc. f. sp. *pisi* (F. R. Jones) W. C. Snyder and H. N. Hans., *Phoma medicaginis* Malbr. and Roum. var. *pinodella* (L. K. Jones) Boerema, *Mycosphaerella pinodes* (Berk. and Bloxam) Vestergr., *Pythium* spp., *Chalara elegans* Nag Raj and Kendrick (synonamorph = *Thielaviopsis basicola* (Berk. and Broome) Ferraris), and *Rhizoctonia solani* Kühn (Hagedorn, (1984; Kraft *et al.* (1981; Tu, (1987). In India, wilt and root rot complex of pea caused by *Fusarium oxysporum* f. sp. *pisi* and *F. solani* f. sp. *pisi* are the most important root pathogens of pea in northern India (Maheshwari and Gandhi, (1998). The disease has been reported from Himachal Pradesh, Bihar, UP, MP and many other states in India. Sometime the disease appeared in combination with *Fusarium solani* and *Rhizoctonia solani* making the disease a complex disease (Dohroo *et al.* (1998). According to Tu ((1987), root rot of field pea has been reported to be caused by *Pythium ultimum* and *Rhizoctonia solani* in South Ontario. In Taiwan, root rot of vegetable pea has been reported to be caused by two species of *Pythium* namely *Pythium aphanidermatun and P. ultimum* in soil less culture system (Lin *et al.*, 2002). Research on root rot pathogens of peas in the Netherlands has confirmed the prevalence of *Fusarium solani, F. oxysporum, Pythium* spp., *Mycosphaerella pinodes* and *Phoma medicaginis* var. *pinodella. Aphanomyces euteiches* and *Thielaviopsis basicola* were identified for the first time as pea pathogens in the Netherlands. An account of major soil borne fungi responsible for pea root rot disease is given below.

Aphanomyces euteiches

The genus *Aphanomyces* belongs to Class Oomycetes (order *Saprolegniales*) that is phylogenetically differentiated from other orders such as the *Peronosporales* and *Pythiales* which have many other important plant pathogens (Petersen and Rosendahl, 2000). *Aphanomyces euteiches* can readily grow on culture media, but in nature, saprophytic growth outside the host is not considered important (Papavizas and Ayers, (1974). The fungus survives as oospores in the soil, and the entire lifecycle is completed in the host roots and surrounding soil. The oospores are 20-35 µm in diameter, which have a thick protective wall and contain energy reserves in the form of a large oil globule. After germination, either the oospores form a mycelium, or a short mycelial strand called a germ sporangium. From a single oospore, hundreds of zoospores can be released through the germ sporangium. The motile zoospores locate and encyst on a host root within minutes, and the cysts are able to germinate and penetrate cortical cells within hours (Papavizas and Ayers, (1974). The mycelium then grows mostly longitudinally and intracellularly through the root tissue and, within a few days of infection, forms large amounts of oogonia, which are fertilised by antheridia and develop new oospores. Live mycelium in infected roots is also capable of releasing new zoospores. There are two recognized races of *A. euteiches*. Isolates are known which may belong to additional races, as yet not defined.

Fusarium oxysporum f.sp. *pisi*

In culture, *Fusarium oxysporum* f.sp. *pisi*, the mycelium is septate producing micro conidia (2.5-4 X 6-15µm) and macro conidia which are typically fusiform, hyaline, multiseptate (3.5-5.5 x 20-22 µm) and as chlyamydospores. There are cultural and pathogenic variability among different races of *Fusarium oxysporum* f.sp. *pisi*. 11 races of *Fusarium oxysporum* f.sp. *pisi* have been reported (Armstrong and Armstrong, (1974) but 1,2,5 and 6 are considered valid by Kraft and Haglund ((1978). Races 1,2,3,4, 5 and 6 are common infecting peas which differ in respect of mycelium, pigmentation, sporulation and pathogenicity (Kraft and Haglund, (1978). There are no reports on the number of races of *Fusarium oxysporum* f.sp. *pisi* available in India.

Fusariun solani f. sp. *pisi*

In culture, *Fusariun solani f. sp. pisi* that is reported to be the individual cause of root rot of pea produce blue green to buff coloured sporodochia. It produces three types of microscopic spores: small, one-celled, elliptical, microconidia; much larger, septate, slightly curved macroconidia; and thick-walled, rounded chlamydospores. Macroconia are mostly triseptate measuring 4.5-5 x 27-40 µm in size, curved and hyaline. Microconidia (9-16 x 2-4 µm) are produced sparsely and chlyamydospores developed terminally or intercalary either singly or in chain (Booth, (1971). The chlamydospores are resistant to unfavorable environmental conditions and can persist in soil for 5 years or more in the absence of peas.

Rhizoctonia solani

The *Rhizoctonia solani* (perfect stage *Thanetophorus cucumeris*) is common in most cultivated soils and attacks hundreds crop plants. It has many strains and anastomosis groups that differ in infection of host and host tissues. *Rhizoctonia solani* may survive indefinitely in soil as a saprophyte and can perpetuate extremes in temperature and soil moisture as small, brown, rounded sclerotia. Under favorable conditions, the sclerotia germinate by producing delicate hyphae that grow through the soil and invade roots directly, through wounds or natural openings when sufficient soil moisture present. The hyphae are branched at right angle and septate nearer to the origin of brances. The *Thanetophorus* sexual state plays little or no part in the disease cycle.

Epidemiology

In India, wilt and root rot complex of Pea caused by *Fusarium oxysporum* f. sp. *pisi* and *F. solani* f. sp. *pisi* is most common. Both the pathogen is seed borne and soil borne. The fungus can survive in conducive soil for more than 10 years. The pathogen is disseminated through contaminated soil or plant fragments by water, wind. In highly wet soil, there is almost no infection. Dry soils are most favourable for the disease. The disease is more common in neutral or alkaline soil than the acid soil. Temperature optimum for the disease development is 24-28 ° C but rapid development occurs at about 21° C. Kumar *et al.* (2003) reported that late sowing of pea (*Pisum sativum*) cv. Arkel minimized the incidence of collar rot (*Fusarium solani* f.sp. *pisi*). Sowing between (19 and 26 November resulted in less disease development. Disease development was high during 18-25 December at the crop age of 48 to 55 days, and 9.1-26.4 ° C temperature, 18.6-93.1 per cent relative humidity and 0-21.3 mm rainfall favoured its development. *Fusarium solani* root rot is most severe at soil temperatures of 26° to 27°C and at moderate soil moistures whereas *Pythium ultimum* is more of a problem in cool (64° to 75°F), wet soils. Peas planted in soil infected with both *Fusarium* and *Pythium ultimum* are more severely attacked than plants infected with only one pathogen at any soil temperature or moisture level.

In other countries, common root rot of pea is caused by *A. eutieches.* It is also a soil borne disease and can survive several years in soil. Spores of the fungus are readily carried over long distances by drainage and splashing water, or during movement of soil from one area to another, and through infected seed. The oospores get mixed with the seed and be introduced into new fields or gardens when such contaminated seed is planted. The zoospores are produced and germinate in abundance between 13° to 20°C with some germination occurring as low as 8°C and as high as 30°C. The *Aphanomyces* fungus also attacks alfalfa, snap beans, cowpea, spring vetch, sweet clover, spinach, sweet pea, and some weed species. According to Jones and Drechsler ((1925), the optimum temperature for infection of susceptible pea plants by *A. eutieches* is between 15-30 ° C. A field soil temperature of 17 ° C is favourable for root rot in pea; however, maximum root infection took place when soil temperature reaches between 24-28 ° C. Almost all the investigators agreed that the root rot disease of pea caused by *A. eutieches* is favoured by high soil moisture level, poor drainage, heavy soil or wet seasons.

All the investigators of pea root rot complex are unanimously agreed that *Rhizoctonia* root rot damages peas at relatively low soil temperatures (18°C) but is most aggressive under warmer conditions (24° to 30°C). *Rhizoctonia* infection and disease development can occur over a wide range of soil moistures. Higher soil temperatures (optimum between 25° to 30°C) and moderate soil moisture, on the other hand, favor *Fusarium* root rot. *Aphanomyces* root rot, most damaging at soil temperatures between 22° to 27°C, is favored by excessive soil moisture. It is most serious when a cool, wet spring is followed by an early, warm, dry summer. Although *P. ultimum* primarily causes a seed rot, damping-off, and seedling root rot, it can cause a watery, soft decay of older plants in wet soils at an optimum temperature of 17° to 23°C.

Genetics

Bhardwaj *et al.* (2001) studied the inheritance of resistance to Fusarium wilt caused by *Fusarium oxysporum* f. sp. *pisi* in garden pea (*Pisum sativum* L.). Analysis of the F1, F2, BC1 and BC2 generations of the crosses involving resistant line DRP-3 and three susceptible parents Ageta-6, Arkel and Sel. 8-1 indicated that resistance in DRP-3 is governed by a single dominant gene. Resistance to *Fusarium* wilt, caused by *Fusarium oxysporum* f.sp. *pisi* race 1, in peas (*Pisum sativum*) is conferred by a single dominant gene, Fw. The gene was located in the pea genome by analyzing progenies from crosses involving genetic markers across all pea linkage groups. Phenotyping of the progenies for reaction to race 1 of the *Fusarium* wilt pathogen was determined by field screening in a "wilt-sick" plot in Pullman, Washington, USA. Fw was shown to be located on linkage group III, about 13 map units from Lap-1 and b and 14 map units from Td. The relatively large distances between these markers and Fw precludes the use of the linked markers in marker-assisted selection for wilt resistance. Additional markers in this region of the pea genome will be required if marker-assisted selection for Fw is to be successful (Grajal-Martin and Muehlbauer, 2002).

Studies was carried out to identify closely linked marker(s) to the Fusarium wilt race 1 resistance gene (Fw) that could be used for marker-assisted selection in applied pea breeding programmes. Eighty recombinant inbred lines (RILs) from the cross of Green Arrow (resistant) and PI 179449 (susceptible) were developed through single-seed descent, and screened for disease reaction in race 1-infested field soil and the greenhouse using single-isolate inoculum. The RILs segregated 38 resistant and 42 susceptible fitting the expected 1:1 segregation ratio for a single dominant gene. Bulk segregant analysis (BSA) was used to screen 64 amplified fragment length polymorphism (AFLP) primer pairs and previously mapped random amplified polymorphic DNA (RAPD) primers to identify candidate markers. Eight AFLP primer pairs and 15 RAPD primers were used to screen the RIL mapping population and generate a linkage map. One AFLP marker, ACG:CAT_222, was within 1.4 cM of the Fw gene. Two other markers, AFLP marker ACC:CTG_159 at 2.6 cM linked to the susceptible allele, and RAPD marker Y15_1050 at 4.6 cM linked to the resistant allele, were also identified. The probability of correctly identifying resistant lines to Fusarium wilt race 1, with DNA marker ACG: CAT-222, is 96 per cent. These markers will be useful for marker assisted breeding in applied pea breeding programmes (McClendon *et al.*, 2002). Neumann and Xue (2003) evaluated about 117 field pea cultivars in Canada for their reactions to the four common races (1, 2, 5, and 6) of *Fusarium oxysporum* f.sp. *pisi* in growth chambers. Based on the visual assessment of foliar wilt symptoms, 49 cultivars were resistant to at least one of the four races, and the remaining 68 cultivars were susceptible to all four races. Of these resistant cultivars, Ascona and 44 other cultivars were resistant to race 1; Impala to race 2; Aladin to races 1 and 2; and Radley and Princess to races 2, 5, and 6. In an effort to standardize the methodology for screening field pea for resistance to the pathogen, other quantitative parameters, including shoot length, vascular discoloration, and shoot and root dry weights, were evaluated on selected cultivars.

Management

Cultural Management

There is no completely satisfactory control for the root-rot diseases of pea once the causal fungus or fungi is/are introduced and becomes prevalent in the soil. Control measures for all root rots are the same. Plant early in fertile, well-prepared, well-drained soil using seed grown in semiarid areas of the Pacific Northwest. Follow a crop rotation scheme in which peas are grown in the same area only once in 5 years or more. If *A. euteiches* is the primary problem, do not include alfalfa, beans, sweet clover, cowpeas, spinach, or vetch in the rotation. Crucifers are suitable for rotation.

At the University of Wisconsin, a method of determining the inoculum potential of Aphanomyces *euteiches prior* to planting (called Field Indexing) has been developed for large scale use by the commercial growers. The soil-assay or "root rot potential" procedure estimates the amount of infective inoculum in soil and the level of disease that can be expected if peas are planted in the field. Growers are advised to avoid planting in moderate to highly infested fields with an index of 75 or higher. There is need to maintain an adequate to high, balanced soil fertility level, based on a soil test. At least 3 tons per acre of rock phosphate, hydrated lime, or calcitic limestone are required to add for effective reduction of losses from root rot. The optimum soil pH is 6.5 to 7.0. Commercial growers should avoid overcrowding, deep planting, over fertilizing, soil compaction, and mechanical damage to the roots and stems.

In India, for effective management of the wilt complex, integrated management practices are to be followed. Sagar and Sugha (2004) reported that soil amendment and repeated cropping with different non-host crops such as wheat, oats, maize and sorghum reduced the population of *Fusarium oxysporum* f.sp. *pisi* and *F. solani* f.sp. *pisi* in the soil and the severity of pea root rot due to these pathogens. *F. solani* f.sp. *pisi* was more sensitive to soil amendment, and wheat and especially maize were more effective than the other two non-host crops. Repeated cropping of non-host species was more effective than a single crop.

Integrated Biological Management

Kumar and Dubey (2001) reported that the seed treatment with captan (1 g kg^{-1})+ *T. harzianum* (10^6 spores ml^{-1} 10 g seed-1) gave good germination, least disease incidence along with highest green pod yield, which was statistically similar with seeds treated with *T. harzianum* alone, thiram + *T. harzianum*, carboxin+ *G. virens* and *G. virens* alone.

Xue (2003) reported that when ACM941 a fungal biocontrol agent and strain of *Clonostachys rosea* (syn. *Gliocladium roseum*) (ATCC 74447) applied to the seed, propagated in the rhizosphere and colonized the seed coat, hypocotyl, and roots as the plant developed and grew. ACM941 significantly reduced the recovery of all fungal pathogens from infected seed, increased in vitro seed germination by 44 per cent and seedling emergence by 22 per cent, and reduced root rot severity by 76 per cent. The effects were similar to those of thiram fungicide, which increased germination and emergence by 33 and 29 per cent, respectively, and reduced root rot severity by 65 per cent. When soil was inoculated with selected pea root rot complex (PRRC) pathogens in a controlled environment, seed treatment with ACM941 significantly increased emergence by 26, 38, 28, 13, and 21 per cent for *F. oxysporum* f. sp. *pisi*, *F. solani* f. sp. *pisi*, *M. pinodes*, *R. solani*, and *S. sclerotiorum*, respectively. Under field conditions from (1995 to (1997, ACM941 increased emergence by 17, 23, 22, 13, and 18 per cent and yield by 15, 6, 28, 6, and (19 per cent for the five respective pathogens. The seed treatment effects of ACM941 on these PRRC pathogens were greater or statistically equivalent to those achieved with thiram. Results of this study suggest that ACM941 is an effective bioagent in controlling PRRC and is an alternative to existing chemical products.

Bio primed pea seeds (cv.Master B) coated with *B. subtilis* and *T. harzianum* has been reported to reduce root rot disease incidence at pre- and post emergence stages if compared with non primed coated seeds with the same biocontrol agents as well as fungicide seed treatment during the two seasons. Bio primed seeds (coated with *B. subtilis* and T. *harzianum*) treatments reduced pea root rot at pre-emergence damping–off stage by 83.3, 72.7 per cent during 2005/2006 and by 84.5, 77.1 per cent during the 2006/2007 season. Meanwhile, seed coating with the same bio agents and seed dressing with Rixolex-T cause 60.6, 57.5, 48.4 per cent and 63.1, 61.4, 56.5 per cent reduction of pre emergence pea root rot caused by *Fusarium solani*, *F. oxysporum*, *Rhizoctonia solani*, *Sclerotoium rolfsii* and *Pythium spp* in Nobaria Province, Egypt. Bio- priming seed treatments can provide a high level of protection against root rot disease of pea plants. It could suggested that bio- priming (combined treatments between seed priming and seed coating with biocontrol agents) may be safely used commercially as substitute of traditional fungicide seed treatments for controlling seed and soil borne plant pathogens (El-Mohamedy and Abd El-Baky, 2008).

Chemical Management

Shalini and Dohroo (2002) reported that in greenhouse experiments, seed treatment with Bavistin, Contaf, Kri-Benomyl and Topas at 0.1 per cent resulted in 100 per cent seed germination and absence of pre emergence rot. Seed treatment with Bavistin resulted in the lowest wilt incidence (5 per cent) and highest disease control (94.37 per cent). In field experiments, Bavistin treatment also resulted in the highest mean seed germination (67.92 per cent), lowest pre-emergence rot (32.08 per cent) and highest yield (635.5 per cent). Bavistin treatment resulted in the lowest wilt incidence (15.34 per cent), which was at par with Kri-Benomyl (15.11 per cent).

Management through Cultivation of Resistant Varieties

Use of resistant varieties is essential for effective management of the disease. Herold, a new pea (*Pisum sativum*) is resistant to wilt (*Fusarium oxysporum* f.sp. *pisi* race 1) and to *Thielaviopsis basicola*. It also has a good level of field resistance to the complex of foot and root rots in Czech Republic in 2002, and in Slovakia and Austria.

Three green pea germplasm W6 26740, W6 26743 and W6 26745 were registered by USDA-ARS, Plant Germplasm Introduction and Testing, Washington State University, Pullman, WA 99164 for resistant to *Fusarium* root rot caused by *Fusarium solani* (Mart.) Sacc. f. sp. *pisi* (F.R. Jones) W.C. Snyder and H.N. Hans.] with acceptable agronomic traits. They will be useful as a resource for developing root rot resistant green pea cultivars. "90-2131" is a pea germplasm release has partial resistance to *Aphanomyces* root rot (caused by *Aphanomyces euteiches* Drechs.), *Fusarium* root rot, and *Fusarium* wilt races 1, 5, and 6. But W6 26740, W6 26743 and W6 26746 are expected to serve as parental lines in the development of cultivars with improved disease resistance primarily to *Fusarium* root rot (Conye *et al.*, 2008).

References

Abda, K. A., Ali, H.Y. and Mansour, M. S. (1992). Phytopathological studies on damping-off and root rot diseases of pea in A.R.E. Egypt. *J. Appl. Sci.* 7(9): 242-261.

Armstrong, G.M and Armstrong, J. K. (1974). Races of *Fusarium oxysporum* f.sp. *pisi:* causal agents of wilt peas. *Phytopathology* 64:849-857.

Bhardwaj, R. K., Kohli, U.K., Gupta, S. K. and Wàlia, D. P. (2001). Inheritance of resistance to *Fusarium* wilt in garden pea (*Pisum sativum* L.). *Crop Improvement* 28(1): 105-108.

Blume, M. C. and Harman, G. E. (1979). *Thielaviopsis basicola:* A component of the pea root rot complex in New Yoik State. *Phytopathology* 69:785-788.

Burke, D. W. and Kraft, J. M. (1974). Response of beans and peas to root pathogens accumulated during monoculture of each crop species. *Phytopathology* 64:546-549.

Chan, M. K. Y. and Close, R. C. (1987). *Aphanomyces* root rot of peas, 1. Evaluation of methods for assessing inoculum density of *Aphanomyces euteiches* in soil. *N.Z. J. Agric. Res.* 30:213-217.

Chang, K.F., Bowness, R., Hwang, S. F., Turnbull, G., Howard, R.J. and Blade, S.F. (2004). The occurrence of field pea diseases in central and southern Alberta in 2003. *Can. Plant Dis. Surv.* 84: 104-106.

Chang, K.F., Bowness, R., Hwang, S.F., Turnbull, G.D., Howard, R.J., Lopetinsky, K., Olson, M. and Bing, D.J. (2005). Pea diseases occurring in central Alberta in 2004. *Can. Plant Dis. Surv.* 85: 89-90.

Chong, G., Banniza, S., Warkentin, T., and Morrall, R. A. A. (2003). Disease survey in field pea in Saskatchewan in 2002. *Can. Plant. Dis. Surv.* 83: 124-125.

Clarkson, J. D. S. (1978). Pathogenicity of *Fusarium* spp. associated with foot rots of peas and beans. *Plant Pathol.* 27:110-117.

Coyne, C. J., Porter, L.D., Inglis, D.A., Grünwald, N.J., McPhee, K.E. and Muehlbauer, F.J. (2008). Registration of W6 26740, W6 26743, and W6 26745 Green Pea Germplasm Resistant to Fusarium Root Rot. *Journal of Plant Registrations* 2:137–139.

Dohroo, N.P.,Verma, S. and Bharat, N. K. (1998). Fusarium wilt and root rot of pea. *Intl. J. Trop. Pl. Dis.* 16(1): 1-20.

El-Mohamedy, R. S.R. and Abd El-Baky, M.M.H. (2008). Evaluation of Different Types of Seed Treatment on Control of Root Rot Disease, Improvement Growth and Yield Quality of Pea Plant in Nobaria Province. *Research Journal of Agriculture* and *Biological Sciences* 4(6): 611-622.

Engquist, G. (1992). Studies on common root rot (*Aphanomyces euteiches*) of peas (*Pisum sativum*) in Sweden. Pages 321-322. In: Proc. Euro. Conf. Grain Legumes, 1st. Euro. Assoc.Grain Legumes.

Grajal-Martin, M. J. and Muehlbauer, F. J. (2002). Genomic location of the Fw gene for resistance to fusarium wilt race 1 in peas. *Journal Heredity* 93(4): 291-293.

Hagedorn, D. J., ed. (1984). *Compendium of Pea Diseases.* American Phytopathological Society, St. Paul, MN.

Hysek, J., Kreuzman, J. and Brozova, J. (2002). Reaction of pea selections to *Fusarium oxysporum* f.sp. *pisi* (races 1, 2, 5, 6) and *Fusarium solani. Plant Protection Science.* 38: 561-564.

Khan, J., Khan, M. and Amin, M. (1998). Distribution and integrated management of root rot of pea in Malakand Division. *Pakisthan J. Biological Sci.* 1(4): 267-270.

Kraft, J. M., Burke, D. W. and Haglund, W. A. (1981). *Fusarium* diseases of beans, peas and lentils. Pages 142-156. *In: Fusarium* Diseases, Biology and Taxonomy. P. E. Nelson, T. A. Toussoun, and R. J. Cook, eds. The Pennsylvania State University Press, University Park.

Kreuzman, J., Drdla, R. and Liska, P. (2003). Pea variety Herold. *Czech J. Gen. Pl. Breed.* 39(1): 29-30.

Kumar, D. and Dubey, S. C. (2001). Management of collar rot of pea by the integration of biological and chemical methods. *Indian Phytopath.* 54(1): 62-66.

Kumar, D. and Dubey, S. C. (2002). Collar rot–a new disease of pea in plateau region of Bihar. *Orissa J. Hort.* 30(2): 108-109.

Kumar, D., Dubey, S. C. and Krishna, G. (2003). Epidemiology of collar rot of pea caused by *Fusarium solani. Journal Research, Birsa AgriculturalUniversity.* 15(2): (199-203.

Lin, Y. S., Huang, J.H. and Gung, Y. (2002).Control of Pythium root rot of vegetable pea seedlings in sopilless culture system. *Plant Pathology Bulletin* 11; 221-228.

Lin, Y.S., Sun, W. and Wang, P.H. (1984). *Fusarium* root rot and wilt of garden peas in Taiwan. *J Agric. Res. China* 33: 395-405.

Maheshwari, S. K., Thooty, T. S. and Gupata, T. S. (1983). Survey of wilt and root rot complex of pea in northern India and the assessment of Losses. *Agricultural Science Digest,* India 3: 139-141.

McClendon, M.T., Inglis, D.A., McPhee, K. E. and Coyne, C.J. (2002). DNA markers linked to fusarium wilt race 1 resistance in pea. *J. American Society Hort. Sci.* 127(4): 602-607.

McLaren, D. L., Conner, R. L., Hausermann, D. L., and Loutchan, K. D. (2005). Diseases of field pea in Manitoba in 2004. *Can. Plant Dis. Surv.* 85: 94-95.

McLaren, D. L., Conner, R. L., Hausermann, D. L., and Penner, W.C. (2006). Diseases of field pea in Manitoba in 2005. *Can. Plant Dis. Surv.* 86: 112-113.

McLaren, D. L., Conner, R. L., Yager, L., and Groom, M. (2004). Diseases of field pea in Manitoba in 2003. *Can. Plant Dis. Surv.* 84: 107-108.

Neumann, S. and Xue, A. G. (2003). Reactions of field pea cultivars to four races of *Fusarium oxysporum* f. sp. *pisi. Can. J. Pl. Sci.* 83(2): 377-379

Olofsson, J. (1967). Root rot of canning and freezing peas in Sweden. *Acta Agric. Scand.* 17:101-107.

Oyarzun, P. and van Loon, J. (1989). *Aphanomyces euteiches* as a component of the complex of foot and root pathogens of peas in Dutch soils. *Neth. J. Plant Pathol.* 95:259- 264.

Papavizas, G. C. and Ayers, W. A. (1974). *Aphanomyces* species and their root diseases in pea and sugarbeet. Technical Bulletin, United States Department of Agriculture (No.1485): 158pp.

Papavizas, G. C. and Ayers, W. A. (1974). *Aphanomyces* species and their root diseases in pea and sugarbeet. U.S. Dep. Agric. Res. Serv. Tech. Bull. No. 1485.

Persson, L., Bødker, L. and Larsson-Wikström, M. (1997). Prevalence and pathogenicity of foot and root rot pathogens of pea in southern Scandinavia. *Plant Dis.* 81:171-174.

Petersen, A. B. and Rosendahl, S. (2000). Phylogeny of the *Peronosporomycetes* (*Oomycota*) based on partial sequences of the large ribosomal subunit (LSU rDNA). *Mycological Research* 104: 1295-1303.

Pfender, W.F. (1984). *Aphanomyces root rot.* Pp. 25-28 *In:* Compendium of Pea Diseases, D. J. Hagedorn (Ed.); American Phytopathological Society, St. Paul, Minnesota, USA.

Ragab, M.M., Aly, M.D.H., Mona, M.M. and Nehals EL. Mougy. (1999). Effect of fungicides, biocides and bioagents on controlling pea root rot diseases. *Egypt. T. Phytopatholoqy* 27: 65-81.

Rauf, B.A, (2000). Seed borne disease problems of legume crops in Pakistan. *Pakistan J. Sci. Indust. Res.* 43: 249-254.

Sagar, V. and Sugha, S. K. 2004. Role of soil amendment and repeated cropping in the management of pea root rot. *Tropical-Science.* 44(1): 1-5

Sneh, B., Pozniak, D. and Salomon, D. (1987).Soil suppressiveness to *Fusarium* wilt of melon, induced by repeated croppings of resistant varieties of melons. *J. Phytopathol.* 120:347-354.

Sundheim, L. (1972). Physiologic specialization in *Aphanomyces euteiches. Physiol. Plant Pathol.* 2:301-306.

Sundheim, L. and Wiggen, K. (1972. *Aphanomyces euteiches* on peas in Norway. Isolation technique, physiologic races, and soil indexing. Sci. Rep. Agric. Univer. Norway 51.

Tu, J. C. (1987). Integrated control of the pea root rot disease complex in Ontario. *Plant Dis.* 71:9-13.

Verma, S. and Dohroo, N.P. (2002). Evaluation of fungicides against *Fusarium oxysporum* f.sp. *pisi* causing wilt of autumn pea in Himachal Pradesh. *Plant Disease Research* 17(2): 261-268.

Xue, A.G. (2003). Biological control of pathogens causing root rot complex in field pea using *Clonostachys rosea* strain ACM941. *Phytopathology* 93(3): 329-335.

Yager, L., Conner, R.L. and McLaren, D.L. 2003. Diseases of field pea in Manitoba in 2002. *Can. Plant Dis. Surv.* 83: 128-129.

Sustainable Disease Management of Agricultural Crops (2011) *Pages* **234–250**
Editors: **S.K. Biswas and S.R. Singh**
Published by: **DAYA PUBLISHING HOUSE, NEW DELHI**

Chapter 17

Important Diseases of Potato (*Solanum tuberosum*) and their Economic Management in Arunachal Pradesh

☆ *A. Kirankumar Singh*[1], *S.P. Singh1, M. Kanwat*[1], *R. Bhagawati*[1] *and S.R. Singh*[2]
[1]*ICAR Research Complex for NEH Region, Arunachal Pradesh Centre, Basar – 791 101*
[2]*Krishi Vigyan Kendra, Belatal, Mahoba – 210 423, U.P.*

The most important diseases of potato which reduces the production of potato drastically in Arunachal Pradesh are discussed and reported here.

Late Blight

Late blight is one of the most important potato diseases in the world. It was the cause of the great Irish Potato Famine of the 1840's and it continues to be a challenge to control in potato producing regions that have climatic conditions ideal for its development. Likewise, in Arunachal Pradesh also this disease reduces the production of potation great extent even though potato is not cultivated in large areas. From most of the areas where potato is cultivated the disease samples were collected and identified the causal organism, *Phytophthora infestans* as most virulent strains which changes adversely the picture of production.

Symptoms

Late blight is appeared on leaves, stems and tubers. Leaf symptoms appear as pale green water soaked lesions. These lesions are usually at the tips or margins of the leaves, but not throughout the growing period. The lesions grow into brown or purplish black lesions which sometimes have a yellow halo which also may not be present in all late blight infections. Lesions also appear on the petioles and stems as black greasy areas. Stem lesions may girdle the stem and kill the foliage above the lesion.

Disease Cycle

There are three conditions that were observed for late blight to occur:

1. Abundant late blight spores (inoculum),
2. A susceptible host (potatoes, tomatoes, correlated species), and
3. The environmental conditions favourable for late blight.

The pathogen survives from one year to the next on infected tubers. These may be curl tubers discarded as storage sheds are emptied for packing or processing, tubers left as volunteers in the field, chips from seed cutting operations or infected tubers planted as seed. All were found to serve as sources of inoculum. It was also observed that wind and air currents spread the spores over a wide area. Late blight spores have been observed to travel over 50-60 km under the favourable conditions. A small amount of inoculum can contaminate a large area very quickly. Because of this fast widespreading nature, if infections in a particular year is observed, it is very much sure that the inoculum will be present that are potential sources of inoculum for the next year growing season.

Environmental conditions must be conducive to disease development before the disease can develop. Humidity needs to be 90 per cent or above for spore development, conditions which can occur frequently inside the potato canopy. Night temperatures of 10-15 °C and day temperatures of 15-20°C are most favourable for disease development.Free moisture must be present on the plant in order for the spores to germinate and infect a new plant. Infection requires cool days to keep evapotranspiration low and frequent rainfall or overhead irrigation or a combination of both to provide

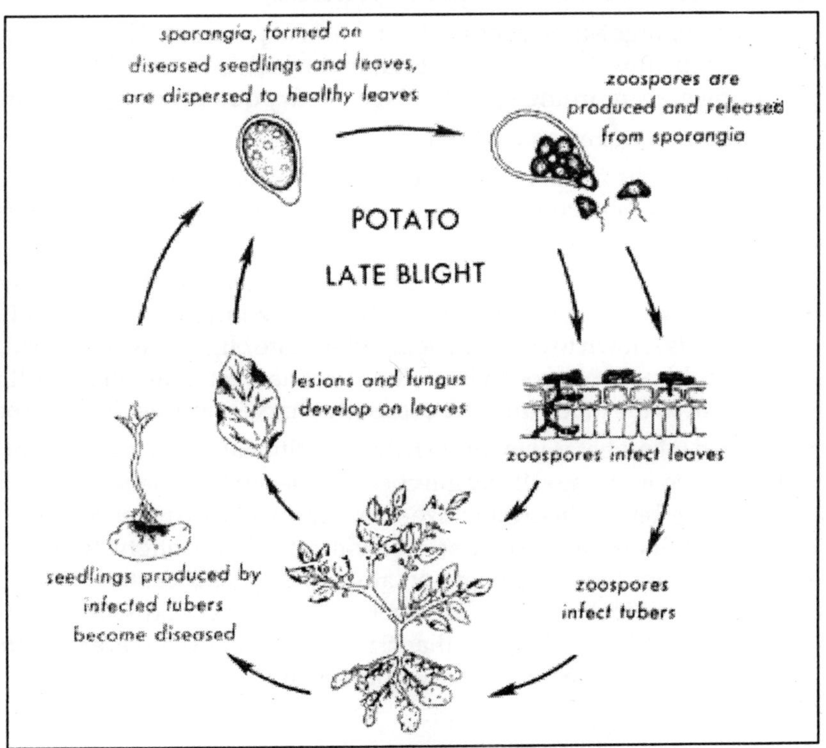

Figure 17.1: Late Blight Disease Cycle

long periods of free moisture over a 3-5 day period of time. To recap, the following environmental conditions are necessary for late blight development:

- ☆ Temperature–below 25°C
- ☆ Humidity–90 per cent and above
- ☆ Moisture–free moisture for 8-12 hours

The fungus can complete many reproductive cycles in a season, accounting for the rapid increase of disease once it becomes established in a field.

Tubers are infected by spores washed from lesions to the soil. Tuber infections are characterized by patches of brown to purple discoloration on the potato skin. Cutting just below the skin reveals a dark, reddish-brown, dry, corky rot.

Management Guidelines

The following guidelines will help reduce the risk of a late blight infection.

Cultural

Cultural practices are the first and foremost line of defense against this disease.

- ☆ Dispose of curll potatoes and seed chips properly by burying to a depth of 2 feet before any potatoes emerge in the spring.
- ☆ Avoid introducing late blight into a field by planting only disease-free seed tubers, preferably Certified seed (from late-blight-free production areas).
- ☆ Avoid frequent or night-time overhead irrigation of potatoes. This practice maintains leaf wetness and high humidity in the plant canopy, which is favourable for the disease *i.e.* try to water during early morning so that foliage can dry quickly during daytime hours.
- ☆ Avoid having waterlogged spots in fields.
- ☆ Hilling will reduce the incidence of tuber infection. The fungus infects tubers by washing through the soil and contacting the tubers. Good soil coverage provides better protection of the potato tubers.
- ☆ Harvest should not be started until vines are completely dead. At *least* one week, and as much as two to three weeks, should pass after vine killing for harvest of fields in an area where late blight is known to occur. In fields where late blight was confirmed, a minimum of two weeks should pass between vine killing and harvest. Late blight will not survive on dead vegetation, so the tubers that are exposed at harvest are less likely to be infected.
- ☆ Remove infected tubers before storage to reduce additional losses from soft rot. Tubers should be dry when placed in storage. If any infection is believed to be present, forced air ventilation through the storage bin can help minimize spread from tuber to tuber. Storage of seed potatoes with small amounts of late blight at 3°C will retard late blight tuber rot. The storage temperatures required for processing potatoes (8-11°C) make storing late blight-infected potato tubers very difficult since these temperatures also favour disease development. Potato lots with excessive tuber rot (greater than 5 per cent total decay) are probably not storable and should be sold or processed directly from the field.

Resistance

- ☆ No cultivar is immune to late blight and most cultivars planted in India are susceptible to late blight. However, some varieties offer partial resistance to this disease. These moderately resistant cultivars could be planted if blight was expected to be a problem. It should be assumed that all varieties grown in this geographic area are susceptible to late blight.

- ✰ A few varieties produced in India are resistant and may not express foliar symptoms of the disease very clearly but presents other hazards to production due to its association with Verticillium fungi.

Monitoring and Forecasting

- ✰ Fields should be monitored at least two times a week. Particular attention also should also be given to monitoring during the two weeks after a cool rainy period looking for late blight in susceptible fields in areas near the rivers.
- ✰ When a late blight infection is reported in a production area, it is important that all the fields in that area be sprayed with a protectant fungicide.
- ✰ Growers and fieldmen are encouraged to be well aware of current weather and disease forecasts. Weather and disease forecasting will be useful for the potato growers to take up the preventive measures in advance.

Chemical

- ✰ When disease is found in the field or is predicted by disease forecasting, fungicides may be applied. Research indicates that fungicide applications are most successful if they start when the canopy begins to close within the row.
- ✰ Apply protectant fungicides. Apply the protectant when foliage is 6 inches high and again just before earthing up. Apply additional fungicides according to the intensity of disease.
- ✰ Protectant fungicides should be used before development of disease in a field. If late blight is present in a field, a combination of protectant and systemic eradicant fungicides should be used.
- ✰ Applications should continue, as needed, throughout the growing season. Complete coverage is critical to the performance of protectant fungicides.
- ✰ The late blight fungus has shown the ability to develop strains that are resistant to some systemic/eradicant fungicides. Resistance to protectant fungicides has never been identified. Because of this threat, eradicant fungicides should always be applied in combination with protectants.

Prevention and Treatment

The first stage in control of the disease is prevention by good field management. Disease free potato seed should be used for planting and potato waste should be burned or treated with herbicides. Disease resistant varieties should be used when possible and farmers should keep abreast of news of outbreaks to select varieties and treatment. The pathogen is at its most virulent in areas with cool, damp climates or where the soil has become overwatered or over-irrigated so good management of soil water content becomes important when an outbreak is reported.

The infection can be treated by repeated spraying with fungicides like:

- ✰ Chlorothalonil
- ✰ Copper preparations such as Bordeaux mixture
- ✰ Mancozeb

- Mancozeb-metalaxyl mixtures
- Maneb
- Metalaxyl
- Ridomyl/Bravo
- TPTH.

Repeated spraying may be necessary and even resistant varieties of potato may need more than one application. Ground spraying is found to be more effective and economical, but aerial spraying is also required for controlling the disease at any cost.

Early Blight

Early blight, caused by *Alternaria solani* (E. and M) Jones and Grout, is a very common disease of potato and is found in most potato growing areas. Although it occurs annually to some degree in most production areas, the timing of appearance and rate of disease progress help determine the impact on the potato crop. The disease occurs over a wide range of climatic conditions and depends in a large part on the frequency of foliage wetting from rainfall, fog, dew, or irrigation, on the nutritional status of foliage and cultivar susceptibility. While losses rarely exceed 20 per cent, if left uncontrolled the disease can be very destructive. In contrast to its name, the disease rarely develops early and usually appears on mature foliage.

Symptoms

Foliar symptoms of early blight first appear as small, irregular to circular dark brown spots on the lower (older) leaves. These may range in size from a pinpoint to an eigth of an inch in diameter. As the spots enlarge, they become restricted by leaf veins and take on an angular shape. Early in the growing season, lesions on young, fully expanded succulent leaves may be larger up to half an inch in diameter and may, due to their size, be confused with late blight lesions. Leaf lesions are relatively easy to identify in the field because lesion development is characterized by a series of dark concentric rings alternating with bands of light tan tissue. A narrow band of chlorotic tissue often surrounds each lesion and extensive chlorosis of infected foliage develops over time. Elongated superficial brown or black lesions may also form on stems and petioles.

By the end of the growing season, the upper leaves of infected potato plants may be peppered with numerous small early blight lesions and subsequently lesions may coalesce to cover a large area of the leaf. Severely infected leaves eventually wither and die but usually remain attached to the plant. Severe infection of foliage by the early to mid-bulking period can result in smaller tubers, yield loss and lower tuber dry matter content.

Tuber symptoms of early blight include circular to irregular lesions that are slightly sunken and often surrounded by a raised purple to dark brown border. The underlying tissues are leathery to corky in texture, dry and usually dark brown. These lesions reduce the quality and marketability of fresh market tubers. Tuber infection also presents a challenge to processors as tuber lesions often require additional peeling to remove the darkened lesions and underlying tissues.

Disease Cycle

Early blight is caused by the fungus *Alternaria solani*. The dark colored spores and mycelium of the pathogen survive between growing seasons in infested plant debris and soil, in infected potato tubers and in overwintering debris of susceptible solanaceous crops and weeds including hairy

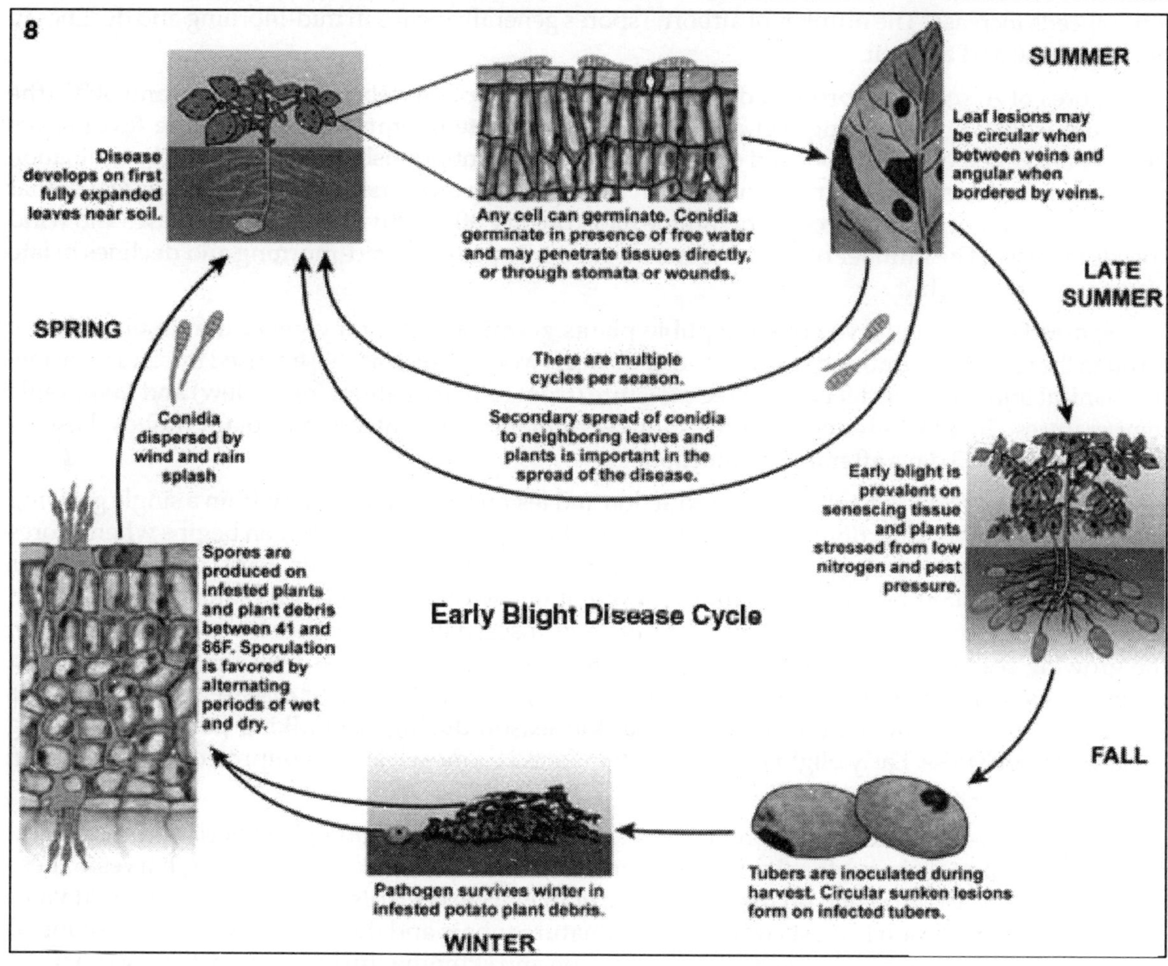

Figure 17.2: Early Blight Disease Cycle

nightshade (*Solanum sarrachoides*). Overwintering spores and mycelia of *A. solani* are melanized (darkly pigmented) and can withstand a wide range of environmental conditions including exposure to sunlight and repeated cycles of drying, freezing and thawing. In spring, spores (conidia) serve as primary inocula to initiate disease. Plants grown in fields or adjacent to fields where potatoes were infected with early blight during the previous season are most prone to infection, since large quantities of overwintering inoculum are likely to be present from the previous crop. Initial inoculum is readily moved within and between fields, as the spores are easily carried by air currents, windblown soil particles, splashing rain and irrigation water.

Spores of *A. solani* are produced on potato plants and plant debris between 5ºC and 30ºC (the optimum is 20ºC). Alternating wet and dry periods with temperatures in this range favor spore production. Few spores are produced on plant tissue that is continuously wet or dry. The dissemination of inoculums follows a diurnal pattern in which the number of airborne spore's increases as leaves that are wet with dew or other sources of night time moisture dry off, relative humidity decreases and wind speeds increase. The number of airborne spores generally peaks in mid-morning and declines in late afternoon and at night.

Spores of *A. solani* are produced on potato plants and plant debris between 5°C and 30°C (the optimum is 20°C). Alternating wet and dry periods with temperatures in this range favor spore production. Few spores are produced on plant tissue that is continuously wet or dry. The dissemination of inoculum follows a diurnal pattern in which the number of airborne spores increases as leaves that are wet with dew or other sources of nighttime moisture dry off, relative humidity decreases and wind speeds increase. The number of airborne spores generally peaks in mid-morning and declines in late afternoon and at night.

Spores landing on leaves of susceptible plants germinate and may penetrate tissues directly through the epidermis, through stomata and or through wounds such as those caused by sand abrasion, mechanical injury or insect feeding. Free moisture (from rain, irrigation, fog or dew) and favourable temperatures (20-30°C) are required for spore germination and infection of plant tissues. Lesions begin to form 2 to 3 days after initial infection.

Many cycles of early blight spore production and lesion formation occur within a single growing season once primary infections are initiated. Secondary spread of the pathogen begins when spores are produced on foliar lesions and carried to neighboring leaves and plants. Early blight is largely a disease of older plant tissues and is more prevalent on senescing tissues on plants that have been subjected to stresses induced by injury, poor nutrition, insect damage, or other types of stress. Early in the growing season the disease develops first on fully expanded leaves near the soil surface and progresses slowly on juvenile tissues near the growing point. The rate of disease spread increases after flowering and can be quite rapid later in the season during the bulking period and during periods of plant stress. Early blight lesions are often found on most leaves of unprotected plants late in the growing season.

In potato tubers, germinated spores penetrate the tuber epidermis through lenticels and mechanical injuries to the skin. Tubers often become contaminated with *A. solani* spores during harvest. These spores may have accumulated on the soil surface or may have been dislodged from desiccated vines during harvest. Infection is most common on immature tubers and those of white- and red-skinned cultivars, since they are highly susceptible to abrasion and skinning during harvest. Course-textured soil and wet harvest conditions also favor infection. In storage, individual lesions may continue to develop but secondary spread does not occur. Infected tubers may shrivel through excessive water loss, depending on storage conditions and disease severity. Early blight lesions on tubers, unlike late blight lesions, are usually not sites of secondary infection by other decay organisms.

Monitoring and Control

Effective management of this disease requires implementation of an integrated disease management approach. The disease is controlled primarily through the use of cultural practices, resistant cultivars and foliar fungicides.

Cultural Control

Cultural practices, such as crop rotation, removal and burning of infected plant debris, and eradication of weed hosts helps reduce the inoculum level for subsequent plantings. Since *A. solani* persists in plant debris in the field from one growing season to the next, rotation with non-host crops (*e.g.* small grains, corn or soybean) reduces the amount of initial inoculum available for disease initiation. Other cultural control measures may include the following.

- ☆ Avoid irrigation in cool cloudy weather and time irrigation to allow plants time to dry before nightfall.
- ☆ Use certified disease free seed.
- ☆ Use tillage practices such as fall plowing that bury plant refuse.
- ☆ To minimize tuber infection after harvest, tubers should be stored under conditions that promote rapid suberization as *A. solani* is unable to infect through intact periderm.

Resistant Cultivars

Cultivars with good levels of field resistance are available, however no immunity to early blight has been found in commercial potato cultivars or in their wild parents. Highly susceptible cultivars should be avoided in locations where early blight is prevalent and disease pressure is high. Field resistance to foliage infection is associated with plant maturity. Thus late maturing cultivars are usually more resistant than early maturing cultivars and therefore, one should avoid planting early and late cultivars in the same or adjacent fields.

Chemical Control

The most common and effective method for the control of early blight is through the application of foliar fungicides. Protectant fungicides recommended for late blight control (*e.g.* maneb, mancozeb, chlorothalonil, and triphenyl tin hydroxide) are also effective against early blight when applied at approximately 7-10 days intervals. The geographical spread of this resistance is not known but applications of strobilurins should be made in combination with tank-mixtures of the fungicides listed above. Other products that have shown efficacy against early blight include azoxystrobin, trifloxystrobin, famoxodone, pyrethamil, fenamidone, and boscalid.

The application of foliar fungicides is not necessary in plants at the vegetative stage, when they are relatively resistant. Accordingly, spraying should commence at the first sign of disease or immediately after bloom. The frequency of subsequent sprays should be determined according to the genotype and age-related resistance of the cultivar. Protectant fungicides should be applied initially at relatively long intervals and subsequently at shorter intervals as the crop ages.

Early season applications of fungicides before secondary inoculum is produced often have minimal or no effect on the spread of the disease. Early blight can be adequately controlled by relatively few fungicide applications if the initial application is properly timed. The first application for early blight control should be timed at 200 P days after emergence. Regular inspection of fields after plants reach 12 inches in height is recommended in order to detect early infections.

Rhizoctonia Stem Canker and Black Scurf of Potato

Rhizoctonia diseases of potato are caused by the fungus *Rhizoctonia solani* Kühn (teleomorph *Thanatephorus cucumeris* (A.B. Frank) Donk) and can be found on all underground parts of the plant at different times during the growing season. *Rhizoctonia solani* has many synonyms and is divided into subgroups called anastomosis groups (AG's), in which isolates are categorized according to the ability of their hyphae to anastomose (fuse) with one another. Three AG's of *R. solani* are prevalent and may be pathogenic to potato at some temperatures, but they generally cause little damage.

Rhizoctonia solani causes black scurf on tubers, and stem and stolon canker on underground stems and stolons, and occurs wherever potatoes are grown. However, *R. solani* causes economically significant damage only in cool wet soils. *Rhizoctonia* is a more consistent problem. Poor stands, stunted plants, reduced tuber number and size, and misshapen tubers are characteristic of diseases caused by *R. solani*.

Symptoms

The symptoms of the disease are found on both above and below ground portions of the plant. Black scurf, is the most conspicuous sign of Rhizoctonia disease. In this phase of the disease the fungus forms dark brown to black hard masses on the surface of the tuber. These are called sclerotia and are resting bodies of the fungus. Sclerotia are superficial and irregularly shaped, ranging from small, flat, barely visible blotches to large, raised lumps. Although these structures adhere tightly to the tuber skin, they do not penetrate or damage the tuber, even in storage. However, they will perpetuate the disease and inhibit the establishment of potato plants if infected tubers are used as seed.

Although black scurf is the most noticeable sign of *Rhizoctonia*, stem canker, is the most damaging of the disease as it occurs underground and often goes unnoticed. Early in the season, the fungus attacks germinating sprouts underground before they emerge from the soil. The sprout may be killed outright if lesions form near the growing tip. Damage at this stage results in delayed emergence and is expressed as poor and uneven stands with weakened plants. Reduction in crop vigor results from expenditure of seed energy used to produce secondary or tertiary sprouts to compensate for damage to primary sprouts. Occasionally, heavily infested potato seed tubers are unable to produce stems. Instead, the tubers will produce stolons with several small tubers. This symptom is referred to as "no top" and can be confused with the same symptom caused by physiologically old seed that has been de-sprouted.

Poor stands may also be mistaken for seed tuber decay, caused by *Fusarium* or soft rot bacteria, unless plants are dug up and examined. *Rhizoctonia* does not cause seed decay, damaging only sprouts and stolons. Poor stands and stunted plants can also be caused by blackleg, a bacterial disease that originates from seed tubers and progresses up stems, causing a wet, sometimes slimy rot. In contrast, *Rhizoctonia* lesions are always dry and usually sunken.

Stolons and roots can also be infected by the pathogen. Early in disease development, stolons, roots and stems have reddish-brown to brown lesions. As lesions mature, they become cankers that are rough, brown and can have craters, cracks or both. Damage varies and can be limited to a superficial brown area that has no discernible effect on plant growth to severe lesions that are large and sunken, as well as necrotic. If cankers are severe they may girdle the stem, interfering with the normal movement of water and carbohydrates throughout the plant.

Late season damage to plants is a direct result of cankers on stolons and stems causing problems with starch translocation. Stolon cankers also affect the shape, size and numbers of tubers produced. If stolons and underground stems are severely infected, the flow of starch from the leaves to the developing tubers is interrupted. This results in small, green tubers, called aerial tubers forming on the stem above the soil. Formation of aerial tubers may indicate that the plant has no tubers of marketable quality below ground. Interruptions in carbohydrate flow may also result in a stunting or rosetting of the plant. A leaf curl, which can be confused with symptoms of the Potato Leaf Roll Virus, has also been reported in severely infected plants.

Disease Cycle

Rhizoctonia diseases are initiated by seedborne or soilborne inoculum. The pathogen overwinters as sclerotia and mycelium on infected tubers, in plant residue, or in infested soils. When infected seed tubers are planted in the spring, the fungus grows from the seed surface to the developing sprout and

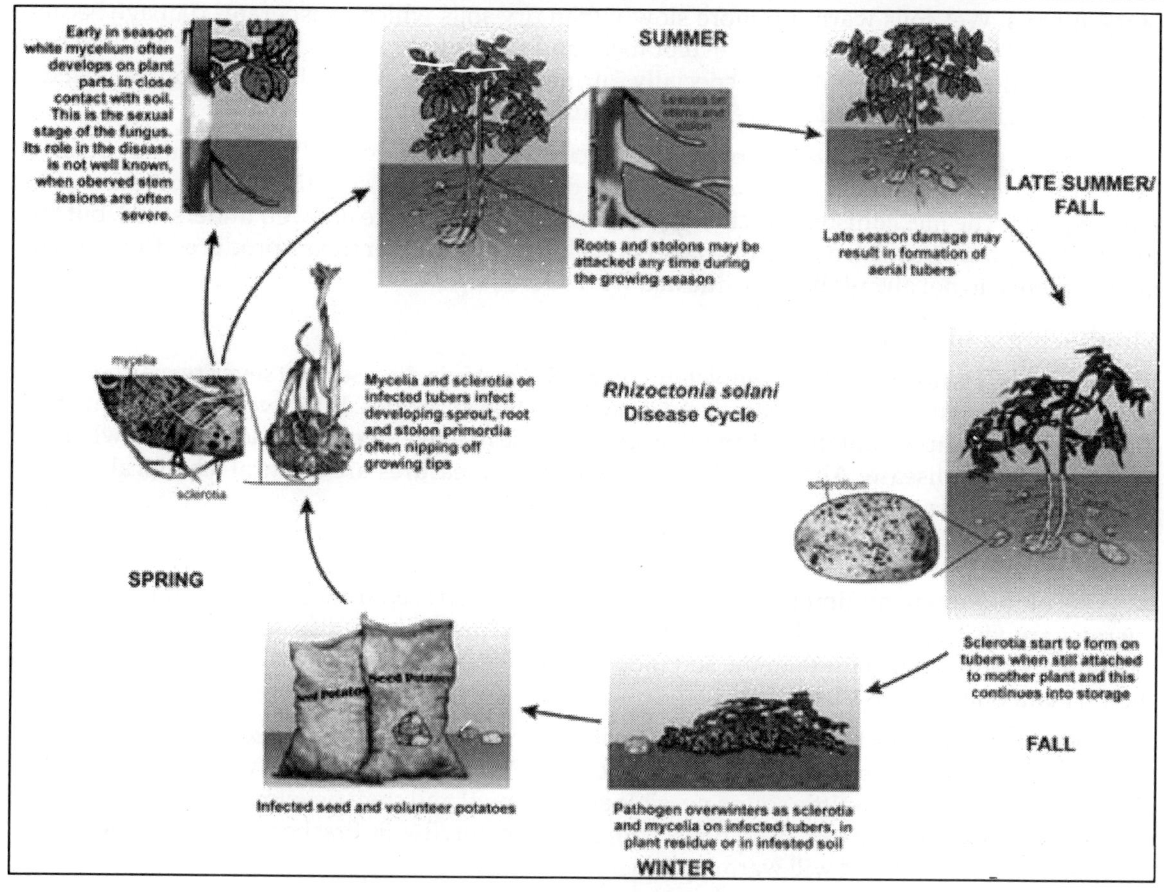

Figure 17.3: *Rhizoctonia solani* Disease Cycle

infection of root primordia, stolon primordia and leaf primordia can occur. Seed inoculum is particularly effective in causing disease because of its close proximity to developing sprouts and stolons.

Mycelia and sclerotia of *R. solani* are living on organic debris, and can cause disease independently of or in conjunction with seedborne inoculum. Soilborne inoculum is potentially as damaging as seedborne inoculum, but it can cause infection only when the plant organs develop in the proximity of the inoculum. Roots and stolons may be attacked at any time during the growing season, although most infections probably occur in the early part of the plant growth cycle. The plant's resistance to stolon infection increases after emergence, eventually limiting expansion of lesions.

Previous research has shown that soil temperature is a critical factor in the initiation of Rhizoctonia disease in potato, with disease severity being positively correlated with the temperature that is most favorable for pathogen growth. The optimal temperature range for the growth of *R. solani* AG-3 is 5 to 25ºC. Thus, plants will be most susceptible to infection when the soil temperatures are within this critical range. Cool temperatures, high soil moisture, fertility and a neutral to acid soil (pH 7 or less) are thought to favor development of Rhizoctonia disease. Damage is most severe at cool temperatures because of reduced rates of emergence and growth of stems and stolons are slow relative to the growth of the fungus. Wet soils warm up more slowly than dry soils which exasperates damage because excessive soil moisture slows plant development and favors fungal growth. However, it has been shown that high soil temperatures, especially during the early stages of plant development tend to minimize the impacts of *R. solani*, even when inoculum is abundant.

Sclerotia begin to form late in the season, principally after vine death. The mechanisms involved in sclerotial development on daughter tubers are different from those acting in the infection of the mother plant. The mechanisms which trigger sclerotial formation are not well understood, but they may involve products related to plant senescence. However, daughter tubers produced from infected mother plants do not always become infested with sclerotia.

Monitoring and Control

Currently it is not possible to completely control Rhizoctonia diseases, but severity may be limited by following a combination of cultural and crop protection strategies. Effective management of this disease requires implementation of an integrated disease management approach and knowledge of each stage of the disease. Although the most important measures are cultural, chemical controls should also be utilized.

Cultural Control

One of the keys to minimizing disease is to plant certified seed free of sclerotia. If more than 20 sclerotia are visible on one side of washed tubers, consider using a different seed source. Tuber inoculum is more important than the soil inoculum as the primary cause of disease. Seed growers should plant only sclerotia-free seed.

Following practices that do not delay emergence in the spring minimizes damage caused to shoots and stolons and lessens the chance for infection. Planting seed tubers in warm soil (above 7°C) and covering them with as little soil as possible speeds spout and stem development and emergence reduces the risk of stem canker. Plant fields with coarse-textured soils first because they are less likely to become waterlogged and will warm up faster.

Rhizoctonia does not compete exceptionally well with other microbes in the soil. Increasing the rate of crop residue decomposition decreases the growth rate of *Rhizoctonia*. Residue decomposition also releases carbon dioxide, which reduces the competitive ability of the pathogen. Since the fungus is not an efficient cellulose decomposer, soil populations are greatly reduced by competing microflora and less disease is observed.

Potatoes should be harvested as soon as skin is set so minimal bruising will occur. The percent of tubers covered with sclerotia increases as the interval between vine kill and harvest is lengthened. Vine removal or burning also reduces the amount of fungus overwintering and thus the amount of inoculum available to infect future potato crops. Do not dump infested tubers on future potato fields as they can become sources of inoculum.

Biological Control

There is growing evidence that a 'bio-fumigation' treatment based on incorporating a mustard cover crop is one way to reduce Rhizoctonia incidence. Mustard residues when incorporated into the soil release cyanide-containing compounds that fumigate the soil, but at the same time they also release carbon and nutrients that are the feedstock for soil organisms. Incorporating green cover crop tissues provides energy that supports the complex web of soil organisms that compete with parasite and disease organisms. Thus mustards, and related 'brassica' plant species such as oil-seed radish, do not leave a soil void of organisms. Instead, these cover crops tend to tip the balance in the favor of beneficial organisms and against parasites and pests.

Our preliminary research indicates that it is important to maximize growth of the cover crop using a high seed rate (7 kg/acre or more) and irrigation to improve establishment if rainfall is insufficient. A tiny seed such as mustard cannot be drilled too deep. It appears to establish well if broadcast and harrowed or irrigated into sandy soil. The bio-fumigation benefits of mustard residues are maximized if they are incorporated at or just before flowering. We suggest that residues be mowed and incorporated while still green. Mustards are rapid growing species and can become a weed in a subsequent crop, so it is important not to let this cover crop produce seed.

We are just beginning to understand the exact mechanisms involved in bio-fumigation using mustard cover crops. Initial results from research indicate that oriental mustard can be used as a cover crop to improve potato root and tuber health. The growth of *Rhizoctonia* was slowed by 90 per cent in soil amended with oriental mustard cover crop tissue compared to bare soil. A field experiment indicated that tubers of the tablestock variety had no observable signs of *Rhizoctonia* when grown after a spring cover crop of oriental mustard. Further research is required to learn more about management practices that optimize the bio-fumigation action of mustard cover crops, but initial results are promising and farmers are encouraged to experiment with brassica cover crops such as oriental and white mustard or oilseed radish to improve soil health.

Chemical Control

Seed Treatment

Several products have been specifically developed for control of seed-borne potato diseases and offer broad-spectrum control for Rhizoctonia, Silver Scurf, Fusarium Dry Rot and to some extent Black Dot (*Colletotrichum coccodes*). These include Tops MZ, Maxim MZ (and other Maxim formulations + Mancozeb) and Moncoat MZ. The general impact of these seed treatments is noted in improved plant stand and crop vigor but occasionally, application of seed treatments in combination with cold and wet soils can result in delayed emergence. The delay is generally transient and the crop normally compensates. The additional benefit of the inclusion of Mancozeb is for prevention of seed-borne late blight.

In-furrow Applied Fungicides

Application of fungicide in-furrow at planting has resulted in significant improvement in control of Rhizoctonia disease of potatoes. Products such as Moncut and Amistar applied in-furrow at planting have given consistent and excellent control of Rhizoctonia diseases of potatoes in trials. However, both seed treatments and in-furrow applications on some occasions have resulted in poor control of Rhizoctonia. This sporadic failure may be due to extensive periods of wet and cold soil shortly after planting or planting in fields with plentiful inoculum. Amistar applied in-furrow has been reported to reduce the symptoms of Black Dot on lower stems and tubers.

Bacterial Soft Rot and Blackleg

Cause

The bacterium, *Erwinia carotovora* sub-sp. *atroseptica*, is usually associated with blackleg and soft rot in storage while *E. c.* subsp. *carotovora* is associated with aerial stem rot, lenticel rot, and soft rot. Moist, cool (below 18°C) conditions enhance blackleg while warmer conditions (21-26°C) are optimal for soft rot. Blackleg can rapidly spread with rain or irrigation and by insects. The principal source of inoculum for blackleg is contaminated seed. The bacteria can be spread among seed pieces by machinery and handling.

Symptoms

- ☆ *Blackleg*–leaves curl up; foliage gradually yellows and dies early. At the stem base, mushy light brown to inky black lesions develop. Aerial tubers also may form on stems. Tuber rot usually begins at the stem end and results in a black, slimy rot.
- ☆ *Aerial stem soft rot*–symptoms are in the upper canopy. Lesions range from light brown to colorless. Stems still get very mushy and hollow and may be filled with mucilaginous slime. Often this disease is a secondary infection associated with stem infections by late blight.
- ☆ *Soft rot of tubers*–on tubers, lesions can be as small as a single eye or involve the entire tuber. The rot is extremely soft and colorless. Although rot of the soft-rot bacterium is relatively odor free, secondary organisms usually cause a foul smell.

Cultural Control

1. Minimize the bacterial contamination in seed tubers (Purchase seed free of bacterial soft rot damage).
2. Sanitize and disinfest seed-handling equipment.
3. Avoid injuring seed tubers and allow cut seed to heal before planting.
4. Crop rotation will help reduce the disease.
5. Avoid planting in overly wet or dry soil.
6. Avoid over-irrigation. More frequent irrigation for a shorter time is less favorable for disease.
7. Do not harvest until tubers are fully mature (after skins "set" and lenticels are closed).
8. Avoid tuber injuries.
9. Provide adequate air flow to promote drying, particularly when tubers first enter storage. Avoid packing and storing wet tubers.

Chemical Control

1. Treat seed with Seed Treatment for Potatoes at 0.45 kg/45 kg seed.
2. Firewall at 100 ppm. Soak cut seed pieces in solution for several minutes. 12-hr reentry.

Use a disinfectant when washing tubers to prevent lenticel infection. Spray a chlorine disinfectant such as sodium hypochlorite at rates to give 75 ppm available chlorine. Check water frequently to maintain proper disinfectant level. Dip treatments lose activity rapidly, and the dip solution then spreads the bacteria.

Black Dot

Black dot, caused by *Colletotrichum coccodes* (Wallr.) Hughes., is a common disease of potato. It is most often observed on tubers but it can affect all parts of the plant. The disease has probably been underestimated in the recent past as the symptoms are similar to more common potato diseases. On potato foliage symptoms are nearly indistinguishable from early blight and on tubers it produces blemishes that are easily mistaken for silver scurf. Although not as serious as other more common potato diseases such as black scurf, silver scurf, or common scab, it can be more devastating as it affects all parts of the plant. Above ground it can infect the vascular system causing wilt, and below ground it can cause severe rotting of roots, shoots and stolons, leading to early plant decline, discolored tubers and reduced yields.

Symptoms

Black dot is named after the abundant black dots that form on tubers. The black dots are microsclerotia that are often just visible to the naked eye. They are not restricted to tubers and can also be found on stolons, roots and stems both above and below ground. Foliar symptoms are not often observed but this may be due to the fact that they bear a resemblance to the small brown to black flecks characteristic of early blight. Stem lesions are more common than foliar symptoms, especially towards the end of the growing season. Stem lesions tend to form around the base of leaf petioles. As with leaf lesions they initially start out as small brown flecks. These gradually coalesce forming lesions which may girdle the stem. As lesions mature they develop circular to irregularly shaped, white to straw-colored centers with wide margins that vary in color from brown to black. Microsclerotia form in the center of the lesions and are often clearly visible against the pale background. As infected tissue senesces, microsclerotia may become abundant covering the entire surface of the stem. Microsclerotia often appear at the base of the plant up to several inches above soil level late in the season and after vine kill.

Lesions on below ground stems and stolons may resemble Rhizoctonia lesions but are darker in appearance. Infection of root cortical tissue causes sloughing of the periderm and may result in severe rotting and early plant death. Microsclerotia and mycelia may also be abundant both internally and externally on roots and stolons.

Tuber symptoms appear as a brownish to gray discoloration over a large portion of the tuber or as circular to irregularly shaped areas. Black dot may develop a silvery sheen during storage, which can be confused with silver scurf. However, black dot tends to show much more irregularly shaped patches with less well-defined margins than silver scurf. Inspection with a hand lens (10x) will quickly differentiate the regularly spaced black dots from the bunched threads of silver scurf.

Disease Cycle

Black dot is caused by the fungus *Colletotrichum coccodes*. The pathogen has a wide host range, occurring on other plants in the Solanaceae family including eggplant, pepper, tomato and weeds such as hairy nightshade (*Solanum sarrachoides*). *Colletotrichum coccodes* readily produces microsclerotia on scenescing plant tissue and the surface of tubers. These structures allow the pathogen to survive for long periods of time in the soil. In Michigan, the pathogen survives between growing seasons as sclerotia in infested plant debris and soil, on infected potato tubers and in overwintering debris of susceptible solanaceous crops and weeds. In the spring, sclerotia develop into acervuli. These are fruiting bodies which produce masses of spores. Spores serve as the primary inocula to initiate disease. Initial inoculum is readily moved within and between fields, as the spores are easily carried by air currents, windblown soil particles, splashing rain and irrigation water. Poor soil drainage and low plant fertility are thought to increase disease.

Spores of *C. coccodes* are produced on potato plants and plant debris between 7 and 35°C. However, in greenhouse studies limited infection occurred below 15°C. As with most *Colletotrichum* species, *C. coccodes* favors temperatures above 20°C, and free moisture (from rain, irrigation, fog or dew) are required for spore germination and infection of plant tissues. Spores landing on susceptible plant tissue germinate and may penetrate tissues directly through the epidermis and or through wounds

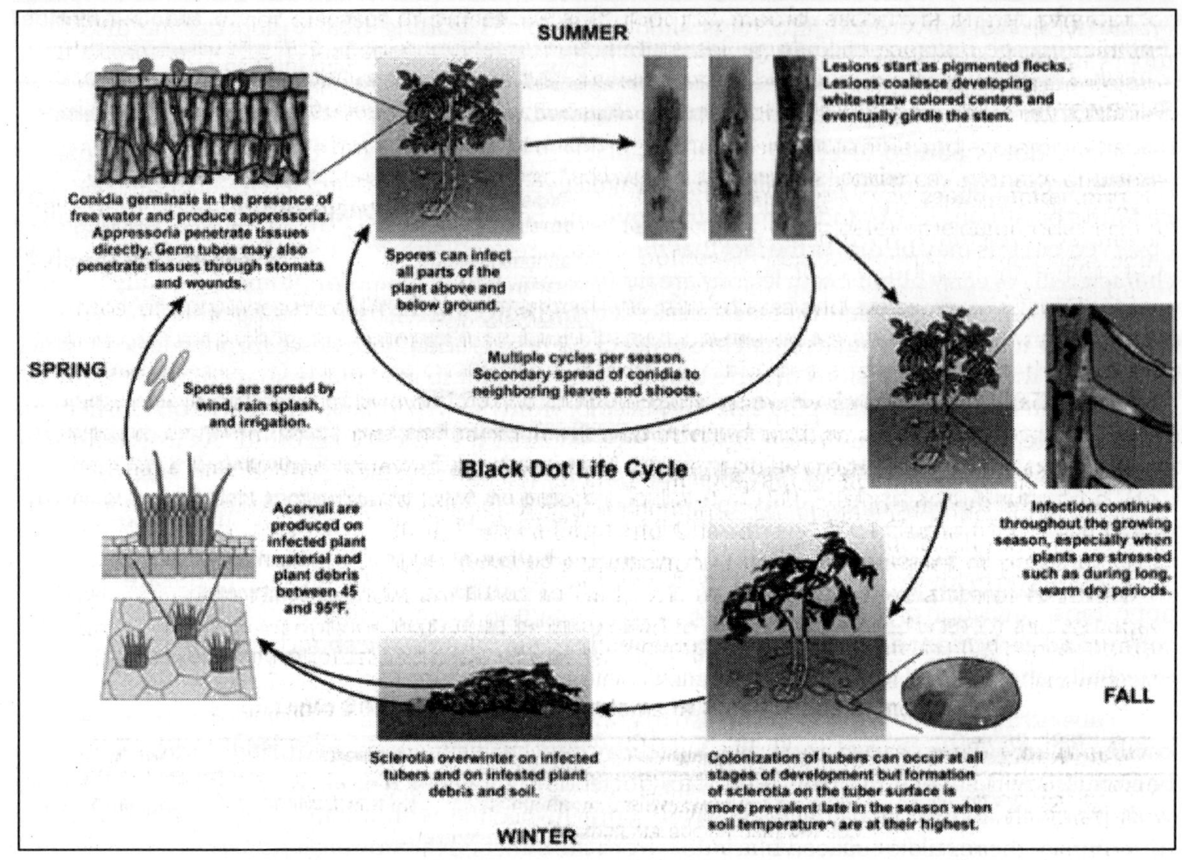

Figure 17.4: Black Dot Disease Cycle

such as those caused by mechanical injury or insect feeding. On tomato leaves, C. *coccodes* has been reported to colonize lesions caused by *Alternaria solani* (early blight) and flea beetles.

Infection of below-ground plant parts continues throughout the growing season, especially when plants are under stress, such as during long, warm dry periods. Production of microsclerotia is greatest at high temperatures and increasing disease incidence has been associated with increasing soil temperatures. Colonization of tubers can occur at all stages of development, but formation of sclerotia on the tuber surface is more prevalent late in the season when soil temperatures are at their highest.

Monitoring and Control

Effective management of this disease requires implementation of an integrated disease management approach. The disease is controlled primarily through the use of cultural practices and foliar fungicides.

Cultural Control

One of the most important approaches to black dot control is to reduce the amount of inoculum in soil through the use of cultural practices such as crop rotation, removal of crop debris, volunteer and cull potatoes from the field, and eradication of weed hosts. Since black dot microsclerotia can persist in the field for up to 2 years, a 3 to 4 year rotation with non-host crops (*e.g.* small grains, soy bean or corn) is often recommended to reduce the amount of inoculum in the soil. The following cultural practices are also suggested to prevent and reduce the incidence of black dot.

Use Certified Disease Free Seed

- ☆ Treat cut seed with a seed treatment (*e.g.* Maxim MZ). Although seed treatments may provide limited control of black dot, they improve plant stand and crop vigor, reducing plant stress which increases susceptibility to black dot.
- ☆ Avoid planting in poorly draining soil if possible.
- ☆ Use good crop production practices, such as timely irrigation and adequate fertilization to reduce crop stress.
- ☆ Use tillage practices such as fall plowing that bury plant refuse and encourage decomposition.
- ☆ Harvest tubers as soon as possible after vine kill.
- ☆ Control temperature and humidity in storage. High temperatures and condensation on the tuber surface promotes disease.

Resistant Cultivars

Currently there are no known commercial varieties with resistance to black dot. However, research has shown that in general late maturing varieties tend to be more vulnerable to yield reductions than early maturing varieties. This may be due to the fact that black dot is a late season disease, and leaving tubers in the ground longer exposes them to more disease pressure.

Chemical Control

In furrow applications of azoxystrobin have been reported to reduce or suppress symptoms of black dot on the stem although under conditions conducive for development of black dot, variable results have been reported. Maxim, Headline and Blocker have also been reported to suppress black dot.

References

Chand, Sudeep (9 September 2009). "Killer genes causes potato famine ". BBC News.

Chemical and Biological Weapons: Possession and Programs Past and Present", *James Martin Center for Nonproliferation Studies*, Middlebury College, April 9, 2002.

Glass, J. R., Johnson, K.B. and Powelson, M.L. (2001). Assessment of Barriers to Prevent the Development of Potato Tuber Blight. *Plant Disease*, 85 (5): 521–528.

Haas, Brian, *et al.* (17 September 2009). Genome sequence and analysis of the Irish potato famine pathogen Phytophthora infestans. *Nature*, 461: 393-398.

James, W.C. (1974). Assessment of Plant Diseases and Losses. *Annual Review of Phytopathology*, 12(1): 27–48.

Koepsell, Paul A. and Pscheidt, Jay W. (1994). *1994 Pacific Northwest Plant Disease Control Handbook*, Corvallis: Oregon State University Press, p. 165.

MacKenzie, D.R. (1981). Scheduling fungicide applications for potato late blight with Blitecast. *Plant Disease*, 65: 394–399.

Moskin, Julia (July 17, 2009). Outbreak of Fungus Threatens Tomato Crop. *The New York Times.*

Mukerji, K.G. (2004). *Fruit and Vegetable Diseases.* Kluwer Academic Publishers, Boston, p. 196.

Novy, R.G. and Love, S.L. *et al.* (2006). Defender: A high-yielding, processing potato cultivar with foliar and tuber resistance to late blight. *American Journal of Potato Research,* 83(1): 9–19.

Reader, John (March 17, 2008). The Fungus That Conquered Europe. *New York Times.*

Song, Junqi, Bradeen, James M., Naess, S. Kristine, Raasch, John A., Wielgus, Susan M., Haberlach, Geraldine, T., Liu, Jia, and Kuang, Hanhui *et al.* (2003). Gene RB cloned from *Solanum bulbocastanum* confers broad spectrum resistance to potato late blight. *PNAS,* 100(16): 9128–9133.

Suffert, Frédéric, Latxague, Émilie and Sache, Ivan (2009). Plant pathogens as agroterrorist weapons: assessment of the threat for European agriculture and forestry. *Food Security,* 1(2): 221–232.

Zwankhuizen, Maarten J., Govers, Francine and Zadoks, Jan C. (1998). Development of potato late blight epidemics: Disease foci, disease gradients, and infection sources. *Phytopathology,* 88(8): 754–763.

Sustainable Disease Management of Agricultural Crops (2011) *Pages* **251–257**
Editors: **S.K. Biswas and S.R. Singh**
Published by: **DAYA PUBLISHING HOUSE, NEW DELHI**

Chapter 18

Diagnosis and Management of Collar Rot of Lentil

☆ ***S.R. Singh[1], R.K. Pandey[2], P.K. Gupta[3], Shashi Mala[4] and Yogita Gharde[5]***

[1]*Krishi Vigyan Kendra, Belatal, Mahoba – 210 423, U.P.*
[2]*Division of ACB and PDM, Amity University, Noida – 201 303, U.P.*
[3]*Division of Plant Pathology, Indian Agricultural Research Institute, New Delhi – 110 012*
[4]*Aligarh Muslim University, Aligarh*
[5]*IASRI, New Delhi – 110 012*

Introduction

Lentil (*Lens culinaris* Medik.) is the second most important *Rabi* legume pulse crop of India and in many parts of the world, cultivated under various production systems (Chaudhary *et al.*, 2009). India alone has nearly 31.5 per cent of the world acreage lentil cultivation. The area under lentil cultivation is 1.50 million hectares with the production of 0.97 million tones and productivity of 652kg/ha (FAO, 2008). Lentil crops are affected by a number of diseases caused by fungi, bacteria, viruses and nematodes. Some diseases are common in most lentil-growing regions worldwide, whereas others are limited to certain production area. The potentially damaging diseases of lentil are wilt, dry root rot, wet root rot, collar rot, rust, ascochyta blight, botrytis grey mould, anthracnose, stemphylium blight, sclerotinia stem rot and downey mildew (Chen, *et al.*, 2009). Among these, collar rot caused by *Sclerotium rolfsii* Sacc. is wide occurrence in lentil growing areas of the country and has been poising a major threat of its cultivation. Under favourable condition collar rot may cause significant economic losses and cause significant reduction in both grain yield and quality. Hence, proper management of the disease is necessary to ensure sustainable productivity and profitability of lentil.

In India, collar rot of lentil was reported for the first time by Butler and Bisby in 1931 from Gwalior region of Madhya Pradesh and in Uttar Pradesh by Pagvi and Upadhhyay, (1976). The disease appears at seedling stage of the crop plant and has been reported to cause 50 per cent loss in farmer's field in Madhya Pradesh, India, Khare *et al.* (1979). Attempts have been made to include the most recent information on the disease; however, the readers are also referred to previous reviews by Khare (1981), Agrawal and Prasad (1997), Tikoo, *et al.* (2005) Davidson *et al.* (2007) and Taylor *et al.* (2007).

Disease Symptoms

The pathogen attacks young seedlings at the collar region. Plants infected at the advanced stage gradually turn yellow pale, drop and dried. On the affected part of the stem, white feathery mycelial growth development. Infected plants were easily pulled out from the soil as the root system poorly developed and side roots were badly destroyed. On the all infected tissues and even the nearly soil, the pathogen produce numerous small radish sclerotia of uniform size that are white when immature, becoming dark brown to black on maturity. Each sclerotium is differentiated in to an outer maladies rind, middle cortex and inner most of loosely arranged hypae (Singh, *et al.*2006).

Etiology of the Pathogen

The fungus of collar rots of lentil *Sclerotium rolfsii* Sacc. produces abundant white fluffy, mycelium.The colonies of fungus were fast growing white with sclerotial margin very smooth, covering full plate within 4-6 days. Mycelium is hyline, branched, septate, sparsely, slender and thin walled with clamp connection at the septum. Hymenium are a dull white radiating mass, profusely branched and forming a network of short, stout broad basidial initials. Basidium is thin walled, hyline, broadly clavate and 14.3-16× 5.2-6.7µ in size. Sclerotium spherical, mustard seed like and light brown initially; becoming dark brown with age. Its perfect stage is *Aethalium rolfsii* (Khare, 1980).

Development of Disease

The development of disease spread mainly through overwintering sclerotia. Spread of sclerotia through moving water, contaminated tools, infected transplant seedlings, infected fruits and vegetables, some hosts and sclerotia mixed with the seed. The pathogen attacks directly tissues of the plant. *Sclerotium rolfsii* secretes oxalic acid and also cellulytic and pectinolytic and other enzymes and it kills and disintegrates tissues before it actually penetrate the host. Pathogen once established in the plants, fungus advances and produces mycelium and sclerotia quite rapidly at high moisture and temperature (Saxena and Khare, 1994)

Management of the Disease

Cultural Management

The control of soil-borne diseases such as the collar rot caused by *Sclerotium rolfsii* Sacc. has always been a problem.Treatment of soil and seed is not sufficient to liminate the disease from field therefore, some cultural practices like crop rotation, mixed cropping and use of nitrogenous fertilizers reduce the incidence of disease.

Crop Rotation

Crop rotation is the best method to eliminating collar rot infection in field conditions. However, since the pathogen survives on deep seated root of the crop plant. The success of crop rotation depend on field condition especially sanitation of the field. Deep ploughing during summer season. Four to five year crop rotation has been found to free the field from the *Sclerotium rolfsii.*

Mixed Cropping

Mixed cropping with non legumes crop provides the most to solve of this problem. Mixed cropping with linseed, mustard and garlic is widely adopted in India. Several experiment conducted over mixed cropping had shown that lower the incidence of collar rot fungus than sole crop. Many reasons were suggested for this incidence of collar rot such as chances of contact between host and pathogen were reduced and root exudates of linseed and mustard act as barrier for the pathogen.

Use of Nitrogenous Fertilizers

The evidence has been adduced on the importance of both level and source of nitrogen in reducing the losses due to the disease in field condition. Thakur and Mukhopadhayay (1972).The high level of nitrogen did not affect sugar recovery from lentil.

The mechanism of nitrogen action, both in form and level was studied in field condition and it was shown that lower level of NO_3 or NH_4 (5Oppm) have inhibitory effect while higher level were directly toxic to sclerotial germination. Effects of nitrate were not dependent on soil pH. A wide variety of nitrogenous substances including inorganics like NO_3 or NH_4, Ca $(No_3)_2$ and organics like chitin, peptone and ammonium acetate affect germination. (Henis and Chet, 1968). Such inhibition of germination was attributed to build up antibiotic producing bacteria in the mycosphere, release of NH_3 that temporarily raised soil pH to above 8.5 and caused direct toxicity to pathogen.

Use of Inorganic Fertilizers

A pot experiment was carried out to determine the influence of inorganic fertilizers on collar rot lentil (*Lens culinaris*) caused by *Sclerotium rolfsii.* The pots were filled with autoclaved soil and inoculated with multiplied on straw + wheat bran @ 10 gram/pot. Urea @ 80 kg/ha, ammonium sulphate @ 50 kg/ha, calcium nitrate @ 40 kg/ha, zinc sulphate @ 50 kg/ha and calcium carbonate @ 40 kg/ha were mixed in inoculated soil of pot before sowing. Amongst all fertilizers, Urea @ 80 kg/ha gave maximum collar rot incidence in comparison to others (Singh *et al.,* 2008).

Soil Solarization

Solar heating of soil (solarization) is a new method for disinfestations, developed in India and tried or adapted in at least 13 countries, since 1976. Moistened soil is covered (trapped) for about one month with transparent polyethylene during the hot season, thereby increasing soil temperatures (Katan, 1981). In many countries, Israel, USA, Greece, Jordan, India, Pakistan etc, soil solarization is used to control several diseases caused by *Verticillium, Rhizoctonia solani, Fusarium oxysporum* f. sp. *vasinfectum* or *melonis, Orobanche, Sclerotium rolfsii, Pratylenchus* spp.and also weeds. Yield was greatly influence soil solarization but, depending on soil infestation. An extended effect in disease control and yield increase in the second year after solarization was found with diseases of cotton (*Fusarium* spp. and *Verticillium*), melons *Fusarium* spp. and others. Biological control, enhanced, in certain cases, by solarization contributes to pathogen control. Computerized models for predicting solar heating were also developed in many countries. Increased plant growth response in various soils without known pathogens was recorded. This was often correlated with release of mineral nutrients and organic substances in the soil solution. Combining solarization with chemical or biological control agents improved and extended the control. Solarization is not effective against certain diseases or in cold regions (Grindrat, 1979). It is a no chemical control method with advantages and limitations, which, under the right circumstances, is effective, simple and economic. Possible negative side effects should be looked for and investigated (Raj and Sharma, 2005).

This method is very most effective to control of several soil borne pathogens especially *Sclerotium rolfsii.*More precise work showed that moistened soil trapped for 19 days with 20–30 μm polythene, heated in the month of May–June sufficiently to sclerotial up to 20–30 cm depths and mortality was completed when the period was extended for another 3-4 weeks. (Elad and Cheet,1980). Subsequently, drip system of irrigation under trap was developed that expedited mortality of fungus sclerotia and increased yield by 42-65 per cent (Grinsten *et al.*, 1979). The mechanism of action may be thermal death or stimulation of antagonistic micro flora in the mycosphere of sclerotia.

Biological Control of the Disease

Biological control of plant disease is any condition or practice under which survival activity of the pathogen is reduced through the agency of any other living organism. A variety of the antagonist have also been identified and their efficacy in managing disease (Abeysinghe, 2009).

Mechanism of Pathogen Control

There are four mechanisms to control plant pathogen.

Competition

Micro organism competes for space, minerals and organic nutrients to proliferate and survive in their natural habitat. Successful biological control of root and bud rot of conifers, *Heterobasidium annosum* by fungal antagonist *Peniophora gigantean* operate on the basis of competition.

Antibiosis

It is defined as antagonism mediated by specific metabolites of microbial origin by lytic agents, enzymes, volatile compound or other toxic substance, As per example *Gliocladium virens* which produces viridian and gliotoxin is responsible for death of sclerotia of *Sclerotium rolfsii.*

Predation

In biological control, predaceous fungi feeding almost exclusively on free living micro organism. Some micro organism capture with an adhesive material which covers the entire surface of hypae.

Parasitism

Some sort of etiological relationship between parasite and host is not rapidly eliminated. Hyhpae of majority of *Trichoderma* spp. coil around hypae of various pathogenic fungi which ultimately causes death of fungi. A model system for assaying soil enzymes was developed to evaluate interaction between the pathogen and antagonist *Tricoderma viride.* Rodriguez-Kabana (1969). It was shown that the high saccharine activity of the pathogen was inhibited enzyme protease produced by the antagonist (Rodriguez-Kabana *et al.*, 1978). Another effect of the antagonist was rise in soil pH that adversely affected the enzymic potentials of the pathogen. Those results shows that antagonists act by the affecting the hydrolytic properties of the pathogen in soil and not by antibiosis. Chet *et al.* (1979) reported that lysis of hypae is occurred due to mediation of, mediated through extra cellular ß-(1-3) glucanase cutinase and luminaries. They also reported that, antibiotic action appeared to be important in control of the southern blight of arhar.

Factors Affecting the Efficacy of Bioagents

There are few report of success of biological control in field condition. Agrawal *et al.* (1977) reported that application of antagonist as a seed dressher is most effective for management of lentil wilt.With several approaches to control integration among these was attempted to improve efficacy of management. In soil *Trichoderma viride* become active with nitrogen. A *Trichoderma harzianum* in wheat bran was more effectively to reduce *Sclerotium rolfsii.* It was also suggested that biological control of this pathogen can be achieved by altering the permeability properties of sclerotia with the help sublethal dosage of CS_2 followed by drying. *Trichoderma harzianum* has emerged as the most antagonists of *Sclerotium rolfsii* (Singh *et al.*, 2007).

Table 18.1: List of Other Important Diseases of Lentil and their Causal Organism Important

Disease Name	*Pathogen*	*Reference*
Alternaria blight	*Alternaria alternate* (Fries) Keissler	Gupta and Das (1964): Kaiser (1992)
Aphanomyces root rot	*Aphanomyces euteiches* C. Drechsler	Lamari and Bernier (1985)
Bacterial leaf sport	*Xanthmonas sp.*	Agrawal and Prasad (1997)
Bacterial root rot	*Pseudomonas radiciperda*	Agrawal and Prasad (1997)
Black root rot	*Fusarium Solani* (Mart.) Sacc.	Karahan and Katrcoglu (1993)
Black Streak Root rot	*Thielaviopsis basicola* (Berk. and Broome) Ferraris	Bowden *et al.* (1985)
Cercospora leaf spot	*Cercospora lensii* Sharma	Sharma *et al.* (1978)
Collar rot	*Sclerotium rolfsii* Sacc.	Khare (1981) Beniwal *et al.* (1993)
Cylindrosporium leaf spot and stem canker	*Cylindrosoprium* sp.	Bellar and Kebabeh (1983)
Downey MIldew	*Perenospora lentis* Gaumann	Mittal (1997)
Dry root rot	*Macrophomina phaseolina* (Tassi) Goidanich	Kaiser (1992); Vishunavat and Katrcoglu (1993)
Helminthosporium leaf spot	*Helminthosporium* sp[a]	Karahan and Katrcoglu (1993)
Leaf yellowing	*Cladosporium herbarum* (Pers.) Link	Kaiser (1992); Vishunavat and Shukla (1979)
Phoma leaf spot	*Phoma medicaginis* Malbr. and Roum	Kaiser (1992)
Pythium root and seedling rot	*Pythium ultimum* Trow, *Pythium* spp.	Paulitz *et al.* (2004)
Reniform nematode	*Rotylenchulus reniformis* Linford and Oliveria	Gaur and Mishra (1990:) Anvar *et al.* (1991)
Wet root rot	*Rhizoctonia solani* Kuhn, telomorph: *Thanatephorus cucumeris* (Frank) Donk	Kaiser (1992); Vishunavat and Shukla, (1979)

a: Pathogenicity not confirmed.

Concluding Remarks

This chapter has described the major aspect of the economically important disease *i.e.* Collar rot of lentil. Certainly, not all diseases have been discussed because of space constraints. Lentil is an important crop for the resources-poor farmers in the developing countries. Therefore, disease control measures must be practicable, affordable and suitable for situation in these countries. In addition, agricultural researchers, breeders, extension worker and growers must continually adjust disease management strategies to cope with changing conditions.

References

Abeysinghe, Saman (2009). The effect of mode of application of *Bacillus subtilis* CA32r on control of *Sclerotium rolfsii* on *Capsicum annuum*. *Arch. of Phytopath and Plant Prot.* 42: (9)835-846.

Agrawal, S.C and Prasad, K.V.V (1997). *Disease of Lentil.* Science Publisher Inc., Enfield, New Hampshire.

Agrawal, S.C., Khare, M.N and Agrawal, P.S (1977). Biological control of *Sclerotium rolfsii* causing collar rot of lentil. *Phytopath.*30:176-179.

Anver, S., Tyagi, S.A., Yadav, A and Alam, M.M (1991). Interaction between *Rotylenchulus reniformis* and *Macrophomina phaeolina* on lentil. *Pakistan Journal of Nematology* 9, 127-130.

Bellar, M and Kababeh, S (1983). A list of disease and injuries and parasitic weed of lentil in Syria, Survey 1997-1980. *LENS Newsletter* 10, 30-31.

Beniwal, S.P.S., Bayaa, B., Weigand, S., Makkouk, K. and Saxena, M.C. (1993). *Field Guide to Lentil Diseases and Insect Pests.* International Centre for Agricultural Research in the Dry Areas (ICARDA), Aleppo, Syria.

Bowden, R. L., Wiese, M.V, Crock, J.E. and Auld, D.L. (1985). Root Rot of Chickpeas and Lentils caused by *Thielaviopsis basicola. Plant disease* 69, 1089-1091.

Butler, E.J. and Bisby, G.R. (1931). The Fungi of India. *Indian Sci. Monograph.* 1: 237

Chaudhary, R. G., Dhar, Vishwa and Singh, R. K. (2009). Association of fungi with wilt complex of lentil at different crop growth stages and moisture regimes. *Archives of Phytopath and Plant Protection*, 42: (4). 340-343

Chen, Weidong. Basandari, A. K, D. Basandari, S. Banniza, B. Bayaa, L. Buchwaldt, J. Davidson, R. Larsen, D. Rubiales and P. W. J. Taylor (2009). The lentil Botany, Production and Uses (eds W. Erskine *et al.*). *CAB International*, 2009. Pp: 262-281.

Chet, I., Hadar, Y., Elad, Y. and Henis, Y. (1979). In: Schippers B. and W. Gams Eds. *Soil borne plant pathogens.* Academic press, London, pp 585-591.

Elad, Y., Katan, J. and Chet.I. (1980). Physical, biological and chemical control integrated for soilborne diseases in potatoes *Phytopatholgy*, 70: 418-422.

Grindrat, (1979). In: Schippers, B. and W. Gams Eds. *Soil borne plant pathogens.* Academic press, London, pp.537-552.

Grinstein, A., Elad, Y., Katan, Razik, A.A., Zeydan, O, and Elad, Y. (1976). *Plant Dis.Rep.*63:1056-1059.

Haleem M.A., Madhwa Raj, M.S. and Rammanna, N. (1971). *Poultry Guide*, 8:13-15.

Henis, Y. and Chet, J. (1968). The effect of nitrogenous amendments on the germinability of sclerotia of *Sclerotium rolfsii* and on their accompanuing microflora. *Phytopatholgy*, 58:209-211.

Katan, J. (1981). Solar heating solarization of soil for control of soilborne pests. *Ann.Rev. Phytopath.*19:211-236.

Khare M.N. (1980). Wilt of lentil. J.N.K.V.V., Jabalpur, Madhya Pradesh, India. p 15

Khare, M.N., Agrawal, S.C. and Jain, A.C. (1979). *Technical Bull.* JNKVV, Jabalpur, M.P., 29pp

Mukopadhyay, A.N. and Thakur, R.P. (1977). Effect of some soil insecticides on Sclerotium root rot of sugarbeet. *Indian J.Agric. Sci.* 47:533-536.

Pagvi, M.S, and Upadhyay, R. (1976). Sci. and Cult.33:75-76.

Papavizas, G.C. and Lewis, J.A. (1979). In: Schippers, B. and W. Gams Eds. *Soil borne plant pathogens.* Academic press, London, pp.483-507.

Paulitz, T.C, Dugan, F., Chen, W. and Grunwald, N.J. (2004). First report of *Pythium irregular* on lentils in the United States. *Plant Disease* 88, 310.

Raj, H.S. and Sharma. D. (2005). Integrated Management of Collar and Root Rot (*Sclerotium rolfsii*). of Strawberry *Acta Horticulturae* Vol. No.: 696

Rodriguez-Kabana, R. (1969). Enzymatic interactions ol *Sclerotium rolfsii* and *Trichoderma viride* in mixed soil culture. *Phytopatholgy*, 59:910-921.

Rodriguez-Kabana, R., Kelley, W.D and Curl, E.A. (1978). *Can.J.Microbial.* 24: 487-490.

Saxena DR, Khare MN. 1994. Influence of soil environment on mortality of lentil due to soil borne pathogens. Ind Phytopathol 47:60–94.

Sharma, N.D., Mishra, R.P. and Jain, A.C. (1978). *Cercospora lensii* sp. Nov.: a new species of Cercospra on lentil. *Current Science* 47, 774-775.

Singh, S.R., Jain, Amit Kumar, Rathi, Brijesh Kr. and Singh, O.P. (2008). Effect of inorganic fertilizers on collar rot of lentil caused by *Sclerotium rolfsii*. *Agri. Sci. Digest.* 28(4): 309.

Singh, S R, R. K. Prajapati, S.S.L. Srivastava Rajesh K. Pandey and Singh, O.P. (2007). Management of Collar Rot of Lentil *Ann. Pl. Protec. Sci.* 15 (1):270-271

Singh, S, R, S.S.L. Srivastava, R.K. Prajapati and R.K. Pandey (2006). Prevalence of collar rot of lentil and its pathogenic virulence. *Ann. Pl. Protec. Sci.*, 14: 488-489.

Taylor, P.W.J., Lindbeck, K., Chen, W. and Ford, R. (2007). Lentil Diseases, in: Yadav, S.S., McNeil, D.L., and Stevenson, P.C. (eds). *Lentil*: An Ancient Crop for Modern Times. Springer, Dordrecht, The Netherlands, pp. 291-313.

Thakur, R.P. and Mukopadhyay, A.N. (1972). *Indian J. Agri.Sci.*42:614-617.

Tikoo, J.L., Sharma, B., Mishra, S.K. and Dikshit, H.K. (2005). Lentil (*Lens culinaris*). in India: Present Status and Future Perspectives. *Indian Journal of Agricultural Sciences* 75, 539-562.

Vishunavat, K. and Shukla, P. (1979). Fungi associated with lentil seeds. *Indian Pthytopathology* 32: 279-280.

Sustainable Disease Management of Agricultural Crops (2011) *Pages* **258–275**
Editors: **S.K. Biswas and S.R. Singh**
Published by: **DAYA PUBLISHING HOUSE, NEW DELHI**

Chapter 19

Diseases of Sunnhemp and their Management

☆ ***S.K. Sarkar and C. Biswas***

[1]*Central Research Institute for Jute and Allied Fibres, Barrackpore, 24-Parganas (N), Kolkata – 700 120*

Sunnhemp (*Crotalaria juncea* L) commonly known as Indian hemp, Madras hemp, brown hemp is a multipurpose Fabacious crop grown widely in India. The fibre obtained from sunnhemp is slightly lignified, light in colour somewhat course, strong and lasting. It is used for various purpose like making ropes, strings, twines, floor mat, fishing nets, handmade paper etc in cottage industries and paper pulp in paper industry. Apart from these industrial values, the plant being a legume is advantageous to grow on poor, fallow or freshly reclaimed soil playing major role as soil builder or renovator as well as deterrent to nematode. One of the major constraints of sunnhemp cultivation is the incidence of diseases especially in monsoon-sown crop. In the present article important diseases of sunnhemp are reviewed.

Sunnhemp Wilt

This disease has been recorded from several countries. Vincens (1921) reported the occurrence of sunnhemp wilt from Tonkin which was described as a collar disease similar to pigeon pea (*Cajanus cajan*) and cotton wilt. He isolated several fungi from the affected plant and found constant presence of *Fusarium* closely allied to *F. udum* Butl. (*F.vasinfectum* Atk) but failed to prove the pathogenecity. Van Hull (1925) reported wilt of *C. usaromoensis* from Dutch East India. Briant and Martyn (1929) recorded it from Trinidad and found that the pathogen was in resemblance with *F. vasinfectum* Atk. infecting pigeon pea. Hean (1947) reported this disease from South Africa. In India this disease was reported by Uppal and Kulkarni (1937) from the then Bombay Province. Mitra (1934) observed this disease and found that the pathogen similar to *F. vasinfectum* Atk. in Pusa, Bihar. Desai *et al.* (1984) reported premature wilting of sunnhemp in *in vitro* conditions. Studies conducted at Sunnhemp Research Station, Pratapgarh, U.P. also showed heavy incidence of vascular wilt in late sown crop (Sarkar *et al.*,

Figure 19.1: Sunnhemp Crop

Figure 19.2: Sunnhemp Wilt

1998). Thakur (1971) reported the occurrence of wilt (*F.udum* Bult.) in *C.verrucosa*. Armstrong and Armstrong (1950, 1951) isolated *Fusarium* from seven *Crotalaria* species from India and USA. He found three races (1,2,3) based on selective pathogenecity and in contrary to other workers, they found that the pathogen did not attack pigeon pea. In spite of appreciable difference in cultural characteristics between Indian and USA isolates, they belong to same pathovar, *F. udum* var. *crotalariae* (Padwick, 1937). In India, it is reported from sunnhemp growing areas of Uttar Pradesh, Bihar, Maharashtra and Madhya Pradesh.

Loss

Generally the incidence was 10-12 per cent (Mitra, 1934) but under favourable conditions the incidence may go as high as 60-80 per cent. Uppal (1937) also reported 88 per cent incidence under green house conditions. More than 60 per cent loss was observed in seedling condition. In seed crop the incidence was much more because of favourable conditions.

Epidemiology

The temperature plays an important role in the development of the disease. Uppal and Kulkarni (1937) reported that the incidence of this disease declined with high temperature at Bombay. Mitra (1934) and Mundkur (1935) recorded similar observation at Pusa. At Sunnhemp Research Station, Pratapgarh, U.P., Sarkar *et al.* (2000) studied the effect of weather parameters (temperature, relative humidity and rain fall) on the crop vis-a-vis incidence of this disease during the months of July to October and found that wilt started at about 62 days after sowing and increased significantly with the age of the crop. Temperature was found to be negatively correlated with the incidence. Relative humidity and maximum temperature were found to be non-significant in disease development. Rainfall during last fortnight of October was found to be conducive for the spread of this disease. The incidence of disease is less in early (mid-April) sown crop. But on the contrary, Saxena (1989) reported that the disease incidence increased with high temperature and high moisture. In low-lying areas wilt incidence was also found to be more.

Symptom

The description of symptoms reported by different workers (Briant and Martyn, 1929, Wright and Leach, 1931 and Mitra, 1934) is more or less similar. The affected plant gradually whither, droops, hang down and later on turn brown, and ultimately dies within a day or two. Usually the whole plant wilts but partial wilting is also noticed. In grown up plants the wilting parts droop at the tips and defoliation starts which consequently die. The sporodochium of the fungi with pinkish tinge are produced on the dead stem or the dead portion of the stem where the infection is confined to one side.

The discolouration of the tissues could be traced to the main tap root or lateral roots. In the early stage the fungus is confined to the lateral roots especially in the tip portion and subsequently attacks the vascular bundle of meristem.

Source of Infection

The fungus survives in the soil as well as in crop residues as facultative parasite (Saxena, 1989). It attacks the plant through the thinner roots and rootlets, even through the cracking in the basal portion of the stem occurs. The pathogen produces enormous spores (both macro and micro condia) in the pink coloured sporodochium that are also capable of infecting the growing crop. The fungus was also noticed on the pod and in many cases in the seed of diseased pod (Mitra, 1934, Sarkar, 2003). The infected discoloured seed also may initiate the infection in the field.

Causal Organism

The disease is caused by *Fusarium udum* (Bult) f.sp *crotalariae* (Kulkarni) earlier named as *F.vasinfectum* Atk. var. *crotalariae, Fusarium lateritium* f.sp. *crotalariae* (Padwick). As the pathogen is a facultative parasite, it survives in the crop stubbles. The pathogen produces pink coloured sporodochia on which enormous micro and macro conidia are produced. Fungal hyphae and spores plug the xylem vessels of the infected part causing the death of the plant. Microconidia 1- celled, hyaline, mostly curved and scattered. Macroconidia is subulate, falcate, narrowed towards either end, 1-3 rarely 4-7 septate and pedicellate. Conidia produced from the tips of phialides are borne on more or less verticillate branced conidiophores forming slimy groups at the tips of the phialides, later scattered on the mycelium to form pale pinkish masses sporodochia on the host usually formed in sporodochia of similar colour. Chlamydospores are 4-10µ in diameter, usually intercalary, ochre yellow, stroma mostly immersed, more or less spread out, plectenchymous, at first pale to pinkish, then salmon orange to cinnabarinous when dry.

Management

Host Resistance

K-12 yellow–a selection from K-12 (black) was found to be largely resistant against wilt (Ghosh *et al.*, 1977, Sarkar *et al.*, 1998, Saxena and Kumar, 1990). Recently, an improved self-compatible strain (Bidhan Shan) developed at Bidhan Chandra Krishi Viswavidyalaya, West Bengal through bulk selection, was found to be highly resistant against diseases (Anon, 2003). Uppal (1937) found that D-IX was highly resistant in decan condition. Considerable work on screening of germplasm against wilt (Bandopadhayay *et al.*, 1982, Kumar *et al.*, 1994 and Sarkar *et al.*, 1998) has been made. But as the crop is highly cross-pollinated in nature, whatever promising germplasm were found to possess resistance became ultimately susceptible. Sarkar *et al.* (1998) tested a number of wild *Crotalaria* species and found that *C. brevidence, C. mucronata, C. verucosa and C. striata* are highly resistant against wilt under natural conditions. This character may be transfered to the cultivated species but so far attempts on interspecific cross between different *Crotalaria* spp. has remained unsuccessful (Kundu, 1964, Kumar *et al.*, 1994).

Seed Treatment

Seed treatment with Benlate @ 3 g ha^{-1} or spraying of Bavistin @ 3 g kg^{-1} was found to reduce the disease considerably. Considerable work on soil application of neem cake along with seed treatment and application of $ZnSO_4$ has been done by Bandopadhyay (1999) at Sunnhemp Research Station, Pratapgarh, U.P and CRIJAF. It was finally recommended that sowing of treated seeds with Bavistin @3 g kg^{-1} and *Rhizobium* and soil application of neem cake @1 q ha^{-1} + 25 kg $ZnSO_4$ ha^{-1} is the best option. In contrast to this, Saxena (1994) tested nine fungicides including Bavistin but the results were not encouraging.

Nutrient

Management of wilt was also attempted with application of NPK but no encouraging result was observed (Saxena, 1989, Sarkar, *et al.*, 1998). Sarkar *et al.* (2000) studied the effect of boron, zinc and iron but no significant decrease in wilt was observed.

Sowing Dates

Sunnhemp can be successfully grown between April to August in North Indian conditions. But it was observed that delay in sowing time increased this disease (Mitra, 1934). In April sowing, the crop almost escape the disease (Sarkar *et al.*, 1998, Sarkar and Tripathi, 2003).

Intercropping

Intercropping with sesame and sorghum reduced the incidence of wilt (Saxena, 1993-1995 and Sarkar, 2000).

Anthracnose of Sunnhemp

Three anthracnose diseases of sunnhemp have been reported. Petch (1917) described this disease from Peradeniya, Srilanka on *Crotalaria striata* and named the pathogen as *Colletotrichum crotalariae* Petch. In southern USA the same fungus was reported on *C. striata* and *C spectabilis* (McKee and Enlow, 1931). Massee (1913) described *Gloeosporium crotalariae* as parasite on young shoot of *C. juncea*. Briant and Martin (1929) described the fungus and named it as *C. curvatum*. Mitra (1937) reported severe incidence of this disease in seedling stage at Pusa.

Symptom

The seedlings are attacked first on the cotyledon and subsequently infection spreads to the stem and growing point. The disease makes an appearance in the form of soft discoloured areas on the cotyledon. Later, brownish spots are formed on all parts of host except underground parts. The affected seedling droops from the point below cotyledon. The affected cotyledons themselves drop from the petiole. The infection spreads downward and acervuli are formed with copious spores on the infected areas within two days. The young seedling when infected generally dies. When older plants are infected the disease is restricted on leaf and stem and the heavily infected leaves fall off.

The spots on older leaves appear on one side of the leaf but gradually enlarge and extend to the opposite side. These spots are grayish brown to dark brown, roundish or irregular. Several spots coalesce and cover the entire leaves or large portion of the leaf. They may also be formed on the midrib. Alongwith the growth of seedlings the plant becomes resistant to the disease.

Epidemiology

Weather condition and age of the plant influence the disease greatly. The cloudy weather accompanied by continuous rain favour the rapid spread of disease in the thickly populated crop. The infection is found to be severe in the seedling stage. Rain splashing helps the spores to spread in adjacent plants (Mitra, 1937).

Pathogen

The disease is caused by *Colletotrichum curvatum* Briant and Martyn. The mycelium is at first localized in the tissues of a lesion either on leaf or stem of a young seedling and does not extend internally to other parts. In severe cases, however, it not only penetrates in the cortical tissues but also invades the vascular bundles. The acervulli are formed in abundance in the epidermis of the diseased area. They are white or faint pinkish in colour and consist of simple erect closely packed conidiophores and setae. From the tip of the conidiophores the conidia are budded off epically one at a time and pile up in a light pinkish heap on the top of acervullus. A mucilaginous secretion holds the conidia together. Conidia are one celled, hyaline, falcate and acute at each end measuring around 15-24 x 3-4 µm. The conidia germinates by producing germ tube, which forms an appresorium to infect the host. The setae are brown to dark brown in colour, septate, tapering towards the tip and swollen at the base. These are found between the conidiophores, measuring 66-140µ x 4-6µ.

Management

Seed Treatrment

Seed borne infection can be checked by disinfecting the seed with suitable fungicides (Mitra, 1937). Sarkar (2003) tested a number of fungicides and found that Thiram or Bavistin are highly effective in reducing this disease.

Spraying

Spraying with Bordeaux mixture (0.5 per cent) at seedling stage also reduces the disease incidence.

Sowing Time

As the disease is favoured by rain and high humidity early sowing in dry season *i.e.* mid-April to mid-May helps to escape this disease, because the crop becomes mature before the onset of monsoon.

Resistance

Amongst the cultivated varieties, K-12 yellow was found to be largely resistant to the disease (Ghosh, 1977). But due to cross-pollinating nature of the crop resistance might have depleted. Dey *et al.* (1990) tested 66 germplasm and found 11 germplasm to be resistant. Recently, a variety named SH-4 has been released, which is resistant against wilt under natural conditions.

Leaf Blight (Figure 19.3 and 19.4)

Sarkar and Pradhan (1999) reported heavy incidence of leaf blight (54 per cent) from Pratapgarh, Uttar Pradesh. The blight started from the margin of the leaf and proceeds inwards. Under moist and warm conditions with intermittent rains, the whole leaf may be blighted. But with the fall in temperature along with receding rains severity was restricted. Often black pinhead like fungal structure is noticed on blighted site. In the early morning, the blighted leaves look greyish and water soaked, which ultimately become brownish with broad yellow margin. Subsequently, the infected leaf becomes weak and droops down from the plant, which gives a sickly appearance of the whole field. This disease is caused by *Macrophomina phaseolina* (Tassi) Goid. It is a soil borne disease. Basuchoudhury and Pal (1982) reported the seed borne nature of the pathogen.

Choanephora Leaf Blight and Tip Rot

Sporadic incidence of this disease was reported by Tripathi *et al.* (1973) from Varanasi, U.P. The authors also observed the disease at Sunnhemp Research Station, Pratapgarh, U.P in highly humid conditions. About 20 per cent incidence of this disease was recorded in the month of August. This disease is characterized by the rotting of terminals. Brown discolouration occurred just below the infection point. Affected portions decay, break and droop. White mycelial growth along with black coloured sexual bodies of the fungus is seen on the affected parts from the tip towards petioles. A characteristic whitish brown discolouration in leaves precedes the disease development. Infected leaves loose their chlorophyll and droop from the stem. The pathogen affects the epidermal and outer cortical layers, and cells in these regions get disintegrated. The disease has also been reported from the wild species like *C. spectabilis, C.sericea* and *C. retusa* (Patel and Patel, 1983, Sarkar, 1998, Bandopadhyay and Mitra, 1994).

Leaf Spot (Figure 19.5)

There are three types of leaf spots reported. These are as follows:

Pleospora Leaf Spot

Pathak and Chauhan (1976) reported this disease from Agra, U.P., occurred in the month of September-October. It is characterized by the development of pale yellow spot measuring 7-15 mm in diameter. The spots turn into black with time.

Phoma Leaf Spot

It was also reported by Pathak and Chauhan (1976). Here the spots appeared as yellow, which turn to brown. Size of the spots is 3-5mm.

Pringsheimia Leaf Spot

Mishra (1976) reported this disease. It is characterized by the development of brown spot (bound by vein and veinlets) on the mature green leaves during the month of September. Microscopic observations showed the presence of minute globose black erumpent perithesia scattered all over the necrotic spots.The perithecia contains a number of pyriform, hyaline asci with a thin but tough wall. The ascus containaing 8 ascospores are hyaline, somewhat thick walled, 4-7 celled, muriformed with longitudinal septum appearing in one or two cells only.

Sunnhemp Mosaic

Raychoudhury (1947) first reported this disease from Delhi and described the symptoms. The first visible symptom appeared within 10 to 12 days after inoculation when mottling was seen on the youngest leaves. As the disease progressed, patches of light and dark green areas became more prominent. Diseased leaves were smaller than the normal. In severe cases of infection, growth of the lamina generally became abnormal. Frequently dark green raised areas on the upper surface were seen with corresponding depression on the lower surface. The infected plant remained shorter in height and therefore, fibre and seed yields were greatly reduced.

Causal Agent

The disease is caused by virus.The structure of the virus was determined under electron microscope (Dasgupta, De and Raychoudhury, 1951). The virus consists of spherical particles measuring 26-40 µm. Anand and Sahambi (1965) studied serological relationship of sunnhemp mosaic virus with other mosaic virus and they found that the sunnhemp mosaic virus produces a flocculent precipitate resembling with bottle gourd mosaic virus, tobacco mosaic virus and cowpea mosaic virus.

Transmission and Host Range

The disease was reported to be transmitted mechanically. Nariani and Chadrashekher (1963) showed the successful transmission through root inoculation. Das and Raychoudhury (1963) found that the hosts of the disease are *C. mucronata, Cyamopsis tetragonoloba, Pisum sativum, Nicotiana tabacum, Datura strumonium* and *Lycopersicon esculenta*. Jensen (1950) reported that sunnhemp mosaic virus of Trinidad can easily be transmitted to cowpea.

The virus movement was studied by Sastry and Vasudeva (1963) and they observed the systemic transmission of the virus. They also noticed that optimum dose of nitrogen and increased dose of phosphorus increased the spread of the virus within the plant but the reverse was true in case of potassium. The concentration of the virus was found to be more in optimum nitrogen application whereas, potash application was found to reduce the concentration of the virus (Sastry and Vasudeva, 1963).

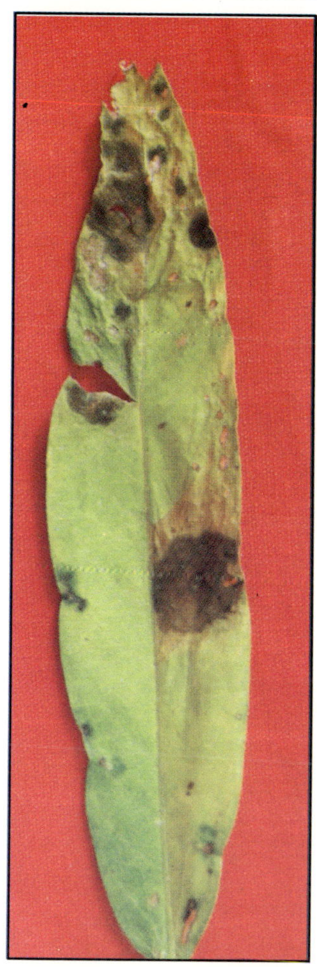

Figure 19.3: Leaf Blight

Figure 19.4: Twig Blight

Epidemiology

Epidemiological studies on the virus have not been studied so far. But at Sunnhemp Research Station, Pratapgarh, U.P. it was observed that the severity of the virus increases with the onset of monsoon. As the virus is reported to be sap transmitted and having wide host range, the disease may be initiated from the other wild or weed host(s). The same observation was also noticed in the farmers' field in and around Pratapgarh. Higher dose of nitrogen increases while potassium application reduces the disease (Sastry and Vasudeva, 1963).

Management

Sowing Time

Time of sowing is very important to escape the disease. Drastic reduction of this disease was noticed in the early sown (mid- April) crop (Sarkar *et al.*, 2003) than the monsoon crop.

Figure 19.5: Leaf Spot

Figure 19.6: Sunnhemp Mosaic

Resistance

Ghosh *et al.* (1977) reported K-12 yellow as a largely resistant variety. Being a cross-pollinated crop resistance might not be observed at present. Recently released SH-4 variety was found to be resistant to sunnhemp mosaic. Wide variation in disease reaction amongst the different germplasm was noticed at Sunnhemp Research Station, Pratapgarh, U.P. as well as at Gobind Ballabh Pant University of Agriculture and Technology, Pantnagar.

Southern Sunnhemp Mosaic

Sunnhemp is extensively grown for fibre and green manure (in South India) in rotation with rice, wheat and sugarcane. Severe incidence of mosaic disease, which caused appreciable loss in fibre yield, seed and green matter, was reported by Capoor (1950, 1962). This mosaic disease is widespread in nature.

Symptom

Initial symptom of the disease appeared within 9-20 days depending on weather conditions. It is characterized by faint discoloured patches appearing first on young leaves. Distinct mosaic with puckering and blistering of leaves developed. Subsequently thin elongated enations running more or less parallel to each other developed on the under surface of the leaves. The characteristics mosaic mottle and varying degree of leaf distortion occurred in advanced stage of the disease. Plants became dwarf with reduced leaves and bear scanty flush of flower resulting in poor pod setting and consequently, reduction in quantity of seed (Capoor, 1962 and Chandra *et al.*, 1975). The incidence of the disease is more in monsoon months when temperature ranges between 22-41°C and relative humidity 80 per cent.

Host Range

The host range of southern sunnhemp mosaic virus was confined mainly to the families, Leguminosae and Solanaceae. Systemic mosaic mottling appeared at about 10-20 days after inoculation in *C. retusa, C. mucronata, C. laburnifolia, C usaramoensis, C.spectabilis, C. lanceolata, Pisum sativum, Vigna sinensis, V. unguiculata, V. cylindricum, V sesquipedalis, Phaseolus vulguris, P latanus, Cassia tora, Cajanus cajan, Solanum nigrum, Datura inoxia* and *Nicandra physaloides*. Local lesion develops on *Nicotiana tabacum, N glutinosa, N. sylvestris, D strumonium, D. aegyptiana,* and *Gompherena globosa*.

Transmission

The virus was readily transmitted by leaf rubbing without the help of any abbrasive or by pinpricking. It was also transmitted by wedge or patch grafting of diseased tissues to healthy sunnhemp plants. However, the virus is not transmitted through seed and soil (Capoor, 1962).

Virus Particles and Physical Properties

The electron monograph revealed that the particles are rod shaped with 300 µm x 18 µm (Anand, 1969 and Nariani *et al.*, 1970). On the basis of cross-inoculation studies, Capoor (1962) concluded that it is a strain of tobacco mosaic virus. Serological studies also proved that it is related with the type strain of TMV. Physical properties of the virus was studied by Capoor (1962) and found that it is highly stable with TIP-95°C, DEP 10^{-7} and aging *in vitro* for 6 years. The virus can withstand complete desiccation.

Sunnhemp Rosette

This disease is characterized by mosaic mottling, crowding, rosetting of leaves, stunted growth and sparse flowering, fruiting and seed setting (Verma and Awasthi, 1976). They reported around 40-60 per cent incidence of this disease at Lucknow, U.P.

Causal Organism

The disease is caused by a rod shaped RNA virus measuring 400 µm x 17 µm. The particle shape although conforms to TMV (Harison *et al.,* 1971) but the length is more as compared to TMV. The physical properties like thermal inactivation point, dilution end point and *in vitro* longevity are 93ºC, $5x10^{-5}$ and 54-56 days at 20-26 ºC, respectively.

Transmision

The virus is transmitted mechanically through sap but it is also seed transmitted (5-10 per cent) in nature if the plants were infected before flowering (Verma and Awasthi, 1976).

Host Range

Unlike other virus diseases of sunnhemp, rosette virus did not produce systemic symptom as in other member of family Leguminaceae, but it produces hypersensitive response in *Cyamopsis tetragonoloba, Vigna sinensis* (Leguminaceae) and *Chenopodium amaranticolour* (Chenopodiaceae).

Management

Practically no information is available to control the disease at the field level. But early sown (mid- April) crop generally escapes the disease at Pratapgarh, U.P. Amongst different varieties, K-12 yellow showed less disease incidence (Ghosh *et al.,* 1977). At present systemic induced resistance is gathering momentum in the field of disease management. In this context, a number of plant extracts like *Cleodendrum aculeatum, Chenopodium murale, Boerhaavia diffusa* (root extract), *Cuscuta reflexa, Solanum melongena* and *Pseuderanthemum bicolour* were found to reduce the disease under laboratory conditions (Verma and Awasthi, 1978, Verma and Khan, 1984; Srivastava and Verma, 1995, Verma and Versha, 1995).

Figure 19.7: Phylloidy

Sunnhemp Phylloidy

Sporadic incidence of this disease is reported from different parts of the country. This disease was first reported by Solomon and Sulochana, 1973 at Chennai after rainy season. Plant once affected with mosaic virus was never affected by this disease. It is characterized by yellowing of apical leaf, followed by big bud formation at the terminal raceme, conversion of floral meristem in the lateral raceme to the vegetative state, leading to buds becoming phylloid and forming dwarf shoot. This disease was also reported from Sri Lanka (Shivanathan *et al.,* 1983A) but they called it a viral disease. At Sunnhemp Research Station, Pratapgarh, U.P. Sharma (1990) reported a similar type of disease and called it as Witches'

Broom, assumed to have been caused by phytoplasma like organism because of its remission after application of tetracycline @ 500 ppm. He observed the disease after 30–60 days of sowing and transmitted mechanically.

Seedling Diseases

Seedling diseases caused by soil borne fungi namely *Pythium butleri, Rhizoctonia solani, R. bataticola, Sclerotium rolfsii, Sclerotinia* and *Colletotrichum curvatum* are mostly responsible for poor stand in the field (Pal and Basuchudhury, 1980, Choudhury and Pal, 1981, 1982, Tripathi *et al.*, 1978, Bandhopadhyay, 1981 and Mitra, 1937). Individual diseases are discussed separately here under.

Wilt

Wilt (*F. udum* f.sp. *crotalariae*) is an important disease known for long time. Chibber (1914) reported this disease for the first time from Mumbai. Later on, Mitra (1934), Mundkur (1935), Uppal *et al.* (1937), Armstrong and Armstrong (1950, 1951) and Padwick (1937) reported this disease from different parts of India and studied several aspects. Briant and Martyn (1929) reported the same disease for the first time from Trinidad. Due to infection, emergence of seedlings was poor and slow, and the affected seedlings remained stunted and weak. Infection was mainly confined to the roots and base of the stem. This disease was characterized by browning of stem at the collar region adjacent to the soil lines, water-soaked cream to brown-coloured lesion developed on cotyledon and hypocotyl. Invaded hypocotyls became thin and soft. The lower parts of tap roots and lateral root systems were completely destroyed resulting into dropping and wilting of infected seedling.

Seedling Blight

Three seedling blights of sunnhemp have been recorded. Petch (1917) described a new species of *Colletotrichum* from Sri Lanka on *Crotolaria striata* and named it *Colletotrichum crotalariae* Petch. This fungus has also been recorded from Southern parts of USA by Mckee and Enlow (1931) on *C. striata* and *C. spectabilis*. Massee (1913) described *Gloeosporium crotalariae* as parasitic on young shoots of *C.juncea*. Briant and Martyn (1929) described another species, *Colletotrichum curvatum*. In India Mitra (1937) reported the seedling blight (*C. curvatum*) of *C. juncea* from Pusa. He mentioned that it caused severe seedling mortality in thickly sown crop at the month of August when the weather is cloudy accompanied by rain. He also studied the different aspects of this disease.

Damping Off

Sporadic incidence of damping off and wilt caused by *Pythium butleri* during the month of August at Varanasi was reported by Tripathi *et al.* (1975). Similarly, Patel and Patel (1980) reported the incidence of damping off caused by *P. aphanidermatum* (Edson) Fitz in *Crotalaria spectabilis* from Gujarat. This disease was not severe, scattered stand of young plants only exhibited symptom. Seedling showed typical damping off in patches in the morning hours. Water soaked lesions appear on the hypocotyls, which gradually encircled it completely, thus causing decay of tender tissues of seedlings. Young plants showed symptom of wilting in healthy dense population. On closer examination of the affected plant, discoloration and necrosis of the stem near collar region was also noticed. Bark of the affected region became water soaked and soft, which gradually moved in both upward and downward directions.

Wilt and Root Rot Caused by *Rhizoctonia, Sclerotium* and *Macrophina*

Rhizoctonia wilt (*Rhizoctonia solani*) was first reported from Pusa (Mitra, 1934). Tripathi (1973) also reported the incidence of *Rhizoctonia* wilt from Varanasi. Similarly Pal and Basuchoudhury

(1980) and Pal (1978) reported some root diseases of *Crotalaria juncea* caused by *S.rolfsii, R.solani* and *P. butleri*. Different pathogenic fungi produced different types of characteristic symptoms on seedlings, which are discussed below.

Rhizoctonia solani

R. solani causes pre- and post-emergence damping-off of seedling. Seeds may rot completely prior to germination. Damping-off was mainly due to colonization in hypocotyl and radicle just after emergence. A reddish brown decay of outer cortical layer appeared on stem base, crown and roots. This decay develops into shrunken reddish brown cankers, which sometimes girdle the stem at or just above the soil line. The cankerous areas may extend in any direction on stem and result in wilt and death of seedlings.

Sclerotium rolfsii

Sclerotium rolfsii causes both pre- and post-emergence rotting of seedling and ungerminated seeds were covered with white mycelial growth of the pathogen. In case of post-emergence infection of seedling, rotting at collar region become apparent resulting in seedling blight. The collar portion and seeds were colonized by white, cottony mycelial growth of fungus alone or with mustard seed like sclerotia.

Macrophomina phaseolina

Macrophomina phaseolina causes seed rot and infects the seedlings, which might show blackish brown discoloration at emerging portion of the hypocotyl at soil line or above. The discoloured area turns into dark brown to black resulting in the death of seedling. In older seedlings the fungus causes a reddish brown discoloration in the vascular tissues of tap root and progresses upto the stem. Small black sclerotia are observed on removing the epidermis of tap root and lower stem.

Management

Seed Treatment

Numerous reports are available on seed treatment of different crops in relation to seedling diseases. But few reports on efficacy of fungicides are available for sunnhemp. Whatever reports are available; these are mostly confined to testing of fungicides against different seedling pathogens under laboratory conditions (Pal, 1978, Tripathi *et. al.*, 1978 and Tripathi, 1973). On the other hand, whatever fungicides are found suitable are either banned or not available in the market. Sarkar (2003) tested five seed treating fungicides, namely Apron, Divident, Mancozeb, Thiram and Bavistin against seedling diseases under field as well as in glass house conditions, and it was observed that maximum reduction of pre- and post-emergence rotting of seedling was in Thiram or Mancozeb treated seeds of all the varieties irrespective of soil inoculation with different pathogens singly or in combination.

Leaf Yellowing

Pradhan and Sarkar (2001) during field survey at Pratapgarh, U.P. reported wide spread occurrence of the disease which is entirely different from known sunnhemp diseases. The crop sown in monsoon months showed heavy incidence of interveinal yellowing. Symptom started in older leaf approximately 50 days after sowing. Afterwards young leaves are also affected. The whole field irrespective of variety was affected which gave a sickly appearance. However, there was no death of plants due to this disease. Remarkably it was observed that mid-April sown crop was almost free from the disease. Drastic reduction in fibre yield was observed in monsoon crop due to this disease. The wide spread

Figure 19.8: Leaf Yellowing

and uniform occurrence of this disease strike two possibilities, *i.e.* either it may be due to virus origin or of nutrient deficiency. Pradhan and Sarkar (2001) studied the effect of boron, zinc, iron, and manganese but found no appreciable recovery of this disease.

References

Anand, G.P.S. and Sahambi, H.S. (1965). Serological relationship of sunnhemp mosaic virus with some other mosaic viruses commonly occurring in Delhi.*Indian Phytopath.* 18: 204-205.

Anand, G.P.S. (1969). Particle shape of southern Sunnhemp mosaic virus. *Curr. Sci*: 706.

Anonymous (2003). Self compatible sunnhemp, Bidhan Shan, (ING No. 01036, IC, 296815). *ICAR News* 9(2): 3.

Armstrong, J.K. and Armstrong, C.M. (1951). Physiological races of Crotalaria wilt (*Fusarium udum* f.sp. *crotalariae*). *Phytopath.* 41: 714.

Armstrong, J.K. and Armstrong, C.M. (1950). The Fusarium wilt of *Crotalaria. Phytopath* 40: 785.

Bandopadhyay, A.K. (1981). Sunnhemp–an allied fibre of great promise. *Nepal Jute,* 38 (Shravan): 19-27.

Bandopadhyay, A.K. and Mitra, G. C. (1994). Incidence of disease caused by Choanephora cucurbitarum (Berk and Lav.) Thaxter in sunnhemp and other *Crotalaria* species. *J. Indi. Bot Soc.* 73.

Bandopadhyay, A.K., Prakash, G and Som, D. (1982). Screening of germplasms of sunnhemp against fusarium wilt disease. *Bangladesh J Bot.* 11(1): 14-16.

Bandopadhyay, A.K. (1990-92). Ann. report, Central Research Institute for Jute and Allied Fibres (ICAR), Barrackpore, West Bengal. Pp161.

Bandopadhyay, A.K. (2000). Annual Report, Report, Central Res. Institute for Jute and Allied Fibres, Barrackpore, West Bengal

Basuchoudhury, K.C. and Pal, A.K. (1982). Infection of sunnhemp seed by *Macrophomina phaseolina*. *Seed Sci. and Technol.* 10(1): 151-53.

Briant, A.K. and Martyn F.B. (1929). Diseases of cover crop. *Tropic. Agr.* 6: 258-60.

Capoor, S.P. (1950). A mosaic disease of sunnhemp in Mumbai. *Curr. Sci.* (19: 22.

Capoor, S.P. (1962). Southern sunnhemp mosaic virus: A strains of tobacco mosaic virus. *Phytopath.*, 52: 393-397.

Chandra, S., Singh, B.P., Nigam, S.K. and Srivastava, K.M. (1975). Effect of some naturally occurring plant products on southern sunnhemp mosaic virus (SSMV). *Current Science* 44: 511-512.

Choudhury, K.C.B. and Pal, A.K. (1981). Infection of sunnhemp seed with *Fusarium* spp. *Seed Sci. and Technol.* 9(3): 729-732.

Das Gupta, N.N., De, M.L. and Raychoudhury, S.P. (1951). Structure of sunnhemp (*Crotalaria juncea* Linn) mosaic virus with electron microscope. *Nature.* 168: 114.

Das, C.R. and Raychaudhury, S.P. (1963). Further studies on host range and properties of sunnhemp mosaic virus.16: 215-221.

Desai, S.A., Siddaramaiah, A.L., Hegde, R.K. and Kulkarni, S. (1984). *Fusarium udum* f.sp *crotalariae* causing pre-mature wilting of sunnhemp in-vitro. *Current-Research*, University of Agricultural Sciences, Bangalore 13: 10/12, 75-76.

Dey, D.K, Banerjee, K, Singh, R.D.N. and Kaiser, S.A.K.M. (1990). Sources of resistance to anthracnose disease of sunnhemp. *Environment and Ecology.* 8(4): 1217-1219

Ghosh, T, Mohan KVJ and Prakash G. (1977). Sunnhemp Variety. Jute Agricultural Research (ICAR), Barrackpore, West Bengal, India.

Harrison *et al.* (1971). Sixteen group of plant viruses. *Virology*, 45: 356-63.

Hean, A.F. (1947). A wilt disease of *Crotalaria juncea* L. found in South Africa. *Science Bull.* Department of Agric. South Africa.15: 255

Jensen (1950). A *Crotalaria* mosaic and its transmission by aphids. *Phytopath.* 40: 512-515

Joshi, L.M. (1960). Preliminary studies on *Uromyces decoratus* Syd., the causal organism of sunnhemp rust. *Indian Phytopath.* 13: 90-95.

Joshi, L.M. (1964). Formation of secondary sporidia in *Uromyces decoratus* Syd. *Indian Phytopath.* 17: 257-258.

Kumar, D. and Prakash G. Collection, Evaluation and maintenance of sunnhemp germplasm. Annual Report, Central Res. Institute for Jute and Allied Fibres, Barrackpore, West Bengal, 1993-95, pp. 108.

Kumar, D., Saxena, P. and Gupta, B.N. (1994). Screening of sunnhemp germplasm against wilt disease caused by *Fusarium udum* Butler f. sp. *crotalariae* Kulkarni, Padwick. *Legume Res.* 17: 109-110.

Kundu, B.C. (1964). Sunnhemp in India. *Soil Crop Sci. Soc.* Florida Proc. 24: 396-404.

Massee, G. (1913). XXXIII addition to the world fauna and flora of the royal Bengal Botanical Gardens, Kew. 14 No 6 (Bull. Misc. Information Royal Bot. Garden Kew).

Mckee, R and Enlow, C.R. (1931). *Crotalaria*, A new legume for the South. U.S. Dept. Agric. Cir. Bull. 137, 30pp.

Mishra, C.B.P. (1976). A new host record of *Pringsheimia* on sunnhemp (*Crotalaria juncea* L.). *Curr. Sci.* 45(12): 468.

Mitra, M. (1934). Wilt disease of sunnhemp. *Indian J. Agric. Sci.*, 4: 701-714.

Mitra, M. (1937). An anthracnose disease of sunnhemp. *Indian J. Agric. Sci.* 7: 443-449.

Mundkur, B.B. (1935). Influence of temperature and maturity on the incidence of sunnhemp and pigeon pea wilt at Pusa. *Indian J. Agric. Sci.*, 5: 609-618.

Nariani, T.K. and Chandrasekhea, B.K. (1963). Transmission of sunnhemp mosaic virus by mechanical root inoculation. *Indian Phyto Path.* 16: 171-73.

Nariani, T.K., Kartha, K.K.and Prakash Nam (1970). Purification of the southern sunnhemp mosaic virus using butanol and differential centrifugation. *Curr. Sci.* 23: 539-541.

Pal, A.K. (1978). Studies on some root diseases of Sunnhemp (*Crotalaria juncea* L.). *PhD. Thesis.* Banaras Hindu University, Varanasi, India.

Pal, A.K. and Basuchudhury, K.C. (1980). Seed mycoflora of sunnhemp (*Crotalaria juncea* L.). *Seed Res.* 8(2): 153-55.

Pal, A.K. and Basuchaudhury, K.C. (1980). Laboratory evaluation of fungicides to control some root diseases of sunnhemp. *Indian J. Mycol. and Plant Pathol.* 16: 212-215.

Patel, B.N., Patel D.J. and Patel R.C. (1983). Choanephora twig blight of *Crotalaria spectabilis. Indian Phytopath.* 36: 764-66.

Patel, D.J. and Patel, B.N. (1980). Pythium damping off of root knot resistant *Crotalaria spectabilis. Indian Phytopath.*, 33: 115

Pathak, P.D. and Chauhan, R.K.S. (1976). A leaf spot disease of *Crotalaria juncea* L. caused by *Stachybotys arta* Corda. *Curr. Sci.* 45: 567.

Petch, T. (1917). Ann Royal Botanic Gardens (Perademya) 6: 239

Pradhan, S.K. and Sarkar, S.K. (2001). Sunnhemp leaf yellowing–a new disease in Uttar Pradesh. *Legume Res.* 24(1): 65-66.

Punithalingam, E. (1968). *Uromyces decoratus*. Commown. Mycol. Inst. Descr. Pathog. Fungi. Bact. 179, 2 pp.

Raychaudhury, S.P. (1947). A note on mosaic virus of sunnhemp (*Crotalaria juncea* L.) and its crystallization. *Curr. Sci.* 16: 26-28.

Sarkar, S.K. (2001). Ann. Rep. Central Research Institute for Jute and Allied Fibres (ICAR), Barrackpore, West Bengal. Pp. 127.

Sarkar, S.K. (2000). Effect of intercropping jower on the incidence of sunnhemp wilt (*Fusarium udum* f. sp. *crotalariae*) *J. Mycol. Pl. Pathol.* 30(1): 41-43.

Sarkar, S.K. (2003). Investigations on seed mycoflora and seedling diseases of sunnhemp (*Crotalaria juncea* L) and their management. *Ph.D. theses* submitted to GBPUA and T, Pantnagar, India.

Sarkar, S.K. and Pradhan, S.K. (1999). Incidence of leaf blight of sunnhemp caused by *Macrophomina phaseolina. J. Mycol. Pl. Pathol.* 29(1): 143.

Sarkar, S.K. and Tripathi, M.K. (2003). Summer crop of sunnhemp escape major pests and diseases. *ICAR News* 9(4): 16.

Sarkar, S.K. and Tripathi, S.N. (2000). Quantitative relationship between plant age, weather parameters and vascular wilt in Sunnhemp. *J. Mycol. Pl.Pathol.*30(2): 274.

Sarkar, S.K., Pradhan, S.K. and Tripathi, S.N. (2000). Influence of boron, zinc and iron on the incidence of sunnhemp wilt (*Fusarium udum* f. sp. *crotalariae*). *J. Mycol. Pl. Pathol.* 30: 116-118.

Sarkar, S.K., Pradhan, S.K. and Prakash, S. (1998). Source of resistance against sunnhemp wilt (*Fusarium udum* f sp. *crotalariae*) and top shoot borer (*Cydia tricentra* Meyr) in wild *Crotalaria* spp. *Legume Res.*21 (3 and 4): 201-204.

Sarkar, S.K., Prakash, S. and Pradhan, S.K. (1998). Influence of dates of sowing, fertilizer levels and varieties on the incidence of wilt (*Fusarium udum* f.sp. *crotalariae*) in Sunnhemp (*Crotalaria juncea* Linn). *Legume-Research.* 21 (3-4): 225-228

Sastry, K.S.M. and Vasudeva, R.S. (1963). Effect of host plant nutrition on the movement of sunnhemp mosaic virus. *Indian Phytopath.* 16: 143-150.

Saxena, P. (1989). Studies on epidemiology of wilt of sunnhemp. Ann. Report, Central Res. Instt. For Jute and Allied Fibres, Barrackpore, West Bengal pp.135.

Saxena, P. (1989). Studies on epidemiology of wilt of sunnhemp. Ann. Report, Central Res. Instt. For Jute and Allied Fibres, Barrackpore, West Bengal pp.

Saxena, P. (1993-95). Studies on epidemiology of wilt of sunnhemp. Ann. Report, Central Res. Instt. For Jute and Allied Fibres, Barrackpore, West Bengal pp.113

Sharma, O.P. (1990). "Witch's Broom" disease of sunnhemp (*Crotalaria juncea* L.) by MLO. *Indian J. Mycol. Pl. Pathol.* 20(3): 264-65.

Shivanathan, P., Shikata, E., Lizhi-Yuan and Maramorosch, K. (1983A). Mycoplasmal disease of *Crotalaria juncea* (sunnhemp) in Sri Lanka. *International-Journal-of-Tropical-Plant-Diseases.* 1: 83-85

Sokhi, S.S. and Singh, S.J. (1979). *Crotalaria medicaginea*, a collateral host of *Uromyces decoratus*, causal agent of rust of sunnhemp (*Crotalaria juncea* L.). *Plant Disease Reptr.* 63: 344.

Solomon, J.J. and Sulochana, C.B. (1973). Sunnhemp phyllody–A virus disease, : *Phytopathologische-Zeitschrift.* 78 (1): 62-68

Srivastava, Neeta and Verma, H.N. (1995). *Chenopodium murale* extract, a potent virus inhibitor. *Indian Phytopath.* 48(2): 177-179.

Thakur, R.N. (1971). Occurrences of Stem rot of *Crotalaria juncea* L. in Jammu and Kashmir. *Indian J. Mycol. Plant Pathol.* 1: 147-48.

Tripathi, H.S. (1973). Studies on some Diseases of Sunnhemp in Varanasi. *M.Sc.Thesis* Submitted to Banaras Hindu Unversity, India.

Tripathi, H.S., Vishwakarma, S.N. and Chaudhury, K.C.B. (1978). Evaluation of fungicides for control of leaf blight and tip rot and damping off and wilt of sunnhemp. *Sci. and Cult.*, 44(1): 30-33.

Tripathi, H.S., Singh, A.K. and Basuchoudhury, K.C. (1975). Two new fungal diseases of sunnhemp from Varanasi. *Indian Phytopath*. 28: 263-65.

Uppal, B.N. and Kulkarni, N.T. (1937). Studies in Fusarium wilt of sunnhemp. I. The Physiology and Biology of *Fusarium vasinfectum* Atk. *Indian J. Agric Sci.*, 7(3): 413-442.

Verma, H.N. and Awasthi, L.P. (1976). Sunnhemp rosette: A new virus disease of sunnhemp.*Curr. Sci.* 45(17): 642-43.

Verma, H.N. and Awasthi, L.P. (1978). Further studies on a rosette virus of *Crotalaria juncea*. *Phytopath. Z.*, 92: 83-87.

Verma, H.N. and Khan, M.M.A.A. (1984). Management of plant virus diseases by *Pseuderanthemum bicolor* leaf extract. *Zeitschrift-fur-Pflanzenkrankheiten-und-Pflanzenschutz*. 91: 266-272.

Verma, H.N. and Varsha (1995). Induction of systemic resistance by leaf extract of *Clerodendrum aculeatum* in sunnhemp against sunnhemp rosette virus. *Indian-Phytopath*. 48 (2): 218-221.

Vincens, M.F. (1921). Bull. Agric. Dev. Inst. Sudu, Saigon, 3: 381-84.

Sustainable Disease Management of Agricultural Crops (2011) *Pages* **276–282**
Editors: **S.K. Biswas and S.R. Singh**
Published by: **DAYA PUBLISHING HOUSE, NEW DELHI**

Chapter 20

Biological Control of Plant Parasitic Nematodes

☆ ***Kusum Dwivedi***

Department of Entomology, C.S. Azad University of Agriculture and Technology, Kanpur – 208 002, U.P.

For the past some years scientist all over the world are engaged in standardizing the pest management strategies by adopting eco-friendly and non-chemical methods including application of biological control agents. Biological control is the reduction of inoculums density of disease producing activity of pathogen or parasite in its active or dominant state, by one or more organisms accomplished naturally or through manipulation of the environment, host or antagonist or by mass introduction of one or more antagonists (Baker and Cook, 1974)

Lohde (1874) suggested for the first time about the possible implication of parasites and predators regulating nematode population in soil. Several organisms like fungi, bacteria, viruses, predatory nematodes and mites have been found to parasitize or prey and some are potential biocontrol agents. However, bacteria and fungi are the major parasites of plant parasitic nematodes.

Fungi

Earlier records of fungi as antagonists of phytonematodes were mentioned by Kuhn (1877). Fungi play an important role and some of them have great potential as biocontrol agents. Nematode destroying fungi are of diverse biology and have constant association with nematodes in the rhizosphere. They range from obligate endoparasitic forms which are zoosporic, to predacious trap forming species and opportunistic fungi that destroy the reproductive structures of nematodes like cysts and eggs. Based on their mode of action, they may be categorized into endoparasitic, predacious, and opportunistic fungi.

Endoparasitic Fungi

They do not produce extensive mycelium and exist in the environment as conidia infecting nematodes by either adhering to the body surface or being ingested. Conidia rapidly germinate and hyphae invade the entire nematode absorbing all the body contents. For example, *Catenaria auxillaris, Nematophthora gynophila, Harposporium, Nematoctonus etc.*

Predacious Fungi

They produce extensive hyphal network in the environment and develop trapping devise at intervals along each hypha. The traps capture nematodes either mechanically or by adhesion, allowing the fungus to penetrate the nematodes rapidly and digest the body contents. There are six different types of traps like, unmodified adhesive hyphae (*Stylopage* spp.), adhesive knobs (*Dactylella* spp.), three-dimensional adhesive nets (*Arthrobotrys* spp.), two-dimensional adhesive nets (Arthrobotrys spp.), non-constricting rings (*Dactylaria condide*) and constricting rings (*Dactylaria brachophaga*). These fungi have been shown to be poor competitors in the natural environment.

Opportunistic Fungi

These can colonize nematode reproductive structures and affect them. Sedentary nematode genera like *Globodera, Heterodera,* and *Meloidogyne* are vulnerable to attack by these fungi either within host plant roots or in soil. Once in contact with cysts and egg masses present in the soil, these fungi grow rapidly and eventually parasitize all eggs in early embryonic developmental stages. This group proved better than earlier two groups and *Paecilomyces lilacinus* and *Vertlicillium chlamydosporium* (*Pochonia chlamydosporium*) are best examples and have been studied by a number of Scientists (Table 20.1). Egg parasites are more effective in reducing nematode populations endophytic in the soil.

Table 20.1: Opportunistic Fungi Used as Bioagents Against Plant Parasitic Nematodes

Antagonists	*Nematodes*	*Crop*
Paecilomyces lilacinus	*Meloidogyne incognita*	Tomato
P. lilacinus	*M. javanica*	Okra
P. lilacinus	*M. incognita* Race 3	Chickpea
P. marguandi	*M. incognita*	Crossandra
P. lilacinus, Verticillium lecanii	*M. incognita*	Crossandra
P. lilacinus, T. harzianum, T. viride	*M. incognita*	Carnation, Gerbera
P.lilacinus, V. chlamydosporium, T. harzianum	*M. Javanica*	Chickpea
P. lilacinus	*M. incognita*	Crossandra
P. lilacinus	*M. incognita*	Tuberose
P. lilacinus, T. harzianum, T. viride	*M.incognita+Fusarium oxysporum f.sp. gladioli*	Gladiolus

Endophytic Fungi

This group includes non pathogenic and root-infecting fungi and mycorrhiza. These fungi are effective against endoparasitic migratory nematodes. They may improve plant growth and can be used as seed/plant material treatment.

Mycorrhization

Mycorrhizal fungi colonize the roots of more than 90 per cent of plant species, to the mutual benefit of both the plant host and the fungus. These symbiotic fungi are present in most terrestrial ecosystems and play a major role in ecosystem process. The most common form is Arbuscular Mycorrhiza (AM) which is found in the majority of commercially important crops and horticultural plants. AM association is beneficial to plants in many fold, since, it increases the area of rhizosphere, microrhizosphere for water and nutrient absorption by plants, decreases disease susceptibility, increases tolerance to adverse environmental conditions, and improves biomass and productivity of plants. Evidence indicates that AM fungi exhibits increased resistance to fungal root disease like wilt and root rots including nematode disease. The most common examples are *Glomus fasciculatum, G. mosseae, G. epigoes., G. intraradices* and *Aucalospora laevis*.

Mycorrhizal fungi have shown antagonistic influence on the population of plant parasitic nematodes which may be due to increased aminoacid contents (especially arginin) in the mycorrhizal roots (Nemec and Meridith, 1981) or presence of additional amount of phophorus in roots. *G. mosseae* was the most effective AM fungi in reducing *M. incognita* population in roots and promoting growth of crossandra (Nagesh *et al.*, 1997). Although most of the research conducted under greenhouse conditions indicated that mycrorrhiza fungi have the potential to nullify the damage due to phytonematodes, but intensive investigations need to be carried out under field conditions to assess their potential utilization as biocontrol agents. Since AM fungi are obligate parasites, they can not be cultured in large scales which hinder their effective utilization in integrated nematode management programme.

Bacteria

Biological control of plant parasitic nematodes by bacteria was first observed by Cobb in 1906 by *Pasteuria* sp. and later by Thorne, 1940. Depending upon their mode of action, the rhizoplane bacteria may be categorized into two groups.

Obligate Parasite

Bacteria those are pathogenic to nematodes or the nematode disease belong to this groups for example *Pasteuria penetrans.*

Pasteuria penetrans

It is an obligate parasite of some plant parasitic nematodes and so for three species of this genus *viz., P. penetrans* (Thorne 1940) Sayre and Starr (1985), *P. thornei* and *P. nishizawae* have been known to parasitize nematodes. *P. penetrous* was first described as a sporozoan, *Dubosqula penetrans.* (Thorne,1940). Finally it was redesignated as *P. penetrans* by Sayre and Starr (1985). It is known to parasitize 205 nematode species belonging to 96 genera from 10 orders of the Phylum Nematoda. It is a mycelial endospore forming bacterium which is an obligate parasite of large number of nematodes. Infection occurs through direct penetration of the cutide by germinating spores that encumber to the surface of the nematodes. The life cycle begins with the vegetative myclial colonies which eventually fill the nematode body with large mass of bacterial endospores. Sporangia are liberated from the body of the nematode upon de composition into the soil. Duration of life cycle of parasite from endospore to endospore varies with species and temperature. Bacteria like *P. penetrans* destroy nematodes by their parasitic behavior.

Rhizosphere Bacteria

Bacteria that colonize the rhizosphere soil under the chemical influence of the root or that of rhizosphere component are commonly referred as rhizobactria (Kloepper *et al.*, 1991). Example-

Agrobacterium radiobactor, Bacillus subtilis, B. thuringiensis, B. sphaericus, Pseudomonas fluorescens, P. chilinolytica and *P. aureojaciens* (Table 20.2)

Table 20.2: Rhizosphere Bacteria Used as Bioagent Against Plant Parasitic Nematodes

Biotic Agent	Nematode
Bacillus subtilis, B. punillis, B. cereus, Pseudomonas spp.	*M. incognita, H. cajani, H. zeae, H. avenae*
B.licheniformis, Pseudomonas mindocina	*M. incognita*
P. fluorescens	*M. incognita*
P. stutzeri, B. thuringiensis	*M. incognita*
B. subtilis, P. fluorescens	*M. incognita*
P. fluorescens	*M. incognita*
P. fluorescens	*Tylenchulus semipenetrans*
P. fluorescens	*M. incognita*
P. fluorescens	*Tylenchulus semipenetrans*
P.chitinolytica	*M.incognita*
Agrobacterium radiobactor	*Globodera* pallida

Rhizobacteria have been divided into two groups according to their mode of action (1) Plant Growth Promoting Rhizobacteria (PGPR) and (2) Plant health promoting rhizobacteria (PHPR) (Sikora, 1988). The rhizobacteria isolated from the root systems of plant exhibited antagonistic activity towards different cyst and root knot nematodes. There are a number of rhizobacteria used for nematode management like, *Agrobacteirum* sp., *Arthrobacterium* sp., *Azotobacter* sp., *Azospirillium* sp., *Bacillus* sp., *Corynebacterium* sp. and *Pseudomonas* sp. But only a few are economically important and found effective against plant parasitic nematodes, like *B. thuringicensis and P. fluorescens.*

Bacillus thuringiensis

A number of strains of *B. thuringiensis* are utilized in the biological control of nematodes. The exotoxin produced by *B. thuringiensis* was found toxic to the population of *Meloidogyne* sp. Dipel and SAN 415 strains of *B. thuringiensis* sub sp. *kurstaki* have been found to suppress populations of both *M. javanica* and *Tylenchulus semipenetrans.* Another isolate of *B. thuringiensis* was effective in reducing galling by root knot nematode on tomato and also reduced the population of *Rotylenchulus reniformis. Bacillus subtilis* and *B.cereus* were found to be antagonistic to *Meloidogyne* spp. (Cannayane *et. al.*,2001), *Heterodera avenae, H.cajani* and *H. zeae* (Gokte and Swarup,1998).

Pseudomonas fluorescens

It is capable of altering root exudates which control nematode behaviour. An interaction between root surface lectins and surface carbohydrates of the nematode may be prerequisite for penetration of nematodes. Fluorescent Pseudomonas has the property to form a ferric-siderophore complex. The bacteria utilizing this complex prevent the availability of iron to other micro organisms. Iron starvation may prevent the survival of other microorganisms including nematodes in the rhizosphere soil. The production of hydrogen cyanide by fluorescent Pseudomonas may also influence plant parasitic nematode population.

Actinomycetes

Avermectins (M K-936) are a new class of macro-cyclic lactones isolated from the soil actinomycetes, *Streptomyces avermitilus*. These compounds are reported to be possessing insecticidal, acaricidal and nematicidal properties (Putter *et al.*, 1981). The Avermectins were commonly distributed in most of the cultivated soil. Avermectins are a family of macrocyclic lactones consist primarily of four major (A1 a, A 2a, B1a, B2a) components. They are grouped into four as A1, A2, B1 and B2 (Laskota and Dybas, 1991). The potent nematicidal, insecticidal and accaricidal activities were established by (Stapley and Woodruff, 1982) Kitasato Institute, Japan. Avermectins have a direct action on neurotransmission. They are antagonists of gama amino butyric acid and has been widely used to control nematode parasites of animals.

Predaceous Nematodes

The predatory nematodes depending upon their mode of feeding and type of feeding apparatus may fall under the following categories;

1. One type feed nematodes by swallowing them whole for example; the species of *Tripyla* and *Monhystera* have a marvelous capacity for swallowing relatively large objects.
2. Another type has the stoma or the anterior part of it widened to form a kind of suction cup. Predators of this kind puncture or slit the body wall of their prey and suck out the internal organs. Certain species of the genus *Diplogaster* and most species of the genus *Mononchus* feed in this manner.
3. A third type of predator is armed with a stylet with that they pierce their prey and feed in the same way as plant parasitic nematodes. The predaceous species of the genus *Aphelenchoides* and most of the many predaceous species found in family Dorylaimidae.

Parasitic Protozoa

Nematodes are infected by sporozoans that may occur as either external or internal parasites. *Duboscqia penetrans* of the sporozoan family Nosematidae, is the commonly studied parasite. Spores and developing sporoblasts of this parasite from the infested nematodes remain in the soil and then attack other nematodes with which they come in contact. In external parasitism the spores attack themselves to the body of nematodes and draw their substance from the nemic body. The internal parasitism is more complicated. The developing sporoblast attack itself to the nematode body and penetrates the cuticle and enters into the body. Here it continues development and become a mature spore and multiply until the body of the host is filled with spores. When the nematode dies and the body disintegrates, the mature spore attacks themselves to other nematodes.

Advantages in Use of Bioagents

- ☆ Eco-friendly, help to achieve pollution free environments and maintain ecological balance.
- ☆ Reduces residues and health hazards.
- ☆ Problems of resurgence by the nematodes can be minimized.
- ☆ Less expensive in comparison to chemicals.
- ☆ Once established they remain effective for long periods especially in perennial crops.

Disadvantages in Use of Bioagents

- ☆ Non-identification of native isolates.

- ☆ Non-compatibility with other management practices.
- ☆ Specificity of bioagents to certain nematode species.
- ☆ Variation in performance due to interaction of other soil microorganisms.
- ☆ Limitations in isolation and mass multiplication of bioagents.
- ☆ Lack of awarness.
- ☆ Lack of standardized techniques for isolation and mass multiplication of bioagents.

In spite of the above limitations the development of bioagents is gaining importance presently with hope to have bright future and will act as an important component in the Integrated Nematode Management Strategies. The concern towards ecofriendly, safe, long lasting methods of crop management through organic farming will involve bioagents as most important component in nematode management practices.

Present Status

Some fungal and bacterial bioagents have effective role in nematode management. However, certain commercial products like Royal 300, a preparation of an isolate of *Arthrobotrys robusta* for the control of *Ditylenchus myceliophagus* on mushrooms, Royal 350, a different isolated of *Arthrobotrys* for control of *Meloidogyne* sp. on tomatoes in France and Biocon from *Paecilomyces lilacinus* against root knot nematodes in Philippines have proved some encouraging results. In India, certain formulations of bioagents are being marketed by some private companies and Institutes like pest control India Ltd. Multiplex and IIHR, Bangalore.

References

Baker and Cook (1974). *Biological control of Plant Pathogen.* Wilt. Freeman and Co., San Francisco. 270 pp.

Cobb, N.A. (1906). Fungus maladies of sugar cane with notes on associated insects and nematodes. (Sec. Edition). Hawaiin sugar planters Assoc. Exp. Sta. Div. Path. Physil. Bull. 5: 163-195.

Cannayane, I., Rajendran, G. and Ramakrishna, S. (2001). Biological potential of *B. cereus* and *B. subtilis* against *Meloidogyne incognita* in tomato. *Crop Research* 21: 167-169.

Gokte, N. and Swarup, G. (1998). On the potential of some bacterial biocides against root knot and cyst nematodes. *Indian Journal of Nematology* 18: 152-153.

Kloepper, J.W., Rodriguez- Kabana, R. Mcinroy, J.A. and Collins, D.J. (1991). Analysis of population and physiological characterization of microorganism in rhizosphere of plant with antagonistic properties to pathogenic nematodes. *Plant and Soil*, 136: 95-102.

Laskota, J. and Dybas, R.A. (1981). Avermectins-A novel class of compounds. Implications for use in Arthropod Pest control. *Annual Review of Entomology*, 36: 91-117.

Lohde, G. (1874). Einige neuen parasitichen plize. Tageblalt der 47 versammlung deusner natureforcher und Aerzter in Breslau: 203-06.

Nagesh, M., Parvatha Reddy, P and Rao, M.S. (1997). Integrated management of *Meloidogyne incognita* on tuberose using *Paecilomyces lilacinus* in combination with plant extracts. *Nematology Medit*. 25: 3-7.

Nemec, S. and Meridith, F.I. (1981). Amino acid content of leaves in mycorrhizal and non-mycorrhizal citrus root-stocks. *Ann. Botany*, 47: 351-58.

Putter, J.G., Mac Connell, F.A, Preiser, F.A., Haidri, A.A., Rishich, S.S. and Dybas, R.A. (1981). A vermectins: Novel class of insecticides, acaricides and nematicides from a soil micro-org. *Eperiention* 37: 963-64.

Sayre, R.M. and Starr, M.P. (1985). Proceedings of Helminthological society of Washington 52: 149-65.

Sikora, R.A. (1988). Interrelationship between plant health promoting rhizobacteria, plant parasitic nematodes and soil micro organisms. Med. Fac. Land bouww. Rijscuniv. Gent. 53/2b, 867-878.

Stapley, E.O., Woodruff, H.B. (1982). Avermectin, anti parasitic lactones produced by *Streptomyces avermitiles* isolated from the soils in Japan. Trends in Antibiotics Research. Japan Antibiotic Research Association. Tokyo. P. 154-170.

Thorne, G. (1940). Doboscqia penetrans n.sp. a parasite of the nematode, *Paratylenchus penetrans* in: Proc. Helminthological Society of Washington, USA. 7: 51-53.

Sustainable Disease Management of Agricultural Crops (2011) *Pages* **283–295**
Editors: **S.K. Biswas and S.R. Singh**
Published by: **DAYA PUBLISHING HOUSE, NEW DELHI**

Chapter 21

Trichoderma spp.: An Important Tools for Management of Diseases as a Biocontrol Agent and its Role in Disease Management

☆ *S.P. Singh*[1], *A.K. Singh*[1], *R. Bhagawati*[1] *and S.R. Singh*[2]
[1]*ICAR Research Complex for NEH Region, Arunachal Pradesh Centre, Basar – 791 101*
[2]*Krishi Vigyan Kendra, Belatal, Mahoba – 210 423, U.P.*

Trichoderma spp. particulary *T. viride*, *T. harzianum*, *T. virens* and *T. koningii* are the most widely used antagonist for the management of plant diseases. They are effective against a wide range of plant pathogens like *Rhizoctonia*, *Sclerotium*, *Macrophomina*, *Fusarium*, *Sclerotinia*, *Pythium*, *Magnaporthe grisea*, *Sarocladium oryzae* etc. They can be applied through seed, root, soil and foliage.

Characteristics of *Trichoderma* spp.

The hyphomycete genus *Trichoderma* is characterized by fast growing hyaline colonies bearing repeatedly branched conidiophores in tufts with divergent, often irregularly bent, flask shaped phialides. Conidiophores may end in sterile appendages with the phialides only borne on lateral branches in some species. Conidia are hyaline of more usually green, smooth walled or roughened. Hyaline chlamydospores are usually present in mycelium of older cultures. If the phialides are stongly convergent the conidial states of otherwise similar cultures are placed in *Gliocladium* while if they are straight and moderately divergent in *Verticillium*.

Trichoderma are best grown on 2 per cent Malt Extract Agar in day light or under NUV, in the dark they quickly loose the capacity to sporulate. For the examination of phialides structures very young cultures (usually 5 days old) must be employed, while conidial roughening is best judged in two week old culture. A selective medium has been developed for the quantitative recovery from soil besides peptone-glucose agar with 17 μg ml^{-1} Rose Bengal and 30 μgml streptomycin sulphate it contains 200 μgml formaldehyde.

T. viride

Conidiophores and their side branches long and slender without sterile hyphal elongation, phialides not crowded, rather slender, colonies yellowish, bright, dull to dark green, floccose or with compact conidiophore tufts. Conidia roughened, 3.6 to 4.8 × 3.5 to 4.5 μm.

T. harzianum

Conidiophores and their side branches long and slender without sterile hyphal elongation, phialides not crowded, rather slender, colonies yellowish, bright, dull to dark green, floccose or with compact conidiophore tufts. Conidia smooth walled. Conidiophores with complicated dendroid branching system, phialides regularly disposed in number of 3 or more. Conidia (2.8 to 3.2 x 2.5 to 2.8 μm.) globose, sub-globose or short obovoid, with length: width ratio of less than 1.25, colony reverse uncoloured, colonies reaching > 9 cm diameter in 5 days at 20 °C on OA.

T. koningii

Conidiophores and their side branches long and slender without sterile hyphal elongation, phialides not crowded, rather slender, colonies yellowish, bright, dull to dark green, floccose or with compact conidiophore tufts. Conidia smooth walled. Conidiophores with complicated dendroid branching system, phialides regularly disposed in number of 3 or more. Conidia ellipsoidal or oblong, often appearing angular, 3.0 to 4.8 x 1.9 to 2.8 μm.

T. virens

Colonies spreading broadly, conidia forming green slimy masses, phialides broadly ampulliform. Fast growing hyaline colonies bearing repeatedly branched conidiophores in tufts with divergent, often irregularly bent, flask-shaped phialides. Conidiophores may end in sterile appendages with the phialides only borne on lateral branches in some species. Conidia are hyaline or more usually, green, smooth-walled or roughened.

T. hamatum

Colonies reaching over 7cm diameter in five days at 20°C on Oat Agar Medium. Conidiophores long and thick, often with sterile hyphal elongations, side branches mostly short and thick bearing crowded, short and plump phialides, colonies white or whitish green to green, generally with compact, green in colour, 3.8-6.0 x 2.2–2.8 μm.

T. polysporum

Colonies slow growing, reaching over 7.0 cm diameter in 10 days at 20 °C on Oat Agar Medium. Conidia hyaline, small, 2.4–3.8 x 1.8–2.2 μm. Conidiophore branches and phialides relatively broad, conidia short-ellipsoidal, smooth walled, compacted into slimy heads.

T. aureoviride

Conidiophores and their side branches long and slender without sterile hyphal elongation, phialides not crowded, colonies yellowish, bright, dull to dark green, conidia smooth walled, conidiophores with a complicated dendroid branching system, phialides regulary disposed in numbers of three or more. Conidia short with a length: width ratio of < 1.5. Conidia obovoid with a truncate base 3.0–4.8 x 2.0 x 3.0 μm, reverse of colonies often discolored.

T. longibrachiatum

Conidiophores and their side branches long and slender without sterile hyphal elongation, phialides not crowded, rather slender, colonies yellowish, bright, dull to dark green, floccose or with compact conidiophore tufts. Conidia smooth walled, conidiophores with a simpler branching system, irregularly laterally disposed, often arising singly, conidia ellipsoidal, phialides usually only slightly alternate at the base, conidia large, partly dark green, 7 μm long, mostly ellipsoidal.

T. pseudokoningii

Phialides usually distinctly alternate at base, conidia smaller, pale green 2.8–4.8 μm long, mostly oblong ellipsoidal.

Selective Media for Isolation of *Trichoderma*

Table 21.1: Composition of *Trichoderma* Selective Media (TSM)

Component	*Quantity*
$MgSO_4.7H_2O$	0.2 g
K_2HPO_4	0.9 g
NH_4NO_3	1.0 g
KCl	0.15 g
Glucose	3.0 g
Agar	20 g
Rose Bengal	0.15 g
Chloramphenicol	0.25 g
Apron (35 SD)	0.015 g
Captan	0.2 g
Distilled water	1000 ml

Table 21.2: Extensive Work Done on Biological Control

Sl.No.	*Biocontrol Agent*	*Pathogen Against which Control Done*	*Crop*	*Application*
1.	*Trichoderma* spp.	*Rhizoctonia solani*	Sesamum	–
2.	*Trichoderma harzianum*	*Chocliobollus sativus*	Wheat	–
3.	*Trichoderma harzianum*	*Puccinia arachidis*	Groundnut	Field application
4.	*Trichoderma harzianum* *Bacillus subtilis*	*M. phaseolina* *Sclerotium rolfsii*	Soybean	Seed Treatment
5.	*Bacillus* spp. *Trichoderma harzianum*	*Alternaria linicola*	Linseed	–
6.	*Trichoderma viridi*	*Sclerotium rolfsii*	Soybean	–
7.	*Trichoderma koningii*	*Rhizoctonia solani*	Cowpea	Spraying
8.	*Trichoderma* spp.	*Fusarium* spp.	Cotton, Watermelon	
9.	*Trichoderma viride*	*F. oxysporum f. sp. ciceri* *R. solani, S. rolfsii*	Chickpea	Soil application

Contd...

Table 21.2–Contd...

Sl.No.	Biocontrol Agent	Pathogen Against which Control Done	Crop	Application
10.	*Gliocladium virens*	*R. solani*	Rice	Soil application
11.	*Trichoderma harzianum* *Gliocladium* spp.	*Aspergillus flavus* *A. niger*	Groundnut	Soil application
12.	*Trichoderma* spp. *Gliocladium virens*	*S. rolfsii*	Rice	–
13.	*Trichoderma harzianum* *T. virens, Gliocladium*	*R. solani*	Chickpea	–
14.	*Trichoderma* spp., *Gliocladium* spp.	*Meloidogyne incognita*	Sunflower	–
15.	*Trichoderma* spp.	*Gaeumannomuces graminis* var. *tritici*	Wheat	–

Trichoderma as an Antagonist

An antagonist is a microorganism that adversely affects another (*e.g.* a target pathogen) living organism in association with it. In biological control of plant pathogens, antagonists are biological control agents, with the potential to interfere the growth and development of plant pathogen. The mechanism of action of an antagonist can be called as antagonism. Antagonism includes:

Direct Action of Biocontrol Agents

1. *Competition*: It is for nutrients, growth factors, space (*e.g.* receptors sites on the cells) oxygen or any other material which are very scarce in supply and are needed by target pathogens also.
2. *Antibiosis*: It is the process of inhibition or destruction of one organism by a metabolic product of another organism.
3. Predation, hyperparasitism, mycoparasitism are other forms of direct exploitation of pathogen by another organism.

Indirect Action of Biocontrol Agents

1. Induction of systemic resistance in plants
2. Plant growth promotion
3. Solubilization and sequestration of plant nutrients.
4. Inactivation of harmful effect of pathogen's enzymes

Antagonism

Mycoparasitism

It is generally studied using dual culture technique.

Dual Culture Method

1. Melted PDA is poured in Petri plates (@ 20 ml per 90 mm plate).

2. Cut 6 mm discs of the antagonist and the test pathogen with the help of a sterilized cork borer from the edge of a 3 day old culture and place them in same plate 5 cm apart.
3. Incubate inoculated plates at 25 ± 2°C.
4. Periodical observations on the growth of the antagonist and test pathogen and the ability of the antagonist to colonize the pathogen are recorded.
5. For hyphal interaction microscopic observation are performed.

Study of Hyphal Interaction

Hyphal interaction between *Trichoderma* spp. and selected test pathogens are observed by employing dual culture technique both under light and scanning electron microscopy.

Light Microscopy

The mycelial mats are lifted gently with the help of needle and put in a drop of cotton blue strain on a microscopic slide. The mycelial mats are then transferred to a clean slide on a drop of lactophenol. Mycelial mats are then separated and spread with the help of needle, covered with cover slip and observed under an Olympus Binocular microscope at different magnification (10 x 40 and x 100). Microphotographs for different stages for hyphal interactions are taken under Olympus Vanox-S microscope using Olympus c-35 ad- camera, OR/OW 35 mm black and white roll.

Scanning Electron Microscopy (SEM)

About 0.5 X 0.5 cm PDA blocks with mycelium mats are cut from the zone of interaction (from dual culture plats) with the help of sterilized scalpel and the blocks are fixed in 3 per cent gluteraldehyde + 2 per cent para-formaldehyde and paste fixed in 1 per cent Osmium tetraoxide in 0.1M Cacodylate buffer at 4°C for 1-2 h. The specimens are then passed through ascending grades of ethanol and ethanol: iso- amyl acetate mixture as per details given below.

30 per cent ethanol- 10 minutes

50 per cent ethanol- 10 minutes

70 per cent ethanol- 10 minutes

90 per cent ethanol- 10 minutes

Absolute alcohol -2 changes for 20 minutes each

Ethanol: iso- amyl acetate

2:1- 2 changes of 30 minutes each

1:1–30 minutes

1:2–30 minutes

100 per cent iso-amyl acetate–changes for 20 minutes each

The dehydrated samples are critically dried in a Balzer's union critical point driver by gradual replacement by iso-amyl acetate with liquid CO_2 it is gradually heated up to 31.9 °C and 71.9 mm Hg atmospheric pressure to achieve complete drying by surpassing the liquid gas interpose.

The critically dried samples are mounted on clean aluminium stubs on double sticking tape and conductivity is made between aluminium stub and samples by applying the silver dag-915 on the margins of the tape. After that the samples are finally coated with Gold-Palladium alloy for 5-15 min in three consecutive series on a sputter coater for uniform coating and to enhance material density and electrical conductivity essential for emission of secondary electrons for fine and perfect image. The samples are observed under Phillips 515 Scanning Electron Microscope at varying magnification and selected areas are photographed using Indu/ORWO 125 Black and White 120 mm roll films.

Sclerotial Parasitization

Capacity of the antagonist to parasitize the sclerotia of pathogen is studied by agar and soil plate method.

Agar Plate Method

One ml fresh (5 days old) conidial suspension (10^6conidia/ml) of antagonist is poured on plain agar plates and uniformly spread by rotary motion. One gram of sclerotia are placed centrally or spread over the antagonist inoculated agar plates and incubated at 28 °C with 12h. light and darkness alternatively.

Periodical observation on colonization of the sclerotia by the antagonist are recorded. The colonized sclerotia are further tested for their viability by surface sterilization with 1 per cent sodium hypochlorite solution for 30 second and washed with distilled sterilized water before placing on PDA plates. Microscopic observations are done for more details.

Light Microscopy

The sclerotia of pathogen are observed regularly under Olympus Vanox-S Microscope using 4 × and 10 × objectives to record colonization by the antagonist.

Scanning Electron Microscopy

Parasitization of sclerotia is studied under Phillips 515 Scanning electron microscopy. Healthy and parasitized sclerotia either as such or cut in to two halves with scalpel and some parasitized sclerotia cleaned with the help of camel hair brush and processed as described under SEM of mycoparasitism. Micropotographs of selected area of the sclerotia whole sclerotia are taken for confirmation.

Soil-Plate Method

Five day old powder preparation of the antagonist is mixed thoroughly with natural soil @ 10 mg/100g soil/plate (90mm diameter) and tap water is sprayed with hand sprayer so as to get the moisture level near to saturation. One gram sclerotia is put on the soil surface and incubated at 28°C. Periodical observations on colonization of the sclerotia by the antagonist is recorded.

Antibiosis

The effect of culture filtrate of *Trichoderma* on a plant pathogenic fungi is studied following the method of Dennis and Webster(1971).

1. Pour 100 ml malt extract (20g/l) broth in 250 ml conical flask and sterilize in an autoclave at 15 lbs psi for 15 minutes.
2. Inoculate three 6 mm disc from 3 day old culture of the antagonist in each flask and incubate at 25±2°C for 7 days with constant shaking.
3. The culture filtrate is passed successively through Whatman No. 1 filter paper, G3 and G5 (bacterial proof) filters using a vacuum pump and the filtrate is collected in a sterilized vacuum flask.

4. Add culture filtrate to melted PDA (at 40 °C @ 6ml/60ml PDA) to obtain final concentration of 10 per cent (v/v)
5. Transfer culture filtrate amended PDA to Petri plates @ 20 ml per 90 mm diam plate). PDA plates inoculated with test pathogens not amended with culture filtrate are maintained as check.
6. Incubate each plate with 6mm disc of the test pathogen.
7. Incubate inoculated plates at 25±2°C
8. Record radial growth at regular intervals.
9. Record radial growth at regular intervals.

Antagonistic activity is calculated as per cent inhibition by using the following formula:

$$I = C - T / C \times 100$$

where,

I: Per cent inhibition,

C: Radial growth in check in mm

T: Radial growth in the treatment in mm

Production of Toxic Volatile Compounds

1. Grow *Trichoderma* on PDA at 25±2°C for 4 and 8 days.
2. Inoculate fresh PDA plate with 6 mm mycelial disc of 3 days old test pathogen.
3. Place pathogen inoculated plate in inverted position over a basal plate containing 4 or 8 days old culture of *Trichoderma*. Both these plates are made airtight by wrapping cellophane tape. Plates with only PDA medium (without antagonist) at the lower side and plate inoculated with mycelial disc of the test pathogen at the upper side served as check.
4. Radial growth of the pathogen is recoded at different intervals. Toxicity of volatile compound is expressed as percent inhibition in growth of the test pathogen.

Production of Extra Cellular Enzymes

Overall cellulolytic activity of the antagonist is investigated by a simple experiment. In this, the antagonist is grown in 250 ml conical flask, containing 100 ml of wheat bran water medium (40g/lit, Ph 5.0), for 6 days under constant shaking at 28 0C. The culture plate is collected after passing through Whatman No. 1 filter paper. Overall cellulolytic activity in this filtrate (crude enzyme preparation) is assayed according to the method described by Mendels (1975).

Composition of Media

Table 21.3

Reagent Require	*Quantity*
Dinitrosalicylic acid (DNSA)	5.0 g
Phenol	1.0 g
NaOH	5.0 g
Na-Metabisulphite	0.25g
Na-K-tartarate	91.0 g
Distilled water	500ml

Preparation of Standard Curve

Five glucose concentrations *viz.* 0.2, 0.4,0.6, 0.8 and 1.0 mg/ml are prepared in 0.1 M citrate buffer of pH 4.8 to 1 ml of each these standard solution, 0.5 ml of buffer and 3.0 ml of DNSA are added. Reagent blank contains 1.5 ml of buffer and 3 ml of DNSA. The solutions are boild in water bath for 10 minutes and cooled to room temperature. The optical density (OD) is measured in a Beckman DU 7 Spectrophotometer at 550nm. The OD values are plotted against the concentration of glucose to obtain the standard curve.

Enzyme Assay

Celluloytic activity determined by measuring the amount of glucose released from filter paper (Whatman No. 1 size 6 cm x 1 cm) by the crude enzyme preparation. To 1 ml of buffer, 0.5 ml of crude enzyme preparation and filter paper strip 96 x 1 cm are added and the reaction mixture is incubated at 50 °C for 1 hours. To this, 3 ml of DNSA is added, boiled for 10 minutes, cooled to room temperature and OD value are recorded. Substrate (filter paper) blank and enzyme blanks are also maintained. Net OD value is calculated as:

Net OD = Observed OD–(OD in filter paper + OD in enzyme blank)

The amount of glucose released is calculated from the standard cure. The enzyme activity is expressed as μ mole of glucose released per ml of enzyme preparation per hour.

Induction of Systemic Resistance

Some isolates/strains of fungal and bacterial biocontrol agents like *Trichoderma Aspergillus, Penicillium, Psedomonas, Serratia, Bacillus,* etc. are clearly potent inducer of induced systemic resistance. Because many biocontrol agents triggering induced systemic resistance can also inhibit growth of the pathogen, directly, their capacity to suppress disease may involve more than one mechanism. Thus, in order to prove that resistance is induced and that it is truly systemic, it must be shown that inducing biocontrol agents are absent from the site of challenge with the pathogen and that challenging pathogen remain spatially separated from the duration of the experiment. Procedures for study of ISR are as follows:

1. Coat the seeds with high number of biocontrol agent and incubate for 35 to 48 hours depending on the type of seeds. Then sowing the seeds in sterilized soil, sterilized water is added in the soil to maintain moisture.
2. Dip the roots of seedlings in the suspension of biocontrol agents and then transplant in autoclaved soil.
3. Pour a suspension of the biocontrol agent on/or mix it with the autoclaved soil.
4. Now seedling is challenged with the pathogen causing disease on the aerial parts of the plant.

Suppression of disease on foliage is indicative of ISR. However, it does not totally rule out the involvement of toxic and systematic metabolite produced by the biocontrol agent.

Plant Growth Promotion is Another Important Mode of Action of Bioagent

1. Control of deleterious root micro-flora including those not causing obvious disease.
2. Direct production of growth stimulating substance *e.g.* hormones and/or
3. Increased nutrient uptake through enhanced root growth.

Growth promoting effect of *Trichoderma* on plants can be tested by applying it as seed treatment or seedling dip @ 4 g formulation (approx. 10^9 spores per gram formulation) per kg seed/litre suspension) Treated seeds are raised in sterilized soil. Comparison of plant growth is done against appropriate control.

Compatibility among Antagonists

Like chemical pesticides, mixed formulation of biopesticides are likely to impart better efficacy and wider applicability with respect to pathogens, crops and environmental conditions. In general fungal antagonists are more effective in acidic soil, whereas bacterial antagonists are better suited for alkaline soil. Therefore, mixed formulation of fungal and bacterial antagonists are likely to be effective under wide range of soil pH. However, for the mixed formulation of antagonists their compatibility becomes a limiting factor. In general they are not compatible. Therefore, compatibility must be tested before making mixed formulation or before mixing prior to application. For the testing of compatibility between two fungal antagonists dual culture method on PDA medium, as described earlier for testing bio-efficacy against fungal pathogens can be used. Dual culture can be used for testing compatibility of *Trichoderma* with bacterial antagonists *e.g. Pseudomonas spp*. However, for dual culture following medium should be used-

Table 21.4

Composition of Media Component	*Quantity*
Potato	100 g
NaCl	2.5 g
Peptone	2.5 g
Yeast extract	0.75 g
Beep extract	0.75 g
Agar	20 g
Water	1000 ml

Compatibility with Agrochemicals

Effect of agrochemicals on radial growth of *Trichoderma* is studied by Poisoned Food Technique (Grover and More,1961). Stock solution/suspension (1000 µg a.i./ml) of each chemicals is prepared in sterilized distilled water. Required amount of stock solution is poured in 250 ml flask containing 100 ml of sterilized melted PDA so as to get final concentrations *e.g.* 1,5 and 10 µg a.i./ml. For calculation of amount of stock solution to be added to PDA to get about concentrations, the formula C1V1=C2V2 is used, where C1 and C2 are the concentrations of stock solution (µg/ml) and concentration desired (µg/ml) respectively. V1 is the volume of stock solution to be added to the unknown volume (V2) of PDA. PDA containing different concentrations of chemicals are poured into sterilized petriplates. After solidification, each petriplates is centrally inoculated with 6 mm mycelial disc of *Trichoderma*. Unamended PDA plates inoculated centrally with *Trichoderma* serve as control. Plates are incubated at 26°C and colony diameter measured when control plates are fully covered with the growth of test fungus. Percent inhibition (I) of growth is calculated by using the formula:

$$I = C-T/C \times 100,$$

where,

I: Inhibition per cent

C: Colony diameter in control

T: Colony diameter in amended medium(mm)

Trichoderma formulations are generally compatible with most of the anti bacterial antibiotics. Among the fungicides, they are compatible with carboxins, oxycarboxin and metalaxyl.

Mass Multiplication and Formulation

One of the greatest obstacles to biological control by introduced antagonists has been the lack or scarcity of methods for mass culturing and delivering the biocontrol agents. The unique problem is developed in biopesticides is that it represents a living system, which must be able to withstand the process of formulation and should remain sufficiently viable for a period until it reaches the farmers. There has been considerable progress in this direction in recent past. *Trichoderma spp.* have been grown on solid substrates like sorghum grains, wheat straw, wheat bran, spent tea leaf waste, wheat bran, saw dust, diatomaceous earth granules impregnated with molasses, *etc.* for their mass multiplication (Baby and Manibhushanrao,1991; Gogoi and Roy,1996; Mukhopadhyay,1994). *Trichoderma* and *Gliocladium* are being mass multiplied by liquid formulation in 50, 100 litre vessels. Fermented biomass of *Trichoderma* and *Gliocladium* consisted mainly of chlamydospores and conidia with some amount of mycellial fragments. *Trichoderma* produced on solid substrate are virulent than those produced by liquid formulation.

Mass Multiplication of *Trichoderma* spp. on Solid Substrate

1. Soak the sorghum grains overnight in water.
2. Decant water and boil these again in 0.2 per cent dextrose solution for 15 minutes.
3. Transfer boiled grains in glass bottles or flasks @ 70 g/250 ml flask and autoclave at 15 lbs psi. for 20 minutes.
4. Inoculate five 6 mm discs of 5-day old *Trichoderma* culture on PDA to each flask and mix by hand shaking.
5. Incubate at 28°C for 3 weeks
6. Dry colonized grains at 40°C for 96 hours and grind with mixture grinder or Willey Mill to fine powder.
7. Count the colony forming units per gram powder.
8. Mix spore powder with commercially available carrier (talc) and carboxy methyle cellulose @ 1 per cent in a ratio so that cfu of formulated product is 10^9 cfu. g^{-1}.

Methods of Application

Seed treatment: @ 3 to 4 g per kg seed depending on seed size. Detail procedure is as fallows:

Seed Coating

The application of biocontrol agent to seed as a coating is the most common method of application. The biocontrol agent is suspended in water and applied as a slurry. A sticker/binder is often used to adhere the biocontrol agent to the seed, in case they are not present in formulation. These include Pelgel, Carboxy Methyl Cellulose (CMC) and Xanthan gum.

Seed Biopriming

Seed biopriming process has potential advantages over simple seed coating. It result in rapid and uniform seedling emergence and may be useful under adverse soil conditions such as low temperature or excess soil moisture. Detailed procedure is as follows:

1. Suspend 4 g *Trichoderma* formulation in 100 ml water in a container.
2. Add 1 kg seed and mix thoroughly.
3. Transfer treated seeds in a tray lined with moist blotting paper. Cover the seeds with moist blotting paper and polythene.
4. Incubate at 25 to 28°C for 24 to 48 hours depending on seed. Fast germinating seeds are incubated for shorter period.

Root Dip

Roots of the seedlings are dipped in the suspension of biocontrol agent before transplanting @ 4 g/litre suspension.

Soil Amendment

Numerous attempts have been made to control several soil borne pathogens by incorporating soil nitrogen substrates colonized by antagonists of pathogen into soil.

Soil Drenching

Drenching will not be feasible as there will be no even distribution of bioagents in the soil. Soil drenching with *Trichoderma herzanium* has given control of stem rot of groundnut caused by *Sclerotium rolfsii*.

Aerial Spraying/Wound Dressing

Trichoderma spp. can be applied as foliar spray to control diseases affecting above ground parts. Biological control of foliar disease is not so developed as biocontrol of soil borne diseases. A most significant example of *Trichoderma* application to aerial plants parts is the biocontrol of wounds on shrubs and trees applied after pruning.

Secondary Metabolites

Trichoderma spp. is a potential biocontrol agent of several soil borne pathogens. *Trichoderma* spp. has been proved to be useful in the control of phytopathogens affecting different crops (Benitez *et al.*,2004). *Trichoderma* spp. *viz., Trichoderma virids, Trichoderma harzianum, Trichoderma koningii, Trichoderma hamatum, etc.* have been in use to produce various biopesticides under different trade and commercial names. It is highly active on root rot, collar rot, stem rot, damping off, wilt, blight/leaf spot of crops like patcholi, coleus, Melissa and pulses, oilseeds, cucurbitaceous crops like cucumber, pumpkin, bottle gourd, ridge gourd, solanaceous crops like tomato, brinjal, chillies, capsicum, cole crops like cabbage, cauliflower, root crop disease like tuber, rhizome rot of potato, ginger, turmeric along with rotten disease of garlic, flowers etc. The product is also effective against sheath rot, sheath blight and bacterial leaf blight of rice (Eziashi *et al.*, 2006).

Trichoderma species produce both volatile and nonvolatile secondary metabolites that adversely affects growth of different fungi (Horvath *et al.*,1995). The biological activity is due to production of antibiotics *viz., Trichodermin* (Kumeda *et al.*, 1994). Neverthless, it appears that each *Trichoderma* isolates behaves differently when faced with the same plant pathogen. The differences observed in the biocontrol would depend on the ability of each isolate to produce antibiotics (Dennis and Webster,1971 a and b) and or to express gene that codify for extracellular fungal cell wall hydrolyzing enzymes such as chitinase,ß-1,3-gluconases and/or proteases, or in the specific isoenzyme pattern expressed by each isolate (Grondona *et al.*,1997).

The accumulation of such secondary metabolites may be induced by changing the condition of culture (Bode *et al.*,2002) or by mutagenesis (Aziz and Smyk,2002). Generally, the biosynthesis of secondary metabolites in bacteria or in fungi is controlled by carbon sources, nitrogen sources and phosphate ions. The nature and concentration of carbon and nitrogen sources regulate secondary metabolism production through the catabolic repression phenomenon (Icgen *et al.*,2002; Lopez *et al.*,2003).

The dominance of *Trichoderma* is achieved by biosynthesizing a wide array of secondary metabolites, transforming a great variety of natural and xenobiotic compounds and producing varied degradative enzymes such as chitinase (Glober,1997). It is clear that *Trichoderma's* growth inhibiting properties on other fungi are probably due to the combined action of cell wall degrading enzymes together with the capacity of *Trichoderma* to produce different secondary metabolites. The important secondary metabolites types isolated from *Trichoderma spp.* emphasizing their biological role that these metabolites play in biological control mechanisms. The types of important fungal metabolites families produced by *Trichoderma* are Anthraquinones, Daucanes, simple pyrones, Koinginins, Trichodermides, Viridins,Viridiofungins, Nitrogen heterocyclic compounds, Trichodenones and cyclopentenone derivatives, Azaphilones, Herzialactones and derivatives; Butenolides, Trichothecenes,Isocyano metabolites, Setin like metabolites, Bisorbicillinoides, Diketopiperazines, Ergosterol derivatives, Peptaibols, Cyclonerodiol derivatives, Statins, Heptelidic acid and derivatives; Acoranes and Miscelanea etc. As indicated by Cardoza *et al.* (2005), the different metabolites exhibiting antibiotic activity in *Trichoderma* can be classified as two main types.

1. Low molecular weight and volatile metabolites which include–simple aromatic compounds, some polyketides such as pyrones and the butenolides, volatile terpenes and the isocyane metabolites. All of these are relatively non-pollar substances with a significant vapour pressure. These "volatile organic compounds" in the soil environment would be expected to travel over distance through systems and thus enhance the status of one organism by affecting the physiology of competitor organisms.
2. The high molecular weight are polar metabolites which, like peptaibols, may exert their activity on direct interactions by contact between *Trichoderma spp.* and their antagonists.

In addition, the study of *Trichoderma spp.* as a source of biologically active metabolites is especially significant and ensures interest in this subject for the years to come.

References

Aziz, N.H. and Smyk, B. (2002). Influence of UV radiation and nitrosamines on the induction of mycotoxins by nontoxigenic molds isolated from feed samples. *Nehrung-Food*, 46:118-121.

Baby, U.I. and Manibhushanrao, K. (1991). Control of rice sheath blight through the integration of fungal antagonists and organic amendments. *Tropical Agriculture*, 70(3):240-244.

Benitez, T., Rincon, A.M., Limon, MC and Condon, AC. (2004). Biocontrol mechanisms of *Trichoderma* strains. *International Microbiology*, 7 (4) 249-260.

Bode, H.B.; Bethe, B.; Hofs, R. and Zeeck, A. (2002). Big effects from small changes possible ways to explore natures chemical diversity. *Chem. Biochem.*, 3: 619-627.

Dennis, C. and Webster, J. (1971). Antagonistic properties of species groups of *Trichoderma harzianum*. *Trans. Br. Mycol. Soc.* 57: 363-369.

Dennis, C. and Webster, J. (1971a). Antagonistic properties of species-groups of *Trichoderma* I. Production of Non volatile antibiotics. *Trans. Br. Mycol. Soc.* 57: 25-39.

Dennis, C. and Webster, J. (1971b). Antagonistic properties of species-groups of *Trichoderma* I. Production of Non volatile antibiotics. *Trans. Br. Mycol. Soc.* 57: 41-48

Gogoi, R. and Roy, A.K. (1996). Effect of soil pH and media on the antagonism of *Aspergillus terreus* to the rice sheath blight fungus. *Indian Phytopath.*49 (1): 32-36.

Grover, R.K. and Moore, D.J. (1961). Adaptation of *Sclerotinia fructicola* and *Scleritinia coax*. *Phytopathology*. 51: 344-401.

Mandels, M. (1975). *Biotech. Bioengg. Sympo.* 5: 81-105.

Mukhopadhyay, A.N. (1994). Biocontrol of soil borne fungi plant pathogens-current status, future prospect and potential limitations. *Indian Phytopath.* 47 (2): 119-126.

Sustainable Disease Management of Agricultural Crops (2011) *Pages* **296–302**
Editors: **S.K. Biswas and S.R. Singh**
Published by: **DAYA PUBLISHING HOUSE, NEW DELHI**

Chapter 22

Disease Management of Important Solanaceous Vegetables

☆ ***Pranabjyoti Sarma[1], R. Bhagawati[1], S.P. Singh[1] and S.R. Singh[2]***
[1]ICAR Research Complex for NEH Region, Arunachal Pradesh Centre, Basar – 791 101
[2]Krishi Vigyan Kendra, Belatal, Mahoba – 210 423, U.P.

Tomato (*Lycopersicon esculentum* Mill.)

Bacterial Wilt

It is one of the most serious diseases of tomato. The disease is caused by *Ralstonia solanacearum*. Disease is characterized by wilting of the plants. Lower leaves may drop before wilting. The vascular system of the plants becomes brown in colour.The bacterium survive in the soil for long period mainly confined to vascular region which later spreads to cortex and pith causing yellowish brown discoloration of tissue. If an infected stem is cut crosswise, it wills look like brown and tiny drops of yellowish ooze may be visible.

Control Measures

1. Planting should be done with disease free healthy seedlings.
2. Use of resistant varieties or immune varieties, *i.e.* Arka Alok, Arka Abhay and Arka Abijit.
3. Crop rotation for 2-3 years with sunflower,potato and eggplant.
4. Application of oil cakes and neem cakes are also recommended.
5. Spray crop with streptocycline@ 0.02 per cent with appearance of disease.

Damping Off

It is also one of the most common diseases which occur mostly during the seedling stage. The disease may be caused by *Pythium ultimum, Phytophthora parasitica* and *Rhizoctonia solani.* The pathogen is seed and soilborne.Symptoms of the disease occur in two phase *ie*,pre and post emergence damping off.Water soaked areas developed at the base of the seedling and the tissues become soft which leads to complete collapse of the plants.

Control Measures

1. Sanitary measure and use of well decompose manure.
2. Seed treatment with Copper oxychloride @ 2 g per kg of seed.
3. Seed bed treatment with Formalin @ 0.5 liter per 10 litre of water at 15 days before sowing.
4. Drenching of nursery bed with Mancozeb @ 2.5 g per liter of water.
5. Avoid excess watering during hot weather.
6. Use of crop protectant *ie.* Thiram @ 2g/kg of seed with *T.harzianum* and *T.viride* @ 4g/kg of seed.
7. Allow the surface of the soil to dry between watering.

Late Blight

The disease is caused by *Phytophthora infestans*. Mostly occurs in sub-tropical and temperate regions of the country. The disease is characterized by brown lesions in different parts of the plants. It may affect the leaves, stems and fruits. On leaves, it appears as pale green, irregular spots on the tips and margins which's moist weather rapidly with central turning necrotic and dark brown or black. On the lower surface of the leaves, a white downy growth of the fungus appears around the dead areas. The disease appears at low temperature and high humidity condition. The pathogen is soil borne and survives in infected seed tubers.

Control Measures

1. Use of disease free tubers for seed.
2. Use of resistant varieties.
3. Avoid planting of tomato just near to potato.
4. Depending on weather conditions, particularly on cloudy weather spraying of Mancozeb @ 2.5 g per litre of water is recommended. Alternatively, Zineb @ 2.5 g per litre of water can be used. Spraying may be repeated at an interval of 10 days depending on the severity of the disease. Care should be taken to cover the lower sides of the leaves while spraying.

Early Blight

This is a fungal disease and caused by *Alternaria solani*. Circular brown spots appear on the leaves and fruits.The spots may not have concentric rings are more prominent in large blotchy spots and give them a target effect. The fruits are also affected by the disease, so the yield loss may go up to 50-60 per cent under severe condition.

Control Measures

1. Follow 3-4 years crop rotation.
2. Dead haulms should be raked together and brun immediately after crop harvest.
3. Seed treatment with Thiram or Captan @ 2.5 g per kg of seed.
4. Spraying of Mancozeb @ 2.5 g per litre of water is recommended.
5. Spray crop with copper fungicide at least 3-4 weeks after planting.

Leaf Curl Disease

This is a viral disease and transmitted by whitefly (*Bemisia tabaci*). Curling of leaves, reduction in leaf size, excessive branching and stunted plant growth are the characteristic symptoms of the disease. An older leaf become thick and breaks easily.

Control Measures

1. Clean cultivation and removal of infected plants
2. Crop rotation with crucifier crop as cabbage, broccoli and turmeric.
3. Spraying of Dimecron @ 1.5 ml per litre of water at 10 days interval to control the whitefly.
4. Application of Carbofuran @ 1.5 kg per hectare at the time of transplanting of seedling.
5. Spray of systemic insecticide *ie*, monocrotophos 36 SL@ 0.4g/lit.
6. Destroy infected plant from field regularly to check vector.

Potato (*Solanum tuberosum* L.)

Late Blight

It is one of the most serious diseases of potato. The disease is caused by *Phytophthora infestans*. Although, the disease appears frequently in the hills but it affects the crops in different parts of the country. It affects the leaves, stem and tubers of the plants. Water soaked spots developed on the leaf which turns brown at later stages. Affected tuber shows reddish brown flesh and sugary texture. The lower surface of the leaves, a white downy growth of the fungus appears the dead areas. The funguses is soil borne and survive in infected potato seed tuber and plant debris.

Control Measures

1. Use disease free seed for cultivation.
2. Seed treatment with 1 per cent Bordeaux mixture
3. Grow resistant varieties like Kufri Jyoti, Kufri Megha for hills and Kufri Sutlej, Kufri Kishan and Kufri Badshah for the plains.
4. Spraying of Metalaxyl @ 2.5 g per litre of water followed by 2 to 3 sprays of Mancozeb @ 2 g per litre of water is recommended for effective control of the disease.

Potato Scab

The disease is caused by *Streptomyces* spp. It is characterized by brownish roughening of tuber skin for which the tubers fetch low price in the market. The lesion are corky tissue arise from abnormal proliferation of tuber periderm when tuber is exposed to pathogen. Individual scab lesions are circular but may into large scbbay areas. In deep pitted scab, a lesion are 1-3 mm in depth and darker than shallow scab.The pathogen is seed and soil borne in nature.

Control Measures

1. Use disease free seed for cultivation.
2. Seed treatment with 3 per cent Boric acid for 30 minutes.
3. Crop rotation with cereals.

Early Blight

This is also one of the common diseases of potato and is found in the hills as well as in the plains. It is caused by *Alternaria solani.* Since the disease appears early in the season, it is known as early blight, in contrast to *Phytophthora infestans* which appears late in the season, to cause blighting of the foliage. Irregular, brown to dark brown spots of different sizes appear on the leaves. Often several spots coalesce to form large patches on the leaves. In severe cases the entire foliage is blighted. Though the fungus is mostly confined to the leaves, sometimes it may affect the tubers near the soil surface, causing brown discoloration. The tuber infection is carried to the storage godowns where it may spread to cause storage-rot, resulting in considerable damage.

Control Measures

1. Follow 3-4 years crop rotation.
2. Use resistant or tolerant cultivars ie, Kufri Naveen, Kufri Sinduri and Kufri jeevan.
3. Dead haulms should be raked and brust immediately after crop harvest.
4. Spraying of Zineb @ 2.5 g per litre of water at 15 days interval is recommended.
5. Spraying of Mancozeb @ 2 g per litre of water when the disease appears.

Bacterial Wilt (Brown Rot)

The disease is caused by *Ralstonia solanacearum*. It is characterized by drooping down of few leaves or entire foliage mostly during noon time.The brown rot refers to the browning of the xylem in the vascular bundles.The browning is often visible of the infected stem as dark patches.The skin of infected tuber is often discoloured.In severly affected tuber the eye buds are blackened. Tuber looks healthy but some tuber may show bacterial exudates in the eyes. Cut stems and tubers shows vascular browning. If the tubers are stored at warm temperature then the infected tubers rot easily.

Control Measures

1. Crop rotation with non solanaceous crops
2. Infected plant residue should be always removed and brunt.
3. Seed tuber should be treated with stretocycline @ 0.02 per cent for 5 minute.
4. Use disease free planting material.
5. In the diseases prone areas use whole tubers for planting instead of pieces of tubers.
6. Application of bleaching powder @ 10 kg per hectare along with fertilizers at the time of planting.

Brinjal (*Solanum melongena* L.)

Phomopsis Blight

This is a fungal disease. The fungus infects the seedlings in the nursusery damping off symptoms. The leaves are infected smaller circular spots appear which become gray to brown with blackish margins. Dark brown spots developed on the leaves and stem. Soft water decayed areas formed on the fruit which turns black at later stages. Ultimately rotting of the fruits occurs. The fungus perpetuates in soil and live on affected tissue.The disease is soil and seed borne in nature.

Control Measures

1. Removal and burning of the infected plants.
2. Seed always obtained from disease free plants should be used for planting.
3. Seed treatment with Thiram or Captaf @ 2 g per kg of seed.
4. Removal and burning of the infected plants.
5. Spraying with Zineb @ 2 g per litre of water at 10 days interval.

Little Leaf

The disease is transmitted by leafhopers. The disease is characterized by occurrence of small yellow leaves and stunted growth. The leaves show a reduction in size and malformed. Disease affected plant are generally shorter bearing a large number of branches. The stimulated axillary buds grow out into axillary shoots which in turn produce secondary and tertiary axillary branches. The petals turn into green and appear leaf like. Affected plants generally do not bear fruits. However; if any fruit is formed it becomes hard and tough and fails to mature.

Control Measures

1. Removal of diseased seedlings and transplanting of healthy seedlings.
2. Destroy the diseased plants from the field.
3. Dipping of seedlings with 0.2 per cent Carbofuran for 24 hours to reduce the insect vector.
4. Use of resistant varieties *ie,* Black Beauty, Brinjal round and Shurti.
5. Spray malathion @2 ml/lit. of water starting with the appearance of the leaf hopper control.

Bacterial Wilt

The disease is caused by *Pseudomonas solanacearum*. Wilting, stunting, yellowing of the foliage and finally collapse of the entire plant are the characteristic symptoms of the disease. Lower leaves may droop first before wilting occurs. The vascular system becomes brown and bacterial ooze comes out from the affected parts. Plant show wilting symptoms at noontime will recover at nights, but die soon.

Control Measures

1. Crop rotation with cruciferous plants.
2. Growing of resistant variety like Pant Samrat.
3. Spraying of 1 per cent Bordeaux mixture.
4. Clean cultivation and destroying of affected plants by burning at the first appearance of the symptoms.

Cercospora Leaf Spot

The disease is caused by *Cercospora solani*. It is characterized by chlorotic lesions, angular to irregular in shape, later turn grayish-brown with profuse sporulation at the centre of the spot. Severely infected leaves drop off prematurely, resulting in reduced fruit yield.

Control Measures

1. Growing of resistant variety like Pant Samrat

2. Spraying of 1 per cent Bordeaux mixture or 2 g Copper oxychloride or 2.5 g Zineb per litre of water effectively controls leaf spots.

Mosaic

This is a viral disease caused by potato virus Y and is transmitted by aphids. Mosaic mottling of the leaves and stunted growth are the characteristic symptoms of the disease. The symptoms in early stages are mild but it becomes severe at later stages. The infected leaves are small, deformed and becomes leathery. Plants show a stunted growth when infected in the early stages.

Control Measures

1. Removal of infected plants.
2. Application of Carbofuran @ 1.5 kg per hectare at the time of transplanting.
3. Spraying of Dimethoate @ 0.05 per cent at 10 days interval.
4. Spraying of phosphomidon@ 0.05 per cent at an interval of 10 days.

Chilli and Capsicum (*Capsicum* spp.)

Damping Off

It is a common disease which appears during nursery stage and caused by *Pythium* spp. It attacks the seedlings near the ground level and cause collapse of the plants.The affected seedlins may topple down. In nursery disease may begin in patches within few days the entire seedlings amy be killed.

Control Measures

1. Seed treatment with Captan or Thiram @ 2 g per kg of seed.
2. Sowing the seeds as thin as possible in raised beds prepared in the open area during summer months.
3. Drenching of seed bed with Captan @ 2 g per litre of water at 15 days interval.
4. Spraying of plants with 1 per cent Bordeaux mixture at monthly interval.
5. Apply *Trichoderma viride*@ 100ppm/m sqare in soil.

Anthracnose

The disease is caused by *Colletotrichum capsici*.The disease symptoms are observed in two phases ie, twig as die back and on ripe fruit as fruit rot. In the infected stem the bark first turn brownish and then turn to shiney white in long and narrow strips containing several blaki dots like frutification. It is also called as fruit rot. Small water soaked spots developed on the fruit which later enlarges. It is a seed-borne disease.

Control Measures

1. The seed from disease free crop only should be used for sowing.
2. Seed treatment with Captan or Thiram @ 2 g per kg of seed.
3. Spraying of Mancozeb or Carbendazim @ 1.5 g per litre of water at 15 days interval.
4. Apply *Trichoderma viride*@ 100ppm/m^2 in soil.

Leaf Curl

The disease is transmitted by whitefly (*Bemisia tabaci*). The disease is characterized by curling of leaves accompanied by puckering and blistering of intervenial areas and thickening of the veins. The axillary buds becomes bunches of small leaves.

Control Measures

1. Removal of infected leaves.
2. Soil application of Carbofuran @ 1.5 kg per hectare followed by 2-3 sprays of Metasystox @ 1.5 ml per litre of water to control the whitefly.
3. Spray of malathion @ 0.04 per cent at 10 days interval.

Mosaic

It is transmitted by aphids and caused mottling of the leaves. The virus induces clearing of veins of younger leaves and in later stages leaf turns yellow followed by irregular pattern of light yellow and green patches. Other important symptoms are leaf distortion, curling, marginal rolling of leaves. The planets become stunted and flower production is ceased.

Control Measures

1. Removal of the infected plants.
2. Growing of resistant varieties like Pusa Jwala, Pusa Sadabahar, Punjab Lal and Pant C-1.
3. Application of Carbofuran @ 1.5 kg per hectare followed by 2-3 sprays of Metasystox @ 2 ml per litre of water to control the insect vector.

References

Handbook of Horticulture (2001) edited by K.L. Chadha and published by Directorate of Information and Publications of Agriculture, ICAR, New Delhi.

Vegetables, Tuber crops and Spices (2001) edited by S. Thamburaj and Narendra Singh and published by Directorate of Information and Publications of Agriculture, ICAR, New Delhi.

Package of Practices of Horticultural Crops of Assam (1994). Published Jointly by Department of Agriculture, Assam and Assam Agricultural University, Jorhat.

www.ikisan.com.

Sustainable Disease Management of Agricultural Crops (2011) *Pages* **303–311**
Editors: **S.K. Biswas and S.R. Singh**
Published by: **DAYA PUBLISHING HOUSE, NEW DELHI**

Chapter 23

Diseases of Isabgol (*Plantago ovata* Forsk.) and their Management

☆ ***Kunal Mandal***

Directorate of Medicinal and Aromatic Plants Research, Boriavi, Anand – 387 310, Gujarat

Isabgol is an irrigated medicinal crop commercially cultivated in India during the winter season (November–March). India enjoys monopoly in production and supply of isabgol to the world market. The crop is grown in the western parts of the country falling under the states of Rajasthan, Gujarat and Madhya Pradesh. It requires less input (irrigation and fertilisers). In fact, it requires half the irrigation that of wheat. Hence, it has become a good choice for the poor farmers. However, some of the diseases pose problems for its successful cultivation.

Downy Mildew

Downy mildew is the most important and destructive disease of this crop. Desai and Desai (1969) first reported this disease from Anand, Gujarat, India. Later the disease was also reported from other parts of the country. Kapoor and Chowdhary (1976) and Rathore and Rathore (1996) reported the disease from Delhi and Rajasthan, respectively. However, this disease was not reported outside India though it was cultivated in Australia, lower desert of North America.

Symptoms

Two distinct types of symptoms–systemic and localised, were observed in the downy mildew of isabgol. Also, we may group the symptoms of vegetative parts and the inflorescence. The symptoms of systemic infection on whole plant appear about 30 days after sowing. The diseased plants turn chlorotic in appearance. Profuse ash coloured downy growth develops on above ground parts of whole plant, mainly the leaves. The leaves later become necrotic, curled, twisted. Rathore and Rathore (1996) observed abnormal spike formation in systemically infected plants.

The initiation of localised symptoms on the vegetative parts starts as pale yellow discoloured patches of leaf (Desai and Desai, 1969). Abundant downy growth is observed on the corresponding lower surface of the leaf. Sporulation is also common on the upper leaf surface. Severe infection covered the whole lamina, which later started drying and curling from the tip. Kapoor and Chowdhary (1976) reported that in case of infection due to *P. alta*, the symptoms were initiated as chlorotic streaks extending along the midribs of leaves with profuse downy growth on the corresponding lower surface. Gradually whole leaf turned necrotic and in case of severe infection, entire plant gave blighted appearance. However, our experience suggests that the causal organisms can not be distinguished on the basis of the disease symptoms as all the causal organisms associated with this disease produce similar types of external symptoms.

Mandal and Geetha (2001) and Mandal *et al.* (2010) reported systemic and localised infection in the floral parts as well. In case of systemic floral infection, spacing between two florets increases compared to healthy one. Long spikes, bearing weak florets become conspicuous and are easily identified from a distance within a population. In extreme cases, the number of florets decreases drastically with a thick inflorescence axis bearing few sterile flowers. Usually, a bunch of flowers towards the apex is infected in case of localised infection. The infected florets display a wide range of symptoms between normal flower opening and failure of flower opening. Filament length reduces significantly due to infection and anthers fail to come out properly out of the corolla tube. Anther does not burst and majority of the pollens turns sterile. However, gynoecium length increases and it retains stigma receptivity after mild infection. When pathogen growth covers the inflorescence, seed formation is hampered and it becomes sterile. Rathore (1997) observed phyllodae like appearance of the floral parts in localised infection.

Yield Loss

The disease causes severe loss of seed yield in isabgol. Patel (1984) and Rathore (1997) estimated yield loss due to this disease from naturally infected and artificially inoculated plants, respectively. In severe infection, reduction in seed yield per plant is more than 80 per cent compared to disease free plants. The disease not only causes quantitative loss in seed yield but also reduces the seed quality. The seeds become smaller in size and test weight is reduced considerably. In addition, percentage of black seed increased with the increase of disease intensity. Interestingly, black seeds from downy mildew infected spikes had higher husk per cent than those of the healthy spikes (Mandal *et al.*, 2010). Loss in seed yield varied from 15 to 50 per cent and that of straw yield from 20 to 45 per cent in the experiments conducted at our farm over the years.

Etiology

Three different fungi were reported to cause downy mildew of isabgol from different parts of India. The disease was first reported from Anand, Gujarat caused by *Peronospora plantaginis* Underwood as the causal agent (Desai and Desai, 1969). Later, Patel (1984) mentioned that Dr. Francis of the then CMI, Kew an authority in this group of fungi, on seeing the diseased material opined that the fungus was near identical to *P. plantaginis*. Recently, Dr. O. Constantinescu of EBC Uppsala University, Sweden, examined the material collected from Anand, Gujarat and confirmed the fungus was *P. plantaginis* (personal communication).

Kapoor and Chowdhury (1976) reported the disease was caused by *P. alta* Fuckel in Delhi and the sample was deposited at HCIO (32000), New Delhi. Rathore and Rathore (1996) reported the pathogen as *P. alta* (IMI No. 332256) from the disease sample of Rajasthan.

Sharma and Pushpendra (1998) observed a new species at Udaipur, Rajasthan, India and named it as *Pseudopeonospora plantaginis* Sharma and Pushpendra since sporangia of this fungus always germinated by zoospores. However, the organism was not published validly and it is considered not existing. Moreover, we re-examined the organism later and found the conidia were germinating with germtubes and some were aborting with no protoplast inside and an opening at the tip. Possibly, observation of the later type resulted proposal by other researchers of new species.

Rao (1968) reviewed the occurrence of *Peronospora* spp. in India. *P. plantaginis* figured in that. However, it was reported on *Plantago amplexicaulis* (Thind, 1942). Originally, *P. alta* and *P. plantaginis* were described on *Plantago aristata* (Fuckel, 1863) and *P. major* (Underwood, 1897), respectively, not on isabgol.

In the infected isabgol leaves, branched, coenocytic hyphae are found within the host tissue. Conidiophores arising out from stomata either singly or in groups are slender, tree like with dichotomous branching. Ultimate branchlets are small, tapering bearing single conidium at the tip. Conidia are thin walled and germtubes always emerge from the side of the conidia, not from the tip. Oospores are smooth, thick walled, globose. Key characters reported by various workers for identification of three species are presented in Table 23.1. However, an identification key for isabgol downy mildew fungi developed after Constantinescue (personal communication) is presented in Table 23.2.

Table 23.1: Morphological Comparison of Two Fungi Causing Downy Mildew of Isabgol

Characters	*P. plantaginis* [a]	*P. alta* [b]
Hyphal width	8.50µm (6.25-12.5µm)	8.8µm (6.5-13µm)
Colour of conidiophore	Grey or pale violet	Grey or pale violet
Ultimate branchlet size	6.81X4.17µm (4.25-12.3X4-4.2 µm)	9.7X2.9 (8.2-21.5X2.5-3.1µm)
Colour of conidia	Sub hyaline	Pale brown to violet
Shape of conidia	Elliptical to subglobose	Ellipsoid
Size of conidia	37X21µm (32-44X17-25µm)	30.2X20.4µm (27.4-34.2X17.8-25.3µm)
Colour of oospore	Yellowish brown	Pale brown to violet
Size of oospore		21-32µm
Haustoria	Cylindrical, long tenuous and curling	Cylindrical, long tenuous and curling

Sources: [a] Patel (1984), [b] Rathore (1997)

Table 23.2: Nomenclature Key for Isabgol Downy Mildew Fungi

a.	Sporangium germinates by zoospores	....*Pseudoperonspora plantaginis*
b.	Conidium germinates by germtube	.............*Peronospora*
ba.	Conidia ovoidal (approx. 22-30X19-24 µm) and yellowish brown in colour when mature	.............*P. alta*
bb.	Conidia ellipsoidal (approx. 30-44X19-23 µm) and dark brown in colour when mature	.............*P. plantaginis*

Histopathology

Conidia germinate directly by producing single, stout germtubes. The germtube penetrates the host tissue through stomatal opening. In case of *P. alta*, infection hyphae were found to arise from substomatal cavities (Rathore, 1997). Infection hyphae branches and produces intercellular, coenocytic, hyaline mycelium sending occasional haustoria into the host cells. Haustoria are peg-like or finger shaped. The sporangiophores emerge through the stomata singly or in groups bearing numerous conidia.

Processes of oospore formation were studied in *P. alta* (Rathore, 1997). Oogonia were formed by swelling of intercellular hyphae which were fertilized by antheridia.

Spore Germination

Effect of Temperature

Under the controlled laboratory condition, conidia or sporangia germination were studied at different temperatures by various researchers (Patel, 1984; Rathore, 1997). Optimum temperature for germination was found to be 15° to 20° C in all three species.

P. alta conidia failed to germinate at or beyond 35°C. Conidia of *P. alta* remained viable up to 11th day when stored at 0° C. However, viability was lost within 24 hrs of storage at 40° C (Rathore, 1997).

Effect of Relative Humidity

Rathore (1997) observed that conidia of *P. alta* germinated on glass slide at or above 30 per cent RH, while highest germination was observed at 100 per cent RH. Higher RH also helped the conidia of *P. alta* to retain viability for longer period. Conidia remained viable up to 9th day at RH value of 100 per cent.

Time of Collection

Maximum conidia germination of *P. alta* was observed when they were collected at 6.00 am. The failure of germination was observed when collection was made at 6.00 pm or later (Rathore, 1997).

Effect of pH

Optimum pH for conidia germination of *P. alta* (Rathore, 1997) was found to be 7.0. Failure of germination was observed at pH 3.0 and below.

Effect of Sugars

Glucose, galactose, mannaose, maltose, sucrose, cellobiose, ribose and raffinose increased conidia germination of *P. alta* while xylose, starch, cellulose, dextrin and rhamnose reduced germination (Rathore, 1997).

Inoculation Technique

Localized Infection

Infected leaves from severely infected field are collected in the afternoon and the remnants of the downy growth is gently removed by brushing off. The leaves are appropriately cut into pieces to fit in the Petriplate and incubated inside moist chambers. The newly formed conidia/sporangia are collected in the sterile water on the next morning with a soft paintbrush. The leaves are inoculated by spraying spores with the help of hand atomizer. Incubation inside humid chamber for initial 24–48 hrs after inoculation favoured establishment of the pathogen (Rathore, 1997).

Systemic Infection

Satisfactory systemic infection could be obtained by adding oospores collected from the previous season's crop, along with the seed during sowing (Dr. M. P. Sharma, personal communication). Diseased leaves containing the oospores are collected from the field and are kept under natural condition in pots. In the next season during sowing, these leaves are ground to powder and then it is used to inoculate the plants under field or pot conditions.

Disease Cycle

The disease is both seed and soil borne in nature. Systemic and localised infections in the floral part of isabgol are reported by Mandal *et al.* (2010). Rathore and Pathak (2002a) observed fungal structures in the seeds. Seed borne inoculum of *P. alta* remains viable for five years. Whereas, under the field conditions, oospores of *P. alta* remained viable up to 3 years in soil. However, oospore viability is reduced when buried deep in the soil (Rathore, 1997). Once the initial infection is established, the secondary spread occurred through conidia. Perennation of the fungus on other host during summer and rainy seasons is not reported. Among the weed flora tested for pathogenecity, *P. alta* infects *Chenopodium album* (Rathore and Pathak, 2002d). However, the fungus did not produce oospore on this host. Hence, this collateral winter season weed host is not very important for survival of the pathogen.

Epidemiology

Environmental Factors

Patel (1984) and Rathore (1997) observed the conidia germination of *P. plantaginis* and *P. alta* respectively, on leaf surface *in vivo*. In case of *P. plantaginis*, the conidia germination pattern showed two peaks–one at early morning (6 am) and second at 8 pm. Patel (1984) tried to explain the observations on the basis of prevailing temperature and humidity. However, the temperature and RH during 8 pm widely differ from that of 6 am. Probably, in nature, the fungus has two peaks for spore maturation, liberation and germination. Highest number of germinating conidia of *P. alta* are observed at 8 am while it is minimum at 2 pm. Conidia germination percent is highly correlated with RH ($r = +0.9488$) and temperature ($r = -0.9328$).

Pathogen

Two different pathogens were found to cause downy mildew of isabgol in India. Information regarding the virulence status of them towards different isabgol lines/varieties is not available. Also, it is not known if there is one or more races of the same pathogen exist.

Under artificial inoculation, disease severity increased when inoculum density was more in case of *P. alta* (Rathore, 1997).

Host

Pot culture study conducted by Rathore (1997) showed that 70–80 days old plants were most susceptible to *P. alta* infection, when spray inoculated with conidia. Whereas 10–30 days old plants remained uninfected to localised infection. Under the field conditions also, the disease appears first about 30 DAS in susceptible varieties with systemic/localised symptoms.

It is observed that in plants grown on acid washed sand supplied with Hogland solution, increase in nitrogen concentration increased disease severity towards *P. alta* (Rathore and Pathak, 2003). Whereas, phosphorous and potassium when applied in excess, decreased disease severity. Among the micronutrients, percent disease index was significantly increased when iron or calcium was absent from the nutrient solution. Under the field conditions also, higher nitrogen fertilisation increased downy mildew susceptibility (Mandal *et al.*, 2008). Higher sugar concentration in the leaves, more foliage, etc are attributed for this.

Disease Progress Under Field Conditions

Rathore (1997) observed that under natural field conditions, the disease followed a typical sigmoid curve with (i) initial slow establishment phase followed by (ii) rapid exponential growth phase ultimately terminating in a (iii) platue.

Linear correlation between per cent disease index and some individual weather parameters like maximum and minimum temperature existed, in case of *P. alta* infection. However, while integrating parameters like maximum temperature, minimum temperature, maximum RH, minimum RH, wind speed, sunshine in a linear regression equation, that did not match well with the disease progress under field conditions (Rathore, 1997).

Disease Management

Cultural Practices

Patel (1984) studied the effect of different sowing dates and seed rates on disease development. The disease pressure was less and the experiment was not conclusive. It was found that disease severity was increased with the increased seed rates. However, difference in seed yield was non-significant.

Rathore and Pathak (2002b) tried several cultural manipulations for disease management. They observed that delay in sowing dates from 15th October to 15th December gradually reduced the time taken for first appearance of disease. Crop sown at 15th November was most severely affected than the earlier and later sown crops. With the increase in plant population from 3-12 lakhs/ha, disease index increased significantly. Also, higher seed yield was obtained when lower seed rate was used. Interaction effect of sowing dates and plant population suggested that highest seed yield with lowest disease severity could be obtained with plant population of 3 lakhs/ha sown on 15th October. Among the different organic amendments tried by them, application of FYM @ 5 tonnes/ha produced minimum downy mildew severity. Elimination of sources of infection (infected seed and infected plant debris) also could reduce the systemic symptoms of downy mildew. However, the seed borne inoculum remained infective up to 5 years of storage whereas soil borne infection reduced to zero within fourth year. Rathore and Chandawat (2003) noticed that downy mildew severity increased with higher irrigation frequency and nitrogen fertilisation. However, reader must bear in mind that both these inputs are essential for higher seed yield. Hence, disease pressure can be reduced by reducing these with the cost of reduced yield (Rathore and Chandawat, 2003; Mandal *et al.*, 2008). Row direction of north-west/south-east developed lowest disease severity. But, farmers may not always find to follow this due to other practical constraints.

Host Resistance

Patel and Desai (1987) screened eight collections of isabgol for resistance against downy mildew. Out of these lines, EC124345 showed lowest disease severity and gave highest seed yield. Sain and Sharma (1999) identified six genotypes as resistant. Rathore and Pathak (2002c) screened 30 lines under pot and field conditions. Among these, none showed immune reaction to downy mildew infection. However, 7 accessions had low disease severity and categorised as resistant against *P. alta* infection. All these entries were free from systemic infection. These accessions also produced higher seed yield.

Among the related species, *P. psyllium*, *P. indica*, *P. patagonica* and *P. drimandii* developed localised infection under artificial inoculation with *P. alta*. The incubation period was lowest for *P. psyllium* and highest for *P. drimandii* (Rathore and Pathak, 2002d). However, at Anand, except *P. ovata* none of the *Plantago* species got infected under natural field conditions.

Chemical Control

Desai and Desai (1969) first tried chemical control of downy mildew of isabgol. They advocated aureofungin application for disease control. Rathore and Pathak (2002) reported that seed dressing with metalaxyl (Apron 35 SD @ 5g/kg seed) significantly reduced number of systemically infected plants compared to control. Patel (1984), Rathore (1996) and Sain and Sharma (1999) observed that three sprays of 0.2 per cent Ridomil (metalaxyl:mancozeb:: 8:64) was best for disease control. Rathore and Pathak (2002e) also noticed that no residue of the fungicide was detected in the plant parts at harvesting by TLC method. Mandal *et al.* (2007) tried systemic (Ridomil) and contact (mancozeb) fungicides singly or in combinations considering farmers' field conditions. It was observed that two sprays of Ridomil or one spray of Ridomil followed by one spray of mancozeb could significantly reduce the disease severity resulting increased seed yield. Mancozeb residue in the seed was at least 10 times lower than the minimum level.

Fusarium Wilt

Fusarium wilt (*Fusarium oxysporum* Schlecht. emend. Snyd. and Hans.) of Isabgol was first reported from Arizona, North America (Russel, 1975). Later it was also observed in Haryana, India (Mehta *et al.*, 1985). The disease became problematic after continuous monoculture (Russel, 1975). The fungus caused pre-emergence damping-off or typical wilting at seed formation stage. In the field, circular wilted patches having dying plants appeared. Wilting started from outer leaves and the leaves turned silvery in colour. Black discolouration of roots starting from the tip was evident. Rotting of cortical roots was observed and internal root tissues became woody. Presence of pathogen hyphae in the xylem tissues of the infected roots could be detected under the microscope. However, fusarium wilt was not a major threat for the Indian isabgol cultivation.

Damping Off

Chastagner *et al.* (1978) reported damping off caused by *Pythium ultimum* Throw from California. Post-emergence (5 to 10 days after sowing) damping-off was the major external manifestation of the disease. In India the crop is grown under water stress conditions hence, this disease does not pose problem for successful cultivation. Moreover, the fungus can be effectively managed by seed treatment with acetyl alanine fungicides such as metalxyl which are recommended for downy mildew.

Leaf Blight

Leaf blight due to *Alternaria alternata* (Fr.) Keissler. (Patel *et al.*, 1984) was reported from Anand, Gujarat, India. The disease appeared late in the season. The leaves turned brownish in colour giving blighted appearance. With the increase in temperature due to climate change, this may become prominent problem for isabgol cultivation. The disease can be managed by fungicidal spray schedule followed for downy mildew disease.

Oozing

The disease was first noticed at Jalore, Barmer and Sirohi districts of Rajasthan during 1990–91 (Rathore, 2003). The main symptoms of the disease are oozing gummy droplets from all above ground parts which turns blackish upon drying. Vegetative remains stunted with crinkled leaves. The plant dies without producing spike. Plants affected after spike formation produce few small spikes which contain small shrivelled seeds. The disease affects all growth stages of the crop–seedling to seed setting. However, maximum frequency is observed before flowering (20–45 days after sowing). The disease is of unknown etiology. Rathore (2003) found accessions RI-90, RI-49 and RI-136 were resistant to this disease. Also, the severity drastically reduces with delay in sowing (November 30).

References

Chastagner G. A., Ogawa J. M. and Sammeta K. P. V. (1978). Cause and control of dampingoff of *Plantago ovata. Plant Disease Report* 62: 929-932.

Desai M. V. and Desai D. B. (1969). Control of downy mildew of isabgol by aureofungin. Hindustan Antibiotic Bulletin 11: 254-257.

Fuckel (1863). *Peronospora alta.* Fungi rehnani no 39.

Kapoor J. N. and Chowdhary P. N. (1976). Notes on Indian microfungi. *Indian Phytopathology* 29: 348-352.

Mandal K. and Geetha K. A. (2001). Floral infection of downy mildew of isabogl. *Journal of Mycology and Plant Pathology* 31: 355-356.

Mandal K., Gajbhiye N. A. and Maiti S. (2007). Fungicidal management of downy mildew of isabgol (*Plantago ovata*) simulating farmers field conditions. *Australian Plant Pathology* 36: 186-190.

Mandal K., Patel P. R., Maiti S. and Kothari I. L. (2010). Induction of male and female sterility in isabgol (*Plantago ovata*) due to floral infection of downy mildew (*Peronospora plantaginis*). *Biologia* 65: 17-22.

Mandal K., Saravanan R. and Maiti S. (2008). Effect of different levels of N, P and K on downy mildew (*Peronospora plantaginis*) and seed yield of isabgol (*Plantago ovata*). *Crop Protection* 27: 988-995.

Mehta N., Madaan R. L. and Thakur D. P. (1985). Record of isabgol wilt from Haryana. *HAU Journal of Research 15*: 473-474.

Patel J. G. (1984). Downy mildew of isabgol (*Plangato ovata* Frosk.). Ph.D. Thesis, B. A. College of Agriculture, Gujarat Agricultural University, Anand.

Patel J. G. and Desai M. V. (1987). Reaction of isabgol varieties/cultures to downy mildew. *Indian Phytopathology* 40: 97-98.

Patel J. G., Patel S. T. and Patel A. J. (1984). New Disease outbreak (leaf blight of *Plantago ovata*). *Indian Phytopathology* 37: 582.

Rao V. G. (1968). *The genus Peronospora Corda in India.* Nova Hedwigia 16: 269-282.

Rathore B. S. (1997). Ph.D. thesis Rajasthan Agricultural University, Bikaner.

Rathore B. S. (2003). Studies on symptomatology and management of oozing in blond *psyllium* through sowing time and host resistance. *Journal of Mycology and Plant Pathology* 33: 498.

Rathore B. S. and Chandawat M. S. (2003). Influence of irrigation, row orientation and nitrogen nutrition on downy mildew of blond *psyllium. Indian Phytopathology* 56: 453-456.

Rathore B. S. and Pathak V. N. (2002). Effect of seed treatment on downy mildew of blonde *psyllium. Journal of Mycology and Plant Pathology* 32: 35-37.

Rathore B. S. and Pathak V. N. (2002a). Infection, symptomatology and histopathology of *Peronospora alta* on blond *psyllium. Journal of Mycology and Plant Pathology* 32: 365-366.

Rathore B. S. and Pathak V. N. (2002b). Influence of planting dates, plant density, organic amendments and sanitation on downy mildew of blond *psyllium. Indian Phytopathology* 55: 269-278.

Rathore B. S. and Pathak V. N. (2002c). Evaluation of blonde psyllium germpalsm against downy mildew. *Indian Phytopathology* 55: 229.

Rathore B. S. and Pathak V. N. (2002d). Studies on the tissue culture and host range of *Peronospora alta* causing downy mildew of blond *psyllium. Journal of Mycology and Plant Pathology* 32: 201-203.

Rathore B. S. and Pathak V. N. (2002e). Studies on uptake, translocation and persistence of metalaxyl in blonde psyllium. *Journal of Mycology and Plant Pathology* 32: 81-85.

Rathore B. S. and Pathak V. N. (2003). Influence of host nutrition on downy mildew on downy mildew of blond psyllium. *Indian Phytopathology* 56: 91-93.

Rathore B. S. and Rathore R. S. (1996). Downy mildew of isabgol in Rajasthan. *PKV Research Journal* 20: 107.

Russell T. E. (1975). Plantago wilt. *Phytopathology* 65: 359-360.

Sain S. K. and Sharma M. P. (1999). Factors affecting development of downy mildw (*Psendoperonospra plantaginis*) of isabgol (*Plantago ovata* Frosk) and its control. *Journal of Mycology and Plant Pathology* 29: 340-349.

Sharma M. P. and Puspendra (1998). A new pathogen causing downy mildew of isabgol (*Plantago ovata* Frosk.) (Abstr.). *Journal of Mycology and Plant Pathology* 28: 74.

Thind K. S. (1942). The genus *Peronospora* in Punjab. *Journal of the Indian Botanical Society* 21: 197-215.

Underwood L. M. (1897). Some new fungi, chiefly from Albama. *Bulletin of Torrey Botanical Club* 24: 81-83.

Sustainable Disease Management of Agricultural Crops (2011) *Pages* **312–332**
Editors: **S.K. Biswas and S.R. Singh**
Published by: **DAYA PUBLISHING HOUSE, NEW DELHI**

Chapter 24

Nematode Problems of Vegetable Crops and their Management

☆ *S.K. Biswas*[1], *R.S. Kamalwanshi*[1], *P. Sinha*[1] *and S. Banik*[2]
[1]*Department of Plant Pathology, C.S. Azad University of Agriculture and Technology, Kanpur – 208 002*
[2]*Division of Plant Pathology, Indian Agricultural Research Institute, Pusa, New Delhi – 110 012*

Vegetables are nutritious and important component of human diet because they supply proteins, vitamins, minerals which are essential for the purpose of human health. The cultivation of vegetables is not only essential part of our Indian farming but also in other countries of the world. The increase of vegetable production was reported to 18 per cent in tropical and 7.7 per cent in developed counting during the period of 1981-85. In India, Vegetable farming occupy 5.5 million hectors of land and producing 48.53 metric tons per annum which is less as compared to developed countries like USA, Japan, and France. The state namely, Bihar, Kerala and Assam occupying maximum acreage of land under the vegetables and India comes second place for vegetable production in the world but its production in every part of the world in reduces to greater extent by nematodes.

Major Nematode Pests

A large number of plants–parasites nematodes have been reported from the rhizosphere and root of vegetables crops by several workers (Table 24.1). Mahajan *et al.* (1986) have compiled a list of 176 nematode species belonging to 72 genera are associated with vegetables crop in India. These nematodes are generally classified into five groups:

1. Root knot nematodes (*Meloidogyne* spp): produces gall on root of many vegetables.
2. Reniform nematodes (*Rotylenchulus reniformis*): losses in the yield of cowpea and tomato.
3. Cyst nematode (*Globodera rostochiensis*): causes a serious problem in Niligiri and Kodai hill of Tamil Nadu.

4. Stunt nematode (*Tylenchorhynchus brassicae)* has been found to reduce the yields of cabbage and cauliflower.
5. Other nematodes:
   (*a*) Lesion nematode (*Pratylenchus delatterer*): reduce the vigour of chilli.
   (*b*) Spiral *Helicotylenchus dihystera* reduce the root growth of chilli, okra, tomato, brinjal and onion have also reported good host.
   (*c*) Lance nematode (*Hoplolaimus indicus*) causes growth reduction in tomato (Gupta and Atwal, 1972).
   (*d*) Stubby root nematode (*Trichodorus allius*) infects onion.
   (*e*) Stem and bulb nematode (*Ditylenchus destructor*) causes dry rot of potato tuber in Shillong.
   (*f*) Mushroom nematode (*Ditylenchus dipsaci, Aphelenchoides compoticola* and *Rhabditoides*)

Table 24.1: List of Major Nematodes Problems in Different Vegetable Crops

Crop	*Type of Nematode*	*Scientific name*	*Yield loss*	*Reference*
Tomato	Root knot	*M. incognita*	61 per cent	Nagnathan, 1984
		M. acrita	50- 80 per cent	Ivanova *et al.*, 1979
	Reniform	*Rotylenchulus reniformis*	42–49 per cent	Subramanyian *et al.*,1990
	Others: (*i*) Root gall	*Paralongidorus buchae*		Zacheo *et al.*, 1985
	(*ii*) Root gall	*Globodera pallida*		Vovlas *et al.*,1986
	(*iii*) Root gall	*Xiphinema basiri* *X. ifacolumi*		Zacheo *et al.*, 1987
Potato	Root knot Cyst	*Meloidogyne* sp. *Globodera* sp.	42.50 per cent 99.50-99.80 per cent	Prasad, 1989 Prasad, 1989
Brinjal	Root knot (major)	*M. incognita*	32–45.7 per cent	Nagnathan, 1984, Darekar Mahase, 1988
		M. javanica	86 per cent	Alseady *et al.*, 1989
	Reniform	*R. reniformis*		
Chilli	Root knot	*M. incognita*	19.7 per cent	Nagnathan, 1984
Onion	Stem and bulb	*Ditylenchus dipsaci*	12 per cent	Fritzsche and Thiele, 1981
	Root knot	*M. arenaria*		
		M. incognita	33.6 per cent	
		M. hapla	60 per cent	McGuidwin, 1987
Garlic	Stem and bulb	*Ditylenchus dipsaci*		
	nematode	*D. destructor*		
		Meloidogyne sp.		
		Pratylenchus sp.		
		Helicotylenchus sp.		
		Trichodorus sp.		
		Aphelenchoides sp.		
Carrot	Cyst nematode	*Heterodera carotae*	50 per cent	Bossis and Mugniery 1989
	Root knot	*M. hapla*	50.95 per cent	

Contd...

Table 24.1–Contd...

Crop	Type of Nematode	Scientific name	Yield loss	Reference
Radish	Cyst Root Knot	*Pratylenchus penetrans* *Heterodera schachtii* *M incognita*		
Cabbage and cauliflower	Cyst	*H. schachtii* *H cruciferae*	42 per cent	Abawi and Mai, 1983
	Root knot	*M. incognita* *M. javanica*		
	Sting	*Belonolaimus* sp.		
	Lance	*Hoplolaimus galeatus*		
	Ring	*Helicotylenchus* sp.		
	Reniform	*Rotylenchulus reniformis.*		
	Lesion	*Pratylenchus* sp.		
Cucumber	Root knot	*M. incognita* *M. javanica*		
	Raniform	*R. reniformis*		
Okra	Root knot	*M. incognita* *M. javanica*	90 per cent 20-40 per cent	Lamberty *et al.*, 1983; Jain *et al.*, 1986
	Raniform	*R. reniformis* *Hoplolaimus aegyptii*	50 per cent at 200 nematodes/plant	
Peas	Root knot Cyst	*Meloidogyne* sp. *Heterodera cajani* *H. goettingiana*	33 per cent	Upadhyaya and Dwivedi, 1987.
Leafy vegetables	Root knot	*M. incognita*		
	Leaf and bulb	*Aphelenchoides. fragariae*		
	Lesion	*Pratylenchus* sp.		
	pin nematode	*Paratylenchus* sp.		
	Awl nematode	*Dolichodorus heterocephalus*		
	Sting	*Belonolaimus longicaudatus*		

Root Knot Nematode (*Meloidogyne* spp.)

The root knot nematode is the most important pest of vegetable crops. There are mainly four important species of root knot nematodes were parasitized on vegetable crops *viz., M. incognita, M. Javanica, M. arenaria* and *M. hapla.*

M. incognita and *M. javanica* are most wide spread in distribution and have a wide host range among vegetables but *M. hapla* attacks potato and *M. arenaria* infect on chilli only. The life cycle of the root-knot nematode is completed within 19th days at 30.6°C and 40-43 days at 21.8°C and can completes 8 generations in a year (Chaudra, 1984); However, Rai and Jain (1988) reported 30 days require to complete one generation and lay an average egg mass 326. The *M. javanica* require 27 to 57 days to complete its life cycle and 2-3 generations for five months (Philis, 1984). The life cycle of the root knot nematode depends on plant age, soil type, temperature and to some extent on nematode race. The duration of life cycle is minimum at temperature of 25-28°C in sandy loam soil on young seedlings.

Interaction of *Meloidogyne* spp. with Other Microorganisms

The association of root-knot nematode with other host plant is often accompanied by other pathogens like bacteria, fungi and viruses and other nematodes in causing disease complex.

Fungi

Interaction of root-knot nematode with Fusarium wilt of cotton was first time reported by Atkinson in 1892. *M. incognita, M. acreta, M. hapla, M. Javanica* have been reported to promote the incidence of *Fusarium* wilt in tomato (Jenkins and Caursen, 1957; Bowman and Bloom 1966; Powell, 1971). In most of the cases *Meloidogyne* spp. have been broken the resistance of many varieties against Fusarium wilt. Simultaneous inoculation with *M. incognita* and *Rhizoctonia bataticola* or *Sclerotium rolfsii* reduce the germination of seed in okra, brinjal and tomato (Shukla and Swarup, 1976; Chhabra and Sharma, 1981). *M. javanica* increased the extent of damage by pre- and post–emergence damping off caused by *Fusarium oxysporum* f.s. *lycopersici, R. solani* and *Pythium debarynum* in tomato (Nath *et al.*, 1984). The incidence and severity of root-rot of brinjal, tomato and okra caused by soil borne fungi such as *R. solani, R. bataticola* and *Phomopsis vaxan* was increased in presence of *M. incognita* (Hazarica and Roy, 1974; Chhabra *et al.*, 1977; Khan *et al.*, 1980; Sharma *et al.*, 1980; Chahal and Chhabra, 1984). The damage to the host plant has generally been seen to increased with the interaction of root-knot nematodes and several fungi like–*Pythium, Rhizoctonia, Sclerotium, Fusarium, Phomopsis, Aspergillus, Verticillium* has been studied. VAM have an antagonistic effect on plant parasitic nematodes. *Glomass mosseae* reduces population of juveniles of nematode in the plants.

Bacteria

Association of root knot nematode and bacteria increase the severity of bacterial wilt, caused by *Pseudomonas solanacearum* (Napier 1978; Goth *et al.*, 1983). Thus combined pathogenic effect of *P. salanacearum* and *M. javanica* on bringal was greater than the independent effect of either of them (Sitaramaiaha and Sinha, 1981). The root-knot nematode has predisposed the bacterial wilt resistance in Pusa purple cluster cultivar of brinjal (Paravatha Reddy *et al.*, 1979). Pani and Das (1972) reported the association of bacterial wilt and *Meloidogyne* spp. on tomato.

Phytoplasma

The combined effect of *M. incognita* and MLO cause little leaf of brinjal was greater in reducing this total plant growth than individual effect of organisms (Dhawan and Sethi, 1977).

Virus

Few viruses have been interact with root-knot nematode, a positive interaction of *M. incognita* and TMV on egg plants, reduce the growth of plant (Goswami and Chanula, 1974). In case of brinjal and bottle gouard, root-knot nematodes development has been enhance by the presence mosaic virus (Mahmood *et al.*, 1974).

Nematode and Nematode Interaction

The root-knot nematode also interact with concomitant nematodes present in the soil (Alam *et al.*, 1975; Sharma and Sethi, 1976). In such interaction, the population of one or both the co-inhabitant may decreases or increases.

Reniform Nematode

It is the second most important pest of vegetable crops after root-knot nematode. It infect tomato brinjal, okra, cowpea, nape bean, pointed gourd and other vegetables. Life cycle of *R. reniformis* normally takes 25 days. The infective stage *i.e.* pre adult female penetrate the root cortex and start feeding and changed its body as kidney shape. After one week of infection, generate eggs in the body of the females and vermiforms male comes out without causing infection. Reproduction mostly amphimixis and rarely pathogenetically. Life cycle of *R. reniformis* is completed on okra 24-29 days (Shiv Kumar and Seshadri, 1971), on brinjal 20-25 days (Singh and Khera, 1978). The number of egg mass may very 39-120 on different vegetable crops (Peacock, 1956; Shiva Kumar and Seshadri, 1972). Population of this nematode reach high in the month of June to October (Prasad *et al.*, 1977). The survival of this nematode in air dried soil (3.3 per cent moisture) is 7 months at 20-25°C (Birch field and Martin, 1967). Sandy loam, black loam and clay loam soil are more effective than sandy soil (Sivakumar and Seshadri, 1972).

Cyst Nematode

Only few species of cyst forming nematodes are problematic in vegetable crops. In India, cyst nematodes are not so important in vegetables crops but potato cyst nematode (*Globodera rostochiensis* and *Globodera pallida*) is the most important pathogen in potato. First time this nematode was reported from Nilgiri and Kodai hills of Tamil Nadu and Monar hill of Kerala (Ramana and Mohan Das. 1988). The potato cyst nematode and the fungi *Rhizoctonia solani* causes root rot complex and in association with *Pseudomonas solanacearium* causes wilt complex in potato. The population of the *G. rostochiensis* has been found to reduce in association of *Glomas faciculatus,* VAM fungi. Cyst nematode is mainly attack cowpea, Indian bean, carrot cabbage snap bean, cauliflower, turnip etc.

The root diffusates of potato attract for the hatching of second stage larvae from the eggs, more actively invade the roots and infect the vascular system which is lie parallel the root. The larvae ultimately develop giant cells from which nematode extract nourishment. After the four moult, the adult larvae converted into globose shape for female and male larvae into filliform and comes out. The second stage larvae took 37-39 days during the summer crop and 40-42 days in autumn crop. Multiplication rate was 7-13 times more in summer crop.

Stunt Nematode

In India, the most common species feeding on vegetables is *Tylenchorhynchus brassicae.* This nematode responsible for poor germination and growth of the cabbage and cauliflower. *T. brassicae* alone did not effect the percentage of emergence of cauliflower seedling, but it has adverse effect with *Rhizoctonia solani* on emergence of seedlings (Khan *et al.*, 1971). The optimum temperature for growth and reproduction is around 30°C with 25-30 per cent moisture (Khan *et al.*, 1969). The nematode to nematode interaction has been noticed that *Tylenchorhynchus dubius* suppressed the population of *Helicotylenchus crenatus* on tomato (Krishnappa and Prasad, 1979).

Losses in Vegetable Caused by Nematodes

The accurate data on the extent of losses caused by nematodes is not available but on the basis of several survey, average yield losses in the world are recorded to be 5 per cent which is higher in tropical and subtropical countries (Taylor and Sasser, 1978). In USA, the vegetable crop suffer 11 per cent annual loss due to nematodes (Anonymous, 1971). In India, unfortunately there is no definite information so far, however, the following yield losses has been recorded in India (Table 24.1).

Management Practices

Cultural Practices

Cultural methods have been adopted in the crops husbandry practices to minimize the nematode population. The following cultural practices are adopted:

Crop Rotation

Crop rotation with certain non hosts, graminaceous crop and certain antagonistic crop for 1-2 years has been found effective in reducing the population of root knot nematodes (Sundresh and Sethy 1977; Jain *et al.*, 1989). Number of years of crop rotation is depend on the initial population and the rate of population decreases. Economically important crop should be selected in crop rotation. There are large, number of varieties of cereals (sorghum millets, maize wheat, rice), cruciferous crops (cabbage cauliflower, knolkhol, mustard), bulbous crop (onion, garlic), oilseed crop (mustard), fibrous crop (cotton) and pulse crop (pigeon pea, cowpea) have been shown to reduce root-knot population in soil. More than 3 years crop rotation of potato with wheat, straw berry, cabbage cauliflower, peas, maize and bean reduces cyst nematode population to a safe level. Likewise, decrease in population of root knot nematodes on tomato, brinjal okra, chillies and sponge gourd occur following marigold, spinach and bottle gourd (Khan *et al.*, 1975). Rotation of cabbage and cauliflower with wheat reduces population *T. brassicae.* Sequence with mustard radish and sesame considerably reduces the population of *Tylenchorhynchus* spp. (Haque and Gaur, 1985).

Fallowing

Keeping land as fallow for a long period of time is uneconomical and even be dangerous for soil erosion. Weeds which are susceptible to nematode should not be allow to grow during the period of fallowing which are also difficult for cultural practices *i.e.* ploughing, harrowing and hoeing etc. Fallow practice decreases the population of *M. incognita* on tomato, brinjal, okra, chillies and spong gourd (Khan *et al.*, 1975).

Organic Amendment

Organic amendment with $CaCN_2$ and Urea Cynamide is effective for controlling root–knot nematodes. The mechanism of control is due to release of HCN or NH_3 which are toxic to nematodes. Potassic fertilizer significantly reduces number of galls in tomato (Gupta and Mukhopadhyay, 1971). Application of potash in combination with phosphorus or nitrogen or potash alone checks the reniform nematode multiplication on okra to a great extent (Siva Kumar and Meerzainuddin, 1974). The efficacy of organic cakes was in the order of mustard >neem > groundnut> castor> mahua. The organic amendment is not only helpful for encouraging useful micro flora but also reduce the pest problem.

Mechanism involve in control of nematodes by use of organic amendment (I) Volatile fatty acids like (Formic acid, acetic acid, propionic acid and butyric acid) ammonia and hydrogen sulphide that are released during the microbial decomposition of organic matter are directly toxic to nematodes. (II) Amino-acid also released during decompositions of organic matter may be responsible for reducing nematode population is soil (Parvatha Reddy; 1975) (III). Microvorous nematode can rapidly reproduce on the abundant materials and decomposable products. (IV). Changes in the physical and chemical condition of the soil may alter the host–nematode relationship.

Organic amendments improve soil conditions for more rapid root growth. This enhances the utilizations of soil nutrients and masks the effect of nematode damage (Parvatha Reddy *et al.*, 1975).

Flooding

Submergence of water decreases oxygen content of the soil and kills the nematode asphyxiation. Chemicals are act as lethal factor for nematodes, such as butyric, propionic acids, hydrogen sulphide and NH_3 often develop in flooded soil and effective against root-knot and stunt nematodes in vegetables.

Trap Cropping

Growing of cowpea help to hatch the egg of *M. incognita* and the larvae easily make entrance within the host but not developed further. Therefore, destruction of crop before the nematode become to mature is efficient trap crops for *M. incognita*. Oat is another efficient trap crop for *Heterodera avenae* (Stone, 1961). Crotalaria, Sun hemp are highly susceptible to invasion by root-knot nematode but resistant to the development of larvae in to adults.

Enemy Plants

There are several plants available in nature which have ability to kill the nematodes, called enemy plant. The following enemy plants are used to control nematodes:

1. *Mustard*: Morgan (1925) first time discovered that cultivation of potato with white mustard in a part of infested soil were less affected by nematodes than potato growing alone. Mustard increase the yield of potato by reducing the severity of nematode attack (Ellemby, 1951). The active principle behind this explanation is that the Allyle isothiocynate is present is mustard which is toxic to nematode.
2. *Pangola grass*: The pyrocatechol, a nematicide compound accumulate in high concentration in the root of pangola grass (*Eragrostic carbula*) in a potent nematicide against *M. javamica, M. hapla, M. arenaria and M. thamasi.*
3. *Marigold*: Intercropping of tomato and okra with *marigold* (*Tagetes erecta* and *T. Patula)* reduces the root-knot nematode population. The population of *Paratylenchus, Tylenchorhynchus, Helicotylenchus, Rotylenchus, Paratylenchus* and *Hoplaimus* was also reduced due to interplant with marigold (Khan *et al.*, 1971, Alam *et al.*, 1977. The active principle is due to secretion of terthenile from the roots of Tegetes.
4. *Asparagus*: Tomato normally a good host of *Trichodorus christiae* is supported by only a low population of asparagus in the same plot. *Asparagus officinalis* would not support to the population of nematode for more than 40-50 days (Rohde and Jenkins, 1958). Glucoside is the active principle which released by asparagus root to reduce the population of *Trichodorus christae.*
5. *Sesame*: Root exudates of sesame (*Sesamum orientale*) have nematicidal property against *M. incognita*–causal agent of root-knot nematode of okra (Atwal and Mangar, 1969). The cultivation of okra with sesame is not only effective against root-knot nematode but also increased the yield of okra as compare to the grown it alone.
6. *Margosa* (*Azadirachta indica*): Margosa is responsible to reduce the population of the *Hoplolaimus indicus, Helicotylenchus indicus, Rotylenchulus reniformis, Tylenchorhynchus brassicae, Tylenchus filiformis* and *M. incognita* due to presence of nematicidal compounds like nimbidin and thionemone.
7. *Cucumber*: The root exudates of bitter cucumber is effective against to the nematodes due to presence of glycones or glycoside. Bitter cucumber is less affected by nematodes than non bitter cucumber which is due to accumulation of cucurbitacin, chemical act as a repellent to *M. incognita.*

Time of Planning

Most of the pathogenic nematodes are inactive during winter month because of low temperature. Early planting of potato and sugar beet reduce the losses caused by nematodes. As for example, planting of potato during third or fourth week of March in Shimla hill would reduces the damage caused by *M. incognita* (Prasad *et al.*, 1983).

Soil Solarization

This is a easy method of nematode management. The soil cover with transparent thin polythene sheath during hot summer increases the upper soil temperature over 5–10°C times more than normal temperature for several hours daily in few weeks which killed nematode fungi bacteria, insect and weeds. The decline of nematode population is corrected to the member of days or hours for which the soil temperature exceed 40°C (Gaur and Dhingra, 1991). This method is expensive, mainly uses in nursery beds orchards and forestry plantation.

Summer Ploughing

Deep ploughing of infested field during summer lead to disturbance and stability of nematode community and also causes mortality by exposing the nematode to solar heat. Three summer ploughing at 10 days interval during May/June is recommended to reduce the population of *M. javanica,Helicotylenchus* spp., *Pratytenchus* spp. Jain and Bhatti, (1987) reported that 2-3 summer ploughing is beneficial for control of nematodes.

Selection of Healthy Propagating Materials

Propagating materials like rhizome, bulb, suckers, seedling, roots, seed etc. should be collected from the healthy plants or reliable source to reduce the nematode infection *e.g.* The potato cyst nematode can be eliminated by selecting nematode free plant material.

Regulatory Methods

Quarantine

Quarantine is an important eco-friendly strategy for management of nematode. It is based on the prevention is better than cure. The principle involved in acting quarantines is exclusion of nematodes into an area which is not infested. The first quarantine law was passed in Rouen (France) in 1890, to prevent the spread of barberry plant, the alternate host of wheat rust, later USA and Australia follow the quarantine law for preventing the spread of barberry plant. The Govt. of India Legislated an act in 1914, known as Destructive Insect and Pest (DIP) act, under which various notifications have been issued from time to time restricting and regulating import of certain plants or plant materials form other countries as well as from one state to another within the country. As an evidence to the spread of plant parasitic nematode through exchange of plant materials, plant quarantined inspection practices have shown interception of many plant nematode. Many economically important nematode like *Heterodera goettingiana, M. incognita, D. dispaci R. similis* (Sethi *et al.*, 1972) and *Globodera rostochiensis* (Renjhen, 1973) have been intercepted from the imported germplasm material. Potato cyst nematode (*G. rostochiensis*) is under quarantine act in Nilgiri hill (India). Infected potato seed materials are not allowed to transport from Nilgiri hill to other part of India.

Seed Certification

Seed certification is the legally sanction method for production of nematode free seed or vegetatively plant propagating materials. During crop growing period, monitoring requires to get the healthy planting material for the propose of certification. Potato seed pieces free from cyst nematode can be grow commercially through seed certification programme.

Physical Method

Physical method is one of the oldest methods and probably a successful control measure of nematode developed so far. The principles behind this is killing of nematodes by heat.

Hot Water Treatment

Hot water treatment is widely used for controlling of nematodes present within the plant tissues before planting and has proved useful in nematode infested seed tuber, bulbs and root of plants (Table 24.2). Nirula and Bassi (1965) have been reported that hot water treatment of potato tubers at 45°C for 48 hrs has been shown to kill about 98.9 per of *M. incognita* without affecting tuber viability. Since most of enzymes are inactivated at a temperature of near to 55°C, death can probably be explain on these basis.

Table 24.2: Time and Temperature Recommended for Control of Nematode by Hot Water Treatment

Nematode	Planting Stock	Time (minute)	Temperature
Meloidogyne spp.	Potato tuber	120.	46.0 to 47.5
	Sweet potato	65	46.7
	Yam tuber	30	51.0
Ditylenchus dipsaci	Onion bulb	120	43.5

Heat Treatment of Soil

Steam sterilization is used in green house bed and small area for the purpose to kill nematode, insect, fungi, bacteria, weed seeds etc. The most important nematodes like potato cyst nematodes, root-knot nematodes of cucumbers and lettuces are controlled by steam heat. In summer, uses of polythene mulched reduce nematode population in soil.

Irradiation

Irradiation of potato tubers with X–rays inhibited sprouting and also development of *G. rostochinesis* (Fassuliotis and Saparrow, 1955). Wood and Goodey (1957) found that x-rays in b/w 48000 to 96000 inactivated *Ditylenchus myceliophagus* in mushroom compost. In field conditions, irradiation method is impracticable because of the length of time needed to irradiate even in small area and damage to plant root. Ultraviolet light also be killed nematode.

Ultrasomic

High frequency vibrations make the nematode to lethal. Kampfe, (1962) used ultrasonic to kill *Heterodera* spp.

Washing Process

Careful washing of soil, adhering to plant material containing plant parasitic nematode reduce the risk of disease spread *e.g.* potato tubers, bulbs.

Chemical Control

Chemical control has been a difficult task in the past but introduction of new chemicals in the market indicated the possibility of exploiting them for better control of nematodes. The primary advantage of chemical control over other methods is that nematodes population is reduced to a very low density within a matter of days after the chemical is applied. The another important benefit of chemical control is that in addition to nematodes, fungi, bacteria and weeds can also be controlled. The chemicals are used as seed and soil treatment and foliar spray.

Soil Treatment in Nursery Bed

Nursery bed treatment is effective preposition for obtaining healthy seedling for transplanted crops and also economizing the quantity of chemicals. Nursery bed treatment with aldicarb and carbofuran @ 2 g a.i./m^2 were effective in increasing seedling growth and reducing root knot population on tomato, brinjal and chili (Rama Krishanan and Bala Shubramanian, 1981; Jain and Bhatti, 1983).

Soil Treatment in Main Field

Generally plant parasitic nematodes of soil born. The control of nematodes can be done by mixing granular chemicals with the soil mechanically. Generally, Aldicarb conbofuram, ethoprophos and phenamiphos @ 1.2 kg/ha were found effective in reducing nematode population in soil of vegetables field (Ahuja, 1983; Singh *et al.*, 1978; Parvatha Reddy, 1985; Singh and Parvatha Raddy, 1981–82; Handa and Mathur, 1981).

The population of *Tylenchorhynchus brassicae* and *T. dubious,* can be reduced from cabbage field by using the soil treatment with aldicarb (0.5 kg/ha.) and carbofuran @ 1.5 kg/ha (Verma *et al.*, 1978). Reddy and Seshadri (1972) reported that thionazin @ 4 kg/ha were effective against *R. remiformis* infecting tomato.

Seed Treatment

Seed treatment with nematicide is one of the most economic method with out any adverse effect on plant growth. Seed treatment gives not only initial protection of seedling but also better plant growth and increase yield. Carbosulphan has been reported to be effective as seed treatment against root-knot nematodes by several workers. Phenamiphos and Aldicarb @ 1 per cent concentration were effective in controlling root-knot nematode, infecting cowpea, french bean and peas (Table 24.3).

Foliar Treatment

Some systemic nematicide like oxamyl may be spread on foliage to affect nematodes at the plant surface or within the tissues, *e.g.* Onion.

Bare Root Dip Treatment

The amount of damage to a crop is usually correlated to the stage of plant growth. Nematode damage is more harmful to seedlings than older plants, because yield loss was greater when seedling were affected with nematode. Hence, root- dip method with nematicide may therefore be reduce nematode damage and proved more economic. Vydate, thionazin trazophas lannate, dimethionate etc. have been shown effective as bare root dip of seedling. Thionazon @ 500 ppm for 15 mins. (Reddy and Seshadri, 1975) Phenamiphos @ 250 to 750 ppm for 30m (Thakar and Patel, 1985), Carbofuram and Oxamyl @ 1000 ppm for 15 to 30 m. (Parvatha Reddy and Singh, 1979), Dimithioate @ 500 ppm for 6 hr, Phosphomidon and Dichlorofenthion @ 1000 ppm for 8 hr. (Jain and Bhatti, 1978), VC- 13@ 1200 ppm for 15 mt (Alam *et al.*, 1973; Saxena, *et al.*, 1974) gave effective control of root-knot nematodes on tomato, chillies, brinjal by bare root dip treatments(Table 24.3). Alam *et al.* (1973) also reported that oxamyl was effected for control of root-knot nematodes infecting brinjal and okra. Bare root dip treatment with aldicarb, carbofuran turbophos @ 500 to 1000 ppm was effective against reniform nematode of brinjal (Prasad and Krishnappa, 1981).

Table 24.3: Chemical Used for Nematodes Management in Vegetables Crops

Crop	*Method of Application*	*Name of Chemicals*	*Dose*	*Reference*
Solanaceous crops (Root knot and Cyst nematodes)	Nursery soil treatment	Metham sodium	60 ml/sq m. 15 day before sowing	Nagnathan 1984
		Carbofuran	10 g/sq. m.	
		Aldicarb	10 g/sq. m.	
		Oxamyl	19.7 g/sq. m.	Ahuja, 1982
		Phenamiphos	2.43 g/sq. m.	
	Root- dip (1 hr.)	Phorate, Fensulphothion	500 ppm	Haque *et al.*, 1984
		Aldicarb carbofuran phosphamidon	500 ppm 1000 ppm	Hussain, 1984
	Seed treatment	Carbofuran or Aldicarb	1 kg/ha	Singh *et al.* (1980)
	Spot treatment	Aldicarb	1 kg/h	
Bulbs crop (onion) (*Ditylenchus dipsaeci*)	Foliar spray	Gentsh Temic Vydate	@ 1.2 kg/ha	
M. javanica		Benfuracarb	10 kg/ha	Greco *et al.*, 1986
D. dipsaci		Phenamiphos (Systemic nematicide)	20-40 kg/ha	Spigel *et al.*, 1983
Garlic		Carbofuran	5.04 a.i/ha kg	Robert and Greathead *et al.*, 1986
D. dipsacei		Fensulphothion	5.04 a.i/ha kg	
	Storage	Aldicarb	@ 2.5 kg/ha	
Carrot *M. hapla*	Farrow	Carbofuran	@1.5 kg/ha	Belair, 1984
M. incognita and hapla	Soil application	Aldicarb	3.3-6.7 kg/ha	Lue *et al.*, 1984
Root–knot nematode	Fumigants	Dichloro di propane, Telone and Vorlex	402, 268, 134 ml/ha	Vrain *et al.*, 1981, Bossi, *et al.*, 1989
		Carbofuran	6 kg/ha	
		Oxamyl	3 kg/ha	
		Aldicarb	6 kg/ha	
Radish *Paratylenchus penetrans*	Soil application	Oxamyl	2 kg/ha	Ohbayashi, 1983
		Cole Crop		
Cabbage *Heterodera schachtii*		*Phenamiphos*	6.7 kg/ha	Abawi and Mai, 1983
Cauliflower Root-knot nematode		*Aldicarb*	2 kg./ha	Maqbool *et al.*, 1985
Heterodera crussiferae	Soil application at depth 15 cm	*Vydate*	20-50 kg/ha	Moel, 1986

Contd...

Table 24.3–Contd...

Crop	Method of Application	Name of Chemicals	Dose	Reference
Cucumber Root–knot nematode	Soil application	*Phenamiphos*	1.7 to 2.5 kg/ha	Turligina *et al.*, 1985
Reniform nematode	seed treatment	Neem oil		Siddiqui and Alam, 1988 and 1990
Okra Reniform nematode	Seed treatment	*Calotropis* and *Euphorbia extract*		
Root knot okra	Seed treatment	Carbofuran	3 per cent	Sivakumar, *et al.*,1976
		Aldicarb	1 kg/ha	
		Carbofuran	2 per cent	
		Aldicarb	6 per cent	
Pea				
Root-knot	Seed treatment	Carbofuran	1.5 to 2.0 kg/ha	Sharma, 1989
Chilli and bottle gourd				
Root-knot		Carbofuran Aldicarb	3 per cent	Kanaswamy and Sivakumar,1981 Anon. 1983
Bitter gourd *Pratylenchus* sp.		Carbofuran Aldicarb and sulfone		Anon. 1985

Host Resistance

The use of resistant varieties is an effective economical and environmentally safe method of reducing yield loss of different vegetables crop against nematodes. In India, screening of germplasm of different vegetables crops against nematodes has been made to identify source of resistance (Table 24.4).

Tables 24.4: Nematodes-Resistant Cultivars of Vegetables Crops

Crop	Nematode	Resistant Cultivars
Tomato	*Meloidogyne* spp.	Nematex, SL-120, NTR-1, SL-12, Punjab NR-7, Mangala, Hawaii-7746, Hawaii-7747, Arka vardaan, Karnataka hybrid
	Rotylenchulus reniformis	Radiant, Ronita, Piersol, SL-120, F-38-E-2, Patriot
Brinjal	*M. incognita*	Giant of Banaras, Black beauty, Gola
	M. javanica	Vijaya hybrid, Multiple purple, Gacha baigan, Ghatikia white, Dingras, PBR-91-2
Chilli	*M. javanica*	CAP-63, Pusa jwala
	R. reniformis	Pusa jwala
Potato	*M. incognita*	Kufri Dewa
	Globodera rostochiensis	Kufri swarna
Okra	*Meloidogyne* spp.	Harichikni, Kanki local green, Vaishali Badhu

Contd...

Table 24.4–Contd...

Crop	Nematode	Resistant Cultivars
Cowpea	*M. incognita*	Barasti Mutant, Iron, New selection, C-152, C-82-1-B, IC-9642-B, Dixie cream, TVU 2430
Carrot	*M. incognita*	Black
French bean	*M. incognita*	Banat, Blue lake stringless, Brown beauty, Cambridge, Bountiful flat, Master piece, Gallaroy
Pumpkin	*M. incognita*	Jaipuri, Dasna
Ridge gourd	*M. incognita*	Panipati, Meerut special
Ash gourd	*M. incognita*	Jaipuri Agra
Watermelon	*M. incognita*	Shahjanpuri
Musk melon	*M. incognita*	Siarsol

Source: P. Parvatha Reddy, 1993

Biological Control

In the present scenario of agricultural research when all nematologist all over the world in one confronted with the non availability of any effective nematicides which is cost effective and environmentally safe, biological control solve their confrontation for the management of nematode at the present time. In the last decayed, nematode management through biocontrol agent is in the forefront of the research and development. Several multination companies and Govt. agencies have been propagating the ideas for commercialization of biocontrol agents. Fungi and bacteria are the most important microbial agents attacking nematodes.

Fungal Agents

The potential of nematode trapping fungi as a biological agents was first examined in Hawaii for control of nematodes on pine apples (Linford *et al.*, 1938). With the advancement of research and development it has wide spread all over the world. Few micro-organisms like *Pasturia penetrans, Paecilomyces lilacinus, Verticillium chlamydosporium, Aspergillis niger, Trichoderma viride, T. harzianum etc.* have shown greater promise against *M. incognita* and several other nematodes (Table 24.5). Few commercial products like Yorker from Agricultural Blotech, based on *P. lilacinus* effective against root-knot nematode have become available in the market. The fungi used for biological of nematodes can be classified as follows:

1. Predacious fungi or trapping fungi
2. Endozoic or endoparasitic
3. Parasitic or opportunistic fungi
4. Indirect influence of fungi from organic additive.

The above fungi increases in the vegetable fields by the manipulation of organic fertilizer and minimize the population *Meliodogyne* spp, *Heterodera* spp, *Rotylenchulus* spp *Pratylenchus* spp. *Hoplolaimus* spp etc. (Singh and Sitaramaiah, 1966; Sen and Das Gupta, 1981).

Predacious Fungi

These fungi also known as nematophagous or trapping fungi like *Arthobotrys errigularis*, protected plants against nematode infection. There are mainly two nematophagous fungi like *Dactyloides* and *Dactylella haptotyla* are very good against root knot and free living nematodes.

Endozoic Fungi

Also called endoparasitic fungi. They produce either simple spore or flagellated spore as infectious agents, instead of extensive hyphal development out side the body of host. *Meria coniospora*, *Nematophthora*, *Catenaria anguillulae* etc. are the best example of endozoic fungi which parasitized on *Globodera*, *M. incognita* and other nematodes of vegetable crops.

Opportunistic Fungi

Several fungi have been found to be parasitized on eggs of plant parasitic nematodes are referred as opportunistic fungi. These fungi are colonize on the reproductive structures of the nematodes like *Meloidogyne Heterodera* and *Globodera* species causes on potato, tomato brinjal etc. Generally *Paecilomyces Verticillium Cylindrocarpon*, *Fusarium*, *Exophiala*, *Torula*, *Gliocladium* and *Phoma* are an example of opportunistic fungi.

Indirect Influence of Fungi from Organic Additive

Amendment of soil with oil cakes of neem, groundnut, mustard and custard was effective is reducing the population of *Hoploluimus*, *Tylenchorhynchus*, *Meloidogyne*, and *Helicotylenchus* spp. on many vegetables crops (Singh and Sitaramaiah, 1966; Misra and Padhi, 1985). Similarly organic amendment with saw dust and rice hull ash also used in nematode management.

Table 24.5: Effect of Some Fungal Antagonists Against Nematode of Vegetable Crops

Crop	*Nematodes*	*Bio-agents*	*Reference*
Potato	*M. incognita*	*P. lilacinus*	P. Parvatha Reddy,1993
Tomato	*M. incognita*	*P. lilacinus*	
Potato	*G. rostochiensis*	*P. lilacinus*	
Tomato and Brinjal	*Rotylenchulus reniformis*	*P. lilacinus*	
Vegetables crops	Root knot and Cyst	*Aspergillus niger*	Khan, *et al.*,1984
		Pasturia penetrans	
		Glomos fasicularum	Sitaramaiah and Sileora,1982

Bacterial Biocontrol Agents

Biological control of nematodes by bacteria is a emerging method of nematode management. There are several bacteria (Table 24.6) which are used for the control of nematodes. The mode of action of the bio agents are parasitism, antibiosis and plant growth promotion.

Plant parasitic nematodes mainly root knot and reniform are the major yield limiting factors of vegetables crops. In India, lot of works have been done for nematodes management, but full package of knowledge on nematodes management is still lacking. Most of the management practices have been reported by several workers are the good gaining of information, not a proper management practices to reduce the nematodes population below the economic threshold level. It is a time to think it again for better future and I hope genetic engineering may be the leading player in the coming era of nematode management.

Table 24.6: Effect of Some Bacterial Antagonists Against Nematode of Vegetable Crops

Corp	*Nematode*	*Bacterial Biocontrol Agents*	*Reference*
Tomato and cucumber	Root-knot and cyst nematodes	PGPR	Zavaleta and Vangundy, 1987
Sugar beet	*Heterodera schachtii*	*Pseudomonas fluorescence*	Oostendorp and Sikora, 1989
Potato	*G. palida*	*A. radiobactor + B. sphebrieus*	Racke and Sikora, 1992
Potato	*G. Palida*	*A radiobactor*	Racke and Sikora, 1992
Bottle guard	*M. javanica*	PGPR (*Azotobacter chroococcum* strain no. NAC-68)	Bansal *et al.*, 1998
Tomato	*M. incognita*	*P. fluoreseens*	Verma *et al.*, 1998 (Proceedings; 79p)
Vegetable crops	Root knot and cyst	*Bacillus thurigiensis*	Sitaramaiah and Sileora,1987

References

Abawi, G.S. and Mai, W.S. (1983). Increase in cabbage yields by phenamiphos treatment of uninfested and *Heterodera schachtii* infested field soil. *Pl. Dis.*, 67: 1343-1346.

Ahuja, S. (1982). Chemical control of root-knot nematode in nursery bed of tomato and egg plant and its effect on yield in the field. *Tropical pest management*, 28: 313-314.

Ahuja, S. (1983). Comparative efficicacy of some nematicides against root-knot nematode, *Meloidogyne incognita* in tomato, okra, brinjal and cauliflower. *Third Nematol Symp.*, H.P. Agri. Univ., solan, P. 43.

Alam, M.M. (1971). *J. Nematd*. Spec. Pub. No. 1: 7.

Alam, M.M., Saxena, S.K. and Khan, A.M. (1973). Control of root-knot nematode, *Meloidogyne incognita* (Kofoid and White, 1919) Chitwood, 1949 on tomato and egg plant with VC-13 and Basamid liquid as bare-root dip. *Indian J. Nematol.*, 3: 154-56.

Alam, M.M., Saxena, S.K. and Khan, A.M. (1975). Influence of interculture of marigold and margosa with some vegetable crops on plant growth and nematode population. *Acta Bot. Indica*, 5: 35-39.

Alsaedy, H.A., Stephan, Z.A. and Girgees, M.M. (1989). Effect of *M. javanica* on egg plant seedling of different age. *Nematol. Medit.*, 17: 31-32.

Anon, (1971). *Estimated crop losses due to plant-parasitic nematode in the United States*. Special Publ No. 1, Suppl. *J. Nematol.*, 7 pp.

Anon (1983). *Proceedings of the fifth Workshop of All India Co-ordinated Research Project on Nematode Pests of Crops and their Control*. Himachal Pradesh Agric. Univ., Solan.

Anon (1985). *Proceedings of the Fourth Biennial Workshop of All India Co-ordinated Research Project on Nematode Pests of Crops and their Control*. Mohanlal Sukhadia Univ., Udaipur, 129 pp.

Atkinson, G.F. (1892). Some diseases of cotton. *Albama Polytech Inst. Agr. Exp. Sta. Bull.* No. 41: 61-65.

Atwal, A.S. and Mangar, A. (1969). Repellent action of root exudates of *Sesamum orientale* against the root-knot nematode, *Meloidogyne incognita* (Heteroderidae: Nematoda) *Indian J. Entomol.*, 31: 286.

Belair, G. (1984). Non fumigant nematicide for the control of northern root-knot nematode in muck growth carrot. *Canadian J. Pl. Sci.*, 64: 143-145.

Birchfield, W. and Martin, W.J. (1967). *Phytopathology*, 57: 804.

Bossis, M. and Mugniery, D. (1989). *Heterodera carotae* John 1950 injuriousness in field in France. *Rev. Nematol.*, 12: 177-180.

Bossis, M., Cavelier, A. and Mugniery, D. (1989). *Heterodera carotae* Johns 1950 advantage and limitation of chemical control. *Rev. Nematol.*, 12: 343-350.

Bowman, P. and Bloom, J.R. (1966). Breaking the resistance of tomato varietes to Fusarium wilt by *Meloidogyne incognita*. *Phytopathology*, 56: 871.

Caudra, R. (1984). Number of generation and life cycle of *Meloidogyne inconita* in *Cuba. Ciencias de la Agricultura*, 20: 3-10.

Chahal, P.P.K. and Chhabra, H.K. (1984). Interaction of *Meloidogyne incognita* with *Rhizoctonia solani* on tomato. *Indian J. Nematol.*, 7: 37-41.

Chhabra, H.K. and Sharma, J.K. (1981). Combined effect of *Meloidogyne incognita* and *Rhizoctonia bataticola* on pre-emergence damping-off of okra and brinjal. *Sci. Cult.*, 47: 256-257.

Chhabra, H.K., Sidhu, A.S. and Singh, I. (1977). *Meloidogyne incognita* and *Rhizoctonia solani* interaction on okra. *Indian J. Nematol.*, 7: 54-57.

Darekar, K.S. and Mhase, N.L. (1988). Assessment of yield loss due to root-knot nematode *M. incognita in tomato*, brinjal and bitter gourd. *Int. Nematol., Netw. Newsl.*, 7: 6.

Dhawan, S.C. and Sethi, C.L. (1977). Inter-relationship between root-knot nematode, *Meloidogyne incognita* and little leaf of brinjal. *Indian Phytopathol.*, 30: 55-63.

Ellemby, C. (1951). Mustard oils and control of the potato root eelworm, *Heterodera rostochiensis* Wollenweber: Further field and laboratory experiments. *Ann. Appl. Biol.*, 38: 859-875.

Fassuliotis, G. (1987). Genetic basis of plant resistance to nematodes In: Vistas on Nematology (eds.) Veech, J.A. and Dickson, D.W.). Society of Nematologist, Incorporation, Hyattsville, Maryland, USA pp. 364-371.

Fritzsche, R. and Thiele, S.I. (1981). Damage by *Ditylenchus dipsaci* to head of onion growth for seeds. Nachridh tenblat fur den Pflanzenchutzdienst in der DDR 35: 211-212.

Gaur, H.S. and Dhingra, A. (1991). Management of *Meloidogyne incognita* and *Rotylenchulus reniformis* in nursery bed by soil solarization and organic amendment. *Rev. de Nematol.*, 14: 189-196

Goswami, B.K. and Chenulu, V.V. (1974). Interavection of root-knot nematode, *Meloidogyne incognita* and tobacco mosaic virus in tomato. *Indian J. Nematol.*, 4: 69-80.

Goth, R.W., Peter, K.V., Sayre, R.M. and Webb, R.E. (1983). *Phytopathology*, 73: 966.

Greco, N., Elia, F. and Brandonisio, A. (1986). Control of *Heterodera carotae, Ditylenchus dipsaci, Meloidogyne javanica* with fumigants and non-fumigant nematicides. *J. Nematol.* 18: 359-364.

Gupta, D.C. and Mukhopadhyaya, M.C. (1971). Effect of N. P. and K. on the root-knot nematode, *Meloidogyne javanica* (Treub) Chitwood. *Sci. Cult.*, 37: 246-247.

Handa, D.K. and Mathur, B.N. (1981). Control of root-knot of chillies with nematicides. *Nematol. Soc. India Symp.*, T.N. Agric. Univ., Coimbatore, p. 48.

Haque, S., Khan, M.W. and Saxena S.K. (1984). Suitability of certain systemic nematicide on dip treatment in relation to penetration and development of *M. incognita* in tomato. *Pakistan J. Nematol.*, 2: 23-27.

Haque, M.M. and Gaur, H.S. (1985). Effect of multiple cropping sequences on the dynamics of nematode population and crop performance. *Indian J. Nematol.*, 15: 262-263.

Hazarika, B.P. and Roy, A.K. (1974). Effect of *Rhizoctonia solani* on the reproduction of *Meloidogyne incognita* on egg plant. *Indian J. Nematol.*, 4: 246-248.

Hussain, S.I., Kumar, R., Khan, T. and Tittoo, A. (1984). Effect of root dip treatment of egg plant seedlings with plant extact, nematicides and oil cakes extract and antihelminthic drugs on plant growth and root-knot nematode development. *Pakistan J. Nematol.*, 2: 79-83.

Ivanova, T.S., Vasileva, T.M. and Dzhurava, L.M. (1979). Dependence of tomato yield on initial densities of *Melodogyne acrita* infective juveniles mortality. *Simpozuima Dushunbe.* pp 40-42.

Jain, B.K. and Bhatti, D.S. (1978). Bare root treatment with systemics for controlling root-knot nematode in tomato transplants. *Indian J. Nematol.*, 8: 19-24.

Jain, B.K. and Bhatti, D.S. (1987). *Tropical pest management*, 33: 122-125

Jain, R.K., and Bhatti, D.S. (1983). Chemical control of *Meloidogyne javanica* in tomato through nursery bed treatment. *Third Nematol. Symp.*, H.P. Agric. Univ., Solan, pp. 37-38.

Jain, R.K.,Bhutani, R.D., Kallo, Bhatti, D.S. and Singh, K. (1989). *Indian Horticulture* (in Press).

Jain, R.K., Paruthi, I.J., Gupta, D.C. and Darekar, B.S. (1986). Appraisal of losses due to root-knot nematode *M. javanica* in okra under field condition. *Tropical pest management*. 32: 341-342.

Jenkins, W. R. and Coursen, B.W. (1957). The effect of root-knot nematodes, *Meloidogyne incognita acrita* and *M. hapla,* on *Fusarium* wilt of tomato. *Pl Dis. Reptr.*, 41: 182-186.

Kampfe, L. (1962). Zur wirkuhg von Ultraschall auf cysten bildende und freilebende Nematoden. *Nematologica* 7: 164-172.

Kanasamy, K. and Sivakumar, C.V. (1981). Effect of *Meloidogyne arenaria* on the germination of chilli (*Capsicum annum* Linn.) seeds. *Nematol. Soc. India Symp.*, T.N. Agric. Univ., Coimbatore, p. 16.

Khan, A.M(1969). Final Tech. Report. Grant No. FG-IN-225, Project No. 17-CR-65- Aligarh Muslim Univ., Aligarh, India

Khan, A.M. Saxena, S.K., Siddiqi, Z.A. and Upadhyay, R.S. (1975). Control of nematodes by crop rotation. *Indian, J. Nematol.*, 5: 214-221.

Khan, A.M., Saxena, S.K. and Khan, M.W. (1971). Interaction of *Rhizoctonia solani* Kuhn and *Tylenchorhynchus brassicae* Siddiqi. 1961 in pre-emergence damping-off of cauliflower seedlings. *Indian J. Nematol.*, 1: 85-86.

Khan, R.M., Saxena S.K. Khan, M.W. and Khan A.M. (1980). Interaction of *Meloidogyne incognita* and *Phomopsis vexans. Indian J. Nematol.*, 10: 242-244.

Khan, T.A., Azam, M.F. and Hussain, S.I. (1984). Effect of fungal filterate of *Aspergillus niger* and *Rhizoctonia solani* on penetration and development of root-knot nematode and the plant growth of tomato var. marigold. *Indian J. Nematol.* 14: 106-109.

Krishnappa, and Prasad, K.S.K. (1979). Effect of concomitant inoculations of *Tylenchorhynhus dubius* and *Helicotylenchus crenatus* on their population build up on tomato. *Mysore J. Agric. Sci.*, 13: 54-56.

Lamberty, F., Boiboi, J.B. and Ciancio (1983). Losses due to *M. incognita* in okra in Liberia. *Nematol Medit.*, 16: 5-6.

Lue, L.P., Lewis, C.C. and Melcher. V.E. (1984). Effect of aldicarb on nematode population and its persistence in carrot and hydroponic solution. *J. Environ. Sci. and Health.*, 19: 343-354.

Linford, M.B., Yap, F. and Oliveira, J. M. (1938). *Soil Science*, 45: 127-141.

Mahajan, R., Mangat, B.P.S. and Nandpuri, K. S. (1986). Plant parasitic and soil nematodes associated with vegetables crops in India-A checklist. Dept of vegetables crops, Landscaping and Floriculture, PAU, Ludhiana, India, pp. 48.

Mahmood, K., Qumar, S., Naqvi, A. and Alam, M.M. (1974). The influence of bottlegourd mosaic virus and brinjal mosaic virus on the development of root-knot caused by *Meloidogyne incognita* (Kofoid and White, 1919) Chitwood, 1949. *Indian Phytopathol.*, 27: 610-611.

Maqbool, M.A., Hashmi, S. and Gaffar, A. (1985). Effect of aldicarb and Carbofuran on the control of root-knot nematode on cauliflower, (*Brassica oleracea*). *Int. Nematol. Netw. Newsl.* 2: 24-25.

McGuidwin, A.E., Bird, G.W., Haynes, A.Z. and Gaze, S.H. (1987). Pathogenicity and population dynamics of *M. hapla* associated with *Allium cepa. Pl. dis.* 71: 446-449.

Mihailescu, N., Lordcachescue, C. and Ivan, M. (1979). Reducing nematode infestation in garlic during storage. *Product Vetel. Hort.*, 28: 19-25.

Misra, R.P. and Padhi, N.N. (1985). Relative efficacy of some nematicides and organic materials in the control of reniform nematode on French bean. *Indian J. Nematol.*, 15: 270-271

Moel, C.D. de. (1986). Cyst nematode and the culture of cauliflower *Groenten en fruit*. 41: 76-77.

Morgan, D.O. (1925). Investigations on eelworm in potatoes in South Lincolnshire. *J. Helminth.*, 3: 185-192.

Nagnathan, T.G. (1984). Chemical control of *M. incognita* in the nursery. *Nematol. Medit.*, 12: 253-254.

Nagnathan, T.G. (1984). Studies on yield loss in vegetable due to *Meloidogyne incognita. South Indian Hort.* 32: 115-116.

Napiere, C. M. (1978). Unpublished graduate thesis, Univ. of Phil., at Los Banos, USA

Nath, R., Khan, M.N., Wanshi, R.S.K. and Dwivedi, R.P. (1984). Influence of root-knot nematode, *Meloidogyne javanica* on pre-and post-emergence damping-off of tomato. *Indian J. Nematol.*, 14: 135-140.

Nirula, K.K. and Bassi, K.K. (1965). Effect of French marigold as a trap crop on the control of root knot nematodes in potato fields. *Indian Potato Journal*, 7(2): 64-68.

Ohbayashi, N. (1983). The control of root lesion nematode *Pratylenchus penetranse* on Japanese radish with DD and Dazomet in combination. *Bull. Kanagwa Hort. Expt.,Stat.*, 30: 81-84.

Oostendorp,M. and Sikora, R. A. (1989). *Revue de Nematologie*, 12: 77-83

Pani, A.K. and Das, S.N. (1972). Studies on etiological complexes in plant diseases. I. Association of root-knot nematodes in bacterial wilt of tomato. *J. Res. Orissa Univ. Agric. Techol.*, 2: 54-59.

Parvatha Reddy, P. (1985). Chemical control of root-knot nematodes infecting peas. *Indian J. Nematol.*, 15: 120-121.

Parvatha Reddy, P. (1993). Nematode management in vegetable crops In: *Advances in Horticulture* Vol. 6-Vegetable crops 2 (eds) K.L.Chadha and G. Kalloo, Malhotra Publishing House, New Delhi, pp. 829-857.

Parvatha Reddy, P., Govindu, H.C. and Setty K.G.H. (1975). Studies on the effect on amino acid on the root knot nematodes *Meloidogyne incognita* infecting tomato. *Indian J. Nematol.,* 5: 36-41.

Parvatha Reddy, P. and Singh, D.B. (1979). Control of root-knot nematode *Meloidogyne incognita* on tomato by chemical bare-root dips. *Indian, J. Nematol.,* 9: 40-43.

Parvatha Reddy, P., Singh, D.B. and Ram Kishun, (1979). Effect of root-knot nematodes on the susceptibility of Pusa Purple Cluster brinjal to bacterial wilt. *Curr. Sci.,* 48: 915-916.

Peacock, F. C. (1956). *Nematologica,* 1: 305-310.

Philips, J. (1984). The development of *M. javanica* on tomato in Cyprus. *Nematologica,* 30: 470-474.

Prasad, K.S.K. (1989). Nematological problems and progress of research on potato in India. *Proc. Fourth Group Meeting on Nematological Problems on Plantation Crops,* Univ. of Agric. Sci., Bangalore, pp. 50-51.

Prasad, K.S.K. and Krishnapa, K. (1981). Effect of bare root-dip treatments with pesticides on *Rotylenchulus reniformmis* affecting brinjal. *Indian J. Nematol.,* 11: 158.

Prasad, K.S.K, Raj, D. and Sharma, R.K. (1983). Effect of planting time on *M. incognita,* infecting potato at Simala Hill Third Nematol. Symp. H.P. Agric. Univ. Solan, pp. 56.

Prasad, K.S.K., Setty, K.G.H. and Krishnapa, K. (1977). *Cur. Res.,* 6: 205-206.

Powell, N.T. (1971). Interaction between nematode and fungi in disease complexes pp. 253-273 in J.G. Horsfall (Ed.) *Ann. Rev. Phytopathol.* Vol. 9, Ann. Of Rev. Ann. Palo. Alto. Calif.

Racke, T. and Sikora, R. A. (1992). *J. Phytopathology,* 134: 198-208.

Rai, B.S. and Jain, R.K. (1988). Comparative biology of root-knot nematode *M. incognita* on cotton G. *hirsutum* and tomato. *L. estculentum. Indian J. Nematol.* 18: 349.

Ramakrishnan, C. and Balasubramanian, P. (1981). Control of *Meloidogyne incognita* in chilli (*Capsicum annum* L.) nursery with nematicides. *Nematol. Soc. India Symp.,* T.N. Agric. Univ., Coimbatore, p. 54.

Ramana, K. V. and Mohandas, C. 1988. Occurrence of potato cyst nematode, *Globodera pallida* (Stone, 1973) in Kerala. *Indian J. Nematol.,* 18: 141.

Reddy, D.D.R. and Seshadri, A.R. (1972). Studies on some systemic nematicides. II. Further studies on the action of thionazin and aldicarb on *Meloidogyne incognita* and *Rotylenchulus reniformis. Indian J. Nematol.,* 2: 182-190.

Reddy, D.D.R. and Seshadri, A.R. (1975). Elimination of root-knot nematode infestation from tomato seedlings by chemical bare-root dips or soil application. *Indian J. Nematol.,* 5: 170-175.

Renjhen, P.L. (1973). *History of Plant Quarantine and its Functioning in India.* Govt. of India, Ministry of Agric., Directorate of Plant. Prot. Quarantine and Storage, Faridabad.

Robert, P.A. and Greathead, A.S. (1986). Control of *Ditylenchus dipsaci* in infected garlic seeds and clove by non fumigants nematicides. *J. Nematol.,* 18: 66-77.

Rohde, R.A. and Jenkins, W.K. (1958). Basis for resistance of *Asparagus officinalis* var. *altilis* L. to the stubby root nematode *Trichodorus christiei* Allen, 1957. *Bull Univ., Md. Agric. Exp. Sta.* A-97, 19 pp.

Saxena, S.K., Alam, M.M. and Khan, A.M. (1974). Chemotherapeutic control of nematodes with Vydate oxamyl. VC-13. and dazomet on chilli plants. *Indian, J. Nematol.,* 4: 235-238.

Sen, K. and Dasgupta, M.K. (1981). Effect of rice hull ash on root knot nematodes in tomato. Trop. Pest Management, 27: 518-519.

Sethi, C.L., Nath, R.P., Mathur, V.K. and Ahuja, S. (1972). Interceptions of plant parasitic nematodes from imported seed/plant materials. *Indian J. Nematol.*, 2: 89-93.

Sharma, G.L. (1989). Estimated losses due to root-knot nematodes, *Meloidogyne incognita* and *M. javanica* in pea crop. *int. Nematol. Network Newsl.*, 6 (1): 28-29.

Sharma, J.K., Singh, I. and Chhabra, H.K. (1980). Observations of the influence of *Meloidogyne incognita* and *Rhizoctonia bataticola* on okra. *Indian J. Nematol.*, 10: 148-151.

Sharma, N.K. and Sethi, C.L. (1976). Reaction of certain cowpea varieties to *Meloidogyne incognita* and *Heterodera cajani. Indian J. Nematol.*, 6: 17-123.

Shukla, V.N. and Swarup, G. (1976). Inter-relationship of the root-knot nematode, *Meloidogyne incognita* (Kofoid and White, 1919) Chitwood, 1949 and *Sclerotium rolfsii* Sacc., on the emergence of tomato seedlings. *bull. Indian Phytopath.* Soc., 6: 52-54.

Siddiqui, M.A. and Alam, M.M. (1988). Efficacy of seed dressing with the extract of neem, Persian lilac against *M. incognita* and *R. reniformos. Nematol. Medit.*, 15: 399-403.

Siddiqui, M.A. and Alam, M.M. (1990). Control of root-knot, reniform and stunt nematode by nimbin seed extract. *Nematol. Medit.*, 18: 19-22.

Singh, D.B. and Parvatha Reddy, P. (1981). Note on chemical control of root-knot nematodes infesting French bean. *Indian J. Agric. Sci.*,51: 534-535.

Singh, D.B. and Parvatha Reddy, P. (1982). Chemical control of *Meloidogyne incognita* infecting cowpea. *Indian J. Nematol.*, 12: 196-197.

Singh, D.B., Rao, V.R. and Parvatha Reddy, P. (1978). Evaluation of nematicides for the control of root-knot nematodes on brinjal. *Indian J. Nematol.*, 8: 64-66.

Singh, R.S. and Sitaramiah, K. (1966). Incidence of root-knot of okra and tomato in oil cake amended soil. *Pl. Dis. Reptr.*, 50: 668-672.

Singh, R.V. and Khera. S. (1978). Culturing and life history studies of *Rotylenchulus reniformis* on brinjal. *Bull. Zool. Survey of India.* 1: 115-128.

Singh, S. P., Ahmad, M., Khan, A.M. and Saxena, S.K. (1980). Effect of seed treatment with certain oil cakes on nematodes and growth of tomato and on rhizosphere population of nematode and fungi. *Nematol. Medit.*, 8: 193-198.

Sitaramaih, K. and Sikora, R.A. (1982). Effect of mycorrhiza fungus *Glonmus fasiculatum* on host parasite relationship of *Rotylenchulus reniformis* in tomato. *Nematologica.* 28: 412-419.

Sitaramaih, K. and Sinha, R.A. (1981). Influence of endomycorrhiza fungus *Glomus massae* on *R. reniformis* penetration and development of bush bean, cucumber and muskmelon. Medelelingcn Van de Faculteit Landbouwwet Rijks univ.

Sitaramaih, K. and Sinha, R.A. (1984). *Indian J. Nematol.*, 14: 175-178.

Sivakumar, C.V. and Meerzainuddin, M. (1974). Influence of N. P. and on the reniform nematode, *Rotylenchulus reniformis,* and its effect on the yield of okra. *Indian J. Nematol.*, 4: 243-244.

Sivakumar, C.V. and Seshadri, A.R. (1971). The life history of the reniform nematode, *Rotylenchulus reniformis* Linford and Oliveira, 1940. *Indian J. Nematol.*, 1: 7-20.

Sivakumar, C.V., and Seshadri, A.R. (1972). Histopathology of infection by the reniform nematode, *Rotylenchulus reniformis* Linford and Oliveira, 1940 on castor, papaya and tomato. *Indian J. Nematol.*, 2: 173-181.

Sivakumar, C.V., Palanisamy, S. and Naganathan, T.G. (1976). Persistence of nematicidal activity in seed treatment of okra with Carbofuran and aldicarb sulfone. *Indian J. Nematol.*, 6: 106-108.

Speigal, Y., Orion, D. and Zehar, R. (1983). Use of systemic nematicides in the control of stem and bulb nematode *D. dipsaci* on onion. *Hassadeh.* 63: 2578-2579.

Stephen, Z.A., Alwan, A.H. and Antoon, B.G. (1988). Effect of planting date on development of root-knot nematode, *M. javanica.* Plant production and per cent of infection of tomato and egg plant. *Iraqi J. Agric. Sci. Zanco.*, 6: 59-68.

Stephen, Z.A., Michbas, A.H. and Shakir, I. (1989). Effect of organic amendment, nematicides and solar heating on root-knot nematode infecting egg plant. *Int. Nematol. Netw. Newsl.*, 6: 35-35.

Stone, L.EW. (1961). Oats as a trap crop for cereal root eelworm. *Pl. Path.*, 10: 164

Subramanyian, S., Rajendran, G. and Vadivellu, S. (1990). Estimation of losses in tomato due to *M. incognita* and *R. reniformis. Indian J. Nematol.*,20-239.

Sundesh, H.N. and Setty, K.G.H. (1977). *Cur. Sci.*, 6: 157-158.

Tanda, A.S. and Atwal, A.S. (1989). Effect of sesame intercropping against root-knot nematode *M. incognita* in okra. Nematologica. 34: 484-492.

Tayler, A.L., and Sasser, J.N. (1978). Biology, identification and control of root-knot nematode (*Meloidogyne* sp.) *Coop. Publi. Pl. Pathol. N.C. State Univ.*

Thakar, N.A. and Patel, C.C. (1985). Control of root-knot nematode, *Meloidogyne javanica* on tomato by chemical bare-root dips. *Indian J. Nematol.*, 15: 110-111.

Turligina, E.S. and Zino eva E.T.V. (1985). Gall nematode in green house soil. *In:* Taksonomiya c biologya fitogelmino. (Edited by Trutygina, E.S.).

Upadhyaya, K.D. and Dwivedi, B.K. (1987). Analysis of crop losses in pea grown due to *M. incognita. Int. Nematol Netw. Newsl.*,4: 6-7.

Verma, M.M., Sharma, H.c. and Pathak, V.N. (1978). Evaluation of systemic granular nematicides against *Tylenchorhynchus* spp. parasitic on cabbage. *Indian J. Nematol.*, 8: 59-60.

Vovlas, N., Moreno, I. and Inseira, R.N. (1986). Histopathoglogy of root gall induced on tomato by *Globodera pallida* and *G. rostochiensis. J. Nematol.* 18: 267-269.

Vrain, T.C., Fournier, Y. and Crete, R. (1981). Carrot yield increased after chemical control of root-knot nematode in organic soil. *Canadian J. Pl. Sci.*, 61: 677-682.

Wood, F. C. and Goodey, J.B. (1957). Effect of gamma-rays irradiation on nematodes infesting cultivated mushroom beds. *Nature*, 180: 760-761.

Zacheo, B.T., Zacheo, Gand and Taylor, C.E.A. (1987). A comparison of histochemical changes induced by *Xiphinema basiri* and *X., ifacolumi* in root of tomato. *Nematol. Medit.*, 15: 235-251.

Zacheo, T.B., Zacheo. G., Lamberti, F. and Chinappen, M. (1985). Gall induction on tomato root tip by *Physiological buchae on ultrastructure aspect. Nematol. Medit.*, 13: 225-238.

Zavaleta-Meija, M. F. and Vangungy, S.D. (1987). *Revista Mexicana de Fitipatologia*,- 5: 38-44.

Sustainable Disease Management of Agricultural Crops (2011) *Pages* **333–353**
Editors: **S.K. Biswas and S.R. Singh**
Published by: **DAYA PUBLISHING HOUSE, NEW DELHI**

Chapter 25

Transgenic Plant and its Role in Plant Disease Management

☆ ***Kahkashan Arzoo, N.K. Pandey, S.K. Biswas, and Mohd. Rajik***
Department of Plant Pathology, C.S. Azad University of Agriculture and Technology, Kanpur – 208 002, U.P.

Considerable progress has been made during last 20 years on isolation, characterization and introduction of novel gene into plants. As per estimate made in 2002, transgenic crops are cultivated world-wide on about 148 million acres of land. Transgenic plants have beneficial traits over normal plant such as disease-resistant, insects-resistant, herbicide tolerance, weed control and many others. Globally transgenic crops were grown during 2003 on nearly 70 million hectare of land. In contrast the coverage in India was <1,00,000 hac compared to 2.8 m ha in China (James 2003). By 2050 it is expected Indian population will reach 1.5 billion and hoped that 30 per cent of Indian population will suffering from malnutrition. Therefore, nutritional security is an important concern for Indian economy by 2050. To meet these challenges we have to manage our resources more precisely and produce more nutritious and healthy crops which is free from injurious diseases and that can only be possible through development of transgenic crop.

Transgenic Plants

The plants obtained through genetic engineering in which a functional foreign gene has been incorporated that generally not present in plant are called transgenic plant.The uptake of foreign DNA/transgene by a plant cell is called transformation.

The type of plant cells employed for transformation must satisfy following criteria:

1. The cells should be competent to take up DNA and allow the expression of transgenes.
2. The cells must be meristematic.
3. The cells should be capable of producing complete plant *i.e.*, it should be totipotant.

Variety of techniques have been used to introduce transgenes into plant cells.The detail of these techniques, are beyond the scope of this chapter we are mainly concern with management of plant disease.

Management of Bacterial Diseases in Plants

Plant pathogenic bacteria have been known since 1882, they are by far the largest group of plant pathogenic prokaryotes causing a variety of plant diseases and causing considerable loss in crop yield.

Genetic engineering techniques have unraveled many options for the management of bacterial diseases in plants. Many new strategies to engineer plants against bacterial pathogens are available. Some of the important strategies are:

Detoxification of Pathogenic Toxins

Many bacteria which causes disease in plant (phytopathogenic bacteria) produce toxins which are directly or indirectly responsible for appearance of symptoms. Toxins are mainly of two types:

Host Specific Toxin

It damages only those plant varieties that are susceptible to the pathogen (Selective in nature).

Non-host Specific Toxin

Such toxin has the capacity to damage the host as well as wide variety of non host plants and living organism (Non selective in nature). (Misaghi,1982).

In case of non-host specific toxin producing bacteria, these bacteria must possess the mechanism to protect itself from toxin. Two strategies for self protection in such bacteria have been documented:

1. Production of an enzyme that modify a toxin to inactive form *e.g.* tabtoxinine-β-lactam
2. Production of toxin insensitive target enzymes.

Production of Bacterial Resistant Genetically Engineered Plants by Resistance Inducing Gene for Signal Transduction

Molecular and biological studies have revealed that bacterial disease resistance in plant is governed by various types of resistance genes and defense genes. The resistance gene R does not code for any toxin but they act as regulatory genes involved in recognition of signals from invading pathogens and subsequent activation of defense genes some examples are as follows.

Pto Gene Mediated Resistance

It confers resistance against *P. syringae* p.v. *tomato* in tomato plant (Martin *et al.*, 1993). Pto itself is a member of tightly clustered family of 5 genes encoding serine/threonine kinase (Martin *et al.*, 1993). Another serine/threonine kinase Pti is also involved in Pto mediated disease resistance (Zhu *et al.*, 1995). The gene Pif bound Pto locus is essential for activation of Pto gene (Loh and Martin 1995; Rommens *et al.*, 1995; Yoneyama, 1998.). Zhu *et al.* (1997) identified 3 genes Pti 9, Pti 5 and Pti 6 encoding proteins that show direct interaction with Pto kinase. Transgenic tomato expressing Pto genes showed resistance in plants against many diseases. (Martin *et al.*, 1993).Pti gene from tomato was cloned and transferred to tobacco and transplant showed resistance to *Pseudomonas syringae* p.v. *tabaci* (Zhau *et al.*, 1995).

Xa-21 Gene Mediated Resistance

Transgenic rice plant expressing cloned Xa-21 gene confer multiple isolated resistance to 29 diverse isolates from 8 countries. It is suggested that Xa-21 gene may be active against diverse pathogen products. It is also possible that Xa-21 may receive signal from pathogen or plant cells and several defense genes may be induced.

Glucose Oxidase Gene Mediated Resistance

The glucose oxidase gene from *Aspergillus niger* catalyses the oxidation of β-glucose by molecular oxygen yielding gluconic acid and H_2O_2. Elevated levels of H_2O_2 induce accumulation of total salicylic acid several fold in the leaf tissue of transgenic plants.

Bacterial soft rot is a destructive disease in many crops. Transgenic potato plants expressing glucose oxidase gene exhibited strong resistance to bacterial soft rot disease caused by *Erwinia caratovora* pv. *caratovora*. The resistance was apparently mediated by elevated levels of H_2O_2 because resistance could be counteracted by exogenously added H_2O_2 degrading catalase.

Arabidopsis

NPRI gene controls onset of Systemic Acquired Resistance and affects Localized Acquired Resistance. It encodes a protein containing ankyrin repeats. Transgenic mutants expressing NPRI gene was more resistant to *P. syringae* p.v. *maculicola* in absence of SAR induction (Lao *et al.*, 1997).

Resistance Via Antimicrobial Peptide

Insect Derived Antimicrobial Peptide

Small peptides with antimicrobial activity *i.e.* the lytic peptides appear to be a major component of antimicrobial defense system of a number of organism like insects, molluscs, amphibians and at least one mammal (Van Hofsten *et al.*, 1985; Casteels *et al.*, 1989; Lee *et al.*,1989). Based on arrangement of amphipathy and high positive charge density of the molecule, the antimicrobial peptides are classed into following:

Cercopins

It is a family of homologous antibacterial peptides of 35-37 residues derived from the giant silk moth (*Hyalophora cercopia*)(Boman and Hulmark, 1991). The antimicrobial activity by cercopin is due to inhibition of a process involving channel formation and subsequent membrane disruption (Christensen *et al.*, 1988).

1. Transgenic tobacco plants that harbored chimeric genes, when were inoculated with *Pseudomonas solanacearum* in their stems, some resistance to wilt was observed but on root inoculation no resistance was observed.
2. Transgenic tobacco plants, containing a modified cercopin gene fused to secretary sequence from barley α-amylase and regulated by promoter and terminator from the potato proteinase inhibitor II, exhibited enhanced disease resistance against tobacco wildfire pathogen *P. syringae pv. tabaci*.
3. The protein extracted from transgenic potato expressing cercopin showed antibacterial activity (Montanelli and Nascari, 1991).

   Several stable analogue of cercopin B such as SB-37, Shiva-1, MB39 have also been synthesized in artificial condition. (Fink *et al.*, 1989; Norelli *et al.*, 1994; Huang *et al.*, 1997).
4. Transgenic tobacco expressing Shiva I showed enhanced resistance against *Ralstonia solanacearum*.

5. Transgenic tobacco expressing MB 39 gene were resistant to *P. syringae* pv. *tabaci*. Resistance was 10 fold compared to control plants (Huang *et al.*, 1997).

Attacins

These are isolated from giant silk moth and act against growing gram negative bacteria but are not directly lytic. They disrupt the structure of outer membrane by interfering with the synthesis of new membrane component (Engsrom *et al.*, 1984).

1. Apple rootstock M-26 susceptible to *Erwinia amylovora* was transformed with gene for attacin E to yield transgenic line T1 which was more resistant than untransformed M-26 but less than most resistant rootstock M-7 (Norelli *et al.*, 1994).

Magainins

The frog's skin possess a group of short peptides which are thought to function as a natural defense mechanism against infection due to their antimicrobial properties. These are known as magainins.

Lactoferin

Lactoferin in an Fe-binding glycoprotein known to have antibacterial properties.

Tachyplesin

A gene encoding lytic peptide tachyplesin has been isolated from horse shoe crab. Transgenic potato expressing the gene showed reduced tuber rot by *E. caratovora* (Allefs *et al.*, 1996).

CTX (Cholera toxin)

It is a multimeric protein consisting of A_1, A_2 and 5 B subunits. Transgenic tobacco expressing the A. subunit of CTX showed greatly reduced susceptibility to bacterial pathogen *P. syringae* p.v. *tabaci* and accumulated high levels of SA.

Plant Derived Antimicrobial Peptides

Thionins are a group of small (5 KDa), cystein rich and basic peptide toxins involved in the resistance of plants to fungal and bacterial pathogen (Bohlmann (1994). These are present in the endosperm and leaves of cereals, in other tissues such as roots and in plant species other than cereals (Bohlmann and Apel 1987; Garcia–Olmedo *et al.*, 1989). Its toxicity has been demonstrated in vitro for several phytopathogenic bacteria and fungi (Fernardez De Caleya *et al.*, 1972; Cammue *et al.*, 1992).

Carmona *et al.* (1993) made chimeric genes in which the coding sequence of an α theorin barley gene, derived from a genomic clone and a wheat α-thionin, derived from a cDNA clone, were placed under the control of the 35 S CaMV promoter.

Synthetic Antimicrobial Peptides

Four synthetic cationic peptides pep 6, pep 7, pep 11 and pep 20 were tested alone or in combination for their antimicrobial activities against economically important phytopathogenic bacteria like *E. caratovora* var *caratovora* and *E. caratovora* var *atroseptica*. The pathogens were found to be inhibited by pep 11 and pep 20. The inhibition was 50 per cent reduction in symptom at 2.0-2.3μm. pep 11 and pep 20 was not phytotoxic to potato plants at 200 μM.

Resistance by Antimicrobial Enzymes

Many genes which govern antibacterial enzymes have been cloned from microorganisms and animals. These enzymes have cell wall degrading activity against bacteria for example, lysozyme from various organisms and phage (Jolles and Jolles 1984).

Lysozymes

These are enzymes having specific hydrolytic activity against the bacterial cell wall peptidoglycan and cleave the glycosidic bond between C-1 of N-acetyl muramic acid and C-4 of N-acetyl glucosamine in peptidoglycon.

The lysozyme genes used for bacterial resistant transgenic plant production are derived from the following:

1. *Hen egg lysozyme*: A hen egg lysozyme has been cloned and inserted on tobacco plant. The expression of such gene Extracts in transgenic plant inhibited growth of several bacteria but reaction of this transgenic plant to bacteria has not yet been studied. (Trudel *et al.*, 1992).
2. Human lysozyme is a powerful lytic enzyme degrading bacterial cell wall.
3. *Bacteriophage*: Bacteriophages lyse the bacterial cell wall by lysozyme (Tsugita 1971).

Bacterial Pectolytic Enzymes

The bacterial pectolytic enzymes degrade the plant cell wall and lead to formation of different sized sugar oligomers. Expression of pectolytic enzymes genes in transgenic plants may lead to resistance by eliciting defense responses.

Table 25.1: Transgenic Plants Conferring Resistance Against Bacterial Pathogens

	Gene	*Source*	*Transgenic Plant*	*Pathogen Examino*	*Reference*
A.	**Bacterial toxin tolerance**				
	ttr	*Pseudomonas*	Tobacco	*P. syringa pv. tobaci*	Azai *et al.* (1989)
	arg K	*P. syringae* p.v. *phaseolicola*	Tobacco	*P. syringae pv phaseolicoca*	De La fuente, Martinez *et al.* (1992) Hatzilou Kas and Panopoulos (1992)
B	**Resistance inducing genes**				
	Pto gene	Tomato	Tobacco	*P. syringae pv. tobaci* (expressing avr Pto)	Thilmong *et al.* (1995) Romonens *et al.* (1995).
	NPRI	*Arabidopsis thaliana*	NPRI defected mutant	*P. syringae pv. maculicola*	Cao *et al.* (1997)
	Glucose oxidase	*Aspergiuns nigar*	Potato	*Ercoinia caratovora*	Wu *et al.* (1996)
C	**Antimicrobial peptides**				
	Certopin	Giant silk moth	Tobacco	*Berkholderia solanacearum*	Jaynes *et al.* (1993)
			Tobacco	*P. syringae pv. tabac*	Huang *et al.* (1997)
	Attacin	Cercopia	Apple	*E. amylovora*	Norelli *et al.* (1994)
	Thionin	Barley	Tobacco	*P. syringae pv. tabaci*	Carmona *et al.* (1993)
				P. syringae pv. syringae	Carmona *et al.* (1994)
	Tachyplerin	South eagg horseshoe	Potato	*E. caratovora*	Allefs *et al.* (1996)
	Cholera toxin	Rab subumit A1	Tobacco	*P. syringae pv. tabaci*	Beffa *et al.* (1995)
D	**Other origin enzyme**				
(1)	Lysozyme	Bacteriophazi T_4	Potato	*E. saratovora*	During *et al.* (1993)
		Human	Tobacco	*P. syringae pv. tabaci*	Nakjima *et al.* (1997)
(2)	Peetate lyase	*E. caratovora*	Potato	*E. caratovora*	Wegener *et al.* (1996)

Management of Viral Disease of Plant

A virus is nucleoprotein entity that multiplies only in living cells and cause disease to economically important crops as well as in other crops of minor importance. Nearly half of known viruses attack and cause diseases in plants. A single virus species can infect a dozens of different species of plants. A plant may sometimes be infected by more than one kind of virus at the same time.

Transgenic plants having resistant to virus have been produced by transforming plants with resistance imparting trans genes.

1. Pathogens for which resistance is to be developed called pathogen derived resistance includes coat protein, replicase protein, movement protein, antisense, protease and ribozyme mediated resistance.
2. Plants include plantibodies genes mediated resistance, R gene, antiviral protein genes.
3. Other sources such as RNA.

Pathogen Derived Resistance

The first virus resistant transgenic plant VRTP was produced in mid 1980s by expressing coat protein gene of TMV Strain U1 in transgenic tobacco plants (Powel-Abel *et. al.*, 1986). Since then many viral genes and virus associated RNAs have been used as transgene to confer resistance in plants.

Both coding and non coding regions of viral genome has been used for developing VRTPs for some groups of viruses, like poty viruses. Resistance in plants have been obtained using almost all types of viral genes as transgene. Coat protein (CP) gene is the most widely used transgene for developing VRTPs followed by replicase genes and movement protein genes:

Coat Protein Mediated Resistance

Coat protein genes from more than 40 RNA/DNA viruses belonging to 17 virus groups have been used to confer resistance to many different plant species (Grumet, 1994; Pappu *et al.*, 1995; Jain *et al.*, 2000; Collaway *et al.*, 2001). Coat protein mediated resistance have been found effective irrespective of virus particle morphology:- rigid rod, flexous, isometric etc.), genome organization and mode of transmission. Jain *et al.* (2000) has shown that complete gene is not always necessary for inducing resistance and genome segment as small as 110 bp long could induce resistance.

Protein Mediated CP Gene Induced Resistance

It occurs when a single copy of transgene is inserted. Reistance so inserted is of moderate levels against a broad range of related viruses and influenced by the levels of protein.

The protein inhibits disassembly of infecting virus and force the assembly/disassembly equilibrium towards assembly. This seems likely for viruses belonging to potex virus, Tenui virus, Tobamo virus groups.

RNA Mediated CP Gene Induced Resistance

It occurs when multiple copies of transgene are inserted (Lomonossoff,1995). Resistance so expressed is of high levels and virus specific and is attributed to lower levels of transcripts. Transgene expression is up to mRNA level with little or no transgenic protein. When mRNA accumulation exceed threshold level, co-suppression (silencing) is initiated, affecting transgene expression and virus replication.

Replicase Protein Mediated Resistance

Replicase protein gene is the second widely used transgene to confer resistance against plant viruses. It was first demonstrated in TMV (Golemboski *et al.*, 1990). It has been successfully used for various RNA/DNA viruses representing from several plant groups (Jones *et al.*, 1998).

Replicase protein mediated resistance provides nearly immune type and highly specific resistance for virus from which the transgene is isolated.

Molecular mechanism of RPMR is not fully understood. However, it seems to be either protein or RNA-medited. Since the concept is relatively new, its commercial application is just being initiated. Potato lines combining replicase protein mediated resistance to PLRV, wheat resistant to Wheat streak mosaic virus (Agrios, G.N. 2006) and Bt. resistance to Colorado potato beetle have been developed (Thomas *et al.*, 1998). Other example is of tomato resistance to tomato yellow leaf curl virus and Cucumber mosaic virus on tomato.

Movement Protein Mediated Resistance

Cell to cell movement of viruses in plants is brought about by movement proteins encoded in virus genomes. Movement proteins have been expressed in transgenic plants which interfere with cell to cell transport of viruses and thus plant is resistant to viruses.

1. Transgenic potato lines expressing mutant PLRV pr17 movement protein exhibited resistance against unrelated viruses like PVX and PVY (Tracke *et al.*, 1996).
2. Transgenic plants expressing the putative movement protein gene NSm from TSWV were resistant to infection by TSWV (Prins and Goldbach, 1998).Some other examples are-resistant raspberry against raspberry bushy dwarf virus, tomato against TMV, dendrobium against Cymbidium mosaic virus.

Satellite RNAs Mediated Resistance

Some viruses have specific satellite RNA molecules, which are consistent viral parasites being dependent on helper virus for multiplication. Some sat-RNAs exacerbate the disease caused by the helper virus, whereas some other sat-RNAs ameliorate the disease. The sat-RNAs that ameliorate the disease have been used for developing VRTPs for resistance to cucumoviruses and nepovirus.

Defective Viral Genome Mediated Resistance

Defective or deleted genomes of some RNA and DNA viruses disrupt the replication of complete genomes of those viruses with which they are associated. There are some evidences that plants expressing such viral subgenomic DNAs show decreased disease severity due to concerned viruses.

Antisense RNA Mediated Resistance

Antisense RNA in its simplest form can be produced by inverting a cDNA copy of an mRNA with respect to the promoter in an expression vector; this yields a full length complementary copy of mRNA sequence. Fragments smaller than full length can also be effective–antisense RNA molecules are thought to interact with mRNA molecules by base pairing to form double stranded RNA.

This approach gives protection by either inhibiting gene expression or viral replication. Although this approach has been used with several RNA and DNA viruses with different levels of success in the case of viruses confined to phloem tissues (Luteoviruses) and the viruses which replicate in the nucleus (Geminiviruses).

Plant Derived Resistance

R-genes Mediated Resistance

Several plant genes (R-genes) having resistance to viruses have been identified (Hulbert *et al.*, 2001); Verma and Mitter, 2001. The 'N' genes for tobacco introduced into tomato gives good protection against TMV and ToMV (Kumar and Baker, 2000).

1. Rx1 and Rx2 genes isolated from *Solanum tuberosum* have been introduced into potato breeding line to confer resistance against PVX (Bendahmane *et al.*, 2000). Several naturally occurring resistant genes are listed in Table 25.2 (Byoung *et al.*, 2005).

Table 25.2

Sl.No.	Gene	Plant	Resistance Mechanism	Viruses
1.	N	*N. tabaccum*	Cell to cell movement	TMV
2.	Rx1	*S. tuberosum*	Replication	PVX
3.	Rx2	*S. tuberosum*	Replication	PVX
4.	S w 5	*S. esculentum*	Cell to cell movement	TSWV
5.	HRT	*A. thaliana*	Cell to cell movement	TCV
6.	RTM1	*A. thaliana*	Systemic movement	TEV
7.	RTM2	*A. thaliana*	Systemic movement	TEV
8.	RCY1	*A. thaliana*	Cell to cell movement	CMV
9.	Tm 22	*S. lycopersicum*	Cell to cell movement	ToMV
10.	Pvr1-pvr2	*C. annum*	Replication	PVY
11.	Pvr1	*C. annum*	Cell to cell movement	PVY
12.	Mol 1	*L. sativa*	Replication	LMV
13.	Sbm 1	*P. sativum*	Replication	PSbMV

Antiviral Principles (AVPs) Mediated Resistance/Ribosome in Activating Protein

Natural resistance to virus infection in non-hosts has been attributed to the presence of antiviral principles or proteins (Varma *et al.*, 1998). This approach can be exploited by genetic engineers for developing resistance against viruses and other pathogens.

1. Tobacco and potato plants transformed with PAP/PAP11 (pokeweed antiviral protein) genes showed resistance against TMV,PVX, PVY and PLRV viruses.

Plantibodies Mediated Resistance

In early 1990's transgenic plants were produced that were genetically engineered to incorporate in their genome, and to express, foriegn genes, such as a mouse gene that produce antibodies against certain plant pathogens. Such antibodies encoded by animal genes but produced in and by the plants are called plantibodies. (Agrios, 2006). Such plant produced antibodies can be full antibody molecule, Fab fragments or specific chain FV (SCFV) fragments produced against specific plant viruses and are targeted against the same viruses (Agrios, 2006). They have been fully expressed in leaves and seeds of some plants and accumulate in intercellular spaces chloroplast and lumen of endoplasmic reticulum.

Resistance can be engineered by the expression of virus specific plantbodies in plant system.

1. Transgenic plants producing antibodies specific to AMCV (Artichoke mottle crinkle virus) coat protein were found resistant to virus infection (Tavladoraki *et al.*, 1993).
2. Similar approach has been attempted to generate transgenic tobacco lines expressing plantibodies to TMV coat protein and TSWV nucleocapsid and glycoprotein (Trancone *et al.*, 1999).
3. Plantibodies to TMV in tobacco decreased infectivity of the virus by 90 per cent (Agrios, 2006)
4. Plantibodies to *Beet necrotic yellow vein virus*, also in tobacco, provides a partial protection against the virus in the early stages of infection and against development of symptoms later on (Agrios, 2006).
5. Plantibodies to stolbus phytoplasma and to corn stunt spiroplasma in tobacco remained free from infection for more than 2 months.

Plantibodies derived resistance appear mostly as a delay in development of disease, and barring a breakthrough, it does not appear that it will become on effective means of plant disease control in the near future.

Ribozyme Mediated Protection

Ribozymes are hybrid RNA molecules consisting of tobacco ring spot virus RNA endoribonuclease catalytic sequences linked to the antisense RNA of specific genes against which they are targeted. The strategy consist of:

1. To produce cDNA of this ribozyme.
2. To integrate it into the host plant genome.

Transgenic tobacco plants expressing ribozymes against TMV showed some resistance to TMV infection. (Singh, 2004). Progress in India for developing resistant transgenic plants against viruses are:

Gemini viruses, poty viruses, Cucumo viruses, Badna viruses, Tobomo viruses, Ilar viruses and Tospo viruses groups which are also major constraints of crop production in India (Verma and Ramachandran, 1994; Bhat *et al.*, 2002). Globally transgenic crops were grown during 2003 on nearly 70 million hectare of land. In contrast the coverage in India was <1,00,000 hac compared to 2.8 mha in China (James 2003).

Management of Fungal Diseases of Plants

Fungi are small, generally microscopic, eukaryotic, usually filamentous, branched, spore-bearing organism that lack chlorophyll. Since from the antiquity of cultivation fungal pathogens are one of the major constraint in crop production. So far we have faced a number of famine which are directly caused by fungal pathogen results widespread loss of life as well as money.

Genetic material can be introduced into plant cells or protoplasts by several methods it may be direct or indirect. In all methods, small or large pieces of DNA are introduced into plant cells or protoplasts may be integrated in the plant chromosomal DNA. Such plants are reffered as 'transgenic'.

When the introduced DNA carries appropriate regulatory genes recognized by the plant cells or is integrated rear appropriate regulatory genes along plant chromosomes, the DNA is "expressed" *i.e.*

it is transcribed into mRNA which is than translated into proteins which is the responsible factors for plant defense. The introduced DNA may be of plant origin or may be derived from a pathogen.

Fungus resistant transgenic plants can be produced by any of the following methods:

1. Expression of gene products that are directly toxic to or which reduce growth of pathogen including pathogenesis related (PR) proteins, such as hydrolytic enzymes (Chitinase, glucanases) and antifungal proteins (osmotin, thaumatin like), as well as antimicrobial peptides (thionins, defensins, lectins), ribosome inactivating proteins and phytoalexine.
2. Expression of gene products that destroy or neutralize a component of the pathogen arsenal, including inhibition of polygalacturonase, oxalic acid and lipase.
3. Expression of gene products that can potentially enhance the structural defenses in plants, including elevated levels of peroxidase and lignin.
4. Expression of gene products that release signals that can regulate plant defenses including production of specific elicttors, H_2O_2, salicylic acid and ethylene.
5. Expression of resistance gene (R) products involved in R/Avr interaction and the hypersensitive response (HR).

Expression of Gene Products that are Directly Toxic to which Reduce Growth of the Pathogen

Expression of Gene Coding for PR-Proteins

PR-proteins are structurally diverse group of plant proteins that are toxic to invading pathogen. PR-proteins are widely distributed in plants in trace amounts but are produced in much greater concentration following pathogen attack or stress.

Several groups of PR-proteins have been classified according to their function, serological relationship, amino acid sequence, molecular weight and certain other properties. PR-proteins are extremely reactive and highly solvable. At least 14 families of PR-proteins are recognized. The better known PR-proteins are:-

1. PR-1–Antioomycetes and antifungal
2. P.R-2–β-1,3-glucanase
3. PR3–Chitinase
4. PR4–Antifungal–harein like proteins
5. PR5–Thaumatin like proteins
6. PR6–Proteinase inhibitors
7. Defensive
8. Thionins
9. Lysozymes
10. Lipoxygenase
11. Osmotin like protein
12. Cystein rich protein
13. Glycine rich

14. Proteinases
15. Chitosanases
16. Peroxidases

Expression of Hydrolytic Enzyme Producing Genes

Most widely used approach has been to over express chitinase and glucanases which belong to the group of PR-proteins (Neuhaus, 1993) and which have been shown to exhibit antifungal activity *in vitro,* (Boller, 1988; Yun *et al.*, 1997). As chitin and glucans are the major components of cell walls of many groups of fungus; the over expression of these enzymes in plant cells was postulated to cause lysis of hyphae, thereby reducing fungal growth (Mauch and Stachelin 1989).

Following expression of different types of chitinases in a range of transgenic plant species, the rate of lesion development and overall size and number of lesions were observed to be reduced upon challenge against many fungal pathogens:

1. Tobacco plants transformed chitinase gene from bean became resistant to infection by soil borne fungus (*Rhizoctonia solani*) but not to infect by fungi belongs omycetes (*Phythium aptanidermatum*) the cell wall of which lacks chitin (Agios, 2006).
2. In other experiment constitutive expression of PR chitinase genes from rice in transgenic rice and cucumber plants made rice plant more resistant to *R. solani* and cucumber to *Botrytis cinerea* (Lin *et al.*, 1995).
3. Similarly tomato expressing both enzymes were more resistant to *Fusarium oxysporum* f.sp. *lycopersici.*
4. Since the characterstics of chitinases from different sources can vary due to substrate binding specificity, pH optimum and localisation in cell, leading to differences.

Expression of Antimicrobial Protein/Peptides/Other Compound Producing Gene

Defensins and thionins are low molecular weight (around 5 k Da) Cysteine-rich peptides (45-54 amino acids in length), found in monocots and dicots that were initially derived from seeds and which have antimicrobial activity (Bohlmann 1987; Brockaet *et al.*, 1995; Evans and Greenland, 1998). It was proposed that these peptides played a role in protecting seeds from infection by pathogens (Bioskaert *et al.*, 1997). Defensins are also found in insects and mammals where they play an important role in curtailing or limiting microbial attack (Rao, 1995).

These peptides may exert antifungal activity by altering membrane permeability and/or inhibiting macromolecule biosynthesis. Thionins may be toxic to plant and animal cells cultures as well (Broekaert *et al.*, 1997).

Chitin binding peptides (knottin type) are 36-40 residues in length and have been recovered from the seeds of some plant species. They contain crystein residues and have been demonstrated to have antifungal activity *in vitro* (Bioekaert *et al.*, 1997). Some examples are as follows:-.

1. The over expression of defensins and thionins in transgenic plant was demonstrated to reduce development of several different pathogens including *Alternaria, Fusarium* and *Plasmodiophora* and provided resistance to Verticillim on potato tuber under field conditions. (Gao *et al.*, 2000).
2. Transgenic tobacco plants expressing PR-1 protein gene were resistant to blue mold Oomycetes *Perenospora tobacina.*

3. Expression of Amianthus have in type peptides and Microbilis knoltin type peptide in transgenic tobacco did not enhance tolerance to *Alternaria longipes* or *Botrytis cenerea* (De Bolle *et al.*, 1996). It was postulated that the presence of catione particularly Ca^{2+} may have inhibited the activity of these peptides *in vivo*. Modifications of amino acid sequence of peptides may enhance the antifungal activity (Evans and Greenland, 1998).
4. An antimicrobial protein homology to lipid transgenic protein was shown to reduce development of *B. cinerea* when expressed in transgenic genonium (Bi *et al.*, 1999).
5. Antimicrobial peptides have been synthesized in the laboratory to produce smaller (10-20 amino acids) molecules that have enhanced potency against fungi (Cary *et al.*, 2000). In addition synthetic cationic peptide chimera (Cercopin-melittin) a broad spectrum antimicrobial activity has been successfully produced (Osusky *et al.*, 2000). When expressed in transgenic potato and tobacco these synthetic peptides provided enhanced resistance against a number of fungal pathogens including (*Colletotrichum sp.*) *Fusarium sp.* and *Phytophthora sp.*

   These synthetic peptides may demonstrate lytic activity against fungal hyphae, inhibit cell wall formation and/or enhance membrane leakage. The ability to create synthetic recombinants and combinatorial variants of peptides that can be rapidly screed in the laboratory should provide additional opportunities to engineer resistance to a range of pathogen simultaneously. Enhancement of specific activities of antifungal enzymes or the creation of variants broad activity using directed molecular evolution (DNA shuffling) has also been proposed as a method to enhance the efficacy of transgenic plants in the future (Lassner and Bedbook, 2001).
6. Transgenic tobacco expressing radish gene encoding antifungal protein 2 (Rs -APP2) which exhibits antifungal activity, showed a high level of resistance to foliar pathogen *Alternaria alternate* tobacco pathotype (Terras *et al.*, 1995).

Expression of Ribosome–Inactivating Proteins (RIP)/Plant Derived Peptide Toxin

RIPs are plant enzymes that have 28S rRNA which have N-glycosidase activity which, depending on their specificity, can inactivate non-specific or foreign ribosomes, thereby shutting down foreign protein synthesis, without interfering with their own ribosomes. RIP genes also show synergism to PR proteins genes when the two are expressed concurrently in the same plant (Leah *et al.*, 1991). Two types of RIRs exist in nature–Type 1 and type 2.

1. The most common cytosolic type I RIP from the endosperm of cereal grains do not act on plant ribosomes but can affect foreign ribosomes such as those of fungi (Stripe *et al.*, 1992; Hortley *et al.*, 1996).
2. Expression of barley seed RIP reduced development of *Rhizoctonia solani* in transgenic tobacco (Logemann *et al.*, 1992) but had little effect on *Bluemeria graminis* in transgenic wheat (Bieri *et al.*, 2000).

   In the latter study, the RIP was targeted to the apoplastic space and may have had less activity against development of intracellular haustoria of the mildew pathogen.
3. It has been demonstrated that combined expression of chitinase and RIP in transgenic tobacco had a more inhibitory effect on *R. solani* development than the individual proteins (Jach *et al.*, 1995).

4. Pokeweed (*Phytolacca americana*) antiviral protein type I RSP activity has been expressed in transgenic tobacco and was shown to reduce development of *Rhizoctonia solani.*

Expression of Phytoalexins

Phytoalexins are toxic antimicrobial substances produced in appreciable amounts in plants only after stimulation by various types of phytopathogenic micro organisms or by chemical or mehanical injury. Phytoalexins are produced by healthy cells adjacent to localized damaged and necrotic cells in response to materials diffusing from damaged cells. Thus phytoalemins inhibit development of a fungus on hypersenstive tissues and are not produced in compatible bio-trophic infections.

More than 3000 chemicals which have phytoalexin like properties isolated from plans belonging to more than 30 families. Some of the better studied phytoalexins include:-

1. Phaseolin–in bean
2. Pisatin–in pea
3. Glyceolin-sin soyabean, alfa alfa and clover.
4. Rishitin–in potato.
5. Gossypol–in cotton
6. Capsidiol–in pepper
7. Mediacarper–in alfa alfa
8. Pheseolin–in French bean

Examples supporting role of phyallexin:

1. A mutant of Arabidopsis deficient in production of indole type phytoalexin 'caumolexin' was shown to be more susceptible to infection by *Alternaria brassicicola* but not to *Botrytis cinerea*. (Thomma, Nelissne *et al.*, 1999).
2. Genes encoding certain phytoalexins such trans resveratrol and medicarpin resulted in delayed development of disease symptom production by number of pathogens on several plant species.

These studies are encouraging in light of the difficulties of engineering the complex biochemical pathways leading to phytoalexins accumulation in plants (Dixon *el al.*, 1996).

Expression of Gene Products that Destroy or Neutralize or Component of the Pathogen Arsenals, including Inhibition of Polygalacturonase, Oxalic Acid and Lipasei

Plant cell wall act as a barrier for pathogen entry. Pathogens have numerous mechanism to overcome this barrier. (Walton, 1994). These include numerous cells wall degrading enzymes and production of toxine like oxalic acid by fungi. Strategies to engineer resistance against pathogen infection targets to inactivate these virulence product of the pathogens.

1. Some plants possess of a glycoprotein in their cell wall known as polygalacturanase inhibitory protein. It can inhibit fungal endopolygalacturanase (Powell *et al.*, 1997; Desideriact *et al.*, 1997). Expression of PGIP in transgenic plants have shown contrasting effects:

   Expression of bean PGIP genes in transgenic tomato did not lead to any resistance against *Fusarium, Botrytis* or *Alternaria* (Desidero *et al.*, 1997). Expression of pea PGIP in transgenic tomato lead to increased resistance against *Botrytis* (Powel, *et al.*, 2000).

It was later shown that bean PGIP different in specificity to fungal polygalacturonase. Therefore appropriate assessment is required before transformation. With PGIP disease is reduced to some extent but cannot be completely prevented.

2. *Expression of genes encoding antibody/plantibodies*: Plantibodies can bind to pathogen virulence product (De Jaeger *et al.*, 2000, Schilberg *et al.*, 2001). The plantibodies can be expressed inter and extracellularly and can bind or inactivate enzymes, toxins or other pathogen factors involved in disease development. Till now there is no published report for expression of antibodies in transgenic plants that has lead to reduction in disease Antilep antibodies inhibited the infection of tomato by *Botrytis* when mixed with spore innoculum by preventing fungal penetration through cuticle (Cornenil *et al.*, 1998). Infection of *Colletotrichum gloeosporioides* on various fruits was inhibited using polyclonal antibodies that bound to fungal pectin lyase (Watted *et al.*, 1997).
3. *Production of phytotoxic metabolite*: Many fungi produce phytotoxic substances like mycotoxin, oxalic acid etc. These metabolites facilitate infection on host plant leading to cell death. Strategies have been developed to degrade these compounds by enzymes expressed by genes encoding them in transgenic plants:

   Expression of trichothecian degrading enzyme from *Fusarium sporotrichioid* in transgenic tobacco reduced plant tissue damage and enhanced seedling emergence in presence of trichothecene (Mutritch *et al.*, 2000). During seed germination in cereals a stable glycoprotein germin was produced. These are produced in response to fungal infection and abiotic stress and have oxalate oxidase like activity. (Dumas *et al.*, 1995; Zhang *et al.*, 195; Beria and Bernia 1997). Activity of substrate oxalate result in production of CO_2 and H_2O_2, H_2O_2 induces defence (Brisson *et al.*, 1994; Mehdy 1994).

Expression of barley oxalate oxidase gene in oilseed rape enhanced tolerance to phytotoxic oxalic acid. (Thompson *et al.*, 1995). Some carry over expression of oxalate oxidase gene in transgenic poplar enhanced resistance to *Septoria* while with oxalate decarboxylase gene enhanced resistance of tomato against *Selerotinia selerotium* which potentially enhance the structure defences in the plant including elevated of levels of peroxidase and lignin.

1. Lignin around the cell at the site of infection can slow pathogen spread (Nicholson and Hammerschmidt 1992). Over expression of tobacco amionic peroxidase gene in tomato enhance lignin level but resistance to fungal pathogen was not enhanced (Lagrimini *et al.*, 1993). Lignin level was enhanced by expression of H_2O_2 generating enzyme glucose oxidase in transgenic potato (Wer *et al.*, 1997) and by expression of IAP in transgenic tobacco (Sitbon *et al.*, 1999). Peroxidase over expression in plants can have negative effect on plant growth and development (Logrimini *et al.*, 1997).

Expression of Gene Products that Release Signals that can Regulate Plant Defenses including the Production of Specific Elicitor Hydrogen Peroxide, Salicylic Acid Ethylene etc.

Elicitors

A signal pathway for defence response in plants including HR, PR-protein and phytoalexin production are triggered by elicitors (Heath 2006; Mc Dowell and Dangel 2000; Shirow and Schslzer–Lefert 2000). Expression of a gene encoding elicitor cryptogein from *Phytophthora cryptogea* in tobacco induced HR as well as several defense genes on challenge innoculation with a range of fungi. The growth of fungi was restricted (Keller *et al.*, 1999). INF-1 elicitor act as Aus factor in tobacco-*Phytophthora infestans* interaction and it resulted in onset of HR (Kamoun *et al.*, 1998).

Expression of gene encoding Avr 9 elicitor from *C. falcatum* in transgenic tomato containing Cf. gene resulted in HR reaction (Hommond–Kosack *et al.*, 1994; Honec *et al.*, 1995). Appearance of lesion resulting from HR in a bacterial proton pump gene (bacteriopsin) from *Halobacterium holobium* activated multiplied defense in transgenic tobacco (Miller *et al.*, 1995). In transgenic potato expression of bacteriopsis enhanced resistance to some pathogen (Abad *et al.*, 1997).

H_2O_2

These are generated by expression of gene encoding glucose oxidase gene and activates the defense response of plants. H_2O_2 inhibit pathogen growth (Wn *et al.*, 1995) and induce PR protein, SR and ethylene (Wu *et al.*, 1997; Chamnangpol *et al.*, 1998) and phytoalexins (Mehdy, 1994).

Expression of increased level of H_2O_2 in transgenic potato, tobacco, cotton reduced disease development against different pathogens–*Rhizoctonia, Verticillium, Phytophthora, Alternaria* etc.

Signal molecules, SA, Jasmonic acid activate plant defense response (Yang *et al.*, 1997; Dong 1995; Reymond and Farmer 1998; Demsey *et al.*, 1999). Over expression of SA in transgenic tobacco was shown to enhance PR protein and provide resistance to fungal pathogen (Verbene *et al.*, 2000). Expression of tobacco catalase, an enzyme with solicylic acid binding activity in transgenic potato enhanced defense gene expression leading to SAR and enhanced tobacco to *Phytophthora infestans* (Yu *et al.*, 1991).

Necrotrophic pathogen attack lead to SA and TP signals in plants. Genetic engineering to alter ethylene and Jasmonic acid production in plants may result in unpredictable effects on disease respone depending on pathogen as well as potential side effect in view of multiple roles played by these signal molecules in plants (O' Dannell *et al.*, 1996; Weiler *et al.*, 1997; Wilkison *et al.*, 1997).

Expression of Resistant Gene (R) Products Involved in R/Avr Interaction and Hypersensitive Response

R-gene products serve as receptor of Avr factors or recognizes the Avr factor through a coreceptor (Staska wicz *et al.*, 1995). Gene for gene interaction trigger one or more signal transduction pathway and that in turn activate defence response in plants to prevent pathogen growth (Hommond–Kosack and Iones 1996; De Wit 1997).

R-gene can be breed into the crop plants to control diseases but this approach has only limited success as many pathogens have evolved methods to counteract plant defence response including. (i)Modification of elicitor proteins by mutation or by deletion of the Avi gene or by down regulation of Avs gene expression. (ii)Secretion of enzymes that detoxify defense components *e.g.* phyloalexin. (iii) Use of ATB binding cassette transporters to mediate the efflux of toxic components. (iv) Secretion of glucanase of inhibition proteins which inhibit the endonucleare activity of host plant. (v) Transforming susceptible plants with cloned R-genes may provide pathogen resistance in some cases for examples.

When susceptible tomato cultivar was transformed with Pto genes, the plant became resistant to bacterial pathogen *Pseudomonas syringae* (Tang *et al.*, 1999).

Once it was thought that major draw back of R-gene mediated resistance is its extreme specifically of action towards a single over gene of a pathogen. However, Pto over expression in plants constitutively activates defense responses and result in general resistance in absence of avr. Pto gene as it also gained resistance against fungal pathogen *C. fulvum* (Tang *et al.*, 1999). Over expression of HRT gene which controls HR to Turnip crinkles virus in Arabidopsis did not confer resistance to *Perenospora tobacina* (Cooley *et al.*, 2006). Expression of cf 9 genes which confer resistance in tomato against races of *C.falcatum*, in transgenic tobacco and potato gave rise to the HR when challenged with Avi 9 peptide (Hammond–Kosok *et al.*, 1998) indicating that cf 9 gene product was produced. Expression of cf 9 gene in oilseed rape enhanced resistance to *Leptospharea maulans* (Menin *et al.*, 2001).

Table 25.3: Transgenic Plants Conferring Resistance Against Fungal Pathogen

	Gene	*Source*	*Transgenic Plant*	*Pathogen examine*	*Reference*
A.	**Defence related proteins**				
(1)	Chitinase	Bean	Tobacco	*Rhizoctonia solani*	Broglie *et al.* (1991)
		Tobacco	Tobacco	*R. solani*	Vierheilig *et al.* (1993)
		Lice	Indica rice	*R. solani*	Lin *et al.* (1995)
		Tomato	Rape	*Cylirdrosporium concentricum, phama lingum, selerotinia sclerotiorum*	Gison *et al.* (1996)
(2)	β1,3 Gluconas	Soybean	Tobacco	*Phytophthora parasitica*	Yoshileawa *et al.* (1993)
				Alternaria alternate	
		Tobacco	Tobacco	*P. tabacina*	Lusso and Kue (1996)
(3)	Chitnase	Rice	Tobacco	*P. cercospora nicotiana*	Zhu *et al.* (1994)
	glucane	Barley	Tobacco		Jach *et al.* (1995)
		Tobacco	Tomato	*F. oxysporum*	Jangedifk *et al.* (1995)
(4)	PR-1	Tobacco	Tobacco	*Perenospora*	Alexander *et al.* (1993)
(5)	Osmotin	Tobacco	Potato	*Phytophthora infestans*	Liu *et al.* (1994)
(6)	Phytoalexin	Grapevine	Tobacco	*Botrytis cinerea*	Hain *et al.* (1993)
		Grapevine	Rice	*Magnapothe grisea*	Stark–lorenzen *et al.* (1997)
(B)	**Resistance inducing gene**				
(1)	Glucose oxidase	*Aspergilus niger*	Potato	*Phytophthora infestans*	Wn *et al.* (1995)
(C)	**Antimicrobial peptides**				
(1)	RIP	Barley seed	Tobacco	*Rhizoctonia solani*	Longemann *et al.* (1992)
(2)	Thionin	*Arabidopsis thaliana*	Arabidopsis	*F. oxysporum f.sp. matthioloe*	Epple *et al.* (1997)
(D)	**Other origin enzymes**				
(1)	Lysozyme	Human	Tobacco	*Erysiphe cichoracearum*	Nakjima *et al.* (1997)
(2)	Chitinase	*Serratia marcescens*	Tobacco	*A. alternata tobacco pathotype*	Suslous *et al.* (1988)
		Rhizo oligosporus	Tobacco	*S. sclerotiorum B. cinerea*	Terakawa *et al.* (1997)

Conclusion

Newly acquired expertise in plant transformation continued discovery of R-gene and novel genes as well as improvement of gene isolation techniques have intensified genetic engineering for resistance against devastating diseases. But transgenic crops are not the cure for all problems. It may have some unforeseen limitation which is yet to be encountered and after all it must qualify political, social and ethical issues.As more and more GM crops come to market, environmental impact, risk analysis and public concern should be explicitly addressed. Being the first generation of new and largely untested area of scientific intervention, perceived benefits of genetic engineering is apparently clearer than the perceived risk.

References

Agrios, G.N. (2006). In: *Plant Pathology*. Academic press. New York. Pp. 242-246.

Alexander, D., Goodman, R.M., Gut-Rella, M., Glascock, C., Weymann, K., Friendrich, L., Maddox, D., Ahl-Goy, P., Luntz, T., Ward, E., and Ryals, J. (1993). "Increased tolerance to two oomycete pathogens in transgenic tobacco expressing pathogenesis related protein la", *Proc. Natl. Acad. Sci. USA*. 90: 7327-7331.

Allefs, J.J., de Jong, E.R., Florack, D.E. A., Hoogendoorn, J., and Stiekema, W.J. (1996). '*Erwinia* soft rot resistance of potato cultivars expressing antimicrobial peptide tachyplesin, I', *Mol. Breed.*, 2: 97-105.

Anzai, H., Yoneyama, K., and Yamaguchi, I. (1989). Transgenic tobacco resistant to a bacterial disease by the detoxification of a pathogenic toxin, *Mol. Gen. Genet.*, 219: 492-494.

Befta, R., Szell, M., Meuwly, P., Pay, A., Vogeli-Lange, R., Metraux, J.P., Neuhaus. G., Meins, F. Jr., and Nagy, F. (1995). 'Cholera toxin elevates pathogen resistance and induces pathogenesis-related gene expression in tobacco'. *EMBO J.*, 14: 5753-5761.

Bendahmane, A., Querci, M., Kenyka, K. and Baulcombe, D.C. (2000). Agrobacterium the isolation of disease resistance genes application to the Rx2 locus in potato. *Plant J.* 21: 73-81.

Bhat, A.I., Jain, R.K., Anil, Kumar and Varma, A. (2002). Serological and capsid protein sequence studies suggest that necrosis disease on sunflower in India is caused by a strain of tobacco streak ilarvirus. *Arch. Virol.* 147: 651-658.

Boller, T. (1988). Ethylene and the regulation of antifungal hydrolases in plants: In *Plant Molecular Cell Biology*, Mifflin, B. Oxford: Oxford University Press. p. 145.

Boman, H.G. and Hulmark, D. (1987). Cell-free immunity in insects. *Annual Review of Microbiology* 41: 103-126.

Broglie, K., Chet, I., Holliday, M., Cressman, R., Biddle, P., Knowlton, C., Mauvais, C.J., and Broglie, R. (1991). 'Transgenic plants with enhanced resistance to the fungal pathogen *Rhizoctonia solani*', *Science*, 254: 1194-1197.

Byoung, C.K., Yeam, I. and Molly, M.J. (2005). Genetics of plant virus resistance. *Annu. Rev. Phytopathol.* 43: 581-621.

Callaway, A., Giesman-Cookmeyer, D., Gillock, E.T., Sel, T.L. and Lommel, S.A. (2001). The multifunctional capsid proteins of plant RNA viruses. *Annu. Rev. Phytopathol.* 39: 419-460.

Cammue, B.P.A., Bolle, M.F.C.D. Terras, F.R.G., Proost, P., Damme, J.V., Rees. S.B., Vanserleyden, J. and Broekaert., W.F. (1992). Isolation and characterization of a novel class of plant antimicrobial peptides from *Mirabilis ialapa* L. seeds. *Journal of Biological Chemistry* 267: 2228-2233.

Cao, H., Glazebrook, J., Clarke, J.D., Volko, S., and Dong, X. (1997). "The Arabidopsis NPRI gene that controls systemic acquired resistance encodes a novel protein containing ankyrin repeats", *Cell,* 88: 57-63.

Carmona, M.J., Molina, A., Fernandez, J.A., Lopez-Fando, J.J., and Garcia-Olmedo, F. (1993). "Expression of the α-thionin gene from barley in tobacco confers enhanced resistance to bacterial pathogens", *Plant J.,* 3: 457-462.

Casteels, P., Ampe. C., Jacobs. E., Vaeck, M. and Tempst. P. (1989). Apidaecins antibacterial peptides from honey-bees *EMBO Journal* 8: 2387-2391.

Christensen B., Fink I. Merrifield R.B. and Mauzeroll D. Toss Channei-forming properties of cecropins and related model compounds incorporated into planer lipid membranes. *Proceeding of the National Academy of Sciences of the United States of America* 85: 5072-5076.

De La Fuente-Martinez, J.M., Mosquerda-Cano, G., Alvarez-Morales, A., and Herrera-Estrella, L. (1992). "Expression of a bacterial phaseolotoxin-resistant ornithyl transcarbamylase in transgenic tobacco confers resistance to *Pseudomonas syringae* pv. *Phaseolicola", Bio/Technology,* 10: 905-909.

During, K., Porsch, P., Flaudung, M., and Lorz, H. (1993). "Transgenic potato plants resistant to the phytopathogenic bacterium *Erwinia carotovora", Plant J.,* 3: 587-598.

Epple, P., Apel, K., and Bohlmann, H. (1995). "An Arabidopsis thaliana thionin gene is inducible via a signal transduction pathway different from that for pathogenesis-related proteins", *Plant Physiol.,* 109: 813-820.

Fernandez De Caleya, R., Gonzalez, B., Garcia-Olmedo, F. and Carbonero, P. (1972). Susceptibility of phytopathogenic bacteria to wheat purothionons in *vitro. Applied Microbiology* 23: 998-1000.

Garcia-Olmedo, F., Rodriguez-Palenzuela, P:, Hernandez-Lucas, C., Ponz, F., Marana, C., Carmona, M.J., Lopez-Fando, J., Fernandez, J.A. and Carbonero, P. (1989). The thionins: a protein family that includes purothionins, viscotoxins and crambis. *Oxford Surveys in Plant Molecular and Cell Biology* 6: 31-60.

Gerlach, W.L., Liewellyn, D. and Haseloff, J. (1987). Construction of a plant disease resistance gene from the satellite RNA of tobacco ringspot virus. *Nature* (*lond*). 328: 802-805.

Grison, R., Grezes-Besset, B., Schneider, M., Lucante, N., Olsen, L., Leguay, J.J., and Toppan, A. (1996). "Field tolerance to fungal pathogens of Brassica napus constitutively expressing a chimeric chitinase gene", *Nat. Biotechnol.,* 14: 643-646.

Grumet, R. (1994). Development of virus resistant plants via genetic engineering. *Plant Breed Rev,* 12: 47-49.

Hain, R., Reit, H., Krause, E., Langebartels, R., Kindle, H., Vornam, B., Wiese, W., Schmelzer, E., Sehreier, P. H., Stocker, R.H., and Stenzel, K. (1993). "Disease resistance results from foreign phytoalexin expression in a novel plant", *Nature,* 361: 153-156.

Herrera L., Estrella and Simpson, J. Genetically engineered resistance to bacterial and fungal pathogens. *World Journal of Microbiology and Biotechnology*. 11, 383-392.

Huang, Y., Nordeen, R.O., Di, M., Owens, L.D., and McBeath, J.H. (1997). "Expression of an engineered cecropin gene cassette in transgenic tobacco plants confers disease resistance to *Pseudomonas syringae* pv. *Tabac"i, Phytopathology,* 87: 494-499.

Hulbert, S.H., Webb, C., Smith, S.M. and Sunqing, (2001). Resistance gene complexes evolution and utilization. *Annu. Rev. Phytopathol.* 39: 285-312.

Jach, G., Gornhardt, B., Mundy, J., Logemann, J., Pinsdorf, E., Leah, R., Schell, J., and Maas, C. (1995). "Enhanced quantitative resistance against fungal disease by combinatorial expression of different barley antifungal proteins in transgenic tobacco", *Plant J.,* 8: 97-109.

Jain R.K., Jyoti, Sharma, Sivakumar, A.S., Sharma, P.K., Byadgi, A.S. and Varma, A. (2004). Variability in the coat protein gene of Papaya ring spot virus isolates from multiple locations in India. *Arch. Virol.* 149: 2435-2442.

James, C. (2003). Preview: Global status of commercialized transgenic crops. ISAAA Brief No. 30, Ithca, NY.

Jaynes, J.M., Nagpala, P., Destefano-Beltran, L., Huang, J.H., Kim, J., Denny, T., and Cetiner, S. (1993). "Expression of a cecropin B lytic peptide analog in transgenic tobacco confers enhanced resistance to bacterial wilt caused by *Pseudomonas solanacearum", Plant Sci.,* 89: 43-53.

Jolles. P. and lolles. J. (1984). Whats new in lysozyme research. *Molecular and Cellular Biochemistry,* 63: 165-189.

Jones, A.L., Johansen, I.E., Bach, I. and Maule, A.J. (1988). Specificity of resistance to pea seed borne mosaic potyvirus in transgenic peas expressing the viral replicase (Nib) gene. *J. Gen. Virol.,* 79: 3129-3137.

Jongedijk, E., Tigelaar, H., van Roekel, J.S.C., Bres-Voloemans, S.A., Dekker, I., van den Elzen, P.J.M., Cornelissen, B.J.C., and Melehers, L.S. (1995). 'Synergistic activity of chitinases and β-1,3-glucanases enhances fungal resistance in transgenic tomato plants, *Euphytica,* 85: 173-180.

K. Yoneyama (1998). Novel Approaches to Disease Control. John Wiley and Sons Ltd.

Kumar, S.P.D. and Baker, B.J. (2000). Alternatively spliced N resistance gene transcripts: their possible role in tobacco mosaic virus resistance. *Proc. Natl. Acad. Sci., USA* 97: 1908-1913.

Lee, J.Y., Boman, A., Sun, C., Andesson, M., Jomvall, H., Mutt, V. and Boman, H.G. (1989). Antibacterial peptides from pig intestine: isolation of a mammalian cecropin. *Proceedings of the National Academy of Sciences of the United States of America* 86: 9159-9162.

Lin, W., Anuratha, C.S., Datta, K., Potrykus, I., Muthukrishnan, S., and Datta, S.K. (1995). 'Genetic engineering of rice for resistance to sheath blight", *Bio/Technology,* 13: 686-691.

Liu, D., Raghothama, K.G., Hasegawa, P.M., and Bressan, R.A. (1994). 'Osmotin overexpression in potato delays development of disease symptoms, *Proc. Natl. Acad. Sci. USA,* 91: 1888-1892.

Logemann, J., Jach, G., Tommerup, H., Mundy, J., and Schell, J. (1992). 'Expression of a barley ribosome-inactivating protein leads to increased fungal protection in transgenic tobacco plants', *Bio/Technology,* 10: 267-283.

Lomonossoff, G.P. (1995). Pathogen derived resistance to plant viruses. *Annu. Rev. Phytopathol.* 33: 323-343.

Lusso, M. and Kuc, J. (1996). The effect of sense and antisense expression of the PR-N gene for β-1,3-glucanase on disease resistance of tobacco to fungi and viruses', *Physiol. Mol. Plant pathol.,* 49: 267-283.

Martin, G.B., Brommonschenkel, S.H., Chunwongse, J. Fray, A., Ganal, M.W., Spivey, R., Wu, T., Earle, E.D. and Tanskley, S.D. (1993). Map-based cloning of a protein kinase gene conferring disease resistance in tomato. *Science* 262: 1431-1435.

Misaghi, I.J. (1982). The role of pathogen-produced toxins in pathogenesis. In. *Physiology and Biochemistry of Plant pathogen Interactions*,. New York: Plenum. pp. 35-61.

Misra, S and Bhargava, A. (2007). Application of Cationic Antimicrobial peptides for management of plant diseases. In: *Biotechnology and Plant Disease Management*. (eds.) Punja, Z.K., De Boer, S.H., Sanfacon, H. CABI Publication Cambridge, pp. 301-320.

Nakajima, H., Muranaka, T., Ishige, F., Akutsu, K., and Oeda, K. (1997). 'Fungal and bacterial disease resistance in transgenic plants expressing human lysozyme', *Plant Cell Rep.*, 16: 674-679.

Norelli, J.L., Aldwinckle, H.S., Destefano-Beltran, L., and Jaynes, J.M. (1994). 'Transgenic' 'Malling 26' apple expressing the attacin E gene has increased resistance to *Erwinia amylovora*', *Euphytica*, 77: 123-128.

Pappu, H.R., Niblett, C.L. and Lee, R.F. (1995). Application of recombinant DNA technology to plant protection: Molecular approaches to engineering virus resistance in crop plant world. *J. Microbiol. Biotechnol.* 1: 426-437.

Pohlmann, H. and Apel, K. (1987). Isolation and characterization of cDNAs coding for leaf-specific thionins closely related to the endosperm specific hordothionin of barley (*Hordeum vulgare* L.). *Molecular and General Genetics* 207: 446-454.

Powell-Abel, P., Nelson, R.S., De B, Hoffman, N, Rogers, S.G., Fraley, R.T. and Beachy, R.N. (1986). Delay of disease development in transgenic plants that express the tobacco mosaic virus coat protein gene. *Science*. 232: 138-143.

Rommens, C.M., Salmeron, J.M., Oldroyd, G.E.D., and Staskawicz, B.J. (1995). 'Intergeneric transfer and functional expression of the tomato disease resistance gene *Pto*', *Plant Cell*, 7: 1537-1544.

Singh, B.D. (2001). In: *Biotechnology*. Kalyani Publishers, New Delhi pp- 248-363

Stark-Lorenzen, P., Nelke, B., Hanbler, G., Muhlbach, H.P., and Thomzik, J.E. (1997). 'Transfer of a grapevine stilbene synthase gene to rice (*Oryza sativa L.*)', *Plant Cell Rep.*, 16: 668-673.

Suslow, T.V., Matsubara, D., Jones, J., Lee, R., and Dunsmui, P. (1988). Effect of expression of bacterial chitinase on tobacco susceptibility to leaf brown spot', *Phytopathology*, 78: 1556.

Tavladoraki, P., Benvenuto, E., Trinca, S., De Martinis, D., Cattaneo, A. and Galeffi, P. (1993). Transgenic plants expressing a functional single chain FV antibody are specifically protected from virus attack. *Nature* (*Lond*). 366: 469-472.

Terakawa, T., Takaya, N., Horiuchi, H., Koike, M., and Takagi, M. (1997). 'A fungal chitinase gene from Rhizopus oligosporus confers antifungal activity to transgenic tobacco', *Plant Cell Rep.*, 16: 439-443.

Thilmony, R.L., Chen, Z., Bressan, R.A., and Martin, G.B. (1995). 'Expression of the tomato Pto gene in tobacco enhances resistance to *Pseudomonas syringae* pv. *Tabaci* expressing *avrPto*', *Plant Cell*, 7: 1529-1536.

Thomas, P.E., Hassan, S., Kaniewski, W.K., Lawson, E.C. and Zalewski, J.C. (1998). A search for evidence of virus/transgene interactions in potatoes transformed with the potato leaf roll virus replicase and coat protein genes. *Mol. Breed.* 4: 407-417.

Tracke, E., Salamini, F. and Rhode, W. (1996). Genetic engineering of potato for broadspectrum protection against virus infection. *Nature Biotechnol.* 14: 1597-1601.

Trudel, J., Potvin, C. and Asselin, A. (1992). Expression of active henegg white lisozyme in transgenic tobacco. *Plant Science,* 87: 55-67.

Tsugita, A. (1971). Phage lysozyme and otherlytic enzymes. In The Enzymes, ed Boyer, P.D.,. New York: Academic Press. pp. 334-341

Van Hofsten, P., Faye, I., Kockum, K., Lee, J., Xanthopulos, K.G. and Boman, H.G. (1985). Molecular cloning, cDNA sequencing and chemical synthesis of cecropin B from *Hyalophora cecropia. Proceedings of the National Academy of Sciences of the United States of America* 82: 2240-2244.

Varma, A. and Mitter, N. (2001). Durable host-plant resistance, a desirable trait for integrated disease management. In: *Rice research for food security and poverty alleviation* (S. Peng and B. Hardy Eds.) International Rice Research Institute, Philippine, 325-344. pp.

Varma, A. and Ramachandran, P. (2001). Replication of plant viruses. In: *Reproductive Biology of Plants* (B.M. Johri and P.S. Srivastava Eds). Narosa Publishing House, New Delhi. 1-21. pp.

Verma, P. S. and Agarwal, V. K. (2009). In: *Genetic Engineering*. S.Chand and Company Ltd, New Delhi.

Vidyashekhran, P.,(2002). In: *Biotechnological applications, molecular manipulation of Bacterial disease resistance.* pp-281-316.

Vierheilig, H., Alt, M., Neuhaus, J.M., Boller, T., and Camejero, V. (1993). "Colonization of transgenic *Nicotiana sylvestris* plant, expressing different forms of *Nicotiana tabacum* chitinase, by the root pathogen *Rhizoctonia solani* and by the mycorrhizal symbiont *Glomus mosseae"*, *Mol. Plant-Microbe Interact.*, 6: 261-264.

Wegener, C., Bartling, S., Olsen, O., Weber, J., and von Wettstein, D. (1996). "Pectate lyase activation in transgenic potatoes confers pre-activation of defense against *Erwiniu carotovor"*, *Physiol. Mol. Plant Pathol.*, 49: 359-376.

Wu, G., Shortt, B.J., Lawrence, E.B., Levine, E.B., Fitzsimmons, K.C., and Shahl, D. M. (1995). "Disease resistance conferred by expression of a gene encoding H2O2-generating glucose oxidase in transgenic potato plants", *Plant Cell,* 7: 1357-1368.

Xing, T. (2007). Signal Transduction Pathway and Disease Resistant Genes and their application to Fungal Diseases. In: *Biotechnology and Plant Disease Management* (eds) Punja, Z.K., De Boer, S.H., Sanfacon, H., CABI Pub, Cambridge, pp-1-15.

Yoshikawa, M., Tsuda, M., and Takeuchi, Y. (1993). 'Resistance to fungal diseases in transgenic tobacco plants expressing the phytoalexin elicitor-releasing factor, β-1,3-endoglucanase, from soybean', *Naturwissenschaften,* 80: 417-420.

Zhu, Q., Maher, E. A., Masoud, S., Dixon, R.A., and Lamb, C.J. (1994). "Enhanced protection against fungal attack by constitutive co-expression of chitinase and glucanase genes in transgenic tobacco", *Bio/Technology,* 12: 807-812.

Sustainable Disease Management of Agricultural Crops (2011) *Pages* **354–367**
Editors: **S.K. Biswas and S.R. Singh**
Published by: **DAYA PUBLISHING HOUSE, NEW DELHI**

Chapter 26

Bacteriophage and its Role in Plant Disease Management

☆ ***N.K. Pandey, Mohd. Rajik, Kahkashan Arzoo, Sunil Kumar and S.K. Biswas***

Department of Plant Pathology, C.S. Azad University of Agriculture and Technology, Kanpur – 208 002, U.P.

Disease caused by various pathogens is the major challenges for food security in the 21st century. It is estimated that out of total crop produced, 31-42 per cent are destroyed by insects, weeds and pathogens (Agrios, 1997). Among the pathogen, bacteria take prime position in crop loss in India. Bacteria also caused various types of post harvest diseases. Various management strategies like cultural, chemicals such as bactericides or plant activators, introgression of plant resistant genes and biological control strategies are proposed form time to time for the management of bacterial diseases Cultural practices are not efficient to manage all types of diseases particularly when disease appears on standing crops. Chemicals lead to various types of environmental hazards and development of pesticide resistant strains of pathogens. Resistant varieties also become susceptible after few years due to introduction of new strains that overcome resistance due to mutation, recombination or genetic differences therefore, now a day, major emphasis is on biological control of plant diseases. Many new approaches have placed greater reliance on biological technologies that would be used effectively in integrated disease management programs (Hall, 1995; Ji and Wilson, 2003; Wilson *et al.*, 1995).

Biological control is a powerful and most economic tool to manage plant diseases. Biological control agents, including antagonistic phyllosphere bacteria, plant growth promoting rhizobacteria and bacteriophages have been successfully used in controlling several plant pathogens. Bacteriophages represent innovative biological control agents because they are highly specific to bacteria with extreme host specificity, unique ultramicroscopic nature and ability to self replicate within the host cells. (Klement, 1959). Bacteriophages have been successfully used for managing several plant diseases, since the submission and acceptance of a bacteriophage patent for plant disease control (Jackson, 1989). Bacteriophage can be used as an organic bactericide and control measures with different modes of action may substantive increase activity in disease management.

What are Bacteriophages ?

The term bacteriophage is derived from two words 'Bacteria' and the Greek word "Phagein" meaning "To eat". The term is commonly used in its shortened form phage. A bacteriophage is any one of a number of viruses that infect bacteria. Bacteriophages are obligate intracellular parasites that multiply inside bacteria by making use of host biosynthetic machinery (*i.e.* viruses that infect bacteria).

Composition and Structure of Bacteriophage

A typical bacteriophage consist of an outer protein coat (capsid) enclosing the genetic material that can be either DNA or RNA but never both. The DNA or RNA may be single stranded or double stranded. The size of genetic material ranges between 5,000 to 5,00,000 nucleosides long which may be either circular or linear. Bacteriophages are much smaller than the bacteria that they destroy. They range from 20 to 200 nm in size.

Bradly (1967) has described different tyes of bacteriophages based on morphological characters which are as follows:

Type A: This type of phage contains a hexagonal head, a rigid tail with contractile sheath and tail fibres *e.g.* ds DNA T-even (T_2, T_4, T_6) phages.

Type B: The phages which contain a hexagonal heads but lacks contractile sheath. Its tail is flexible and may or may not have tail fibres *e.g.* ds DNA phages (T_1, T_5 phages).

Type C: It is characterized by hexagonal head and a tail fibre, eg. ds DNA phages *e.g.* T_3, T_4.

Type D: It contains a head which is made up of capsomeres but lacks tail, *e.g.* ssDNA phages (ø×174).

Type E: This type consist of a head which is made up of small capsomeres but contains no tail eg. ss RNA phages (F_2, MS_2 phages)

Type F: It is a filamentous phage, for eg. ssDNA phages (fd, Fl phages) recently a group G is discovered which has a lipid containing envelops and has no detectable casid, *e.g.* a ds DNA phage MV-L2. Types A, B and C are unique bacteriophages. Type D and E are also found in plants and animals and type F is found in some plant viruses too.

Dubey and Maheshwary have classified bacteriophages into following morphological type:

The ssDNA Bacteriophages

1. Icosahedral phages–ø×174, øR, St-1, 6 SR, BR 2, U3 and G eg. G_4, G_6, G_{13}, G_{16}. All are like ø×174.
2. Helical (Filamentous): -
   (*i*) The F1 group–They are F specific phages and assort to the tip of type sex pilus *e.g. E. coli* phages (fd, fl, M_{13})
   (*ii*) If group–they are absorbed to I type sex pilus specified by R factors eg. If_1, If_2 etc.
   (*iii*) The third group is specific to strains carrying RF_1 sex factors.

The ds DNA Phages

1. E-even phages of *E. coli* eg. T_2, T_4, T_6.

2. T- odd phages of *E. coli* eg. T_1, T_3, T_5, T_7
3. The other *E.coli* phages *e.g.* P_1, P_2, Mv, ø80
4. The phages of *Bacillus subtilis e.g.* PBSX, PBSI, PBS_2
5. The phages of *Salmonella e.g.* P_1, P_2
6. The phages of *Shigella e.g.* P_2
7. The phages of *Haemophilus e.g.* HP_1
8. The phages of *Pseudomonas* et PM_2

The ss RNA Phages

1. Group I: *E.coli* phages such as f_2, MS_2, M_{12}, R_{17}, fr etc.
2. Group II: The QB phages.

The ds RNA Phages

1. The ø6 bacteriophage.

Classification

The bacterial virus subcommittee of the International Committee on Taxonomy of viruses works on the classification and nomenclature of bacteriophages. ICTV has classified phages into 13 families based on morphology and nuclei acid which are given in following table.

Table 26.1: ICTV Classification of Phages

Order	*Family*	*Morphology*	*Nucleic acid*
Caudovirales	Mycoviridac	Non enveloped, contractile tail.	Linear ds DNA.
	Siphoviridae	Non enveloped, long, non contractile tail.	Linear ds DNA.
	Podoviridae	Non enveloped, Short, non contractile tail.	Linear ds DNA.
	Tectiviridae	Non enveloped, isometric	Linear ds DNA.
	Corticoviridae	Non enveloped, isometric	Circular ds DNA
	Lipothrixviridae	Enveloped, rod shaped	Linear ds DNA
	Plasmaviridae	Enveloped, pleomorphic	Circular ds DNA
	Rudiviridae	Non enveloped rod shaped	Linear ds DNA
	Fuselloviridae	Non enveloped lemon shaped	Circular ds DNA
	Inoviridae	Non enveloped filamentous	Circular ss DNA
	Microviridae	Non enveloped isometric	Circular ds DNA
	Leviviridae	Non enveloped isometric	Linear ss RNA
	Cystoviridae	Enveloped spherical	Segmented ds RNA

History of Bacteriophage

History of Bacteriophage is given in Table 26.2.

Table 26.1: History of Bacteriophage

Year	Name	Contribution
1896	Ernest Hanbury Hankin	Reported that something in water of Ganga and Jamuna rivers in India had marked antibacterial action against cholera and could pass through a very fine porcelain filter.
1915	Fr-derick Twort (British bacteriologist)	Superintendent of Brown Institution of London, discovered a small agent that infected and killed bacteria. He considered it either a stage in life cycle of bacteria, an enzyme produced by bacteria themselves or a virus that grew on and destroyed the bacteria.
1917	Felin d' Herelle	Working on Pasteur Institute in Paris announced the discovery of an invisible antagonistic microbe of the dysentery bacillus. He found this entity to be a virus parasitic on bacteria.
	D' Herelle	Called this virus as bacteriophage and also recorded recovery of non suffering from dysentery to good health by bacteriophages
1921	Brunogh and Maisin	Reported the control of *Bacillus anthracis* and *Staphylococcus* infections with bacteriophage treatment.
1924	Mallman and Hamstreet	Observed the control of *Xanthomonas compestris* pv *compestruis* from diseased cabbage tissue to be controlled by filtrate of liquid collected from decomposed cabbage.
1925	Coons and Kotila	Observed the control of *Erwinia caratovora* subsp. *Caratovora* in carrot by its homologous phage.
1925	Kohila and Coons	Observed bacteriophage from soil sample active against the causal agent of blackleg disease of potato
1926	Moore E.S.	Discovered phages as plant disease control agents.
1934	Messey	Suggested that phages are prime factor in limiting severity of bacterial blight of cotton under field condition.
1935	Thomes	Reported reduced disease incidence (18-1.4 per cent) of stewarts wilt of corn by seed treatment with phage.
1963	Okabe	Claimed the phages to be ineffective in controlling disease development.
1992	Goto	Claimed the practical use of phages for plant disease control to be unsuccessful.
2006	U. K. Ministry of Defense	Took responsibility for a G-8 funded global partnership priority Eliana Project as a retrospective study to explore the potential of bacteriophages for 21st century.

Isolation of Phages

The routine way for isolating of phage is as follows:

1. The bacterial pathogen which is used as a phage detector is cultured first.
2. The material that may eassy the phage is then crushed in a mortar with a small amount of water and is then centrifuged.
3. The superetant fluid is then filtered through an ultrafilter and finally examined to determine whether the phage is present using high populations of detecting isolates.

Sources of Phages

The phages are produced only from the lysed cells of the host bacteria. So phages should be found wherever their host bacteria may be present for *e.g.*

- ☆ Lesions on diseased plants are sources of phages for the pathogens of those diseases.
- ☆ Soils are sources of phages for the pathogens of soil infecting diseases.

☆ Irrigation waters are sources of phages for the pathogens of diseased plants grown in or adjacent to the irrigation canal or ditch.

However many examples are there to explain sources of phages without host bacteria in nature. for *e.g.*

☆ Anderson (1928) has reported the isolation of phage for *X pruni* from the soil under infected peach tree but not from diseased leaves.

☆ Matsumato and Okabe (1937): isolated a phage for *X. citi* from soil.

☆ Massey (1931) isolated the phages for X. *malvacerum* from fallow or cultivated land that bore infected cotton crops and from water of the Blue Nile.

☆ Kent(1937) has demonstrated the presence of a phage of *A. tumefacins* in the healthy part of the tomato stem about six inches away from the gall.

☆ Kauffman (1928) has isolated a phage of *A. tumafaciens* from animal feces.

☆ Rosbug and Parrack (1955) have isolated a phage of X. *malvacerum* from two year old dry leaves but not from leaves of naturally infected cotton plants.

Table 26.3: Kinds of Phages of Plant Pathogen

Sl.No.	*Origin of Phage*	*Host bacterium*	*Reference*
1.	*E. caratovora* (ginger)	*E. caratovara, A. cloacae, E.coli*	Terada (1956)
2.	*P.solanacearum*	*P. solanacearum, A. cloacae, E. caratovora*	Terada (1956)
3.	*A. radiciola*	*A. tumefaciene, A. radicicola*	Issailsky (1927) Roslycky *et al.* (1962)
4.	*E. caratovora*	*E. caratovora, A. tunefaciene*	Coons and Kotila (1925)
5.	*P. cannabina* Ca P1 *P. mori* Mo P1	*P. mori, P. larymans, P. morepru-norun,* *P. phaceolicola*	Klement (1959)
6.	*P. cannabira* Ca P2	*P. cannabina, P. marginalis, P. mori, P. tomato,* *P. syingae, P. syringae*	Klement (1959)
7.	*P. tabaci* phage 2	*P. angulata, P. tabaci. P. syringae,* *P. coronafaciens, P. phaseolicola*	Fulton (1961)
8.	*P. pisi*	*P. lactrymane, P. phaseolicola, P. syringae,* *P. mori-pwnosum*	Klement and Hevesi (1959)
9.	*E. amylovora*	*E. amylovatre, E. lathyri, E. ananas*	Billing *et al.* (1961)
10.	*P. mori MhP1*	*P. mori, P. atrofaciens, P. morsp-runoun*	Klement (1959)
11.	*C. flaccumfaciene*	*C.flaccumfaciene, C. peinsettlae*	Klement *et al.* (1959)
12.	*X. phaseoli fuscane*	*X.phaseoli fuscane*	Klement (1959)
13.	*X. tranalucens hordei*	*X. tranolucens* f. sp. *hordei*	Sutton and Ratzne-lson (1959)
14.	*A. tumefaciene*	*A. Tamefaciene*	Kauffman (1928) Muncie and Patel (1930)

Strategies to Control of Phage Longevity

1. Use of formulations that attenuate sunlight damage.
2. Evening application to minimize U.V. radiation. (Balogh *et al.*, 2003).

3. Use of phage integrated with SAR inducers. (Obradovic *et al.*, 2005).
4. Use of phage integrated with antagonistic phyllosphere bacteria (Svircev *et al.*, 2006).
5. Using target bacteria for phage propagation in the target environment.

Table 26.4: Comparative Study of Bacteriophages with Antibiotics

	Bacteriophages		*Antibiotics*
1.	Very specific, affects only target bacteria.	2.	Non specific in action and targets not only pathogenic microorganism but also normal microflora.
2.	Dysbiosis *i.e.* chances of development of secondary infection avoided.	2.	Can lead to development of serious secondary infection.
3.	No side effects.	3.	Multiple side effects.
4.	It is able to self reproduced so long as corresponding host bacteria is present in the environment therefore need of reported application in greatly reduced.	4.	Repeated administration after needed.
5.	Since phage is targeted to receptors on bacterial membrane or capsule which are important virulence determinants, development of phage resistance usually means changes in those structures and may therefore lead of attenuation of strain in virulence.	5.	Antibiotic resistant bacterial strains remain pathogenesis.
6.	Selection of a new phage against a phage resistant bacteria in a rapid process and frequently can be accomplished in days.	6.	Developing new antibiotics against a drug resisted bacteria is a time consuming process and may take several years.
7.	Production is simple and relatively in expensive.	7.	Lots of labour and money needed
8.	Because of high specificity of phages the disease causing bacterium has to be identified before phage therapy can be started.	8.	Can be used without-knowing the exact characteristics disease causing bacteria.

Replication of Bacteriophages/Phage Multiplication Cycle

On the basic of method of replication, the phages can be divided into lytic and lycogenic cycle.

Lytic Cycle

Lytic of virulent phages are the phages which can only multiply on bacteria and kill the cell by lysis at the end of life cycle. The T-phages like T_1 to T_7 (contain ds DNA as their genetic material) cause the lysis and death of their host cells (*E.coli*). As soon as the cell is destroyed, new phages can find new hosts, lytic phages are suitable for phage theray.

The lytic cycle consists of following steps:

Adsorption

This is the first step of infection of a host bacterial cell. The phage particle undergoes a chance collision at a chemically complementary site on the bacterial surface, than adheres to that site by means of its tail fibres. The specific receptors on the surface of bacteria help the phage to adhere on the surface of the cell which is mainly by lipopolysaccharides, teichoic acid, proteins or even flagella.

Penetration

The phage injects its DNA into the bacterial cell, the tail sheath contracts and the core in driven through the wall to the membrane. Penetration may be mechanical or enzymatic. Phage T_4 packages a bit of lysozyme in the bacterial cell wall for insertions of the tail core. The DNA is injected into the periplasm of the bacterium and how it penetrates the membrane, it is not known with clearity.

Synthesis of Early Proteins

A part of phage DNA undergoes transcription and translation initially to make a set of proteins that are needed to replicate the phage DNA. The early proteins produced are repair enzyme to repair the hole in bacterial cell wall, DNAase enzyme degrades the host DNA into precursors of phage DNA and a virus specific DNA polymerase that will copy and replicate phage DNA.

Synthesis of Phage Nucleic Acid

Several copies of viral nucleic acids (DNA in case of T-even bacteriophages) are synthesized after synthesis of early proteins. Walter Fiess (1972) (University and Ghent, Belgium) was the first to establish the complete nucleotide sequence of gene and of viral genome of bacteriophages and MS_2.

Synthesis of Late Proteins

Each of the several replicated copies of phage DNA can be used for transcription and translocation of a second set of proteins called late proteins, the late protein are mainly structural protein that make up the capsomeres and the various components of tail assembly. Lysozyme is also a late protein that are packed in a tail of a phage and are used to escape from host cell during last step for the replication process.

Assembly

The protein capsomeres assemble into heads and "seal in" a copy of phage DNA. The tail and accessory structures assemble and incorporate a bit of lysozyme in the tail plate.

Release of Virions

T-even bacteriophages are released by the cell lysis. Other phages may be released by extrusion or by budding. Lysis by tailed phages is achieved by an enzyme called endolysin or lysozyme which attacks and breaks down the cell wall peptidoglycon. An altogether different phage type, the filamentous phages, continuously form new virus particles and released from the host cell. Released visions are described as free and unless defective are capable of infecting new virus particles. Budding is associated with certain mycoplasma phages. The number of particles released per infected bacteria may be as high as 1000.

Assay of Lytic Phage

Plaque assay–A plaque is a clear result of lysis of bacteria. Each plaque arises from single infectious phage. The infection particle that give rise to plaque is Plaque Forming unit (PFU).

Lysogeny or Temperate Phage Replication

In lysogenic replication, the phage DNA actually integrates into the host chromosome and is replicated along with the host chromosome and passed in to the daughter cells. The phage DNA in repressed or quiescent state is called a prophage. The cell harboring a prophage is not adversely affected by the presence of the prophage and the presence of prophage and lysogenic state may persist indefinitely. The cell harboring a prophage is termed as lysogen. λ-Phage is the prototype phage of lysogeny of laysogenic cycle consist of following steps:

Circularization of Phage Chromosome

λ-Phage's DNA is a double stranded linear molecule with small single stranded regions at the 5′ ends. These single stranded ends are complementary (Cohesive ends) so that the base can easily pair and produce a circular molecule. In the cell, the free ends of the circle can be ligated to form a covalently closed circle.

Site Specific Recombination

A recombination event, catalysed by a phage coded enzyme occurs between a particular site or the circularized phage DNA and a particular site on the host chromosome. The result is integration of phage DNA into the host chromosome.

Repression of Phage Genome

A phage coded protein called a repression is formed which binds to a particular site on the phage DNA called the operator and shuts off transcription of most phage genes except the repressor gene. The result is a stable repressed phage genome which is integrated into the host chromosome. Each temperate phage represses its own DNA only and not from other phage so that suppression is very specific immunity to super infection with same phage events leading to termination of lysogeny. Anytime a lysogenic bacterium is exposed to adverse condition, the lysogenic state can be terminated. The process is called induction conditions which favour the termination of the lysogenic state, include desiccation, exposure to UV or ionizing radiation, exposure to mutagenic chemicals etc. Adverse conditions lead to production of proteases (sec A protein) which destroy the repressor protein. This is turn lead to expression of genes, reversal of integration process and lytic multiplication.

Phage Therapy

Phage therapy is the therapeutic use of bacteriophages to treat pathogenic bacterial infection. Phages were discovered as antibacterial agents and were put to use as such soon after their discovery. But their use was limited in west after discovery of broad spectrum antibiotics. In the former Soviet Union, however phages were used in 1940's as an alternative to antibiotics.

Practical Use of Bacteriophages for Disease Control

Ubiquitous distribution and occurrence of specific phages in bacteria infected plant tissues irrespective of disease severity suggests that bacteriophages may check multiplication of sensitive phytopathogenic bacteria (Hayward, 1964; Okabe and Goto 1963). Although bacteriophages may be important factor in the ecology of their host bacteria (Hayward, 1964; Anderson, 1957). The epidemiological role of phage is less understood. The role of lytic bacteriophages in certain bacterial plant disease has been primarily of experimental interest only (Civerola, 1572; Goto, 1969; Goto *et al.*, 1971; Koizum *et al.*, 1969; Okabe and Goto 1963; Tagam, 1959).

The use of phages for control of plant diseases caused by phytopahogenic bacteria was reviewed by Okabi and Goto (1963). Various phages have been applied as useful tools in many diverse studies such as the identification of bacterial species, the differentiation of strains in bacterial species, the detection of pathogens hibernating in fields or on plants, the counting of bacterial density and the forecasting of disease development.

Application of phage for phage therapy: It can be done by two ways:

1. Infection of phage in diseased plant tissue.
2. Soaking of seeds in phage suspension.

Table 26.5: Works Done on Use of Phages for Plant Disease Control

Use of Bacteriophage	*Disease Name*	*Causal Organism*	*References*
Bacteriophage	Bacterial spot of tomato	*X.c. pv vesicatoria*	(Flaterty *et al.*, 2000).
Bacteriophage	Bacterial spot of peach	*X.c.* pv *pruni*	Civerolo and Keil, 1969; Saccardi *et al.*, 1993
Bacteriophage	Geranium bacterial blight	X c. pv *pelargoni*	Flaherty *et al.*, 2001
Bacteriophage	Bacterial leaf spot of tomato Xanthomonas leaf blight of onion	*X.c. vesicoloria,* X. *performans* *X.axonopodis* p.v. *allii*	(Balogh *et al.*, 2003; Flaherty *et al.*, 2000) Lang *et al.*, 2007; Obsadovic *et al.*, 2004; obradovic *et. al.*, 2005).
Plant growth promoting Rhizobacteria, two systemic acquired resistance inducers (harpin and acebenzolar-S-methyl), unformulated bacteriophage (Agiphage) and two antagonistic bacteria	Bacterial spot in tomato	*X.c.* pv *vesicatoria*	Obradovic *et al.* (2005)
Bacteriphages in combination of acebenzolar-s-methyl	Bacterial spot of tomato	*X.c.* pv *vesicatoria*	Obradovic *et al.* (2005)
PCF and cascrate formulated phages	Bacterial spot of tomato	*X.c.* pv *vesicatoria*	Balogh *et al.* (2003)
Bacteriophage alone as well as in combination with copper bactericides	Bacterial spot of citrus Citrus canker	*X. anonopodis* pv *citrulmelo*) *X.c.* pv *citri*	Balogh *et al.* (2008),
Phages	Many phytopathogenic bacteria		Okabe and Goto, 1963

Benefits of Phage Therapy

1. Bacterophages are much more specific than antibiotics that are in clinical use.
2. It is harmless to eukaryotic host undergoing therapy.
3. It does not effect the beneficial normal flora of the host.
4. It has no or very little side effects.
5. It can be used to treat those bacterial diseases which do not respond to antibiotics.
6. They are especially successful where bacteria have constructed a biofilm composed of a polysaccharide matrix that antibiotics cannot penetrate.
7. Phages are self replicating.
8. They are natural component of bioshere.
9. Phage proportion is relatively inexpensive and fairly easy to produce.
10. Readily degraded
11. Relatively safe, no residue effect.

Demerits

1. Bacteriophages have extremely short residual activity in the phyllosphere.

2. Need of very high dose of phage initially.
3. Effective in controlling bacterial population only if present before bacterial inoculation but ineffective after ingression of bacteria. (Civerolo and Keil, 1969).
4. Efficacy of phages may be influenced by relative fitness of phage resistant mutants.
5. Efficacy is reduced due to low rate of diffusion.
6. Efficacy is influenced by timing of phage application.
7. Efficacy is reduced due to harmful environmental factors like CO_2, rain, temperature and sunlight (U.V. radient), high or Low pH.

Model Bacteriophages

The bacteriophages which extensively studied are Phage, T_2, T_4, (169 to 170 bp, 200nm long), T_7, T_{12}, T_{17}, M_{13}, MS_2, Phage (23-25 nm in size), G_4, P_1, P_2, P_4 phage, ø×174, ø 6 Phage, 29 phage, N4phage, 186phage that are commonly known as model of bacteriophage.

Lysogenic Phages Not Safe for Phage Therapy

Lysogenic phages integrate their DNA into their host DNA and create a prophage. Prophages can escape from original host (by cutting not only its DNA back, but possibly some genes of host bacterium as well) and can integrate into a different one (the process is called transduction). Such phages are inappropriate candidates for phage therapy because of their mode of action and because they can lead to transfer of virulence gene and those mediating resistance to antibiotics.

Others Recent Uses of Bacteriophages

Bacteriophages are recently used as biocontrol agent for controlling bacterial diseases in plants. There are some useful properties of bacteriophage that are used in various bacterial diseases to identify and minimize disease. The salient other uses of phages are as follows:

1. Bacteriophages are useful tools for identification of bacterial species.
2. They are useful tools for differentiation of strains in bacterial species.
3. They are useful tools for detection of pathogen libernating in fields or on plants.
4. They are useful for counting bacterial density.
5. They are useful tools for for casting disease development.
6. Bacteriophages have been used as important tools for molecular cloning in biotechnology.

Commercialization of Bacteriophages

The first commercial company to produce phages for control of bacterial plant disease was Agriphi, Inc, established by L.E. Jackson. This company received the first EPA registration to use phage in agriculture, wild type phages and H-mutants are used to minimize development of resistant bacterial strains. Only lytic phages isolated from plant parts (leaves, stem etc) soil, water etc are utilized. Bacteriophages mixtures developed from diverse host bacterial strains Agriphage, omnilytics, salt lake city, U.T. But there are limitations in commercialization of bacteriophages, which are as follows:

1. Failure to select specific phages against the target bacteria in vitro before using them *in vivo* models.
2. Lack of availability and/or reliability of bacterial laboratories for carefully identifying the pathogens involved and thus making selection of specific phage impossible.

3. Use of single phages in inflections which involved mixtures of different bacteria.
4. Failure to understand the heterogenicity and mode of action of phage (eg difference between lytic and lysogenic phages in respect to their ability to act as therapeutic agents).

Integrated Management Strategies

Disease control is best achieved by using on integrated management approach by combining bacteriophages with proper cultural practices, chemicals such as bacterial or plant activators and biological control strategies. The phage based integrated management stratergies of bacterial plant disease management can be achieved by these approaches:

1. Combination of phages and biological control agents.
2. Combination of phages and SAR inducers
3. Combination of phages and bacterial antagonism.

Combining bacteriophages with a plant activator may be effective management strategies for Xanthomonas leaf bligh of onion. Bacteriophages with copper hydroxide or mancozeb and bacteriophages with Acebenzotar-S-Methyl (ASM) reduced disease severity by 31 per cent and 50 per cent respectively in *X. axonopodis* pv *allii.*

Combined application of bacteriophages with bacterial biocontrol agents which served both as propagating host for the phage and as antagonist of the target bacterium are a successful strategy in controlling fire blight of pear and tobacco bacterial wilt.

Janka *et al.*, reported that tobacco bacterial wilt incited by *Ralstonia solanacearum* is reduced by co-application of an avirulent strain of *R. solanacearum* and its phage that in active against both the virulent and avirulent strains.

Conclusion

Among the various strategies developed for plant disease control, biological control in cheapest method and is of great demand in present era. Various types of biocontrol agents for control of bacterial diseases are developed. Among this bacteriophges *i.e.* viruses that eat bacteria or that parasitize bacteria are potential biocontrol agents. Bacteriophages are recently used as biocontrol agents of bacterial diseases of plants. There are some useful properties of bacteriophages that are used in various bacterial diseases to identify minimize diseases. In India, there is an urgent to explore the possibility of integrating bacteriophages for managing bacterial diseases.

References

Anderson, E. S. (1957). The relations of bacteriophages to bacterial ecology. P 189-217. In R.E.O. Williams and C. C. Spicer [ed.] 7th Sympos. *Soc. Gen. Micro. Biol.* Combridge, England.

Balogh, B., Jones, J. B., Momol, M. T., Olson, S. M., Obradonic, A., King, P., and Jackson, L.E. (2003). Improved efficacy of newly formulated bacteriophages for management of bacterial spot on tomato. *Plant Dis.* 87: 949-954.

Balogh, B., Jones, J.B., Mommal, M. T. and Olson, S. M. (2004). Persistence of bacteriophages as biocontrol agents in the tomato conoky, Pages 299-302 In: Proc. Ist is on tomato disease; June 21-24. 2004, Orlando, Fl. M. T. Momol, P. Ji and I. B. Jones, eds. *Acta Hortic. ISHS*. 695.

Bergamin Filho, A. and Kimati, H. (1981). Estoidos sobre um bacteriofage isolated xanthomonas comprises. 11 seu imprego no control de *X. compestris, X.vesicotoria*. Sum. *Phytopathol.* 7: 35-43.

Billing, G., Baker. L.A. E., Crosse, J.E. and Garrett, C. M. E. (1961) *J. Applied Bacteriol.*, 24, 195-21.

Bologh, B., Conteros, B.I., Stall, R. E. and Jones, J. B. (2008). Control of citrus canker and citrus bacterial spot with bacteriophages. *Plant dis.* 92: 1048–1052.

Bradley, D. E. (1967). Ultrastructure of bacteriophages and bacteriocins. *Bacterial Rev.* 31: 230-314.

Civerola, E.L. (1972). Introduction between bacteria and bacteriophages on plant surfaces and plant tissue p. 25-37. In: H. P. Maas Geesteranustedj Third Int. conf. Plant–Pathogenic Bacteria Proc. Centre Ag. Publ. Doc. (Pudoc), Wageringen. The Netherlands.

Civerolo, E. L. and Kerl, M. L. (1969). Inhibitor of bacterial spot of peach foliage by *Xanthomonas* prin bacteriophage. *Phytopathology* 59: 1966-1967.

Coons, G. H., and Kotila, J.E. (1925). *Phytopathol. Soc. Japan*, 15, 357-70.

Coons. G. H. and Kotila, J. E. (1925). Investigation on the blackleg disease to potato. Mich. Agric. Exp. Stn. Tech. Bull. 3-29.

Fiers, W. (1972). Complete nucleotide–sequence of bacteriophage MS2-RNA primary and secondary structure of replicase genes, *Nature*, 260: 500-507.

Flaherty, J. E., Harbaugh, B. K., Jones, J. B., Somedi, G. C. and Jackson, L. E. (2001). H-mutant bacteriophages as a potential biocontrol of bacterial blight of geranium *Hort. Science* 36: 98-100.

Flaherty, J. E., Jones, J. B., Harbangh, B. K., Somodi, G. C. and Jackson, L.E. (2000). Control of bacterial spot on tomato in the green house and field with H-mutant bacteriophges. *Hort. Science* 35: 882-884.

Flaherty, J. E., Jones, J. B., Harbangt, B. K., Somodi, G. C. and Jackson, L. E. (2000). Control of bacterial spot on tomato in the green house and field with H-mutant bacteriophages. *How science* 25: 882-884.

Fulton, R. W. (1950). *Phytopathology*, 40: 936-49.

Goto, M. (1952). Fundamentals of Bacterial Plant Pathology sen Diego: Academic

Goto, M. (1969). Ecology of phage bacteria interaction of *Xanthomonas oryzae* (Uyeda et Ishiyama) Dowson. Bull. For Agr. Shizuoka Univ. 19: 31-67.

Goto, Mee Frank, F. G. and Ou, S. H. (1971). A study on the phage bacteria relationship of *Xanthomonas translucens* f. ap. *oyzicola*. *Ann. Phytophathol.* Sco. Japan 37: 249-258.

Hall, R. (1995). Challenges and prospects of integrated pest management pages 1-24 in: Novel approaches to integrated pest management R. Reuveni. ed. Lewis Pblishers, CRC Press, Inc., Baca Raton, FL.

Hayward, A. C. (1964). Bacteriophage sensitivity and biochemical group of *Xanthomonas malvaceasum* Genmicrobs 35: 287-298.

Israilsky, W. P. (1927) *Centre. Bacteriol. Parasitank.*, 71: 302-11.

Jackson, L.E. (1989). Bacteriophage prevention and control of harmful plant bacteria. Vs Patent. Patent No. 4, 828, 999.

Jhomas, R. C. (1935). A bacteriophage in relation to slewents disease of clove. *Phytopathol.* 25: 371-372.

Ji, P. and Wilson, M. (2003). Enhancement of population size of biological control agents and efficacy in control of bacterial speak of tomato through salicylate and ammonium sulphate amendments. *Appl. Environ. Microbiol.* 69: 1290-1294.

Kauffman, F. (1928) *Z. Krebsforsch.*, 28: 109-20.

Keel, H. L and Wilson, R. A. (1963). Control of peach bacterial spot *Xanthomonas pruni* bacteriophage. *Phytopathology* 53: 746-747(Abst.)

Kent, G. C. (1937). *Phytopathology*, 27: 871-902.

Klement, Z. (1959). Some new specific bacteriophage for plant pathogenic *Xanthomanas* spp. *Nature.* 184: 1248 -1249.

Klement, Z. (1959). *Nature*, Lond., 184: 1248-49.

Klement, Z., and Hevesi, M. (1959). In: *Omagiu lui T. Savulescu (Hungry)*, 347-53.

Klement, Z., and Lovas, B. (1959). *Phytopatholoy*, 49: 107-12.

Koizum, M. S., Yamamoto, S and Yamada, J. (1969). Ecological studies on citrus canker caused by *Xanthomnas citri.* I, some ecological studies on bacteriophages of causal bacterium in lesions of various citrus leaves. Bull. *Host. Res. Stn.* Japan. 5: 105-117.

Kotil, J. E. and Coons, G. H. (1925). Investigation on the blockleg disease to potato. Mich. Agric. Exp. Stn. Tech. Bill. 3-29.

Lang, J. M., Gent, D. H. and Schwartz, H. F. (2007). Management of Xanthomonas leaf blight of onion with bacteriophages and a Plant activator. *Plant Dis.* 91: 871-878.

Mallmann, W. L. and Hemstreet, C. J. (1924). Isolation of an inhibitory substance from plants *Agric. Res.* 599–602.

Massey. R.E. (1931) *Empire Cotton Growing Rev.*, 8: 187-213.

Mc Guine, M.R., Shasha, B. S., Lewis, L.C., Nelson, T.C. (1994). Residual activity of granular starch-encapsulated *Bacillus thuringiensis J. Econ. Entomol.* 87: 631-637.

Moose, E. S. (1926). d'Herelle's bacteriophage in relation to plant parasites. South Ajr. J. Sci. Sci. Univ. Fl. Online.

Obradoric, A., Jones, J. B., Momal, M. I., Olson, S. M., Jackson, L. E., Baloygl, B., Guven, K. and Iriaste, F. B. (2005). Integration of biological control agents and systemic acquired resistance inducers against bacterial spot on tomato *Plant Dis.* 89: 712-716.

Obradovic, A., Jones, J. B., Momol, M.T., Balogh, B., and Olson, S. M. (2009). Management of tomato bacterial spot in the field by foliar application of bacteriophages and SAR inducers. *Plant Dis.* 88: 736-740.

Okah, N. and Goto, M. (1963). Bacteriophages of plant pathogens *Ann. Rev. Phytopathology* 1: 397-418.

Rosberg, D.W., and Barrack, A. L. (1955) *Phytopathology*, 45: 49-51.

Roslycky, E. B., Allen. O. N., and McCoy, E. (1962). *Can. J. Microbiol.*, 8: 71-78.

Stolp, 17, stan, M. P. and Baigent, N. L. (19). Problems in speciation of phytopathogenic *Pseudomonas* and *Xanthomonas. Ann. Rev. phytopathol* 3: 231-264.

Sulton, M.D., and Katzenelson, H. (1953). *Can. J. Bot.*, 31: 201-5.

Svireen, A. M., Lehman, S.M., Kim, W. S., Barzez, E., Schneder, K. E. and Cash. A.J. (2006). Control of fire blight pathogen with bacteriophages. In: Proc 1st Int. Sympor. Biol. Control bacterial Plant Dis. W. Xeller and C. Ullrich, eds. Sacheim/Dasmstadt, Germany. pp 259-261

Tagami, Y. (1959). Relationship between the occurrence of bacterial leaf blight and rice and phages in rice growing season plant 13: 389-394.

Tanaka, H., Negishi, M. and Macda, H (1990). Control of tobacco, bacterial wilt by an avirulent strain of *Pseudomonas solanaceaum* M4S and its bacteriophages *Ann. Phytopathol* 56: 243-246.

Tereda, M. (1927). *Studies on Bacterial Viruses* (Naya publishing Co., Tokyo, 153 pp.,).

Wilson, M., Savka, M. A, Hwang, I. Farrand S. K. and Lindow S. E. (1995). Alterted epiphytic colonization mannityl opine producing tranogenic tobacco plants by a mannityl opine-catabolyzing strain of *Pseudomonas syringae*. *Appl. Enviorn. Microbial.* 61: 2151-2158.

Sustainable Disease Management of Agricultural Crops (2011) *Pages* **368–376**
Editors: **S.K. Biswas and S.R. Singh**
Published by: **DAYA PUBLISHING HOUSE, NEW DELHI**

Chapter 27

Important Diseases of Button Mushroom (*Agaricus spp.*)

☆ ***Ved Ratan***

Department of Plant Pathology, C.S. Azad University of Agriculture and Technology, Kanpur – 208 002, U.P.

Introduction

Presently, the scenario of mushroom cultivation in India is different from all the mushroom growing countries of the world with respect to substrate, casing mixture, system of cultivation and occurrence of diseases (Sharma, 1993).Among the cultivated mushroom button and oyster mushroom cultivation is now picking up very fast in central and southern parts of India and the growers are using a variety of substrates prepared by different methods. These mushrooms are being cultivated under natural conditions in ordinary crop rooms or sheds without environmental control. Like any other living organism, mushroom is attacked by several pests and diseases. A large number of biotic and abiotic factors are responsible for lowering down both productivity and quality. The most common biotic causes are parasistic and antagonistic fungi, bacteria, viruses, namatodes, mites etc.

.During the crop period some moulds are commonly observed during cultivation which, under certain conditions may cause of complete failure of crop. However, very meger information's are available on pest and diseases of mushroom. Therefore the management of important diseases of button mushroom are being discussed here:

Fungal Diseases

Dry Bubble

Causal Organism

Vertificillium fungicola.

Other Names

Verticillium diseases, fungus spot, brown spot, La mole.

The first report of the heavy incidence of this disease in India was from mushroom farm located at Chail and Taradevi (Seth *et al.*, 1973).

Symptoms

In initial stage whitish mycelial growth can be noticed on the casing soil which has a tendency to turn greyish yellow. If infection takes place in an early (pin-head) stage, the typical onion shaped mushrooms are produced; the stem is thicker than the cap, which is often not distinguishable from the stem. In the case of sever infection, cracked and deformed mushroom can be seen. On fully developed sporophores, it produces localized light brown depressed spots. Adjacent spots coalesce and form irregular brown blotches like those of bacterial blotch but lighter in colour and somewhat sunken at the centre. Diseased caps shrink in blothched areas, turn leathery, dry and cracking.

Epidemiology

The infected casing soil is the main source of *Verticillium.* Although the possibility of spores also is being carried by dirty picking containers. The spread of dry bubble is quicker than that of wet bubble, because of smaller spore size and quicker dispersal through various agencies. The fungus is soil-borne and spores can survive in moist soil for one year. The pathogen also perpetuates through resting mycelium and in spent compost. High humidity, lack of proper air circulation, delayed picking (of diseased fruit bodies) and temperature above 16°C favour its development and spread (Sohi, 1988). It becomes more common when cropping is extended beyond 60 days.

Control

1. Use of sterilized casing soil, proper disposal of spent compost, control of insects and better hygiene is essential to avoid primary infection.
2. Spraying with Dithane Z- 78 (0.2-0.5 per cent) nine days after casing, Topsin-M, and chlorothalonil, have been found very effective in checking dry bubble.
3. Roughing of diseased fruit bodies and spraying of 2.0 per cent commercial formalin on affected patches.
4.. Strict hygienic measures; use of properly pasteurized compost and 'sterilized casing mixtures are also important for the management of dry bubble.

Cobweb

Causal Organism

Clabodotryum dendroides.

Other Names

Mildew, soft decay, *Hypomyces* mildew disease, *Dactylium* disease.

In India, it was first recorded in Chail and Shimla (HP) (Seth, 1977) and later from Solan and Kasauli with natural incidence ranging from 8.17-18.83 per cent in 1982.

Symptoms

In the beginning on the surface of casing soil it first seen as circular, small, white patches of the mycelium on the surface of the casing soil. As the disease progresses, a fluffy white mould grows over the mushroom. Eventually, mushrooms affected by cobweb disease turn brown, begin to rot and die-off. In severe condition, mushrooms changing from a fluffy cobweb to a dense mat of mycelium. The white colour can turn pink or even red with age. One symptom which is generally not associated with the disease is cap spotting. The spots can be brown or pinkish brown.

Epidemiology

High relative humidity and temperature encourage the disease. *Cladobotryum* is spread mainly by conidia which are the only known spores produced by the fungus. The pathogen is a soil inhabiting fungus and is normally introduced into the crop by farm workers, soil contaminants, mycelium, spores, on crop debris. It is easily transmitted by the spores those are very easily spread by air movement and have been experimentally trapped a long way from the source. Usually fourth or fifth flush, where the casing is contaminated with the mycelial fragments, the disease develops on the pinheads and can cause severe losses.

Control

The following steps can be taken for the effective management of this disease:

1. Proper sterilization of casing soil is necessary. Sterilizing the casing mixture even at 50°C for 4 hours has been found sufficient to completely kill the pathogen.
2. Regular cleaning of crop beds
3. During picking the temperature and relative humidity should not go beyond 65°F and 90 per cent respectively.
4. If hygiene measures breakdown, fungicides can be applied to the surface of casing. carbendazim, Chlorothalonil can be applied in the initial casing mix or sprayed on the bed between flushes.

Wet Bubble

Causal Organism

Mycogon pernisiosa.

Other Names

Mycogone disease, White mould and La mole etc.

This disease was first reported from few mushroom farms of village Pinglen in J&K State (Kaul *at al.*, 1978).

Symptoms

Wet bubbles ranges from one half the size of a golf-ball to as large as a grape fruit.When young pin heads are infected with *Mycogone,* they develop monstrous shapes which often do not resemble with mushrooms. The short, curly: pure white fluffy mould of *Mycogone* can easily be observed on and around the infected sporocarp. After few days the colour of infected area becomes creamy brown due to the formation of dark coloured chlamydospores.

Epidemiology

The main source of infection is casing soil, air, infested trays and spent compost. Chlamydospores may survive for a long time (upto 3 years) in casing soil.

Control

Good hygiene, use of properly sterilized casing soil and spraying with Dithane Z-78 (0.3 per cent) at weekly intervals gives good control.

Green Moulds

Causal Organism

Trichoderma spp. (*Trichoderma viride, T. hamatum, T. harzianum, T. koni*), *Penicillium cyclopium, Aspergillus* spp.

Symptom

A number of *Trichoderma spp.* have been reported to be associated with green mould symptoms in compost and casing soil in the spawn bottles and on grains after spawning. In addition some of the other species are also attack. The sporocarp producing dark brown or greenish lesions/spots/ discolouration along with the pileus as well as, the stripe. *T. viride* produces reddish brown discoloration along with the stipe by killing the tissues through the toxins produced by the mould fungus. *T. koningii* grows as a cottony weft of grayish mycelium over the casing surface. *Tricoderma hamatum* is frequently growing during spawn-run, but it also causes brownish spots on mushroom caps under high humid condition. When young pin heads are affected, spots are enlarged and cracking of caps becomes more conspicuous.

Epidemiology

As stated above, green moulds generally appear in compost rich in carbohydrates and deficient in nitrogen. If the compost is tampled too hard in the beds or if the filling weight is too high), which make the peak heating process difficult. Frequent use of formalin possibly in too high concentrations tends to promote the development of *Trichoderma.* High relative air humidity accompanied by a low pH in the casing soil also promotes the development of mould.

Control

The following steps can be adopt to avoid green moulds:

1. Very good hygiene condition;
2. Sterilizing the supplements before use and mixing them thoroughly, preferably after spawn-run.
3. Proper pasteurization and conditioning of compost;
4. Using the correct concentration of formalin (maximum 2 per cent).
5. Weekly sprays of Dithane Z-78 (0.2 per cent) or Bavistin (0.1 per cent) or treatment with Zineb dust or Calcium hypochlorite (15 per cent) have given effective control of the disease (Sohi, 1986, 1988; Bhardwaj and Seth, 1983).

False Truffle

Causal Organism

Dichliomyces microsporus.

Common Name

Truffle disease.

This is most dreaded competitor in mushroom bed. It was first reported by Lambert (1930) from Ohio, U.S.A. In India, Sohi *et.al.* (1965) observed false truffles causing sever losses to mushroom crop during 1929 when the compost in the beds reached beyond 22–24°C.

Symptoms

The color of the mycelium is white to creamy yellow at a later stage. The sporocarps vary appreciably in size ranging form 0.5 to 3 cm in diameter. At maturity, these turn pink, dry and reddish, and finally disintegrated into a powdery mass emitting chlorine like odor. The fungus does not allow the mushroom mycelium to grow and compost turns dull brown. The spawn in affected patches turns soggy and disappears.

Epidemiology

Ascospores develop in the truffles in 3 to 6 weeks and are released when the truffle disintegrates. The major source of infection are casing soil and surviving ascospores/mycelium in wooden trays from the previous crops. Ascospores can survive for long period (up to 5 years) in soil and spent compost (up to 6 months) as the major source of primary inoculum. Temperature and stage of infection are important factors for determining the severity of the diseases. False truffle seems to depend either on mushroom metabolites or on depletion of inhibitory factor by mushroom mycelium It is mainly a disease of A. *bitotquis* where ever crop is raised at 25°c but it also develops very fast in *A. bisporus* when the temperature rise 20°C during crop period.

Control

1. Compost should be prepared on a concrete floor and never on uncovered soil. Because during composting there is rise in temperature which activates the ascospores present in the soil.
2. Pasteurization and conditioning of the compost should be carried out carefully.
3. Temperature above 26-27°C during spawn run and after casing should be avoided. During cropping, temperature should be kept below 18°C. Under such condition, it is practically impossible to grow *A. bitorquis* but disease can be managed effectively in *A. bisporus.*
4. Casing soils known to harbour traces of spores should not be used. Young truffles must be picked and burried before the fruit bodies turn brown and spores are ripe. Woodwork, *trays* or side-boards of shelf-beds should be treated with a solution of sodium pentachlorophenolate at the end of the crop. Air-drying of wood work for 2-3 months also eradicate the pathogen.
5. Initial infection can be checked by treating the affected patches with 2 per cent formaldehyde solution.

Brown Plaster Mould

Causal Organism

Papulospora byssina

This disease was first reported on horse dung compost from Missouri (Motson,1917). In India, it was reported by Munjal and Seth (1974) causing 90-92 per cent yield loss in A. *bisporus* in Chail and Delhi and complete crop failures of oyster mushroom in Kasauli (H.P.) (Dar and Seth, 1981).Loss in number and weight of fruit bodies on artificial inoculation of mould has been found 7.7-53.5 per cent and 3.0-50.7 per cent, respectively (Sharma, 1990, Sharma and Vijay, 1993).

Symptoms

It was first noticed as whitish mycelial growth on the exposed surface of compost and casing soil in trays as well as on sides in bags due to moisture condensation. This develops further into large dense patches gradually changing colour through shades of tan, light brown to cinnamon brown; ultimately becoming rust coloured. No mushroom mycelium grows on places where plaster mould occurs.

Epidemiology

Primary infection comes through workers, air-borne bulbils, containers and causing silete. Its development is favored by wet, soggy and improper compost preparation. Higher temperature during spawn-run and cropping favors the disease development. In wet, greasy compost which had not received enough oxygen during fermentation, development of the disease is quicker.

Control

1. Strict hygienic measures should be followed.
2. Composting should be carried carefully.
3. Using sufficient gypsum and not too much water.
4. Peak heating should be of sufficient duration and at proper temperatures.
5. The compost should not be too wet before or after peak heating. Localized treatment of infected patches with 2-4 per cent Formalin has been recommended for its control. Spraying of systemic fungicides, namely, Baustin, Tops in-M or Vitavea at 0.1 per cent concentration has also been recommended.

Bacterial Disease

Bacterial diseases have been reported from all over the world. In India, only bacterial blotch has been reported on white button mushroom which is discussed below.

Bacterial Blotch

Causal Organism

Pseudomonas fluorescens biotype G (*P. tolassii, Phytomonas tolassii, Bacterium tolasii*)

This disease was first described by Tolaas (1915) from America. Later on, the disease has been reported from all the major mushroom growing countries of the world. In India it was first reported in 1976.

Symptoms

The pathogen induces lesions on mushroom tissues that are pale yellow initially but later on they become golden yellow or chocolate brown. This discoloration is superficial not more than 2 to 3 mm deep and the underlying, mushroom tissue may appear to be water-soaked or grayish yellow. Blotch can appear on mushroom of its any growth stage including its early stage to harvesting stage. If moisture conditions are favorable the spots become large and coalesce, sometimes covering the entire mushroom cap. Typical spotting is observed at or near the edge of mushroom cap, at the contact points between two mushroom caps.

Causal Organism

Pseudomonas fluorescens and *P. tolaasii*

Epidemiology

Casing soil and air-borne dust are the primary source of blotch pathogen into a mushroom house. Occurrence of the disease is associated with the rise of bacterial population on the mushroom cap rather than on the population in casing. Bacteria splashed into a mushroom surface will reproduce in moist conditions especially when moisture or free water film persists for more than 3 hours after watering. Once the pathogen has been introduced at the farm, it may survive between crops on the surfaces, in debris, on tools and on various structures.

Control

The following practices should be adopted for the management this disease:-

Ecological

As has been stated above that the disease development depends upon the free water film on the mushroom surface and any alteration in the management of environment which results in regulating free water film on the mushroom surface will also automatically alter the disease development. Temperatures above 20°C and of more than 85 per cent R.H. should be avoided. Additional ventilation and circulation after watering can ensure the quick drying of mushrooms.

Biological

Isolates of *P. fluorescence* and other antagonistic bacteria have resulted in 30-60 per cent control of bacterial blotch. Spraying of the casing soil with a mixture of *P.fluorescence* and bacteriophage has resulted in more than 80 per cent control of blotch symptoms (Guillaumes *et al.*, 1988: Munsch *et al.*, 1991).

Chemical

Application of Terramycin at 9mg/sq.feet, streptomycin (200 ppm), oxytetracycline (300 ppm), has been found effective in lowering down the disease.

Physical

Pasteurization of casing soils by steam/air mixture and short wave length irradiation have been reported effective in eliminating the bacterial pathogen but over-heating should be avoided otherwise biological vaccum will be created and successive invasion of moulds would be very high.

Viral Diseases

Gandy and Hollings (1962) were the first to demonstrate the association.of three types of 'viruses in mushrooms. Since then, viral diseases have been reported throughout the world on different mushrooms. In India, virus and virus-like disease have been reported on button mushroom (Tewari and Singh, 1984, 1985; Goltapeh and Kapoor, 1990). Transmission and control measures of button mushroom viruses are discussed below.

Causal Organism

Several viruses of different shapes and sizes have been reported. In India, viruses have been found associated with a virus disease of button mushroom (Tewari and Singh, 1984, 1985). Virus-like particles measuring 29 nm in diameter have also been reported on button mushroom (Goltapeh and Kapoor, 1990).

Symptoms

The various names coined for mushroom viruses give some indication of the diversity and variation of symptoms induced by viruses in mushrooms. Symptom expression depends upon virus concentration; time of infection, strain of the spawn used and cultural and environmental conditions. The common symptoms observed are as below:

1. Pin heads appear late and sometimes below the surface of casing layer. If the sporophores appear above the casing layer, their pilei are already opened.
2. Mycelium isolated from diseased sporophores on agar shows a slow and degenerated growth as compared with healthy mycelium.
3. Symptoms on sporophores are highly variable. The following abnormalities can be found separately or in combination:
   (*a*) Watery stipes or streaking in the stipe.
   (*b*) Dwarfing and slow development of the pin heads.
   (*c*) Elongated and slightly bent stipes.
   (*d*) Early maturity with Off-white color of the caps,

Epidemiology

There is no information's about the incidence of viruses in mushrooms and losses caused by them in India but studies carried out and found that when virus inoculations were done 0 to 12 days after spawning, the extent of 'loss varied from 37.5 to 95.5 per cent over un-inoculated (Dieleman van Zaayen, 1972).

Control

Hygienic measures are to be followed strictly. When virus-infected cultures are grown at 33°C for 2 weeks then sub cultured and retumed to 25°C, the newly obtained cultures are generally free from viruses.

Abiotic Disorders

Management of environment is of great significance in mushroom cultivation and any deviation from the optimum requirements may lead to various kinds of abnormalities. Since a major proportion of button mushrooms is being produced under natural climatic conditions in India, the following important abiotic disorders are quite frequently observed:

Storma: Sectors, Sectoring etc.

Stroma is noticeable as aggregations of mushroom mycelium on the surface of spawned compost or the casing. It seen like aerial patches of white mycelium form a dense tissue layer on the substrate surface, which can be easily peeled from the surface of compost or casing.

Stroma and sectors are related to the genetic character of the spawn but are sometimes induced if spawn is mishandled or exposed to harmful petroleum-based fumes of chemicals or certain detergents during preparation, storage, transit at the farm.

Management

1. Avoid the use of chemical and petroleum products during cropping.
   (*a*) Once the spawn open it should be utilized.
   (*b*) Cleaning of crop room.

Weepers: Stinkers, Leakers

Mushrooms described as being 'weepers' typically extude considerable amount of water from mushroom cap. When small water droplets exude from stem or cap, the mushrooms are called leakers. The distinction between a 'leaker' and 'weeper' is that the water droplets remain as droplets on the leaker mushrooms while it actually falls or flows from a weeper. Weepers are usually noticed since they are quite unusual. Water collects on the casing surface beneath a weeper and the area develops a putrid odour becoming a 'stinker'.

Factors that induce a mushroom to become a weeper are not known but low moisture compost(less than 64 per cent) coupled with high moisture casing is where weepers are frequently seen. The combination of these two conditions often foster weeper mushrooms prior to and during the first break.

References

Bhardwaj, S.C. and Seth, P.K. (1983). Studies on green mould *Trichoderma viridi* (Pres. Ex. S. F. Gray)and its control.*Indian j. Mush.*, 9: 44-51.

Dar G.M.and Seth P. K. (1981) Studies on brown plaster mould (*Papulospora bassina* Hots.) and its control. *Indian J. Mush.* 7: 60-83.

Gandy, D.G.and HollingsM. (1962). Die back of mushroom: a disease associated with virus. Rep. *G.C.R.I.*, 1961. pp. 103-108.

Goltapeh, E.M. and Kapoor, J. N. (1980). VLP, s in white button mushroom in north India. *Indian Phytopath.* 43: 254 (abstr.).

Guiliaumes.J., Haudean, G., Geamin, R., and Olivier, J.M. (1988). *Amelioration de la lutte biologique centre P. tolaasii par utilization de bacteriophage. Bull.OEPP/EPPO.*, 18: 77-82.

Guleria, D.S. (1976). A note on the occurrence of brown blotch of cultivated mushroom in India. *Indian J. Mush.*2 (1): 25.

Munjal, R.L. and Seth, P.K. (1974).Brown plaster mould-a new disease in white button mushroom.*Indian. Hort.*18(4): 13.

Munsch. P., Oliever J.M. Houdean, G. (1991). Experimental control of bacterial blotch by bacteriophage. *Mush. Sci.*13(1): 389-396.

Seth, P.K. (1977). Pathogens and competitor of *Agaricus bisporus* and their control.*Indian J. Mush.*3: 31-40.

Seth P. K., Kumar, S. and Shandilya, T. R. (1973). Combating dry bubble of mushroom. *Indian Hort.* 18(2): 17-18.

Sharma. S.R. (1993). Present scenario of mushroom cultivation in India- Abroad prospective. *Agric. Situation in India.*47(2): 825-834.

Sohi, H.S. (1986). Disease and competitors mould associated with mushroom culture and their control.*N.C.M.T., Chambaghat Solan, Ext. Bull.* No. 2.12 p.

Sohi, H.S. (1988). Diseases of white button mushroom (*Agaricus bisporus*) in India and their control. *Indian J. Mycol. Plan Pathol.* 18: 10-18.

Tewari, R.P. and Singh, S. J. (1984).Mushroom virus diseases in India. *Mushroom J.*, 142 pp.354-355.

Tewari, R.P. and Singh, S. J. (1985). Studies on a virus disease of white button mushroom [*Agaricus bisporus* (Lange) Sing.] in India. *Indian J, Virol.*, 1: : 35-41.

Sustainable Disease Management of Agricultural Crops (2011) *Pages* **377–383**
Editors: **S.K. Biswas and S.R. Singh**
Published by: **DAYA PUBLISHING HOUSE, NEW DELHI**

Chapter 28

Fusarium Wilt Diseases and their Management

☆ *P. Kishore Varma*[1] *and C.N.L. Reddy*[2]

[1]*Assistant Professor, Department of Plant Pathology, Agricultural College, Aswaraopet – 507 301, Khammam, Andhra Pradesh*

[2]*Assistant Professor, Department of Plant Pathology, Sericultural College, Chintamani – 563 125, Chikkaballapura, Karnataka*

Vascular wilts caused by fungal and bacterial pathogens *viz. Fusarium, Verticillium, Acremonium, Pseudomonas, Ralstonia and Burkholderia* are major constraints in the production of many cultivable crop species. Among these, wilt incited by *Fusarium* species is very important as they cause heavy economic losses in yield. Eventhough, different formae speciales of the fungus *Fusarium oxysporum* are morphologically similar, they are generally host-specific.

The most economically important crops which are prone to Fusarium wilt are arhar, cotton, castor, chickpea, watermelon, tomato and safflower. Most of the Fusarial wilts are similar in their symptomatology, biology, epidemiology and management. Consequently, Fusarium wilt of Arhar is described as an example of the group. Fusarium wilt of redgram caused by *Fusarium udum* Butler is the most devastating disease of redgram which is distributed worldwide. The annual losses from India due to wilt have been estimated at US $ 71 million (Reddy *et al.*, 1993). Further, in India it is widely distributed in Andhra Pradesh, Maharashtra, Madhya Pradesh, Tamil Nadu, Uttar Pradesh and Bihar.

Symptomatology

Symptoms incited by wilt Fusaria are dependent on several interacting factors, including the susceptibility of the host, amount of primary inoculum in the soil, environmental conditions and nutritional status of the soil. The appearance of the disease starts from early stages of plant growth (4-6 week old plant) up to flowering and podding stage. Initial symptoms often include a dull green appearance of the lower leaves that precedes a loss of turgor pressure and wilting. Wilting is followed by chlorosis of the leaves and finally leads to necrosis.

The prominent symptoms appear when the plant put forth flowers and pods. Plants die in patches under conditions of sufficiently high inoculum density in conjunction with a susceptible host (Figure 28.1). The characteristic symptom of Fusarium wilt is the discoloration of the vascular system (xylem), which can be observed readily in longitudinal or cross section of roots or stems. In redgram and safflower, a purple or brown necrotic band may be visible on the lower stem and extend along the length of the stem.

Partial wilting of the plant is a definite indication of Fusarium wilt and distinguishes from *Phytophthora* blight that kills the whole plant. Further, partial wilt is associated with lateral root infection, while total wilt is due to tap root infection.

Cotton is also susceptible to a vascular disease caused by *Verticillium dahliae* that may be confused with *Fusarium* wilt, as the symptoms are similar. Fusarium wilt can be distinguished from the *Verticillium* wilt by partial wilt and continuous browning of vascular tissues. In case of *Verticillium* wilt, pinkish to pinkish brown discolouration of the woody tissue is observed which may be continuous or interrupted (Tiger stripe). Higher temperatures appear to impede Fusarial wilt infection often resulting in plants that are yellowed and stunted but not wilted.

Mechanism of Wilting

The pathogen produces abundant mycelium, micro- and macro-conidia or polysaccharides in the vascular bundles (xylem) resulting in clogging of vascular bundles. The pathogen is also known to produce a vivotoxin, fusaric acid, *in vitro* and *in vivo*. However, the extent to which the fungal toxins are involved in pathogenesis is not known. In addition, the enzymes secreted by the pathogen breakdown the parenchyma cells resulting in various types of gums and gels leading to the blockage of xylem vessels. The breakdown products are further oxidized leading to browning of vascular bundles. The host parenchyma cells defend through the production of tyloses clogging the vascular bundles ahead of pathogen invasion; thereby restricting the subsequent invasion of the pathogen resulting in wilt. Once the *Fusarium* incites the disease, it is usually restricted to the xylem and generally emerges only after death of the plant.

Pathogen Biology

(Sub-Div: Deuteromycotina; Class: Hyphomycetes; Order: Tuberculariales; Family: Tuberculariaceae)

The fungus produces septate and branched hyphae on which short, hyaline, simple or branched conidiophores arise bearing a whorl of spore producing cells called phialides. *Fusarium udum* produces two types of conidia, *viz.*, macroconidia and microconidia. Macroconidia are long, crescent (fusiform) shaped, multiseptate, typically having three to five cells and are produced in large numbers, often aggregated into asexual fruiting bodies known as sporodochia. Macroconidia have the ability to form chlamydospores and help in the survival during off-season, besides it may also infect roots. Chlamydospores are the primary means of survival and typically form under conditions of suboptimal growth for the fungus or on death of the host (Rangaswami and Mahadevan, 2004).

Fusarium oxysporum forms two types of chlamydospores; one within the macroconidium and the other within the mycelium. Those formed from mycelia tend to occur singly or in pairs and may be either intercalary (within the mycelium) or terminal (occurring at the ends). Disease results from infection of the root or hypocotyls by hyphae resulting from either the macroconidia or chlamydospores. Once infection has occurred, the fungus penetrates and grows within the xylem. Abundant microconidia are produced in the xylem vessels which are carried upstream along the transpiration stream. These

Figure 28.1: Arhar (cv. LRG 30) Plants Showing Advanced Symptoms of Wilt

Figure 28.2: Safflower Plant Showing Partial Wilt

microconidia are generally single celled, oval and are produced on short conidiophores and probably play a little role in initial infection in the field. The secondary spread of the pathogen to short distances occurs by means of irrigation water and contaminated farm equipment and over long distances primarily in infected planting material or in the soil carried with them.

Fusarium oxysporum is a soil-borne fungus with both saprophytic and pathogenic members. Pathogenic strains (formae speciales) are somewhat specialized and are defined on the basis of the host they predominantly infect (Table 28.1).

Table 28.1: Host Specificity of Pathogenic Strains of *Fusarium oxysporum*

Sl.No.	Causal Organism	Crop
1.	*Fusarium udum*	Redgram
2.	*F. oxysporum* f.sp. *vasinfectum*	Cotton
3.	*F. oxysporum* f.sp. *ricini*	Castor
4.	*F. oxysporum* f.sp. *carthami*	Safflower
5.	*F. oxysporum* f.sp. *ciceris*	Chickpea
6.	*F. oxysporum* f.sp. *lycopercisi*	Tomato
7.	*F. oxysporum* f.sp. *cubense*	Banana
8.	*F. oxysporum* f.sp. *niveum*	Watermelon
9.	*F. oxysporum* f.sp. *chrysanthemi*	Chrysanthemum
10.	*F. oxysporum* f.sp. *callistephi*	Aster

Figure 28.3: Rhizome of Banana Showing Abundant Sporulation of *Fusarium*

Many are further divided into pathological races based on their pathogenicity to a set of differential host cultivars, each with a unique set of resistance genes.

Disease Cycle

The disease cycle for all *Fusarium* wilts is almost similar. *Fusarium* survives in the soil in a dormant state for 8 to 20 years as chlamydospores and is capable of inciting initial infection of seedlings (Reddy *et al.*, 1993). Under conditions of optimal growth of the fungus, chlamydospores germinate and produce infection hyphae that penetrate the root cortex through the crevices between tap root and lateral roots. Infection may be enhanced by wounds or damage to the roots caused by the nematodes or farm operations. Upon colonization of the root cortex, the fungus invades the xylem tissue producing abundant mycelia and microconidia which are carried up the xylem in the transpiration stream. In response to the mycelial growth and sporulation in the xylem tissue, the parenchyma cells adjacent to xylem produces balloon like structures called tyloses ahead of the pathogen invasion; thereby preventing systemic colonization of the pathogen in resistant varieties. However, in susceptible plants, production of tyloses is slow and do not completely block the vessel, ultimately allowing the fungus to colonize. The formation of tyloses in some vessels and not in others partially explains the partial wilt often seen in redgram, cotton and Safflower (Figure 28.2). Further colonization of the tissue results in necrosis and death of the plant and produces the characteristic brown discoloration of the vascular tissue.

Upon plant death or in the late stage of the disease, the fungus produces thick mats of white mycelia and abundant macroconidia (Figure 28.3) out side the host which are converted into chlamydospores, which remain dormant until next season.

Epidemiology

Wilt diseases incited by Fusaria are generally considered to be monocyclic as these don't produce propagules capable of secondary spread above ground level until very late in the crop season. Movement of infested soil by means of farm implements during field preparation, inter-cultivation and uncontrolled irrigation (flooding) may result in spread of the disease in a field. Many *Fusarium* wilt pathogens are capable of being externally seedborne, although the extent of the transmission varies from host to host. *Fusarium* wilts are generally most severe in light, sandy, slightly acidic soils when temperatures are between 24-27° C (Agrios, 2005).

Disease Management

Resistant cultivars are of prime importance in the management of Fusarium wilts; however, there are a number of options that can help in reducing the severity of the disease.

Host Resistance

Use of resistant varieties for the management of *Fusarium* wilt is the best and most economical method of control. However, it is difficult to get complete resistance to all races available in any given locality. Hence, deployment of available resistant germplasm based on the race flora existing in a locality is very important for the management of Fusarial wilts. Grow resistant/tolerant varieties of pigeonpea like ICPL 87119, ICP 8863, ICPL 85063, ICPL 84031, etc., in wilt prone areas (Reddy *et al.*, 1993).

Field Sanitation

The pathogen is known to survive in the infected stubbles which serve as repository of inoculum.

Hence, collection and subsequent destruction of the infected stubbles helps in reducing the initial inoculum, and thereby the disease incidence.

Crop Rotation

The same crop should never be grown in succession on the same land as successive cropping can cause the pathogen to build up to high levels leading to mass mortality (Singh, 2008). As the pathogen survives as chlamydospores in the soil for several years, prolonged crop rotation with non-host crops like tobacco, sorghum and castor is generally recommended.

Mixed Cropping

Mixed cropping with sorghum is generally recommended to reduce the lateral spread of the disease. The disease incidence of Arhar wilt was highly suppressed under mixed cropping with *Crotolaria medicagina* (Upadhyay *et al.*, 1981). In cotton, mixed cropping with non-host plants is known to lower the soil temperature below 20°C by providing shade and thereby the disease incidence.

Soil Fumigation

Soil fumigants like methyl bromide, chloropicrin and metham-sodium (vapam) have shown moderate reductions in disease severity under field conditions. However, fumigation with these chemicals is not an economically viable option and is seldom successful, primarily because of the resistant nature of the chlamydospores, re-colonization of the soil by the pathogen, or because of improper application of the fumigants (Hopkins and Elmstorm, 1979).

Soil Solarization

Soil solarization or slow soil pasteurization is the hydro-thermal soil heating accomplished by covering moist soil with transparent polyethylene sheets as soil mulch during summer months for 4-6 weeks. Soil solarization traps the energy from the sun and heat the soil to temperatures that are lethal to the pathogen. It has been used effectively in some locations around the world for high value crops; however, it is costly and laborious. Soil solarization can lower populations of wilt fusaria in soil sufficiently to delay the onset of wilt symptoms as well as in reducing disease incidence. However, it does not eliminate the disease and its effectiveness is limited to areas of the world where appropriate climatic conditions exist (Martyn, 1986; Martyn and Hartz, 1986).

Seed Treatment

Seed treatment is the important management practice for seed borne wilt Fusaria (Singh, 1998). In general seed treatment with systemic fungicide like carbendazim at 2g/kg seed or Carboxin at 4 g/kg seed is recommended for seed disinfestations. However, seed treatment with antagonists like *Trichoderma viride* (4-6g/kg seed) and *Pseudomonas fluorescens* (10g/kg seed) is also recommended to manage the disease.

Soil Amendments

Application of organic amendments like farm yard manure, oil cakes, green manures etc., effectively manages the wilt diseases caused by *Fusarium*, *Verticillium* and *Acremonium* (Upadhyay and Bharat Rai, 1981). Organic amendments serve as a substrate for the multiplication of antagonistic micro-organisms in soil and helps in suppression of wilt Fusaria. Application of farm yard manure at the rate of 10 tonnes per hectare is generally recommended for the management of soil borne diseases. In addition to this, application of neem cake at 2.5 tonnes/ha is also recommended for the management of nematodes which aggravates the disease. Green manure crops like sunhemp and dhaincha can be grown and incorporated at the flowering stage one month before planting the main crop.

Biological Control

Potential of different types of bioagents has been examined for the management of Fusarial wilts in a number of crop plants. While many demonstrate promise in laboratory and greenhouse trials, few have shown significant control at the field level. Antagonistic or antibiotic producing soil fungi and bacteria, *viz.*, *Trichoderma* spp., *Gliocladium* spp., *Pseudomonas* spp. and *Bacillus* spp. are tested for the management of wilt fusaria (Upadhyay, 1992). *Bacillus subtilis*, a gram positive bacterium capable of producing endospores was found promising in the management of *Fusarium udum*.

Integrated Disease Management

With the exception of host resistance, it is likely that no one technique will result in the complete control of *Fusarium* wilts. However, by integrating various approaches such as chemical, biological, resistant varieties, *etc.*, can be used to manage the disease. The development of resistant varieties and combined application of bioagents and fungicides is considered as more practicable approach for the management of *Fusarium* wilts. An integration of management practices like seed treatment with carbendazim (2kg/kg seed) followed by soil application of *Pseudomonas fluorescens* and *Trichoderma viride* (2.5kg/ha in FYM@50kg/ha) was found to reduce wilt incidence under field conditions (Mahesh *et al.*, 2010).

References

Agrios, G.N. (2005). *Plant Pathology*. Elsevier Academic Press, New York, pp. 952

Hopkins, D.L. and G.W. Elmstrom. (1979). Evaluation of soil fumigants and application methods for the control of Fusarium wilt of watermelon. *Pl. Dis. Repor.* 63:1003-1006.

Mahesh, M., Saifulla, M., Srinivasa, S. and Shashidhar, K. (2010). Integrated management of Pigeonpea wilt caused by *Fusarium udum* BUTLER. EJBS, 2(1): 1-6

Martyn, R.D. (1986). Use of soil solarization to control *Fusarium* wilt of watermelon. *Pl. Dis.* 79:762-766.

Martyn, R.D. and T.K. Hartz. (1986). Use of soil solarization to control *Fusarium* wilt of watermelon. *Pl. Dis.* 70:762-766.

Rangaswami, G. and Mahadevan, K. (2004). Diseases of Crop Plants in India. Prentice Hall of India Pvt. Ltd., New Delhi, pp. 507

Reddy, M.V., Raju, T.N., Sharma, S.B., Nene, Y.L. and Mc Donald, D. (1993). Handbook of Pigeonpea diseases. (In En. Summaries in En, Fr.) Information bulletin No. 42, Patancheru 502 324, Andhra Pradesh, India: International Crops Research Institute for the Semi-Arid Tropics. 99.64

Singh, R. (1998). Chemical control of wilt disease of Pigeonpea. *Korean Journal of Mycology*. 26: 416-423

Singh, R. S. (2008). *Plant Diseases*. Oxford and IBH Publishing Co. Pvt. Ltd., New Delhi, pp. 721

Upadhyay, R.S. (1992). Ecology and biological control of *Fusarium udum* in relation to soil antagonistic microorganisms. *J. Pl. Protect.Tropics* 9: 1-9

Upadhyay, R.S. and Bharat Rai. (1981). Effect of cultural practices and soil treatment on incidence of wilt disease of pigeonpea. *Plant and Soil* 62 (2): 309-312.

Sustainable Disease Management of Agricultural Crops (2011) *Pages* **384–400**
Editors: **S.K. Biswas and S.R. Singh**
Published by: **DAYA PUBLISHING HOUSE, NEW DELHI**

Chapter 29

Soil Solarization: A Novel Tool for Soil Borne Diseases and Pest Management

☆ ***Prem Naresh, Abu Manzer, S.K. Biswas and N.K. Pandey***
Department of Plant Pathology, C.S. Azad University of Agriculture and Technology, Kanpur – 208 002, U.P.

Plants are the major source of energy hence called primary producer. The consumer herbivore, carnivore and omnivore depend on the plants for their survival *viz.* shelter, nourishment, furniture and even oxygen supply. India as well as world population is increasing day by day and per capita food availability becoming decrease, due to shrunken of land and low crop production. Plants are the major source of energy to men and animals which are affected by so many abiotic and biotic factors that inhibit their normal growth and development resulting low production and productivity.Plant disease make the difference between happy life and life hunted by hunger and even death due to starvation. Among different plant diseases soil borne diseases are very harmful to plant because they encounter from seedling emergence till seed maturity. It also makes the soil sick. The major soil borne pathogen fungi (*Pithium, Phytophthora,Fusarium, Rhizoctonia, Sclerotium, Sclerotinia* etc.) Some bacteria and nematodes etc. They are most versatile type and found round the year as well as till long year which make difficulties to manage soil borne disease.Thus soil solarization is an effective and innovative method which not only minimizes soil borne disease but also increase yield quality of the crop. Therefore soil solarization techniques are to be developed to manage this harmful organisms with the aims to eradicate or reduce the existing inoculum in the soil, before to planting/sowing, and finally to achieve an economic reduction in disease incidence for at least one season.

Important Characteristics of Soil Solarization

☆ It is a hydrothermal process.

☆ The covering polythene is transparent.

☆ Soil should be saturated *i.e.* it must be wet (not flooded).

- ☆ Solarised soil is comparatively hotter in nature than unsolarized soil (Only applied in hot area not in cool)
- ☆ Producing high soil temperature (after solarization) which are mostly lethal to soil pathogen.
- ☆ It can be used only for soil borne pathogen not for other like (air borne, seed borne etc) pathogens.

What is Soil Solarization

A sacred hyms from the Atharava Veda says: *Ut surya Div. Eti Puro Rakshansi Nijurvan 6.52.1* that means the sun was worshipped as one of the gods by ancient Aryans. They realized the antimicrobial property of solar rays and had used them to control plant diseases and keeping water and grains free from germs. This concept comes in the form of soil solarization in the field of agriculture with special reference to management practices, particularly for soil borne disease.

Thus soil solarization is a hydrothermal process in which covering moist soil with transparent polythene sheet as soil mulch during hot summer for 4-6 weeks, to producing high soil temperature *i.e.* lethal to soil borne pathogen. According to Lamberti and Greco (1991), soil solarization is a newly developed technique for the control of soil borne pathogen. Soil solarization is also termed as solar heating, plastic or polyethylene trappings, polyethylene or plastic mulching of soil and solar pasteurization. Since this method involves repeated daily heating at relatively mild temperatures the term solar pasteurization is justified (Katan, 1981).

History of Soil Solarization

- ☆ Rising sun destroys the numerous germs causing diseases. Almost the same expressions have this holy mantra from Atharva Veda: *Udyan Adityah Krimin Hantu....... Rashmibhi* 2.32.1
- ☆ Renowed Israeli phytopathologist Dr. Katan and his co-workers I[st] time developed a solar heating approach for soil disinfestations to control of plant pathogenic pests, diseases and weeds (Katan *et al.*, 1976).
- ☆ In 1939- Polyethylene (also known as polythene) one of the most important plastic materials used commercially in agriculture.
- ☆ Ist report on use of solar energy for the control of soil–borne diseases was published by Katan *et al.* (1976).
- ☆ Katan *et al.* (1976) term hydrothermal soil heating used for accomplished by covering moist soil with polyethylene sheet as soil mulch during summer months for 4-6 weeks.
- ☆ Katan (1980) use the term solar pasteurization.
- ☆ Barbercheck *et al.* (1986) and Katan *et al.* (1983) used successfully used soil solarization to reduce population of plant parasitic nematode, various plant pathogenic fungi and weeds.
- ☆ Pullman *et al.* (1981) successfully controlled some soil borne fungal diseases.

Method of Soil Solarization

At mid summer *i.e.* when maximum solar radiation can be obtained, the field is first tilled and irrigated when the moisture level is around the field capacity (saturated soil). Then the field is covered by a clear transparent polyethylene sheet. The sheet is compacted into the soil and left as such for 3-6 weeks. At the end of this period, the sheet is removed and the field is available for use. The method of soil solarization as described by Katan *et al.* (1976 and 1981). They also pointed out the following criteria have to be followed for soil solarization.

1. The edges of the polythene should be inserted in furrows.
2. Organic matter should be added to soil before starting of solarisation.
3. The sheets are compacted in to the soil left as such for 4-6 weeks.
4. The land to be solarized should be thoroughly tilled and leveled to minimize such protrusion as clods stubble and stones to prevent tearing the polythene.
5. Irrigation (40-50) mm should be given at mid summer when maximum solar radiation can be obtained before laying of the sheet.
6. Clear, transparent polythene sheet of 25-100. µm thickness should be applied immediately after irrigation. It is best to apply the sheet at dawn when it is least windy. End of this period the sheet should be removed.

Principle of Soil Solarization

During solarization the plastic sheets trap solar energy and prevent heat loss as caused by evaporation and conviction, thus creating green house effect (GHE). Soil solarization involves trapping of solar heat/energy through polyethylene covering to raise the soil temperature to the level where it becomes lethal to temperature sensitive to mesophillic soil pathogenic organisms. Polythene covering of soil induces green-house effect and raises soil temperature. Solar heating involves the use of heat as a lethal agent for pest control through the use of traps for capturing solar energy by means of transparent polyethylene soil mulches (Katan, 1981).

In Mediterranean type climates temperatures as high as or higher than 60 °C are achieved in the surface layers of soil generally representing increase of 10-12 °C over uncovered soil. Report show that many nematode and fungal pathogens are killed by temperatures within the range of 45°C to 55°C (Pullman *et al.*, 1981). Consequently, the principal effect of solarization is to raise soil temperatures to levels which are lethal to soil borne pathogens.

Katan (1980) developed the following principles.

1. The accumulation of heat in polyethylene mulched soil by transmission of short wave solar radiation (280-2500 nm) but prevention the loss of long wave radiation (5000 to 35000 nm) from the soil.
2. Maintenance of high soil moisture which improve the thermal conductivity of the soil and retention of heat by high latent heat of water vapor.
3. Increase in thermal conductivity by hydration of pathogenic organisms.
4. Low thermal sensitivity in and fast re-colonization by antagonistic micro-organisms, thus shifting the biological equilibrium in favour of the natural enemies of plant pathogen.
5. Acceleration of microbial and physico-chemical reaction in the soil resulting in the accumulation of gases some of which are toxic and the release of microbial ions some of which are also toxic to pathogen while others serve as nutrients for or induce resistance in the subsequent crops.
6. Thermal inactivation of pathogen, rendering them more vulnerable to antagonists and reduces their infectivity and longevity.

7. Condensation of water vapor on the under side of the polyethylene sheet in the form of planoconvex droplets, reducing heat loss and concentrating the solar radiation.
8. Concomitant control of weeds (which otherwise may act as secondary host for the nematodes) and other pathogen, some of which forms disease complexes with the target pathogen.

Factors Affecting the Efficacy of Solarization

The several parameters which by and large determine the effectiveness of solarization-

Colour of the Sheet

Transparent polyethylene (TP) is more effective than black sheets. Usmani and Gaffar(1981) recorded highest temperature build up under transparent polyethylene.

Thickness of Covering Sheet

Thinner sheets give better result than thicker ones (Katan, 1981). Thickness of 25-40 µm generates maximum soil temperature than higher thickness. But this transparent polythene sheet may not get tear by the pressure of water vapors and gasses produce under it.

Thinner sheets are not suitable for reuse because the emerging weed puncture the sheet. A thickness of 50-100 µm is most desirable, in case of reused point of view. The old polythene sheet produces more heat than new one, because the water droplets accumulated under the new sheet reflect more radiation than the dusts and water fullness on the old one (Avissar *et al.*, 1986).

Chemical Nature of the Covering Sheet

Poly vinyl chloride (P.V.C.) to be an effective mulching sheet. It gives 3-7°C grater increase in soil temperature than normal polyethylene (Garibaldi, 1987). Patel *et al.* (1990) found that Low density polyethylene (LDPE) when used as mulching sheet gives two fold increases in tobacco seedling against root-knot nematode. But the cost of PVC is high and it is also not easily available in market.These are the main draw back of PVC for it use by the farmers in soil solarization process.

Number of Layers

The number of layers of plastic sheet is alone important factors for increasing soil temperature The Garibaldi (1987) found that double layer bubble plastic gave soil temperature 1-2°C higher than those obtained with P.V.C. The double layer of clear plastic mulch give 3 °C higher temperature than single layer plastic sheet.

Photodegradable and Biodegradable Additives

Biodegradable plastic which incorporates water soluble biodegradable plasticizers and Low density polyethylene (LDPE) into starch ethylene acrylic acid copolymer films increases the rate of degradation without adversely affecting plastic properties. Because starch is biodegradable this plastic disintegrates into small particles much faster than conventional plastic (De Luca *et al.*, 1994; Manera *et al.*, 1999).

Pre-irrigation

High moisture content in soil not only retains large amount of heat, but also activates the pathogen and hence make them more vulnerable to the high lethal temperature. The nursery beds were flood irrigated before putting the polyethylene mulch to increase the thermal sensitivity of viable propagules of different pathogenic fungi (Raj *et al.*, 1995).

Period of Solarization

Duration of soil solarization also plays an important role in plant disease management. For getting satisfactory nematode/disease control, the optimum period of solarization are essential, it depends upon solar radiation intensity, locality, type of mulch, water holding capacity etc. In general 2-8 weeks of solarization is optimum for control of soil borne pathogens.

Soil Solarization Combination with Other Method of Control

Solarization is a part of Integrated Pest Management (IPM) Programme (Nageswari, 2001). The chemical or biological control agents can easily act upon work or inactivated pathogen by the effect of sub lethal heat. Soil solarization acts through physical, chemical and biological mode and improves the plant growth response by conserving moisture and alternation by mineral and organic matter solubility.

The polyethylene sheet prevents the losses of gas produced by the pesticide fumigates and then promote their effective dosage. Walker and Watchel (1989) demonstrated that solarization increase the infections of Juveniles of *Globodera rostochiensis* by *Posteuria penetrans*. Soil solarization (SS) increase the level of available mineral nutrients specially ammonium (NH_4) nitrogen. Ammonia giving adversely affects the certain survival of soil borne pathogen.

Soil Solarization in Nursery and Pots

Soil solarization has a big draw back in respect of purchasing of large polyethylene sheet, involving of laborious practice and keeping the field for long time without any crop. It can be over come when nurseries and plots are solarized. Kamra and Gaur (1995) reported that a small amount of soil (1 kg) kept under polythene bags is more effective in solarization process and also reduces the duration of time. According to them this sealed bags reduces the *Meloidogyne juvenilla* and upto 100 per cent within 5 days of solarization. This method is very much useful and can be explored for minimising nematode fungi, bacteria from the field of vegetables, ornamentals, horticultural and plantation crops.

After studying the above factors the following recommendation are made to bring about effective solar heating of soil (Katan, 1981).

1. Transparent, not black polyethylene should be used since it transmits most of the solar radiation that heat the soil.
2. Soil mulching should be carried out during the period of high temperature and intense solar radiation (April-June).
3. Soil should be kept wet during mulching to increase thermal sensitivity of resting structures and improve heat conduction (Katan *et al.*, 1976).
4. The thinnest polyethylene trap possible (25-30 μm) is recommended since it is both cheaper and some what more effective in heating (Horowitz *et al.*,1980) due to better radiation transmittance than the thicker one.
5. The mulching period should be sufficiently extended, usually four weeks or longer in order to achieve pathogen control at all desired depths because temperature at the deeper soil layers are lower than at the upper once *e.g.* mortality rate of *Sclerotium rolfsii* sclerotia at depths of 5 and 20 cm were 100 per cent and 25 per cent after 19 days of mulching and 100 per cent and 80 per cent after 21 additional days respectively (Elad *et al.*, 1980).

Significance

Plant diseases incur 26 per cent yield losses in different crops (Singh. 1998). Soil solarization has been used to effectively control certain soil borne pathogens *viz.* root knot nematode *Fusarium oxysporum, Rhizoctonia, S. rolfsii* and some pre and post emergence damping off caused by species of *Pythium,* (Patel *et al.*, 1991).

Table 29.1: Effect of Soil Solarization on Pre and Post Emergence Damping Off on Different Vegetable Crops (Harender *et al.*, 1997)

Crops	% Seed Germination		% Seedling Survival (30 daysold)	
	Solarized	Non-solarized	Solarized	Non-solarized
Brinjal	85.6	49.6	84.6	40.3
Broccoli	87.3	69.0	85.6	56.0
Cabbage	87.6	61.3	85.0	51.6
Capsicum	75.3	45.0	73.0	36.3
Cauliflower	80.6	54.0	71.6	45.0
Chilli	72.3	42.3	71.0	34.3
Chinese sarson	89.0	47.0	88.0	36.0
Lettuce	88.3	59.3	85.6	50.0
Onion	73.3	45.6	72.6	35.3
Tomato	88.0	56.3	86.6	44.3

According to (Harender *et al.*, 1997), higher seed germination was recorded in Chinese sarson lettuce and tomato in solarized nursery beds. In general seed germination was more in vegetables which were raised in solarized nursery beds. Thus incidence of post emergence damping off was recorded lower in solarized nursery beds in different vegetable crops.

Table 29.2: Effect of Soil Solarization on Seedling Vigour of Different Vegetable Crops

Crops	% Increase in Growth Vigours Over Plants Grown in Non-solarized Soils	
	Shoot Length	Root Length
Brinjal	27.5	24.5
Broccoli	20.6	33.3
Cabbage	25.3	21.9
Capsicum	20.2	26.9
Cauliflower	33.7	25.0
Chilli	20.2	22.8
Chinese sarson	20.5	28.5
Lettuce	18.4	23.4
Onion	18.6	27.5
Tomato	29.6	22.6

In case seedlings raised in solarized nursery beds had better vigour (shoot and root lenghts) than seedling raised in non solarized beds. Highest increase in shoot length in solarized beds over non solarized beds was recorded in cauliflower (33.7 per cent) tomato 29.6 per cent and brinjal 27.5 per cent.However, increase in root length in solarized nursery beds over non solarized beds was recorded in broccoli (33.3 per cent) followed by Chinese sarson (28.5 per cent) and onion (27.5 per cent) Harender *et al.* (1997).

Table 29.3: Example of Some Soil Borne Pathogen Controlled by Solarization

Pathogen/Fungal	*Crop*	*Reference*
Fusarium oxysporum	Wilt of citrus	Raj and Gupta (1995)
S. rolfsii	Southern blight of tomato	Chellemi and Olson (1994)
F.o. f. sp. carthami	Safflower wilt	Sastry and Chattopadhyay (1999)
Macrophamina phaseolina	Stem root rot of sesame	Chattopadhyay and Sastry (1996)
S. rolfsii	Sclerotial root rot of chickpea	Tiwari *et al.* (1997)
F.o f. sp ricini	Wilt of castor	Raoof and Rao (1997)
M. phaseolina	Dry root rot of cruciferal	Lodha *et al.* (1996)
Rhizoctonia solani	Cucumber root rot	Keinath (1995)
Pythium aphanidermatum	Damping off of horticulture crop	Abdul *et al.* (1995)
F.o. f. sp ciceri	Gram wilt	Rao and Krishnappa (1995)
Phytophthara spp.	Blight of solanaceous crop	Abdul *et al.* (1995)
F.o. f. sp. vasinfectum	Cotton wilt	Katan *et al.* (1983)
Verticillium sp.	Verticillium wilt of potato	Gumiel *et al.* (1999)
Bacterial		
Pseudomonas solanaceassum	Potato	Chellemi and Olson (1994)
Agrobacterium tumefacience	Fruit trees	Raio *et al.* (1997)
Nematode		
Helicotylenchus sp. Heterodera sp., Pratylenchus spp.	Vegetable crop	Sharma and Nene (1990)
Meloidogyne spp.	Root knot nematode of *Bidi* tobacco	Patel *et al.* (1995)
Heterodera avenae	Nematode of wheat	Patel *et al.* (1995)

Mechanism of Disease Control

The reduction in disease incidence in plant grown in solarized soils, as with any soil treatment results from the effects exerted on each of the three living components *viz.*, host, pathogen and surrounding beneficial microorganisms as well as on the physical and chemical environments which in turn affect the activity and inter relationship of the organisms (Devay and Katan, 1991). The following mechanism may occur in controlling the disease (Katan, 1981).

1. Thermal inactivation (Physical effect)
2. Biological control
3. Chemical like volatile compound
4. Other mechanism *viz.* soil anarobiosis.

Thermal Inactivation

Whenever organisms are subjected to moist heat at temperatures exceeding the maximum for growth, their viability is reduced. Thermal death rate (TDR) of a population of an organism depends on both the temperature level and exposure time, which are inversely related. At a given temperature and exposure time, mortality rate is related to the inherent heat sensitivity of the organisms and to the prevailing environmental condition. It depends on both temperature level and exposure time. *P. cinnamon* complete activation was ceased after exposure of 90 °C for 4.5 min. (Bensen, 1978). Spores of *Armillaria mellea* were not viable after expose of 4-7 hours at 41°C (Munnecke *et al.*, 1976). Exposing moistened inoculum of *Fusarium oxysporum* f. sp. *lycoperseci* and *Verticillium dahliae* for 1 hour at 50°C resulted in 95 per cent eradication. Wet heat was more effective than dry heat. In the presence of water less energy is required to unfold the peptide chain of protein resulting in a decreased heat resistance (Precht *et al.*, 1973).Weakened propagules may possess lower inoculums potential and shorter longevity due to slower germination or growth.

Biological Control

Biological control may affect the pathogen by increasing its vulnerability to soil micro organisms or increasing the activity of soil micro organism toward pathogen or plant which will finally lead to reduction in disease incidence, pathogen survivability or both. After or during solarization, biological control may operate at any stage of pathogen survival or disease development. The following biological control mechanisms may be created or stimulated by solarization (any disinfestations method) (Katan, 1981).

1. Affect the inoculum existing in soil through-microbial killing, parasitism or lysis by antagonists, antibiosis or competition.
2. Suppressing the inoculum introduced from deeper layers
3. The effect on the host due to cross protection.

Chemicals Like Volatiles in Disease Control

Volatiles in the soils are involved in key processes such as fungistasis and biological control (Lewis *et al.*, 1975; Papvizar etal., 1980; Pavlica *et al.*, 1978; and Zakaria *etal., 1980*). Amonia (NH_3) for example is apparently involved in the suppression of *Fusarium* by oilseed amendment and the suppression of *Aphanomyces enteiches* by cruciferous amendments was related to the formation of volatile-Sulpher-containing compounds (Lewis and Papavizas, 1975). Arya (1990) tried three organic amendments leaves of Neem, Ecalyptus and Calotropis were added to the soil before solarization, maximum effect was recorded with use of neem leaves.

Soil Anarobiosis for Microbe Content

The fields irrigating is commonly practiced before soil solarization. It causes anaerobic (oxygen-free or anoxic) condition. Excess water greatly affects the activities of soil microorganism (Drew and Lynch, 1980). With the on set of anaerobial soil condition it might be expected that fungal pathogen which are generally considered to be obligate aerobes would die both through lack of oxygen (O_2) and by a build up in facultative aerobes acting as antagonists.

Beneficial Effects

Increased Growth Response (IGR)

Plant growth in disinfected soil is enhanced as compared to untreated infested soil which is might be due to free of pathogens in soil. Besides these, the following points have also been suggested for explaining increase growth rate in disinfested soils.

- ☆ Increased availability of macro and micro elements in the soil solution.
- ☆ Elimination of minor pathogens or parasites.
- ☆ Destruction of phytotoxic substances produced by pathogen or adverse soil condition in the soil.
- ☆ Release of growth regulators like substances.
- ☆ Stimulate the activities of mycorhizae or other beneficial microorganism (Altman, J., 1970; Broadbent *et al.*, 1977 and Newhall, 1955).

Increase in dry matter production and seed yield in pigeon pea have also been found where crop is grown in solarized soil (Chauhan *et al.*, 1988). Disease control was usually accompanied by an increase in yield to various extents. The yield of onion is also increased by 28 per cent in solarized soil (Katan, 1981).

Table 29.4: Example of Increase Crop Yield Due to Soil Solarization

Crop	*Pathogen*	*Increase over control*	*Reference*
Potato	*Verticillium dahliae* and *Pratylenchus thornei*	35	Gristein *et al.*, 1979
Cotton	*Verticillium dahliae*	60	Pullman *et al.*, 1981
Cotton	*Fusarium oxysporum* f.sp. *vasinfectum*	40-70	Katan, 1980(a)
Peanut	*Slerotium rolfsii*	42-64-a and 105-146-b	
Onion	*Pyronchaeta terrestris*	60-125	Katan, 1980(b)
Tomato	*P. lycoperssici*	100-300	Tjamos *et al.*, 1980
Tomato	*P. lycoperssici*	245	Katan *et al.*, unpublished
Tomato	*P. lycoperssici*	140-300	Tammietti *et al.*, 1980
Safflower	*Verticillium dahliae*	113	Pullman, 1979
Eggplant	*Verticillium dahliae*	215	Katan *et al.*, 1976
Carrot	*Orobanche aegyptica*	c	Jacobsohn *et al.*, 1980

a: Total yield; b: Grade a of the yield; c: The yield is the untreated control was zero as compared to ca 70 tons/ha in the solarized treatment.

Higher concentration of soluble organic matter and minerals were found in extract of solarized soil. But it is still unknown whether these soluble organic substances play a role in growth stimulation or not. Katan (1995) believes that secretion of more amino acids by roots of plants, grown in solarized soil is the responsible factors for plant growth.

Change in *Rhizobium* Population

Soil solarization did not significantly affect rhizobial population or nodulation of pigeonpea and chickpea in the non-irrigated treatment, but it caused a significant reduction in these parameters after irrigation (Chauhan *et al.*, 1988). However, shoot dry matter was not reduced, perhaps due to compensatory effect of solarization in enhancing nitrate concentration in these treatments.

Control of Weeds

Soil solarization is an effective method of weed control, lasting in some cases for a whole year or even longer. In general, most of the annual and many perennial weeds are effectively controlled by soil solarization.The weeds are belongs to gramminae are more sensitive to soil solarization (Katan, 1981). The possible mechanisms of weed control are direct killing of weed seeds by heat or indirect microbial killing of seed, weakened by sublethal heating, killing of seeds stimulated to germinate in moistened mulched soil follwed by killing germinated seeding in heated soil. Reduction in growth of weed has also been recorded by Arya and Mathew (1993). Solarization significantly decreased the weeds growth at 80.6 and 90.5 per cent in weed count and dry weight, respectively as compared to unsolarized control. (Rao *et al.*, 1995).

Increase of Soil Temperature

Temperature is higher in above 10 cm of surface soil. During night the soil temperature slowly decreased according to atmospheric condition. The fluctuation of daily increases and decreases in soil temperature is vital for success of solarization. In Israel, the soil temperature of mulched soil recorded at depths of 5 cm. is 45-50°C where as at 20 cm.,it was 39-45°C. The plots covered with polythene sheet markedly had markedly higher temperatures (45.9°C) as compared to unsolarized control (38.2°C). During solarization every day soil temperature become increased by about 10-15°C at day time which is not in case of bare soil in mulched soil. Soil solarization increases soil temperature.The temperature above 45°Cis lethal for most of microorganism was also recorded during solarization period. (Rao *et al.*, 1995).

Control of Soil Insects

Harmful soil insects may be reduced after solarization. Priliminary data showed that population of the soil mite *Rhizoglyphus robini* are drastically reduced by solar heating (Katan, 1981).

Need of soil solarization: A pest control practice must be economically feasible for to be used. Thus it is beneficial to evaluate the practice on a cost/benefit analysis. There are at least seven areas where opportunities exist to use soil solarization economically which are as follows:

- ☆ In a crop where there is no available pesticide due to lack of registration, availability, crop tolerance, hazard of application from a pesticide or expense.
- ☆ In a crop where pests are not manageable by other means.
- ☆ In certain circumstances where more than one pest problem exists.
- ☆ Where a crop is growing without synthetic pesticide.
- ☆ Where solarization can change the cropping sequence or maintain yield on smaller areas.
- ☆ In a crop where early seedling vigour and rapid growth is an advantage.
- ☆ Competition in markets where organic produces are competing with conventionally produced pesticide treated areas.

Important Criteria for Soil Solarization

To ensure more effective solarization, the following guidelines are proposed–Solarization should be conducted for at least 6 weeks during the hottest part of the year (April-June).

1. The land should be leveled properly to prevent tearing of polyethylene sheets before mulching.
2. Soil must be wet during solarization.Irrigation has been proved beneficial for it. If necessary, light irrigation (40-50 mm) irrigation should be given prior to lying of the sheets.
3. Soil should be kept moist during mulching to increase thermal sensitivity of resting structure and improve heat condition. (Katan *et al.*, 1976).
4. Clear, transparent sheet of 25-100 mm thickness rather than high thickness sheet should be used; experimental results show that thinner sheet is more effective.
5. Polythene sheeting should be applied at dawn when it is least windy.
6. Two edges of polythene sheets should be inserted in furrow. All free edges should be buried to prevent exchange of escape heated air or soil moisture.
7. It is necessary to allow for a buffer zone of at least 0.5 m around the edges of solarized area.
8. Any damage by creation of hole on plastic sheets should be scaled immediately during solarization.
9. If necessary, visit at mulched plot should be done through watching by feet or with smooth soled shoes.
10. To prevent tearing of polyethylene sheets, it is recommended to keep heavy weights on polythene sheeting.
11. Irrigation water should not be allowed to flow from contaminated plots to solarized plot at the time of crop growing season.
12. Solarization can be integrated with pesticides organic amendments and bio agents to get maximum benefit organic matter should be added to soil before starting solarization.

Merits of Soil Solarization

1. It is non-chemical method which is not hazardous to the user and does not involve substances toxic to the consumer, to the host plant or to other organism.
2. It is less expensive than other methods.
3. It is easy method and that can be easily transmitted to the ordinary farmers.
4. It can be applied in larger areas by machines or manually in small plots. Thus it is suitable for both developed and developing countries.
5. It may have long term effect since effective disease control lasted for more than one season.
6. The method has characteristics of an integrated disease pest control (Baker and Cook,1974)
7. This method has been used successfully to reduce population of plant parasitic nematodes various plant pathogenic bacteria, weed and certain soil born fungal disease (Barber check *et al.*, 1986; Katan *etal*, 1983) (Ashworth,1979).
8. Soil solarization is a simple cheap and highly effective method for controlling plant pest, diseases, weeds and environmental poisoning.
9. Soil solarization is a natural hydrothermal process of disinfecting soil from plant pest *i.e.* accomplished by passive solar heating.

10. Soil solarization occurs through a combination of physical, chemical and biological mechanism and is compatible with another disinfestations method to process integrated pest management.
11. Though, it is used commercially on small scale as a substitute for synthetic chemical toxicants, but its use is increasing as methyl bromide, the major chemical fumigant is phased out due to its zone depleting properties.
12. Soil solarization has been mainly used for disinfecting soils in areas where air temperature is very high during summer.
13. The solar heating of soil is a novel agricultural practice in which disease is controlled by soil disinfestations. Its aims is to eradicate or reduce the inoculum existing in the soil prior to planting.
14. Solar heating involves the use of heat as a lethal agent for pest control including disease. Through the use of traps for capturing solar energy by means on transparent polyethylene soil mulches.
15. Renowned Israeli Phytopathologist Dr. Katan has pointed out that farmers of the Deccan plateau in India have long exploited a form of solar heating of soil by ploughing the soil so as to expose sub soil prior to the hot summer period (April-June), when maximum daily temperature exceeds 40°C and then leaving it fallow.
16. The solar heating method for disease control is similar in principle to that of artificial soil heating by steam or other mean, which are usually carried out at 60-100°C (Baker and Cook, 1974).

Demerits of Soil Solarization

1. The main draw back of soil solarization is used in region specific which means that it can be particularly used in tropical climate areas but not in temperate region.
2. It can be applied in field as well as in nursery, but the particular place must be free from crop for about one month or more.
3. In certain cases it is ineffective to control of some diseases.
4. Repeated application of the process, there is a chance for development of heat tolerant pathogen among the population.
5. It is effective only to control soil borne disease but not seed or air borne disease.
6. Some times it indiscriminately killed antagonists organism due to excessive heat resulted decreases in plant growth.
7. In some cases, the cost of solarization exceeds in many less expensive field crops.
8. Solar heat may affect target pathogens and at the same time reduced the population of beneficial soil micro organism like fungi bacteria, insects etc in soil.

Conclusion

We have passed more than 100 years of induction of soil disinfestations and over 50 years of Sanford's classical biological control our hungry world is still crying out for reducing crop losses caused by soil borne plant pathogens. We are continuously facing comforted difficulties on appearance of new physiological races of the pathogens which are being resistant to pesticide. Thus, we have acquired modesty and innovative method of disease management *i.e.* soil solarization. But single methods cannot give long last protection against disease and keeping in mind the term. "Sustainable Agriculture" soil solarization can be included in the IPM programme for long lasting ecofriendly and over all agricultural growth and development.

References

Abdul Wajid, S. M., M. M. Shenoi and S. S. Sreenivas (1995). Seed bed soil solarization as a component of integrated disease management in FCV tobacco nurseries of Karnataka. *Tob. Res.* 21: 58-65.

Adams, P. B. (1971). Effect of soil temperature and soil amendments on Thieloviopsis root rot of sesame. *Phytopathology* 61: 93-97.

Al- Radad. A. M. M. (1979). *Soil disinfestations by plastic trapping*. M. Sc. thesis Univ. Jordan. 95pp.

Altman, J. (1970). Increased and decreased plant growth responses resulting from soil fumigation. In: *Root Diseases and pathogens Soilborne*. (ed) T. A. Tousson, R. V. Bega, R. E. Nelson, PP. 216-21. Berkeley: Uni. Calif. Press.

Arya, A. and Mathew, D. S. (1993). 'Studies on rhizosphere microflora of pigeon pea: qualitative and quantitative incidence of microorganism after solarization'. *Indian Phytopath.* 46: 151-54.

Arya, A (1990). Integrated management of wilt disease of pigeon pea'. Final tech. report DST(SP/YS/L4187).

Ashworth, L. J. (1979). Polyethylene tarping of soil in a pistachio nut gorve for control of *Verticilium dahliae*. *Phytopathology* 69: 913.

Avissor, R., Mahrer Y., Margulies L. and Katan J. (1986). *Soil Sci Soc. America J.* 50: 202-205.

Baker, K. F., Cook, R. J. (1974). *Biological Control of Plant Pathogens*. San Francisco. Freeman. 433PP.

Barbercheck, M. E. and Brambsen, S. L. (1986). Effect of soil solarization on plant parasitic nematodes and *Phytophthora cinnamoni* in South Africa. *Plant Diseases* 70: 945-50.

Benson, D. M. (1978). 'Thermal inactivation of Phytophthora cinnamomi for control of Fraserfir root rot, *Phytopathology. 69*: 1373-76.

Bradbent, P., Baker, K. F., Franks, N., Halland, J. (1977). Effect of Bacillus spp. on increased growth of seedling in steamed and in non treated soil. *Phytopathology* 67: 1027-34.

Chattopadhyay, C. and R. K. Sastry (1996). Management sesame stem root rot through soil solarization-An Ecofriendly Strategy. *IndianJ. Mycol. Pl. Pathol.* 26: 123.

Chauhan, Y. S., Nene, Y. L., Johansen, C., Haware, M. P., Saxena, N. P. Singh, S., Sharma, S. B. Sahrawat, K. L., Burford, J. R., Rupela, O. P., Rao, J.V.D.K. and Sithanantham, S. (1988). Effect of soil solarization on pigeon pea and chickpea, ICRISAT Research Bulletin. Andhra Pradesh, No. 11.

ChauhanY. S., Nene, Y. L., Johansen, C., Haware, M. P., Saxena, N. P., Singh, S., Sharma, S. B., Sahrawat, K. L., Burford, J. R., Rupela, O. P., Kumar, Rao, J.V.D.K. and Sithanatham, S. (1981). Effect of soil solarization on pigeon pea and chickpea. Research Bulletin- II, ICRISAT, Patancheru, 16 pp.

ChauhanY. S. , Nene, Y. L. , Johansen, C. , Haware, M. P. , Saxena, N. P. , Singh, S. , Sharma, S. B. , Sahrawat, K. L. , Burford, J. R. , Rupela, O. P. , Rao, J. V. D. K. and Sithanatham, S. (1981). Effect of soil solarization on pigeon pea and chickpea. ICRISAT Research Bulletin, Andhra Pradesh, No. II, 1988.

Chellemi, D. O. , S. M. Olson and D. J. Mitchelley(1994). Effect of soil solarization and fumigation on survival of soil born pathogens of tomato in Northern Florida. *Plant Dis.* 78: 1167-72.

De-Luca V. Manera C. Mazza S. , Immirzi B. , Malinconico M. and Martuscelli E. (1994). Innovative biodegradable plastic films to enhance efficiency of soil solarization in agriculture. In: *Proceed. of the 13th International Congress on Plastic in Agriculture held in Verona,* Italy 8-11 March, 1: 12.

Devay, J. E. and J. Katan. (1991). Mechanism of pathogen control in solarized soil. In: *Soil solarization.* CRC, Press. Boca Raton, FL, USA. pp87-101.

Drew, M. C. and Lynch, J. M. (1980). 'Soil anaerobiosis, microorganism and root function'. *Ann. Rew. Phytopathol.* 18: 37-66.

Elad, Y. , Katan. J, Chet. I. (1980). Physical biological and chemical control in integrated for soil borne diseases in potatoes. *Phytopathology* 70: 418-22.

Gamliel, A. , Kritzman, G, Alon Y. P, Becker E, Zilberg V. and Heiman O (1999). Soil solarization using sprayable plastic polymers to control soil borne pathogen in field crops (Abstracts) 14th IPPC. Meeting. Jerusalem P92.

Garibaldi A. (1987). *Colture Protette* 16: 25-28.

Grinstein, A. , Orion, D. , Greenberger, A. , Katan, J. (1979). Solar heating of the soil for the control of *Verticillium dahliae* and *Pratylenchus thornei* in potatoes. In *Soil born Plant Pathogene* pp. 431-38, ed. B Schippers, W. Gams. London, New York, San Francisco: Academic. 686pp.

Harender, R. , Sharma, N. K. and Gupta, V. K. (1995). Effect of siol solarization on population of *Fusarium solani.* Biological and cultural test for control of Plant Diseases. *Indian Phytopath.* 10: 55.

Harender Rraj, M. L. Bharadwaj and N. K. Sharma (1997). Soil solarization for the control of damping off of different vegetable crops in the nursery. *Indian Phytopath.* 50 (4): 524-528

Harender, R. , Bharadwaj, M. L and Sharma N. K (1997). Soil solarization for the control of damping off of different vegetable crops in the nursery. *Indian Phytopath* 50(4): 524-28.

Horowitz, M. , Regev, Y. (1980). Mulching with plastic sheets as a method of weed control. *Hassadeh* 60: 395-99 (In Hebrew) In press.

Jacobsohn, R. , Greenberger, A. , Katan, J. , Levi, M. , Alon, H. (1980). Control of Egyptian broomrape (*Orobanche aegyptiaca*) and other weeds by means of solar heating of the soil by polyethylene mulching. *Weed Sci.* 28: 312-16

Karma A. and Gaur H. S. (1995). *Ann. Pl. Protec. Sci.* 3: 7492.

Katan J. (1981). Solar heating (solarization) of soil for control of soil born pest. *Ann. Rev. Phytopathol.* 19: 211-36.

Katan J. , Fisher G. and Gristein A (1983). Short and long term effect of Isoil solarization and crop sequence on Fusarium wilt and yield of cotton in ISRAEL. *Phytopathology*, 73: 1215-19.

Katan, J. (1981). Solar heating (solarization) of soil for control of soil borne pests. *Ann. Rev. Phytopathol.* 19: 211-226.

Katan, J. (1995). Soil solarization: A non–chemical tool in plant protection (Abstr.) *Indian J. Mycol. and Pl. Pathol.* 25: 45-47.

Katan, J. Green berger A. , Alon. H. and Gristein, A. (1976). Solar heating by polyethylene mulching for the control of disease caused by soil born pathogen. *Phytopathology* 66: 683-688.

Katan, J. (1980). Solar pasteurization of soil for diseases control: Status and prospects. *PlantDis.* 64: 450-54.

Katan, J. (1983). Soil solarization. *Acta. Hort.* 152: 227-36.

Katan, J. , Fishler, G. , Grinstein, A. (1980)-a. Solar heating of the soil and other methods for the control of *Fusarium*, additional soilborne pathogens and weeds in cotton. See Ref. 37, pp. 80-81.

Katan, J. , Greenberger, A. Alon, H. and Grinstien, A. (1976). Solar heating by polethylene mulching for the control of diseases caused by soil borne pathogens. *Phytopathology* 66: 683-88.

Katan, J. , Rotem, I. , Finkel, Y. , Daniel, J. (1980)-b Solar heating of the soil for the control of pink root and other soil borne diseases in onions. *Phytoparasitica* 8: 39-50.

Keinath, A. P. (1995). Reductions in inoculum density of *Rhizoctonia solani* and control of bellyrot on pickling cucumber with solarization. *Plant Dis.* 79: 1213-19.

Lamberti F. and Greco N. (1991). FAO Pl. Production and Protection. 109: 167-172.

Lewis, J. A. and Papvizar, G. C. (1975). 'Effect of sulpher containing volatile compounds and vapours from cabbage decomposition on *Aphanomyces euteiches.' Phytopathology* 61: 208-214.

Lodha S. , Sharma, S.K. and Agrawal R. K (1996). Solarization and natural heating of irrigated soil amended with cruciferous residues for improved control of *Macrophamina phasiolina. Plant Ppathol.* 26: 186-90.

Manera C. , Margiotta S. and Prcuno P. (1999). *Colture Protette* 28: 59-64.

Munnecke, D. E. , Wilber, W. and Darley, E. F. (1976). 'Effect of heating or drying on *Armilaria mellea* or *Trichoderma viride* and the relation to survival of *A. mellea* in soil. '*Phytopathology*. 66: 1363-68.

Nene, Y. L. (1986). 'Opportunities for research on disease of pulse crops'. *Jeersannidhi award lectur. Indian Phytopath Soc.* 13 February 1986, Bhagalpur.

Newhall, A. G. (1955). Disinfestation of soil by heating, flooding and fumigation. *Bot. Rev.* 21: 189-250.

Papavizar, G. C. and Lumsden, R. D. (1980). Biological control of soil bornefungal propagules. *Ann. Rev. Phytopathol.* 18: 389-413.

Patel, D. J. , Patel, H. R. , Patel, S. K. and Makwana, M. G. (1990). Soil solarization through clear polythenetraping for management of root knot disease in a tobacco nursery. In: *Proc. Of XIth Int. Cong. on the use of Plastic in Agric.*, held on 29th February-2ndMarch, 1990.

Patel, D. J. , Patel, H. R. , Patel, S. K. and Makwana, M. G. (1991). Soil solarization through clear polythenetraping for management of root knot disease in a tobacco nursery. *Tab. Sci.* 35: 63-64.

Patel, H. R. , Makwana M. G and Patel B. N (1995). The control of nematodes and weeds by soil solarization in tobacco nurseries, effect of the film thickness and of the covering duration, *Plasticulture* 107: 23-28.

Pavlica, D. , A. , Hora, T. S. , Bradshaw, J. J. , Skogerboe, R. K. , Baker, R. (1978). Volatiles from soil influencing activities of soil fungi. *Phytopathology* 68: 758-65.

Precht, H. Chistophersen, J. , Hensel, H. and Larcher, W. (1973). Temperature and life. Burlin: Springer.

Pullman, G. S. DeVay, J. E. , Garber, R. H. Weinhold, A. R. (1981). Soil Solarization for the control of *Verticillium* wilt of cotton and the reduction of soil borne populations of *Verticillium dahliae, Pythium spp. , Rhizoctonia solani* and *Thielaviopsis basicola. Phytopathology* 71: In press

Pullman, G. S. (1979). *Effectiveness of soil solarization and soil flooding for control of soil–borne diseases of Gossypium hirsutumL. In relation to population dynamics of pathogens and mechanisms of propagule death.* PhD thesis. Uni. Calif. , Davis. 95pp

Pullman, G. S. , De VayJ. E. and Garger, R. H. (1981). Soil solarization and thermal death: a logarithmic relationship between time and temperature for four soil borne plant pathogens. *Phytopathology* 71: 959-64.

Pullman, G. S. , Devay, J. E. and Gorber, R. H. (1981). Soil solarization and thermal death, A logarithmic relationship between time and temperature for four soil borne plant pathogens. *Phytopathology* 71: 959-964.

Raj H. , Kapoor, I. J. (1993). Soil solarization for the control of tomato wilt pathogen (Schl.). *Z. Pflkrankh. Pflschutz.* 100: 652-661.

Rao V. K. and Krishnappa K. (1995). Soil solarization for the control of soil borne pathogen complex with special reference to *Meloidogyne ingonita* and *Fusarium oxysporum* fsp *ciceri. Indian Phytopath.* 48(3): 300-303.

Rao, V. K. and K. Krishnappa(1995). Soil solarization for the control of soil borne pathogen complex with special reference to *Meloidogyne incognita* and *Fusarium oxysporum* f. sp. ciceri. *Indian Phytopath.* 48: 300-303.

Raoof, M. A and T. G. Nageshwar Rao (1997). Effect of soil solarization on caster wilt. *Indian J. Pl. Protect.* 25: 154-59.

Roio, A. , A. Zoina and L. W. Moore. (1997). The effect of solar heating of soil on natural and inoculated agrobacteria. *Plant Pathol.* 46: 320-28.

Sastry, R. K. and C. Chattopdhyay (1999). Effect of soil solarization on *Fusarium oxysporum* f. sp *carthami* population in endemic soils. *Indian Phytopath.* 52: 51-55.

Sharma, S. B. and Y. L. Nene (1990). Effect of soil solarization on nematodes parasitic to chick pea and pigeonpea. *Supple. J. Nematol.* 22: 658-64.

Singh R. S. (1998). *Plant Diseases.* Oxford and I B H publication company New Delhi.

Tamietti, G. , Garibaldi, A. (1980). Control of corky root in tomato by solar heating of the soil in greenhouse on Riviera Ligure. *La Difesa delle Piente* (In Italian)

Tiwari, R. K. S. , Parihar, S. S and Chaure N. K (1997). Soil solarization for the control of *S. rolfsii* causing sclerotial root rot of chickpea. *Plant Protect.* 25: 142-45.

Tjamos, E. C. , Faridis, A. (1980). Control of soilborne pathogens by solar heating in plastic houses. See Ref. 37, pp. 82-84

Usmani S. M. H. and Ghaffar A. (1981). Polyethylene mulching of soil to reduce viability of *Sclerotium oryzae*. *Soil. Biol. and Biochem*. 14: 203-206.

Walker, G. E. and Watchel M. F. (1989). *Nematologica* 34: 477-483.

Zakaria, M. A. , Lock wood, J. L. and Filonow, A. B. (1980). Reduction in Fusarium population density in soil by volatile degradaton products of oil seed meal amendments. *Phytopathology* 70: 495-99.

Index